T0310139

Fog Computing: Theory and Practice

WILEY SERIES ON PARALLEL AND DISTRIBUTED COMPUTING

Series Editor: Albert Y. Zomaya

Parallel and Distributed Simulation Systems / Richard Fujimoto

Mobile Processing in Distributed and Open Environments / Peter Sapaty

Introduction to Parallel Algorithms / C. Xavier and S. S. Iyengar

Solutions to Parallel and Distributed Computing Problems: Lessons from Biological Sciences / Albert Y. Zomaya, Fikret Ercal, and Stephan Olariu (Editors)

Parallel and Distributed Computing: A Survey of Models, Paradigms, and Approaches / Claudia Leopold

Fundamentals of Distributed Object Systems: A CORBA Perspective / Zahir Tari and Omran Bukhres

Pipelined Processor Farms: Structured Design for Embedded Parallel Systems / Martin Fleury and Andrew Downton

Handbook of Wireless Networks and Mobile Computing / Ivan Stojmenović (Editor)

Internet-Based Workflow Management: Toward a Semantic Web / Dan C. Marinescu

Parallel Computing on Heterogeneous Networks / Alexey L. Lastovetsky

Performance Evaluation and Characterization of Parallel and Distributed Computing Tools / Salim Hariri and Manish Parashar

Distributed Computing: Fundamentals, Simulations and Advanced Topics, 2nd Edition / Hagit Attiya and Jennifer Welch

Smart Environments: Technology, Protocols, and Applications / Diane Cook and Sajal Das

Fundamentals of Computer Organization and Architecture / Mostafa Abd-El-Barr and Hesham El-Rewini

Advanced Computer Architecture and Parallel Processing / Hesham El-Rewini and Mostafa Abd-El-Barr

UPC: Distributed Shared Memory Programming / Tarek El-Ghazawi, William Carlson, Thomas Sterling, and Katherine Yelick

Handbook of Sensor Networks: Algorithms and Architectures / Ivan Stojmenović (Editor)

Parallel Metaheuristics: A New Class of Algorithms / Enrique Alba (Editor)

Design and Analysis of Distributed Algorithms / Nicola Santoro

Task Scheduling for Parallel Systems / Oliver Sinnen

Computing for Numerical Methods Using Visual C++ / Shaharuddin Salleh, Albert Y. Zomaya, and Sakhinah A. Bakar

Architecture-Independent Programming for Wireless Sensor Networks / Amol B. Bakshi and Viktor K. Prasanna

High-Performance Parallel Database Processing and Grid Databases / David Taniar, Clement Leung, Wenny Rahayu, and Sushant Goel

Algorithms and Protocols for Wireless and Mobile Ad Hoc Networks / Azzedine Boukerche (Editor)

Algorithms and Protocols for Wireless Sensor Networks / Azzedine Boukerche (Editor)

Optimization Techniques for Solving Complex Problems / Enrique Alba, Christian Blum, Pedro Isasi, Coromoto León, and Juan Antonio Gómez (Editors)

Emerging Wireless LANs, Wireless PANs, and Wireless MANs: IEEE 802.11, IEEE 802.15, IEEE 802.16 Wireless Standard Family / Yang Xiao and Yi Pan (Editors)

High-Performance Heterogeneous Computing / Alexey L. Lastovetsky and Jack Dongarra

Mobile Intelligence / Laurence T. Yang, Augustinus Borgy Waluyo, Jianhua Ma, Ling Tan, and Bala Srinivasan (Editors)

Advanced Computational Infrastructures for Parallel and Distributed Adaptive Applications / Manish Parashar and Xiaolin Li (Editors)

Market-Oriented Grid and Utility Computing / Rajkumar Buyya and Kris Bubendorfer (Editors)

Cloud Computing Principles and Paradigms / Rajkumar Buyya, James Broberg, and Andrzej Goscinski

Energy-Efficient Distributed Computing Systems / Albert Y. Zomaya and Young Choon Lee (Editors)

Fog Computing: Theory and Practice

Edited by

Assad Abbas
COMSATS University Islamabad, Pakistan

Samee U. Khan
North Dakota State University, USA

Albert Y. Zomaya
University of Sydney, Australia

This edition first published 2020.

Registered Offices
John Wiley & Sons, Inc., 111 River Street, Hoboken, NJ 07030, USA

Editorial Office
111 River Street, Hoboken, NJ 07030, USA

For details of our global editorial offices, customer services, and more information about Wiley products visit us at www.wiley.com.

Wiley also publishes its books in a variety of electronic formats and by print-on-demand. Some content that appears in standard print versions of this book may not be available in other formats.

Library of Congress Cataloging-in-Publication data applied for

ISBN: 9781119551690

Cover Design: Wiley
Cover Image: © Khing Choy/Shutterstock

Set in 9.5/12.5pt STIXTwoText by SPi Global, Chennai, India

Printed in the United States of America

V10018106_031820

Contents

List of Contributors

Assad Abbas
COMSATS University Islamabad
Islamabad Campus
Pakistan

Mansoor Ahmed
COMSATS University Islamabad
Islamabad Campus
Pakistan

Isam Mashhour Al Jawarneh
Department of Computer Science and Engineering
University of Bologna
Italy

Imran Ali Khan
COMSATS University Islamabad
Abbottabad Campus
Pakistan

Mazhar Ali
COMSATS University Islamabad
Abbottabad Campus
Pakistan

Ahmad Ali
COMSATS University Islamabad
Islamabad Campus
Pakistan

Parastoo Alinia
Washington State University
Pullman, WA
United States

Delaram Amiri
Department of Electrical Engineering and Computer Science
University of California–Irvine
United States

Arman Anzanpour
Department of Future Technologies
University of Turku
Finland

Cosmin Avasalcai
Vienna University of Technology
Vienna

Kamran Sattar Awaisi
COMSATS University Islamabad
Islamabad Campus
Pakistan

Iman Azimi
Department of Future Technologies
University of Turku
Finland

Thais V. Batista
Federal University of Rio Grande do Norte
Brazil

Micah Beck
University of Tennessee
Knoxville, TN
United States

Pete Beckman
Argonne National Laboratory
Lamont, IL
United States

Paolo Bellavista
Department of Computer Science and Engineering
University of Bologna
Italy

Javier Berrocal
Department of Computer and Telematics Systems Engineering
University of Extremadura, Cáceres
Spain

Prasad Calyam
Department of Electrical Engineering and Computer Science
University of Missouri–Columbia
United States

Chii Chang
School of Computing and Information Systems
University of Melbourne
Australia

Ahmed Chebaane
Landshut University of Applied Sciences
Landshut
Germany

Dmitrii Chemodanov
Department of Electrical Engineering and Computer Science
University of Missouri–Columbia
United States

Antonio Corradi
Department of Computer Science and Engineering
University of Bologna
Italy

Sajal K. Das
Department of Computer Science
Missouri University of Science and Technology
United States

Flavia C. Delicato
Federal University of Rio de Janeiro
Brazil

Panagiotis D. Diamantoulakis
Electrical and Computer Engineering Department
Aristotle University of Thessaloniki
Thessaloniki
Greece

Jack Dongarra
University of Tennessee
Knoxville, TN
United States

and

Oak Ridge National Laboratory
Oakridge, TN
United States

Schahram Dustdar
Vienna University of Technology
Vienna

Nikil Dutt
School of Information and Computer Sciences
University of California–Irvine
United States

Biyi Fang
Michigan State University
East Lansing, MI
United States

Nicola Ferrier
Argonne National Laboratory
Lamont, IL
United States

Miodrag Forcan
Faculty of Electrical Engineering
University of East Sarajevo
East Sarajevo
Bosnia and Herzegovina

Luca Foschini
Department of Computer Science and Engineering
University of Bologna
Italy

Geoffrey Fox
Indiana University
Bloomington, IN
United States

Hassan Ghasemzadeh
Washington State University
Pullman, WA
United States

Usman Habib
National University of Computer and Emerging Sciences
Peshawar
Pakistan

Amnir Hadachi
Institute of Computer Science
University of Tartu
Estonia

Muhammad Imran
COMSATS University Islamabad
Islamabad Campus
Pakistan

George K. Karagiannidis
Electrical and Computer Engineering Department
Aristotle University of Thessaloniki
Thessaloniki
Greece

Osman Khalid
COMSATS University Islamabad
Abbottabad Campus
Pakistan

Muhammad Usman Shahid Khan
COMSATS University Islamabad
Abbottabad Campus
Pakistan

Asad Khan
National University of Computer and Emerging Sciences
Peshawar
Pakistan

Muazzam A. Khan
National University of Science and Technology (NUST)
Pakistan

Samee U. Khan
North Dakota State University
United States

Hasan Ali Khattak
COMSATS University Islamabad
Islamabad Campus
Pakistan

Abdelmajid Khelil
Landshut University of Applied Sciences
Landshut
Germany

Nicholas D. Lane
Oxford University
UK

Marco Levorato
School of Information and Computer Sciences
University of California–Irvine
United States

Pasi Liljeberg
Department of Future Technologies
University of Turku
Finland

Mirjana Maksimović
Faculty of Electrical Engineering
University of East Sarajevo
East Sarajevo
Bosnia and Herzegovina

Asad Waqar Malik
National University of Science and Technology (NUST), Pakistan
Department of Information System
Faculty of Computer Science & Information Technology
University of Malaya
Malaysia

Jakob Mass
Institute of Computer Science
University of Tartu
Estonia

Diomidis S. Michalopoulos
Nokia Bell Labs
Munich
Germany

Terry Moore
University of Tennessee
Knoxville, TN
United States

Shuja Mughal
COMSATS University Islamabad
Islamabad Campus
Pakistan

Ilir Murturi
Vienna University of Technology
Vienna

Kannappan Palaniappan
Department of Electrical Engineering and Computer Science
University of Missouri–Columbia
United States

Paulo F. Pires
Federal University of Rio de Janeiro
Brazil

Tariq Qayyum
National University of Science and Technology (NUST), Pakistan

Amir M. Rahmani
School of Information and Computer Sciences, and School of Nursing
University of California–Irvine
United States

Rao Naveed Bin Rais
College of Engineering and Information Technology
Ajman University
Ajman
UAE

Dan Reed
University of Utah
Salt Lake City, UT
United States

Aluizio F. Rocha Neto
Federal University of Rio Grande do Norte
Brazil

Yuanchao Shu
Microsoft Research
Redmond, WA
United States

Satish Narayana Srirama
Institute of Computer Science
University of Tartu
Estonia

Neeraj Suri
TU Darmstadt
Darmstadt
Germany

Stergios A. Tegos
Electrical and Computer Engineering Department
Aristotle University of Thessaloniki
Thessaloniki
Greece

Inayat ur Rehman
COMSATS University Islamabad
Islamabad Campus
Pakistan

Hafeez Ur Rehman
National University of Computer and Emerging Sciences
Peshawar
Pakistan

Ling Wang
Department of Automation
Tsinghua University
Beijing
China

Ning Wang
Department of Computer Science
Rowan University
Glassboro, NJ
United States

Jie Wu
Center for Networked Computing
Temple University
Philadelphia, PA
United States

Chu-ge Wu
Department of Automation
Tsinghua University
Beijing
China

Hui Xu
AInnovation
Beijing
China

Shen Yan
Michigan State University
East Lansing, MI
United States

Alessandro Zanni
Department of Computer Science and Engineering
University of Bologna
Italy

Xiao Zeng
Michigan State University
East Lansing, MI
United States

Faen Zhang
AInnovation
Beijing
China

Mi Zhang
Michigan State University
East Lansing, MI
United States

Acronyms

ADAS	advanced driving assistance systems
AAL	ambient assisted living
APIs	application programming interfaces
AI	artificial intelligence
AR	augmented reality
BTS	base transceiver station
CSA	Channel Switch Announcement
CloVR	Cloud Virtual Resource
CoAP	Constrained Application Protocol
CaaS	context as a service
CAPWAP	Control and Provisioning of Wireless Access Point
CNN	Convolutional Neural Network
DNNs	Deep Neural Networks
DTN	delay tolerant network
DBF	distance-based forwarding
edgeOS	edge operating system
ETSI	European Telecommunications Standards Institute
WSN	wireless sensor network
E2C	Elastic Compute Cloud
FLOPs	floating-point operations
FCW	forward collision warning
GRA	Grey relational analysis
HMM	Hidden Markov model
HWMP	Hybrid Wireless Mesh Protocol
IaaS	infrastructure as a service
iFog	indie fog
ITU	International Telecommunication Union
IoD	Internet of drones
IoMaT	Internet of marine things

IoMT	Internet of medical things
IoT	Internet of Things
IoVs	Internet of vehicles
ISP	Internet service provider
IFT	Iterative Feature Transformation
LV-Fog	land vehicular fog computing
LoRa	long range
LTE	long-term evolution
LTE advanced	long-term evolution-advanced
LPWANs	low-power wide-area networks
Marine Fog	marine fog computing
MDP	Markov decision process
MQTT	Message Queuing Telemetry Transport
MANET	mobile ad hoc network
MFC	mobile fog computing
MVCs	mobile vehicular cloudlets
MCS	monitoring and control server
MAC	multiple access control
NIST	National Institute of Standards and Techology
NFC	near-field communication
NCS	Neural Compute Stick
NOMA	nonorthogonal multiple access
ONF	Open Network Foundation
QoE	quality of experience
RFID	radio-frequency identification
RSUs	roadside units
RoT	roots of trust
SLA	service level agreement
S3	Simple Storage Service
SNP	single nucleotide polymorphism
SNS	social network services
SaaS	software as a service
SDN	software-defined network/ing
STORMSeq	Scalable Tools for Open-Source Read Mapping
SOC	system-on-chip
TOSCA	Topology and Orchestration Specification for Cloud Applications
UAVs	unmanned aerial vehicles
UAV-Fog	unmanned aerial vehicular fog computing
UE	user equipment
UE-fog	user equipment-based fog computing

VAT	Variant Annotation Tool
VFC	vehicular fog computing
V2D	vehicle-to-device
V2V	vehicle-to-vehicle
VHF	very high frequency
VM	virtual machine
WAVEs	wireless access in vehicular environments
WPAN	wireless personal area network
WSNs	wireless sensor networks
WiMAX	Worldwide Interoperability for Microwave Access

Part I

Fog Computing Systems and Architectures

1

Mobile Fog Computing

Chii Chang, Amnir Hadachi, Jakob Mass, and Satish Narayana Srirama

Institute of Computer Science, University of Tartu, Estonia

1.1 Introduction

The Internet of Things (IoT) paradigm motivates various next-generation applications in the domains of smart home, smart city, smart agriculture, smart manufacturing, smart mobility, and so forth [1], where the online systems are capable of managing physical objects, such as home appliances, public facilities, farming equipment or production line machines via the Internet. Moreover, mobile objects, such as land vehicles (e.g. cars, trucks, buses, etc.), maritime transports (e.g. ships, boats, vessels, etc.), unmanned aerial vehicles (UAVs; e.g. drones), and user equipment (UE; e.g. smartphones, tablets, mobile Internet terminals, etc.), have become the indispensable elements in IoT to assist a broad range of mobile IoT applications.

Mobile IoT applications emphasize the connectivity and the interoperability among the IoT infrastructure and the mobile objects. For example, in an Internet of Vehicles (IoVs) application [2], the IoT-based smart traffic infrastructure provides the connected roadside units (RSUs) that assist the smart cars to exchange the current traffic situation of the city center toward reducing the chance of traffic accidents and issues. As another example, classic disaster recovery activities of a city require numerous manned operations to monitor the disaster conditions, which involve high risk for human workers. Conversely, by integrating an Internet of Drones (IoD) [3], the smart city government can dispatch a number of drones to monitor and to execute the tasks without sending human workers to the frontline. Unexceptionally, mobile IoT also has benefited maritime activities in terms of improving the information exchange among the vessels and the central maritime management system, hastening the overall process speed of fishery or marine scientific activities [4].

Fog Computing: Theory and Practice, First Edition.
Edited by Assad Abbas, Samee U. Khan, and Albert Y. Zomaya.

Besides the public applications, mobile IoT plays an important role in personal applications, such as Internet of Medical Things (IoMT) applications [5], which utilize both inbuilt sensors of the UE (e.g. smartphone) and the UE-connected body sensors attached on the patient to collect health-related data and forward the data to the central system of the hospital via the mobile Internet connection of the UE.

Explicitly, the mobile IoT applications described above are time-critical applications that require rapid responses. However, the classic IoT system architecture, which relies on the distant central management system to perform the decision making, has faced its limitation to achieve the timely response due to latency issues deriving from the dynamic network condition between the front-end IoT devices and the back-end central server. Furthermore, the large number of connected mobile IoT devices have raised the challenges of mobile Big Data [6] that increase the burden of the central server and hence, lead to bottleneck issues. In order to improve the agility and to achieve the goal of ultra-low latency, researchers have introduced fog computing architecture [1].

Fog computing architecture (the fog) distributes the tasks from the distant central management system in the cloud to the intermediate nodes (e.g. routers, switches, hubs, etc.), which contain computational resources, to reduce the latency caused by transmitting messages between the front-end IoT devices and the back-end cloud. Specifically, the fog provides five basic mechanisms: storage, compute, acceleration, networking, and control toward enhancing IoT systems in five subjects: security, cognition, agility, low latency, and efficiency [1]. For example, in IoV application, the central server can migrate the best route determination function from the cloud to the roadside fog nodes to assist the travel of the connected vehicles. As another example, in an outdoor-based IoMT application, the hospital system can distribute the health measurement function and the alarm function to the UE in order to perform timely determination of the patient's health condition and to perform an alarm to catch the proximal passengers' attention when the patient is having an incident.

Here, we use the term *mobile fog computing* (MFC) to describe the fog-assisted mobile IoT applications.

MFC brings numerous advantages to mobile IoT in terms of rapidness, ultra-low latency, substitutability and sustainability, efficiency, and self-awareness. However, the dynamic nature of MFC environment raises many challenges in terms of resource and network heterogeneity, the mobility of the participative entities, the cost of operation, and so forth. In general, the static fog computing frameworks designed for applications, such as the smart home or smart factory would not fully address the MFC-specific challenges because they have different perspectives from the involved entities and the topology. For example, a classic fog computing framework, which may involve a thin mobile client-side application for smartphone users, would not consider how to provide a reliable fog service to the high-speed moving vehicles. Moreover, the classic fog computing framework also would not consider how to provide a reliable fog service to vessels at sea where the telecommunication base stations are not available, and the satellite Internet is too expensive.

The goal of this chapter is to provide an introduction and guidance to MFC developments. Specifically, different from the existing works [7, 8] related to MFC, this chapter discusses MFC in four major application domains: land vehicular fog computing (LV-Fog), marine fog computing (Marine Fog), unmanned aerial vehicular fog computing (UAV-Fog), and user equipment-based fog computing (UE-fog).

The rest of the chapter is organized as follows: Section 1.2 clarifies the term *MFC*. Section 1.3 breaks MFC down into four application domains and describes their characteristics. Section 1.4 enlists the wireless communication technologies used in the mentioned application domains, while Section 1.5 proposes a taxonomy of nonfunctional requirements for MFC. Open research challenges, both domain-oriented and generic, are identified in Section 1.6, and finally, Section 1.7 concludes this chapter.

1.2 Mobile Fog Computing and Related Models

In this chapter, MFC has its specific definition and it is not an exchangeable terminology with the other similar terms, such as mobile cloud computing (MCC) or multi-access (mobile) edge computing (MEC). In order to clarify the meaning of MFC, one needs to understand the aspects of the parties who introduced or adapted the terminologies. Commonly, MCC refers to a system that assists mobile devices (e.g. smartphones) to offload their resource-intensive computational tasks to either distant cloud [7] or to the proximity-based cloudlet [8].

Fundamentally, MEC is an European Telecommunications Standards Institute (ETSI) standard aimed to introduce an open standard for telecommunication service providers to integrate and to provide infrastructure as a service (IaaS), platform as a service (PaaS), or software as a service (SaaS) cloud services from the industrial integrated routers or switches of their cellular base stations. Explicitly, MEC is an implementation approach rather than a software architecture model. Further, as stated by ETSI, ETSI and OpenFog are collaborating to enable the MEC standard and the OpenFog Reference Architecture standard (IEEE 1934) to complement each other [9].

Today, researchers of industry and academia have been broadly using *edge computing* as the exchangeable term with fog computing. However, National Institute of Standards and Technology (NIST) and the document of IEEE 1934 standard for fog computing reference architecture, which derives from OpenFog Consortium, have specifically explained the differences between fog computing and edge computing. Accordingly, "the Edge computing is the network layer encompassing the end-devices and their users, to provide, for example, local computing capability on a sensor, metering or some other devices that are network-accessible" [10]. Further, based on the literature in edge computing domain, which include cloudlet-based computing models [11], one can explicitly identify that edge computing is loosely a *bottom-up* model. Specifically, an edge computing-integrated system emphasizes task offloading from the end-devices to the nearby cloudlet resources, which are capable of providing Virtual Machine (VM) or containers engine (e.g. Docker[1])-based service to the other nodes within the same subnet.

On the other hand, fog computing is a hierarchical *top-down* model in which the system specifically tackles the problem about how to utilize the intermediate networking nodes between the central cloud and the end-devices to improve the overall performance and efficiency. Commonly, such intermediate nodes are Internet gateways such as routers, switches, hubs (e.g. an adaptor that interconnects Bluetooth-based device to IP network). Moreover, a fog node is capable of providing five basic services – storage, compute, acceleration, networking, and control [1]. Correspondingly, when a cloudlet or an IoT device is providing gateway mechanism to the other nodes and they are capable of providing some or all of the basic fog services, we also consider them as fog nodes.

By extending the notion above, MFC is the subset of fog computing that addresses mobility-awareness. Specifically, MFC involves two types – infrastructural fog (iFog)-assisted mobile application and mobile fog node (mFog)-assisted application. In summary, iFog-assisted mobile application enhances the performance of a cloud-centric mobile application by migrating the processes from the central cloud to the stationary fog nodes (e.g. the cellular base station) that are currently connecting with the mobile IoT devices. On the other hand, mFog-assisted applications host fog services on mobile gateways (e.g. in-vehicle gateway devices or smartphones) that interconnect other devices (e.g. onboard sensors, body sensors, etc.) or other things (e.g. proximal cars, people, sensors, etc.) to the cloud.

1.3 The Needs of Mobile Fog Computing

MFC encompasses four application domains: land vehicular applications, marine applications, unmanned aerial vehicular applications, and UE-based applications. Specifically, each domain involves both iFog- and mFog-based architecture. Ideally, the approaches of iFog aim to provide generic solutions that are applicable

1 www.docker.com.

to all the MFC domains where the infrastructure is applicable. On the other hand, mFog-based approaches aim to overcome the challenges in which the iFog is inapplicable or is unable to resolve effectively. To clarify the terminologies used in the rest of the chapter, iFog denotes infrastructural fog node and mFog represents the generic term of mobile fog nodes. Moreover, we further classify mFog to four types: LV-Fog, Marine Fog, UAV-Fog, and user equipment-based fog (UE-fog) corresponding to the mobile fog node hosted on a land vehicle, a vessel, a UAV, and a UE (e.g. smartphone, tablet, etc.).

1.3.1 Infrastructural Mobile Fog Computing

1.3.1.1 Road Crash Avoidance

The number of vehicles on the road is increasing every year as well as the number of road accidents [12]. In order to reduce or avoid the collision accident, academic and industrial researchers have been working on improving the safety aspect of the vehicles. Specifically, the advancement in communication technologies has allowed the development of advanced driver-assistance systems (ADAS), which has emerged as an active manner of preventing car crashes. ADAS has made many achievements through the development of systems that include rear-end collision avoidance and forward collision warning (FCW). The vehicle-to-vehicle (V2V) communication plays a big role in ADAS systems and it is manifested in ensuring that the controllers on board the vehicles (i.e. onboard unit, OBU) are capable of communicating with other vehicles for the purpose of negotiating maneuvers in the intersections and applying automatic control when it is necessary to avoid collisions [13]. The success of these systems relies a lot on the reliability of the communication. Therefore, many models of V2V communication have been investigated. Some of them focus on probabilities and analytic approaches in modeling the communication message reception while others adapt Markovian methods to assess the performance and reliability of the safety-critical data broadcasting in IEEE 802.11p vehicular network [14]. Vehicular ad-hoc network (VANET) has also contributed to integrate and improve the car-following model or platooning, which reduces the risks of collisions and makes the driving experience safer [15]. Explicitly, today's smart vehicular network systems have applied the fog computing mechanisms that utilize the cloud-connected OBUs of the vehicle to process the data from the onboard sensors toward exchanging context information in the vehicular network and participating in the intelligent transport systems.

1.3.1.2 Marine Data Acquisition

Today, marine data acquisition and cartography systems can achieve low-cost data acquisition and processing by composing IoT, mobile ad hoc network (MANET), and delay-tolerant networking (DTN) technologies. Specifically, sea vessels, which equip multiple sensors, can utilize International Telecommunication Union (ITU) standards-based very high frequency (VHF) data exchange system to route the sensory data to the gateway node (i.e. cellular base station) at the shore via the ship-based MANET. Afterwards, the gateway can relay the data to the central cloud. In general, such an architecture may produce many duplicated sensory transmission readings due to the redundant data transmitted from different ships. In order to remove such duplication and to improve the efficiency, the system can deploy fog computing service at the gateway nodes to preprocess the sensory data toward preventing the gateways sending duplicated data to the central server [4].

1.3.1.3 Forest Fire Detection

Emerged smart UAVs, which are relatively inexpensive and can be flexibly dispatched to a large area under different weather conditions, both during day and night, without human involvement are the ideal devices to handle forest fire detection and firefighting missions. Specifically, with onboard image detection mechanism and mobile Internet connectivity, UAVs can provide real-time event reporting to the distant central management system. Further, in order to extend the sustainability of the image-based sensing mission, the system can distribute the computational image detection program to the proximal iFog hosted on cellular base stations and made accessible via standard communication technologies, such as Long-Term Evolution (LTE), SigFox, NB-IoT, etc. Hence, the UAVs can use their battery power only for flying and sensing tasks [16] (Figure 1.1).

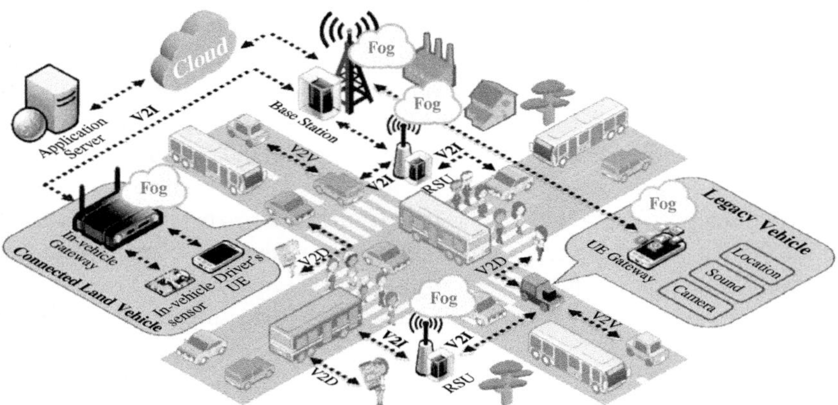

Figure 1.1 Land-vehicular fog computing examples. (*See color plate section for the color representation of this figure*)

1.3.1.4 Mobile Ambient Assisted Living

Today's UE devices, such as smartphones, have numerous inbuilt sensor components. For example, the modern mobile operating systems (e.g. Android OS) have provided numerous software components that are capable of integrating both internal and external sensors to support mobile Ambient Assisted Living (AAL) applications such as real-time health monitoring and observing the surrounding environments of the user to avoid dangers. Fundamentally, classic mobile AAL applications rely on the distant cloud to process the sensory data in order to identify situations. However, such an approach is often unable to provide rapid response due to communication latency issues. Therefore, utilizing proximal fog service derived either from the MEC-supported cellular base station or individual or small business-provided Indie Fog [17] has become an ideal solution to enhance the agility of mobile AAL applications [18].

1.3.2 Land Vehicular Fog

The development of vehicular networking has improved safety and control on the roads. Especially, LV-Fog nodes have emerged as a solution to introduce computational power and reliable connectivity to transportation systems at the level of Vehicle-to-Infrastructure (V2I), V2V, and Vehicle-to-Device (V2D) communications [19]. These networks are shaped around moving vehicles, pedestrians equipped with mobile devices, and road network infrastructure units. Further, these aspects have facilitated the introduction of real-time situational/context awareness by allowing the vehicle to collect or process data about their surroundings and share these insights with the central traffic control management units or other vehicles and devices in a cooperative manner.

To perform such activities, there is a need for adequate computing resources at the edge for performing time-critical and data-intensive jobs [20] and face all the challenges related to data collection and dissemination, data storage, mobility-influenced changing network structure, resource management, energy, and data analysis [21, 22].

Most of the techniques proposed to solve these challenges are focusing on merging the computation power between vehicular cloud and vehicular networks [19]. This combination allows usage of both vehicles' OBUs and RSUs as communication entities. Another side that has been investigated is the issues related to latency and quality optimization of the tasks in the Vehicular Fog Computing (VFC) [20], and it was formulated by presenting the task as a bi-objective minimization problem, where the trade-off is preserved between the latency and quality loss. Furthermore, handling the mobility complexity that massively affects the network structure is addressed by using mobility patterns of the moving vehicles and devices to perform a periodic load balancing in the fog servers [23] or distance-based forwarding (DBF) protocol [19]. The energy management and computational power for data analysis are controlled by distributing the load among the network entities to make use of all the available resources based on CR-based access protocol [22].

Moreover, the design of the Media Access Control (MAC) layer protocol in the vehicular networks is essential for improving the network performance, especially in V2V communication. V2V enables cooperative tasks among the vehicles and introduces cooperative communication, such as:

Dynamic fog service for next generation mobile applications. The emergence of new mobile applications, such as augmented reality (AR) and virtual reality, have brought a new level of experience that is greedy for more computational power. However, the traditional approach of a distant cloud-driver is incapable of achieving with good performance due to latency. Therefore, introducing *Metropolitan vehicle-based cloudlet*, which is a form of mobile fog node model, solves the latency issue by dynamically placing the fog at the areas with high demand. Furthermore, by adopting a collaborative task offloading mechanism, the vehicle-based mobile fog nodes are capable of effectively distributing the processes across all the participative nodes, based on their encounter conditions [24].

Federated intelligent transportation. Traffic jams start to have a considerable negative impact by wasting time, fuel, capital, and polluting the environment due to the nonstop increase in the number of vehicles on the roads [25]. Fortunately, cloud-driven smart vehicles have emerged as a facilitator to overcome the problem. The solution resides in considering the serviceability level of mobile vehicular cloudlets (MVCs), which are a form of the mobile fog node model, based on the real-world large-scale traces of mobility of urban vehicles collected by onboard computers. Based on the peer-to-peer communication network, vehicles can further improve the traffic experience by exchanging real-time information and providing assistance to the manned or unmanned vehicles [26].

Vehicular opportunistic computation offloading. Public transportation service vehicles, such as buses and trams, which commonly have fixed routes and time schedules, can be the mobile fog nodes for the other mobile application devices inside the proximal encountered vehicles that need to execute time-sensitive and computation-intensive tasks, such as augmented reality (AR) processes used for the advanced driver assistance systems and applications [27].

1.3.3 Marine Fog

Integrating IoT to existing legacy marine systems can provide rapid information exchange. Initially, classic marine communication systems utilize VHF radio to obtain ship identification, ship location, position, destination, moving speed, and so forth. Unfortunately, VHF can provide only 9.6 kbps, which is insufficient to provide marine sensory data streaming [28]. Alternatively, ships can utilize the new satellite Internet to deliver their data, which is capable of achieving 432 kbps. However, satellite communication is not affordable for small and medium-size businesses since a simple voice service can cost USD $13.75 per minute [28]. In order to overcome the issue, researchers have introduced fog computing and networking-integrated marine communication systems for the Internet of marine t hings (IoMaT) [4]. Here, we term such a fog computing model Marine Fog.

By integrating a virtualization or containerization technology-based fog server with onboard equipment, vessels are capable of realizing a software-defined network (SDN) that allows the vessels to (re)configure a message routing path dynamically. Afterwards, by utilizing Marine Fog–based SDN mechanism, vessels can easily establish an ad hoc–based DTN, which caches data at the onboard Marine Fog node until the vessel encounters the next Marine Fog node, for delivering sensory data from the data source to the base stations toward relaying the data to the cloud. Moreover, an advanced Marine Fog node within the network may also perform data preprocessing in order to further reduce the transmission latency [29] (Figure 1.2).

Existing wireless sensor network (WSN) architecture in marine monitoring uses sea buoys as sink nodes, capable of communicating with nearby sensor nodes (other buoys, vessels) directly (e.g. using ZigBee), as well as via the cellular Internet network [30]. By introducing the previously mentioned virtualization, the WSN architecture could be extended to be used in Marine Fog. However, this approach amplifies the need for energy-harvesting technology at the buoys.

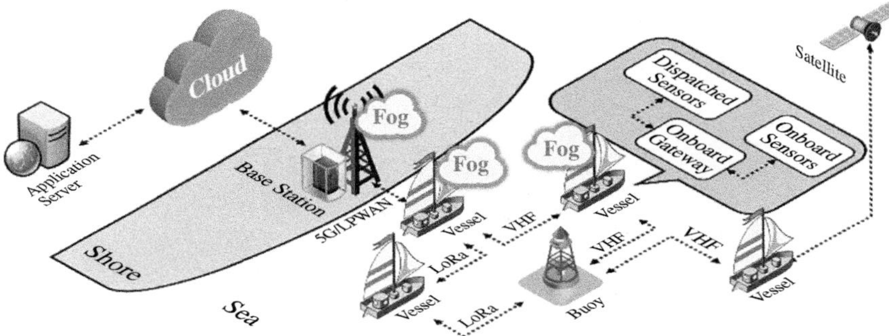

Figure 1.2 Maritime fog computing examples.

1.3.4 Unmanned Aerial Vehicular Fog

UAVs, which are also referred to as drones, can be employed in a broad range of applications due to their unrestricted geo-location feature. Hence, they have become one of the major elements in the IoT ecosystem. For example, smart city application can utilize UAVs as mobile sensors that can be sent to critical environments that are dangerous for humans (e.g. a forest fire) [16]. Specifically, by utilizing UAVs, the system is capable of establishing a UAV ad hoc network that is capable of performing a wide-ranged geolocation-based sensory data streaming network dynamically. Moreover, by mounting fog nodes on UAVs to perform sensory data analysis, the system can further improve the agility of identifying the emergency situations.

Below, we describe the features of the UAV-Fog [31], with corresponding examples indicated in Figure 1.3.

Fast deployment. Modern UAVs are capable of carrying on tasks programmatically without human interference. Further, a system can dispatch a large number of UAVs to perform a temporary mission in an area where the manned vehicles are unable to reach or unable to effectively perform the tasks. For example, UAV-Fogs can assist wildfire problems in Portugal, Spain, and Australia [32].

Scalability. The rapid growth of large-scale IoT applications requires more network infrastructure and fog computing resources in order to compensate for ultra-low latency. However, investing in the base infrastructure in certain areas is not cost-efficient. Hence, instead of developing the infrastructural IoT and fog network, the service provider can deploy UAVFog nodes to those areas. For example, in order to support the sensory data streaming performed by underwater vehicles, UAV-Fog nodes can fly above the water surface in order to route the data stream to the base station at the shore [31].

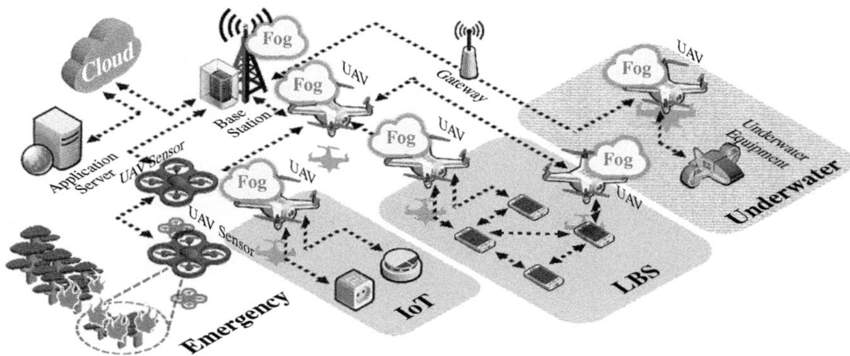

Figure 1.3 UAV fog computing examples.

Flexibility. UAV-Fog nodes can equip heterogeneous capabilities to support various applications. For example, an Olympic event in a city lasts 16 days. During the contests, a large number of visitors are gathering in the city and many of them are using Social Network Services (SNS) to disseminate information (e.g. text, image, video posts) related to the event. However, the city's network infrastructure may not have sufficient capacity to provide the high-quality experience for the SNS users due to traffic overload. In order to support the best quality of experience (QoE) for the SNS users, SNS providers may deploy UAV-Fog nodes to the city to provide a temporary location-based social network (LBSN) mechanism that directly routes the content (e.g. Twitter posts, YouTube video stream, etc.) within the city when the content provider and the receiver are within the city.

Cost-effective. The content described in previous paragraphs has indicated that employing UAV-Fog nodes is a cost-effective solution for many applications that require only a temporary enhancement for computational or networking needs. For example, wildfires in Australia often occur in areas where the network infrastructure is unavailable. Hence, establishing an infrastructural IoT-based smart monitoring system at such an area is unrealistic. Second, many cities in the world are unable to provide fundamental infrastructure for the rapid growth of IoT applications. Instead of waiting for the hardware service provider to complete the infrastructure, the IoT software service provider can simply deploy more UAV-Fog nodes to the areas that require more resources. Finally, many cities often spent a large amount of money on network infrastructure for temporary events, which is cost-inefficient. Although it is possible to send the manned land vehicular-based nodes (e.g. mobile base stations) to support the need, compared to unmanned UAVs, the manned solutions require payroll for human workers and extra petrol or electricity, since the movement of land vehicles is constrained based on the roads.

1.3.5 User Equipment-Based Fog

UE represents the end-users' terminal devices (e.g. smartphones) that are connecting to an Internet Service Provider (ISP)'s network. Initially, UEs are thin clients in the IoT ecosystem. However, recent research efforts have utilized UEs as one of the major service provisioning elements. Accordingly, the following use cases illustrate the usefulness of fog-integrated UEs (Figure 1.4).

1.3.5.1 Healthcare

Healthcare is one of the top-five application domains in fog computing that has potential market value up to $2737 billion by year 2022 [33]. Specifically, IoMT applications, which commonly rely on the central cloud for managing data and

Figure 1.4 UE fog computing examples. (*See color plate section for the color representation of this figure*)

performing decision-making, are now distributing certain tasks to the intermediate gateways. In particular, IoMT applications broadly utilize UEs (e.g. smartphones, tablets, etc.) as the gateways of wearable body sensors, in order to let the IoMT servers (e.g. hospital) acquire the sensory data anytime, anywhere from the patients. Further, considering the need of agile sensory data stream processing when the patients are in outdoor areas, which may not be able to maintain a high-quality network connection, utilizing the infrastructural fog (e.g. hosted at the cellular base station or ISP's Wi-Fi access points [APs]) can highly improve the overall agility of the data processing and identification of emergency situations. Moreover, considering modern UEs have powerful central processing units (CPUs) and decent storage spaces when the infrastructural fog is unavailable, the processes can also migrate to UEs to support the service continuity [34].

1.3.5.2 Content Delivery

Besides utilizing UEs as computing-based fog servers, UE-based fog computing nodes (UE-fog nodes) are also the ideal networking service providers. For example, replays of soccer matches, which catch important moments of the game, haver become an indispensable element of the sports game for the fans both inside or outside the stadium. Fundamentally, the audience members download the replay video to their UEs via the Internet. However, considering that there can be a large number of requests coming from the audience, the local wi-fi or cellular base stations may not be able to provide sufficient speed, especially when the audience demands high-quality or even ultra-high-quality videos. In order to address this problem, the application can host fog servers in the UEs of the audience to support the SDN mechanism, which allows the service provider to establish a local ad-hoc content caching group among the fans in the stadium and its surrounding areas toward reducing the burden of both the ISP and the soccer replays service provider [35].

1.3.5.3 Crowd Sensing

Integrating UEs into the fog can raise many advanced people-centric applications that are directly related to people's daily lives. For example, in the lost child situation, local government or police can collaborate with a local ISP to request the SDN mechanism-supported UE-fog nodes, which are hosted on the smartphones of the people on-site, to assist the lost child situation by utilizing the inbuilt cameras and audio recorder of the UEs as the sensors, then utilize some of the more powerful UE-fog nodes as the super-peers to route the data to the local cloud of the ISP, which is accessible by the government and the police, toward hastening the entire process. Moreover, the UE-fog nodes that have high-performance computational power can also provide context as a service (CaaS) instead of routing the raw sensory data [36]. Specifically, CaaS mechanism–supported UE-fog nodes allow the requesters to deploy their own data processing algorithm to UE-fog nodes toward preprocessing the raw sensory data and hastening the overall process.

1.4 Communication Technologies

In this section, we give an overview of communication technologies used by mobile things and mobile fog nodes in each of the major application domains. Unsurprisingly, in existing literature Wi-Fi technology is most prevalent, due to the ready availability of devices with 802.11 support: routers, smartphones, single board computers. Wi-fi is suitable for mFog thanks to the easy mobility of Wi-Fi AP technology. On the other hand, 4G technology, e.g. long-term evolution (LTE), which needs static base stations is the dominant option for iFog.

The fast movement of nodes among road network environment has obliged the existing protocol for the physical layer to consider multipath fading and Doppler frequency shifts. Therefore, the trend is to use very high-frequency radio waves, such as micro or millimeter waves.

1.4.1 IEEE 802.11

The IEEE 802.11 set of specifications, commonly referred to as Wi-Fi, is the most widely used wireless communication technology found in fog computing. Since Wi-Fi infrastructure is widely deployed in homes, offices, public spaces, and so forth, it is the natural choice for fog-thing and fog-UE communication, with the fog node hosting the Wi-Fi AP.

The standards *802.11n* and *802.11ac*, for instance, have a typical data rate of 200 and 400–700 Mbps respectively, the typical range of 802.11 routers in the 2.4 GHz band can be up to 50 m [37]. The next generation *802.11ax* promises to enhance data rates further, but it is unlikely that significantly higher coverage range will be achieved in practice, due to the recent versions using the 2.4 and 5 GHz bands.

Interestingly, advertising capabilities of 802.11 can be improved by including additional information (e.g. fog node capability and status) in the 802.11 advertising beacons [38].

The mentioned signal coverage ranges are suitable in domains where the client device mobility speed is low; consider, for example, pedestrians in UE-fog. Additionally, the smartphones already employ the technology. Existing UE-fog research that does not include real-world technology choice and simply consider the data rate aligns with the capabilities of wi-fi. For example, even 6.9 Gbps rates [39] are theoretically supported by *802.11ac*. In terms of existing research prototypes, laptop hotspots are a common choice to establish the wi-fi AP [40–42]. Since laptop hotspots generally operate with *802.11n* technology, it is important to consider the newer standards in future fog prototypes.

In UAV-fog, following the mFog concept, a wi-fi AP could reside at UAV node or, alternatively, static *802.11ac* APs may act as sink nodes supporting a group of UAVs [16].

IEEE 802.11p, a.k.a. wireless access in vehicular environments (WAVEs) is adapted for the wireless environment with vehicles. In addition, they are designed in such manner that they are very suitable for single hop broadcast V2V communications; however, this technology suffers from an issue related to scalability, reliability, and unbounded delays due to its contention-based distributed medium access control mechanism [23, 43].

For maritime use cases, the coverage ranges offered by Wi-Fi are generally not suitable. However, Wi-Fi is useful in vessels for onboard networks where the clients are crew and passengers, but such scenarios can be categorized rather as UE-fog.

1.4.2 4G, 5G Standards

Currently operating cellular networks target the requirements of the 4G standard (also known as IMT-Advanced), specified by the International Telecommunication Union (ITU) in 2008 [44]. For instance, the requirements suggest 100 Mbps data rates for clients moving at high speeds (e.g. in a train) and 1 Gbps for stationary situations.

Among technology standards accepted as fulfilling the 4G requirements are IEEE *802.16m* (WiMAX v2) and 3 GPP Long Term Evolution-Advanced

(LTE Advanced), the latter of which has seen far bigger deployment and thus is the common option for existing mobile fog.

To supersede 4G, the ITU is defining requirements for the 5G networks, also called IMT-2020. The 2017 draft of technical performance requirements [45] notes peak data rates of 20 and 10 Gbps for downlink and uplink, respectively, and specifies channel link data rates for four different mobility classes. 5G also specifically targets supporting cases where the density of devices is large, growing from 4G's 10^5–10^6 devices km^{-2} [46].

5G networks are enablers for smart collaborative vehicular network architecture since they provide possibilities for fulfilling the requirements of reliability, handover, and throughput of future vehicular networks [47]. LTE D2D-based VANET has proven to be suitable for the safety-critical IoV applications, thanks to their effectiveness in coping with high mobility and precise geo-messaging [43].

The MEC paradigm has introduced the handover and migration of VMs to the cellular base stations for supporting the UE [48–50], however, the similar idea potentially applies to the other mobile fog domains.

In maritime fog systems, the shore-located cellular base stations can be leveraged to also act as sink nodes [4, 51], gathering sensor data from the vessels. Multiple access techniques, such as nonorthogonal multiple access (NOMA) offered by 5G, are considered useful for UAV cloudlets to maximize efficiency [52]. However, generally 4G/5G coverage is available in more urban areas, so for marine communication at sea or UAV deployments in remote areas, alternatives such as satellite communication need to be considered.

1.4.3 WPAN, Short-Range Technologies

From the perspective of the mobile thing, wireless personal area network (WPAN) technologies such as ZigBee (802.15.4) and Bluetooth (801.15.1) are suitable for lower bandwidth and lower energy communication needs, such as interacting with IoT devices or exchanging metadata.

The shorter range, while unfit for marine scenarios, can be applied in UAV-Fog since use cases,such as supporting land/marine vehicle, mandate that the UAV itself will adjust its location to stay close to the peers [31].

The traditional Wi-Fi AP-based infrastructure can be expanded using Wi-Fi direct. Here the client devices form a local Wi-Fi direct group, reducing the load on the AP by locally disseminating the data [35, 53] and advertising device services [54].

Bluetooth and Bluetooth Low Energy are common choices for UE when mediating data from other devices to a fog node (e.g. Wi-Fi AP-based), for instance, forwarding sensor data from Medical IoT devices [5].

Near-field communication (NFC) radio-frequency identification (RFID)-based technology, such as NFC may add an additional layer of physical security due to the extremely low signal range, for instance in UE-based Mobile-to-Mobile computation off-loading [55].

1.4.4 LPWAN, Other Medium- and Long-Range Technologies

VHF marine radio VHF is the internationally used technology for marine radio in the frequency range 156—163 MHz. Typical range of VHF is reported to be up to 70 nautical miles from a land-based station [56], while ship-to-ship signal range is below 40 km [57]. However, VHF is limited in supported data rate, which is below 30 kbps. While VHF may be sufficient to transmit sensory readings from ships to shore-deployed fog nodes [4] for aggregation and forwarding, the low data rate is unsuitable for agile transfer of larger data (e.g. video streams or VM images).

Satellite systems offer higher speeds compared to VHF and provide the greatest signal coverage, which is an important factor considering the distances involved in the marine domain. Yet, due to their high cost, these are a viable option only for larger vessels [57, 58].

Low-power wide-area networks (LPWANs) long range (LoRa) has been receiving attention lately as an energy-efficient, long-range wireless technology. In the mF2C project [29], the physical layer-LoRa, accompanied with LoRaWAN at the data link layer is used for ship-to-ship communication, while [59] use it for ship-to-shore communication in harbors. LoRa with LoRaWAN can cover up to 15 km in rural areas with a data rate up to 37.5 Kbps [60], making it a lower-energy alternative to VHF.

Sigfox and NB-IoT can be considered as competitors to LoRaWAN. While LoRaWAN and Sigfox operate in unlicensed bands, NB-IoT operates in licensed frequency bands.

Dedicated short-range communications (DSRC) is another technology defined especially for VANET, which are one-way or two-way short-range to medium-range wireless communications designed for allowing V2V and V2I communications. It is characterized by its frequency of 75 MHz licensed spectrum in 5.9 GHz band, which is provided by Federal Communications Commission (FCC) in the United States [20, 26].

1.5 Nonfunctional Requirements

An MFC system needs to address a number of nonfunctional requirements in order to achieve the basic quality of service (QoS) principles. In order to provide a comprehensive guide to the developers, this section describes the nonfunctional requirements in five aspects – heterogeneity, context-awareness, tenant, provider, and security.

Figure 1.5 summarizes the elements of the five aspects of the nonfunctional requirements. In general, the five aspects are corelated to one another. For example, heterogeneity is a common factor that needs to be considered when the fog tenant plans to choose which fog server to use to deploy their applications. Similarly, context awareness, which represents the runtime factors, is also influencing the decision of when and where to deploy and execute the tasks. Furthermore, the tenant-side end-device or end-user (i.e. tenant-side clients) is influencing the decision of how the provider of fog servers manage the fog nodes. On the other hand, the QoS of the provider's fog nodes also influences the decision for the tenant to choose the right fog node for tenant-side clients. In addition, although all the four MFC domains involve the five aspects, the complexity level of each aspect in a different domain can be quite different. For example, an application scheduling scheme designed for LV-fog may require significant adjustment when the developers intend to apply the scheme to UAV-fog because the heterogeneity level and the context factors are very different between the two domains.

As indicated above, we distinguish the tenant and the provider in which the *providers* represent the owners who own the fog server machines and are providing the fog infrastructure and PaaS to application service providers known as the *tenants* in a multitenancy manner [61]. For example, a telecommunication company such as Vodafone may host fog servers on their base transceiver station (BTS) and enable the accessibility of the fog server via 4G/5G/5G NR (https://www.qualcomm.com/invention/5g/5g-nr) communication. Therefore, application service providers can tenant the fog servers and deploy fog-based applications on the

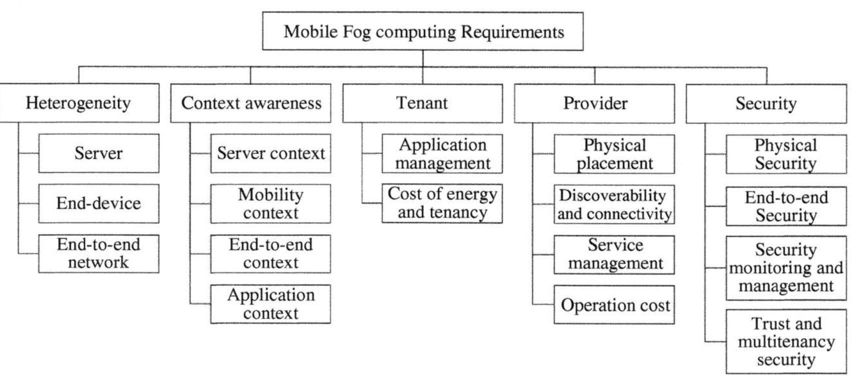

Figure 1.5 A taxonomy of non-functional requirements of mobile fog computing.

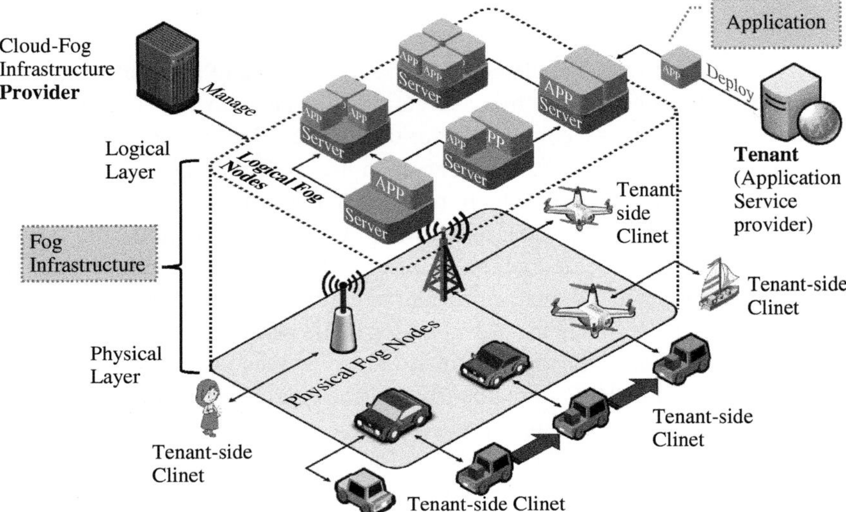

Figure 1.6 Fog infrastructure service provider, fog service tenant, and tenant-side clients.

servers toward serving the tenant-side clients. Correspondingly, Figure 1.6 illustrates the relationships among the involved entities.

Note that although, in a general perspective, fog servers are expected to support multitenancy [61], in many cases, service providers may provide the fog nodes in a highly integrated manner in which a single provider manages both the underlying fog servers and the application services. For example, it would be a common case that an indie fog [17]-based UE-fog service provider who follows the common standards to provide the micro-fog services from their own customer premises equipment (CPE) would manage both the fog service software and the host hardware. As another example, Marine Fog systems would deploy the integrated fog nodes on vessels based on the isolated platform for marine activities in order to prevent security issues.

Based on the state-of-the-art literature in both iFog and mFog across the four MFC domains, we explain the elements of the five aspects and what needs to be addressed in order to achieve the QoS in MFC.

1.5.1 Heterogeneity

There are three types of *heterogeneity*: server heterogeneity, end-device heterogeneity, and end-to-end heterogeneity.

1.5.1.1 Server Heterogeneity

Different to the common IaaS/PaaS-based cloud services, which are virtual resources, fog services are hosted on resource constrained physical equipment that has limited computational power and networking performance. Therefore, when tenants intend to deploy their applications, they need to consider the compatibility, connectivity, interoperability, and reliability of the fog servers. Specifically, we can classify the heterogeneity of the fog servers into two aspects: hardware type and software type.

- *Hardware type.* Represents the hardware component specification and configuration. In detail, the provider should clearly provide information on the hardware in terms of the computational resource specifications, such as CPU model code and speed, RAM model code and speed, read/write speed of storage, independent or integrated GPU, vision processing unit (VPU), field-programmable gate array (FPGA), application-specific integrated circuit (ASIC), AI accelerator, etc.; the available networking resources specification, such as IEEE 802.11a/b/g/n/ac, Bluetooth LE, IEEE802.15.4, LoRa, NB-IoT, etc.; extra components such as inbuilt or connected sensors that are accessible via the API provided by the fog server. Furthermore, if the fog server is hosting on a mobile Fog node, the provider should also provide the corresponding mobility-related information, such as the route of its moving path, the moving speed, and so forth.
- *Software type.* Denotes the software application deployment platform supported by the fog server. For example, the fog server may support VM-based service which allows a flexible configuration. Alternatively, the fog server may provide a containerization-based platform service, such as Docker (https://www.docker.com). Whereas, for the resource-constraint devices, which can serve only a limited number of requests, the provider may configure a FaaS-based platform that allows the tenant to deploy functions on the fog servers to provide microservice to tenant-side clients instead of the completed applications.

1.5.1.2 End-Device Heterogeneity

The heterogeneity in hardware specification and the software specification of the end-device/user influences the overall QoS and QoE that tenant can provide. In order to achieve the best QoS and QoE, tenants need to choose fog nodes that are most compliant and most efficient for their end-devices. Specifically, end-device heterogeneity involves hardware specification in terms of the type of the device (e.g. smartphone, mobile sensor, drone, etc.), and computational power and network capabilities, which influence how the end-device/user interact with the tenant's fog applications deployed on the provider's fog server. Further, from the software specification aspect, the tenants need to consider what software they can install at the end-device/user-side and how well the client-side software may perform when it interacts with the fog nodes?

1.5.1.3 End-to-End Network Heterogeneity

The end-to-end network heterogeneity involves the three aspects listed here, and all of them would influence the performance of the other nonfunctional requirements.

- *Device-to-fog (D2F).* Represents the radio network and the communication protocol used between the fog server and the end-device. In general, D2F has a broad range of networking options and hence, increase the complexity from the perspective of both fog server provider and the tenant. For instance, radio network options encompass IEEE 802.11 series, IEEE 802.15.4, Bluetooth, 3G/4G cellular Internet, LoRa, SigFox, NB-IoT, LTE-M, 5G New Radio (5G NR), and so forth, which indicates that the provider may need to support multiple radio networks in order to fulfill multitenancy. On the other hand, tenants also need to consider the available radio network provided by the fog server in order to optimize their applications.
- *Fog-to-fog (F2F).* Represents the communication network between fog nodes in both horizontal and vertical manner. For example, Figure 1.7 illustrates the vertical communication between two LV-Fog nodes, while both LV-Fog nodes also rely on an iFog node to interconnect them to the cloud. Explicitly, the F2F network would highly influence the performance of the tenant-side application,

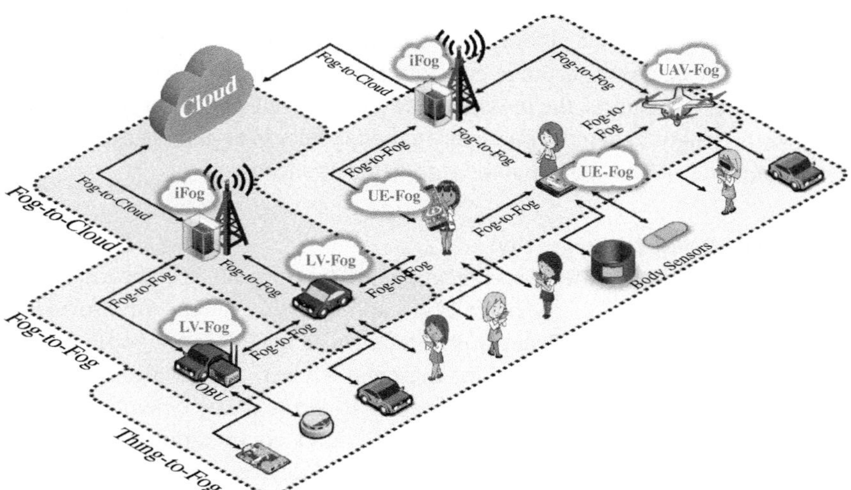

Figure 1.7 The three types of end-to-end networking.

especially when the application requires a certain process or decision making from the cloud.

- *Fog-to-cloud (F2C).* Represents the communication network between the fog node and the cloud, which has a similar influence as F2F to the tenant-side applications. In particular, depending on the underlying infrastructure, geo-location of the cloud and the intercontinental routing path between the fog node and the cloud, the latency between the end-device and the cloud can be very different. Therefore, besides the D2F and F2F, the tenant also needs to consider the F2C network, especially when the application is unable to operate fully in a self-managed manner.

1.5.2 Context-Awareness

The context represents the runtime factors that influence the server operation and the applications. In general, the context of MFC involves the follows:

1.5.2.1 Server Context

Server context influences the serviceability and sustainability of the fog server. Specifically, it involves the following runtime factors:

- *Current task in queue and the task types* [24]. Regardless of the end-to-end communication latency between the tenant-side client and the fog server, the task in queue and the task types are the factors that influence the response time of a tenant-side application deployed on the fog server. In particular, if a fog server has a large number of resource intensive computational tasks in the queue, its response will be much slower than a fog server, which has a decent number of simple tasks in the queue. In addition, the complexity of the task is a relative factor depended on the performance of the fog server.
- *Energy* [62], which is a specific context factor for mobile fog nodes that rely on battery power. For example, UE-fog node and UAV-fog node are commonly operated based on battery power. Hence, tenants need to identify the sustainability of the mobile fog nodes before they decide to deploy the application on them for serving the tenant-side clients.

1.5.2.2 Mobility Context

Mobility context includes a number of movement-related factors that influence the accessibility and connectivity between the tenant-side client and the fog node. In particular, mobility context includes the following elements [27, 62, 63]:

- *Current location and destination,* which represents the current physical geo-location and the destination of the tenant-side client and the fog node, which are the basic parameters to identify the potential encounter rate between the two nodes, based on measuring the possible movement routes.

- *Movement,* which includes the moving frequency (i.e. how often the entity is moving), maximum and average moving speed, acceleration rate (i.e. moving speed of a specific time), moving direction, moving path, and route. Specifically, the system requires continual up-to-date information of these factors in order to adjust the mobility context reasoning.

1.5.2.3 End-to-end Context

The end-to-end context represents the factors that influence the communication performance between the tenant-side client and the fog server. Essentially, a system can measure the end-to-end context based on the server context and mobility context. Hence, one can consider that the end-to-end context is a second level context. In summary, end-to-end context involves the following factors [26, 64]:

- *Signal strength.* A dynamic factor at run-time in many cases, such as LV-fog and UE-fog environments where the density of the wireless network objects is high and hence, they can interfere the signal strength of one another. Therefore, tenants need to consider the signal strength when they intend to measure the connectivity between the fog server and the tenant-side client.
- *Encounter chance and Inter-contact time.* Represent the possibility of the fog server and the tenant-side client to successfully communicate and how long they can maintain the connection for the current stage and next stage. Specifically, it influences the decision of whether the tenants should deploy the application on the fog server for their clients or not, and whether the tenant-side clients can utilize the fog servers for task distribution or not.

1.5.2.4 Application Context

Application context influences the server-side performance and also the tenant-side decision regarding where the application should perform the task. In some cases, the process involves handling large amounts of sensory data locally at the tenant-side client, which could be more energy efficient than distributing the task to a fog server, because the involved data size is too large for the tenant-side client to transmit it to the fog server effectively. In summary, application context involves the following elements [28, 64]:

- *Request data type and the amount of data.* Determines the complexity of the request from the perspective of the computational power of the tenant-side client and the fog server.
- *Deadline of the task and task priority.* Determines the required time to complete the task derived from the tenant-side client. In general, it is a part of the service level agreement (SLA), which specifies that the fog server should adjust the tasks in its queue when it needs to prioritze some tasks.

- *Energy consumption of the client*. Which represents one of the major costs of the tenant-side client, when the tenant-side application demands that the client relay the data to its encountered fog server for processing. As mentioned previously, in some cases local processing at the tenant-side client can be more energy-efficient. Hence, the tenant needs to consider the optimal energy efficiency for the clients in the application management.

1.5.3 Tenant

The nonfunctional requirements of tenants, who are responsible to manage the fog applications, aim to achieve the adaptability of runtime fog applications that involve both application services deployed on the fog servers and the application clients operating on the end-devices. In order to comply with the dynamic MFC environment, tenants need to address the adaptability at each phase of the application management. In general, the management of fog applications, which fundamentally are process management of IoT applications [65], encompasses three basic phases: (re)design, implement/configure and run/adjust.

1.5.3.1 Application Management

(Re)design phase involves the software architecture design and the application process modeling design. Specifically, in MFC, tenants need to consider the adaptivity of their software architecture in terms of self-awareness and deployability. Self-awareness assures that the applications have corresponding mechanisms to identify the situation of the runtime application (e.g. the movement of the end-device or the fog node) and are capable of optimizing the process model automatically with a minimal dependence on the distant central management system. In order to fulfill this requirement, the architecture design may need to support decomposition mechanisms that allow the applications to move the processes of applications or even portions of a process (i.e. tasks) from one node to another node dynamically at runtime.

Implement/configure phase represents the stage that transforms the design abstraction to the executable software and deploying the software to the involved fog nodes and end-devices. In contrast to the classic static IoT systems which do not need to frequently adjust the location of application services, in MFC, based on the runtime context factors of the fog nodes and end-devices, the tenant needs to support the rapid (re)deployment mechanism in order to allow the fog applications to move the processes among the fog nodes toward optimizing the agility of the fog applications. Essentially, the rapid (re)deployment mechanism requires a compatible technical support from the fog infrastructure providers, considering the heterogeneity and dynamic context factors of the fog nodes, the tenants also need to develop the optimal decision-making schemes specifically for their applications in order to deploy the applications to the fog nodes in an optimal manner.

Run/adjust phase represents the runtime application and capabilities of autonomous adjustment based on contextual factors. In general, MFC applications need to support three basic mechanisms and consider two cost factors:

- *Task allocation.* Represents where the tenants should (re)deploy and execute their tasks. In general, unless the entire system utilizes only iFog nodes, MFC applications are rarely operating tasks at fixed locations. Therefore, while the mFog nodes or the tenant-side clients are moving, the fog applications need to continuously determine the next available fog nodes for the clients based on the contextual factors and the specifications (see previous descriptions) in order to rapidly allocate and deploy the tasks to the fog nodes.
- *Task migration.* Has a slight similarity to task allocation in term of (re)deploying tasks at fog nodes. However, task migration can involve much more complexities. For example, in a stream data processing-based fog application, when the client encounters the next fog node while the previous process is in progress, the previous fog node may try to complete the task and then intent to save the process state and wrap the result, the process state information together with the application software (i.e. in case the application is not preinstalled at all the fog nodes) as a task migration package toward sending the task migration to the next encountered fog node of the client. However, there is a chance that the routing path to the new fog node of the client does not exist, which leads to failure of the task migration procedure. Certainly, the example has illustrated only one of the failures in task migration. In order to support the adaptability at the run/adjust phase, the tenants need to consider all the contextual factors and heterogeneity issues in supporting adaptive task migration.
- *Task scheduling.* Represents the timing of any action at the run/adjust phase. In general, based on the application domain, task scheduling can involve different actions including the schedule of task allocation or task migration. For example, in Marine Fog [28], the system needs to identify the best time to route the marine sensory data among the ad hoc network nodes in order to deliver the most important information on time. For example, in LV-Fog, each vehicle needs to perform local measurements in order to identify the encounter and the intercontact time between itself and the incoming vehicle, toward performing rapid information exchange [64].

1.5.3.2 Cost of Energy and Tenancy

Besides the process and task management aspects, tenants need to consider the cost of energy and tenancy. First, *energy cost* commonly refers to the energy consumption of battery-powered end-devices. Here, the tenants should optimize the application design and the runtime processes in order to minimize the energy consumption of the end-devices toward extending the sustainability of the overall processes.

Second, the cost of tenancy indicates the cost-performance trade-off of the application. Specifically, in some cases, the tenants intend to achieve the best agility in terms of task allocation, execution, and migration, they would pre-deploy the application at every single fog node that potentially will be encountered by the end-devices. For example, in the AAL-based application, the tenant might have deployed the application at all the fog nodes on the potential moving path of the patient in order to perform proactive fog-driven sensory data reasoning [18]. However, such an approach may demand a high tenancy cost for the tenant, especially when we consider that the patient (the end-device) may not encounter some of the fog nodes.

1.5.4 Provider

The multi-tenancy-supported fog server providers are responsible to provide adaptive application deployment platforms for the tenants and to cater reliable accessibility to the tenant-side clients.

In order to achieve high QoS for the fog servers, the providers need to address the following aspects.

1.5.4.1 Physical Placement

The physical placement represents where the providers should deploy their fog servers. Commonly, in the case of iFog, the provider may enable fog servers on all the possible nodes (e.g. cellular base stations) and rely on the underlying communication technologies (see Section 1.4) to support the accessibility. On the other hand, in case of mFog [24, 63], providers need to identify the best geo-location to place the mobile fog nodes in order to provide the best QoS to the end-devices and also to support the cost-efficiency of the operation. For example, in UAV-Fog, the provider may choose the locations for the mobile fog nodes based on the density of the end-devices, the signal coverage of the fog node, the distance between the fog server and the end-devices, and the other context factors described in the previous content related to context-awareness. In general, the primary goal of physical placement is to achieve the lowest latency in terms of request/response time, application service handover time, and application task migration time.

1.5.4.2 Server Discoverability and Connectivity

Server discoverability is a specific requirement for the multitenancy fog services, and it involves two phases.

- *Multitenancy fog service provider discovery*. Presents the phase when the tenants intend to discover feasible fog service providers for deploying their applications. Commonly, based on the experience of the cloud service business model, it is likely that the tenants would discover the providers via the indexing services (e.g. Google searching). Alternatively, the providers may establish a federated service registry for the service discovery. Furthermore, the provider may follow the open standard-based service description mechanism or interface (e.g. ETSI – MEC standard) to describe their fog services toward helping the tenants to discover the service that matches to their requirements.
- *Runtime fog server discovery*. Presents the runtime service discovery phase for the fog applications. In general, the fog applications hosted on the fog servers, need to perform seamless interaction with the end-devices on the move. Besides the mobility schemes that help the tenants to identify the movement of the end-devices, tenants need a corresponding mechanism that can help the end-devices to continuously discover and to connect to the new fog servers automatically, without any inference from the end-users. Therefore, the fog servers need to support the corresponding API that allows the tenants to configure the application process/task handover and migration mechanism among the fog nodes. Commonly, if such an API support is not available, tenants have to enable the corresponding mechanisms from the higher layer of the application, which may result in an inefficient tenancy cost and operational performance.

1.5.4.3 Operation Management

In general, fog servers have limited resources in which they can serve a fairly limited number of tenant-side applications at each time slot. Therefore, fog servers require dynamic and optimal mechanisms to support their serviceability. Here, we list the basic elements involved in the operation management of fog servers.

- *Load balancing of request and traffic*. Commonly, fog servers are connecting with one another vertically or horizontally. Therefore, it is possible to establish a cluster computing group among the fog nodes connected in 0-hop range toward enhancing the overall computational capability. Besides the computation-related loads, since fog nodes are fundamentally Internet gateway devices, the heavy network traffic can always affect their serviceability. In order to overcome the traffic-related issues of fog servers, the provider may configure multilayered caching mechanism that utilizes the fog nodes in the hierarchy to reduce the burden [66].

- *Server allocation, server scheduling, and server migration.* Three corelated elements, especially for the resource constraint mobile fog node (constraint mFog), such as UE-fog and UAV-fog nodes. To explain, a provider may deploy a specific type of fog server on the constraint mFog device for a domain-specific application in a specific period of time. Whereas, the rest of the time, the device does not operate the fog server at all. For example, in an indie fog environment [17], the owner of a smartphone (UE)-fog may configure the device to serve the context reasoning–based fog server [36] only when the owner is carrying the device in outdoor areas and the battery level of the device is over 50%. Further, the owner can also configure that, when the battery level of the device is between 51 and 70%, it will redirect/migrate the request to another authorized fog node. Similarly, the notion described here is applicable to other MFC domains, such as LV-fog [24] and UAV-fog [63].

1.5.4.4 Operation Cost

Operating fog servers can be costly for the providers especially when the providers are unable to identify what tenants really demand. For example, stream data filtering is a common method used in fog computing and the corresponding program can be quite simple in comparison to the scientific programs operated on the cloud. However, in the classic approach, the tenant may need to wrap the simple program to a package that runs on the resource-intensive VM environment because the provider was following the classic cloud service deployment approach to providing the fog server. Explicitly, the provider has inefficiently increased the burden of the fog node while providing excess service to tenants who demanded only a simple method. Therefore, the provider should consider what are the most cost-efficient service types that should be supported by the fog servers. Beside the service type, the providers of the battery-powered mobile fog nodes need to specifically address the energy-efficient of the fog servers in order to improve their sustainability. For instance, although providers can easily replace UAV-Fog nodes, considering the extra latency derived from the process/task handover and migration while replacing the UAV-Fog nodes, frequently replacing UAV-Fog nodes will reduce the QoE for the tenant-side clients.

1.5.5 Security

In large-scale MFC systems, the classic perimeter-based security approach will not suffice, security strategies in MFC must account for various factors: physical, end-to-end, and also monitoring and management.

1.5.5.1 Physical Security

Since the fog nodes will be deployed in the wild (e.g. road-side infrastructure in LV-Fog), physical exposure is a more serious threat than in conventional enterprise or cloud computing. The devices need antitamper mechanisms that prevent, detect, and respond to intrusions, while simultaneously considering how to allow maintenance operations without compromising these mechanisms [61].

1.5.5.2 End-to-End Security

End-to-end security is concerned with the security capabilities of each device within the MFC, spanning different layers of the fog architecture and devices therein.

- *Execution environments*. The devices need to include capable software and hardware components solely dedicated to performing security functions (so-called roots of trust (RoT). These components should, on one hand, be isolated from the rest of the platform while also verifying the functions performed by the platform.
  Based on RoT-s, the nodes must have the capability to provide trusted execution environments. In the case of virtualized environments, this can be achieved through virtual trusted platform modules.
- *Network security*. According to the OpenFog security requirements, fog nodes should provide the security services defined by the ITU X.800 recommendation by using standard-based secure transport protocols.
  Some nodes in the MFC system can provide security services on the network through network function virtualization (NFV) and SDN, for example, deep packet inspection.
- *Data Security* protection of data must be taken care of in all the mediums in which data may lie or move: in system memory, in persistent storage or data exchanged over the network.

1.5.5.3 Security Monitoring and Management

The system must be capable of observing security state in the network and reacting to new threats via monitoring and management mechanisms. Security management should allow definition and updating of security policies and propagation of the policies over the network in real-time.

In addition to policy management, identity and credential management is a requirement. In addition to the necessary registration and credential storage functions, an MFC system must handle the challenge of authentication and access control in situations with intermittent connectivity, e.g. negotiating session keys when crossing different trust domains while ensuring data integrity and privacy [67]. Security monitoring, on the other hand, should collect log traces while ensuring their integrity.

The OpenFog security requirements define at least two logically separate security domains: (1) policies concerning the collection of Fog entities within the system that can interact with one another and (2) for policies regarding the individual services and applications being executed and provided on the platform.

1.5.5.4 Trust Management and Multitenancy Security

Managing trust. The information flows in MFC raise the issue of how to determine which nodes in the network are trustworthy for a particular client and request? Trust management frameworks need to manage trust assessment of both devices and applications and be able to adjust to the real-time updates of trust-related data, such as social relationships and execution results [68]. Some scenarios may also require decentralized trust management achievable with the help of blockchain technology; however, this aspect has issues with latency [69].

Multitenancy. As the fog architecture dictates that fog platforms can host services for multiple parties, this poses the issue of how to ensure isolation in the runtime environment and ensure that only data that was intended to be shared is available across the instances [61, 70]. Secure isolation of tenant and user space continues to be a challenging requirement, as vulnerabilities allowing adversaries to access memory of other tenants VMs without permission [71] have surfaced, even in mass-produced CPUs.

1.6 Open Challenges

Fog computing has been introduced to overcome many challenges that cloud computing was incapable of handling, such as latency-sensitivity, connectivity between the large set of cloud-based applications, etc. Therefore, many researchers investigated and proposed different architectures for introducing fog computing into the traditional cloud computing to create an extension of the existing design. This action resolved into solving some of the old obstacles and generate new perspectives and ambitions that led to new challenges.

1.6.1 Challenges in Land Vehicular Fog Computing

Introducing the cloud computing into VANET was redemption since it provided a solution to most challenges that traditional design of VANET faced, such as decreased flexibility, scalability, poor connectivity, and inadequate intelligence [72]. However, the new generation of VANET has introduced new requirement with respect to the high mobility, low latency, real-time applications, and reliable connectivity. For all these reasons, adding fog computing to the equation has emerged as a potential solution. Based on the existing research it seems that there are still some obstacles to be dealt with, for example, the management of fog server in geographically distributed fog nodes. The difficulty relies on the positioning of each edge server in a given area, which is important for fog approach and it considers many parameters, such as vehicles density, traffic status data, and processing load on the servers. Another challenge is related to management of neighboring fog servers with respect to communication and access, which can be affected by the environment. Finally, the assessment and handling dissemination of real-time critical message can be problematic, since design should be capable of distinguishing between a true or false event [73].

1.6.2 Challenges in Marine Fog Computing

Existing works [4, 28] have proposed the frameworks for improving the efficiency of the communication and for application management in the hybrid MFC environment that consist of both iFog and mFog nodes. Essentially, comparing to the UAV-Fog nodes and the UE-fog nodes, the Marine Fog nodes do not have the resource-constraint issue in terms of computation and data storage. However, because the operating marine environment lacks infrastructure, the communication between the Marine Fog nodes and the distant central cloud faces the latency issue derived from the limitation of the underline network topology and technology. For example, the bandwidth of the common VHF radio used in maritime communication can reach only 28.8 kbps and the range of the infrastructure-less 4G/5G LTE device-to-device communication is unable to fulfill the need of a marine environment. In order to overcome the fundamental communication issue, developers may consider integrating UAV-fog nodes [31] in which the system can deploy the UAV-fog nodes between vessels to form a mesh network toward dynamically supporting better bandwidth and more stable connection. However, UAVs have a limited available operation time slot because they are battery-powered. The system needs to introduce an adaptive scheduling scheme, physical location placement scheme. Further, the system needs to dynamically adjust the movement of the UAV-fog nodes based on the interconnected vessels in order to seamlessly maintain the mesh network.

1.6.3 Challenges in Unmanned Aerial Vehicular Fog Computing

Current works in UAV mFog [32, 52] were focusing on the underlying system design and communication mechanisms. Although UAV-Fog nodes have many potential applications due to the features in terms of fast deployment, scalability, flexibility, and cost-efficiency [31], integrating UAV-Fog nodes to pervasive computing systems or IoT systems raises many new challenges besides the underlying communication mechanisms. Specifically, existing works have not fully addressed the requirements for both tenant and provider. For example, although an existing work [63] has proposed schemes that enable UAV-Fog nodes to perform data-driven service handover, which transfers the client data from one UAV-Fog node to another, this scheme was designed for a domain-specific application in which the author assumes the system has preinstalled the application to all the UAV-Fog nodes and hence, at runtime, the UAV-Fog nodes need only to transfer the client data in order to support the mobility. On the other hand, considering the multitenancy fog service model, preinstalling applications to the UAV-Fog nodes for all the tenants will cause a high burden to the storage size, especially when the application involves large size files. Therefore, UAV-fog nodes require the mechanism that supports rapid and dynamic application management in terms of task allocation/placement and task migration.

1.6.4 Challenges in User Equipment-based Fog Computing

There exist a large number of frameworks designed for supporting mobile UE from iFog. However, existing works rarely address the challenges in UE-fog nodes. The use cases described in the previous section indicate that systems which integrate UE-fog nodes require a dynamic program deployment mechanism. For example, in the advanced crowd sensing use case, which utilize UE-fog nodes to provide the interpreted context information derived from the collected sensory data, the UE-fog nodes need to provide the corresponding service that allows the clients to deploy the program code of the context reasoning algorithm on the UE-fog nodes. Explicitly, considering the UE-fog nodes have constraint resources, they are unable to support the common VM or containers engine-based service for the dynamic program deployment. Instead, developers generally would develop the standalone solutions which leave the interoperability as an unsolved problem in UE-fog. In order to address such an issue, developers may consider integrating an open standard–based service interface or to develop a specific mobile fog node description language based on the extension of existing cloud service-based standard, such as OASIS Topology and Orchestration Specification for Cloud Applications (TOSCA).

1.6.5 General Challenges

1.6.5.1 Testbed Tool

The complexity of the MFC topology, which encompasses the hierarchical, vertical, and horizontal interconnections among the cloud, the iFog, the mFog, and the end-devices, has increased the challenges in system design and validation. Commonly, the developers of the distributed computing system, such as cloud services, pervasive services, or mobile services, have a broad range of simulation options (e.g. CloudSim [http://www.cloudbus.org/cloudsim], the ONE simulator [https://www.netlab.tkk.fi/tutkimus/dtn/theone], and iFogSim [https://github.com/Cloudslab/iFogSim], etc.) to validate their system designs. However, existing simulation tools are insufficient to validate many MFC systems because they are unable to address all the elements of MFC. For example, although the cloud service–based simulation tools are capable of simulating the hardware heterogeneity, they do not include mobility-related factors. For another example, while the mobile service-based simulation tools are capable of simulating the movement of entities, they do not have corresponding mechanisms to simulate the complete MFC network that contains the hierarchical and the vertical interconnection among the entities. Finally, iFogSim is capable of simulating the stationary fog nodes but it does not provide the mechanism to simulate the mobile fog nodes. Consequently, integrating the existing tools to develop a comprehensive MFC testbed becomes a critical challenge.

1.6.5.2 Autonomous Runtime Adjustment and Rapid Redeployment

To achieve optimal operation in MFC, the system demands autonomous adjustment and rapid redeployment based on context-awareness and the real-time system process analysis. In particular, considering an MFC system with the large-scale deployment of mFog and iFog nodes, manned optimization becomes impractical and inefficient. In order to overcome such an issue, the system needs to introduce a certain level of self-aware mechanisms to the fog nodes. Specifically, at an early stage, the system manager can preconfigure the basic knowledge to the fog nodes that help the fog nodes to identify the situation at runtime and to adjust or to redeploy the fog service. While the system continuously operates, the fog nodes should support edge intelligence mechanism in which the fog nodes together with the back-end cloud can study the historical records of the operation in order to identify the weak parts and to perform adjustment and redeployment automatically. For example, by enabling edge intelligence on the UAV-Fog nodes, the UAV-Fog nodes are capable of learning when and where to adjust their location, when and where they should migrate or redeploy their services, or when they should reserve their computational resources in order to provide the best QoE to the tenant-side clients.

1.6.5.3 Scheduling of Fog Applications

A few works have addressed fog application scheduling for hybrid MFC environments that consist of both iFog and mFog nodes. However, existing frameworks for mFog either designed for a specific purpose, such as for data routing in SDN [28], or for process/task distribution [20, 64], have not considered all the contextual factors (see Section 1.5.2) and the heterogeneity factors (see Section 1.5.1). Specifically, besides the movement of the entities, the tenants' application scheduling scheme should consider that the fog servers have the different computational and networking performance at different time slot due to the hardware specifications and the runtime context factors. In other words, developers need to have insight into the processing delay by considering all the factors toward proposing the optimal application scheduling scheme.

1.6.5.4 Scalable Resource Management of Fog Providers

In general, fog nodes have limited resources to serve tenants because they are fundamentally the independent network gateway devices that do not interconnect with one another in a short range like the server pools in the cloud. In other words, introducing computational scalability in MFC faces the network latency challenge. Commonly, providers of fog servers may manage multiple fog servers that are interconnected vertically within the hierarchy or are interconnected horizontally in a peer-to-peer manner. However, since the primary objective of fog servers in MFC is to serve the tenant-side clients, the distances between the fog servers are rarely within the range that is capable of achieving ultra-low latency. Therefore, the classic cloud-based scalability scheme is incompatible in MFC and hence, scalability becomes an unsolved challenge, especially for mFog environments. In order to address the challenge in scalability, the developers may consider developing a hybrid framework that integrates opportunistic computing, SDN, and context-aware software architecture toward enabling an adaptive fog service topology that can be orchestrating the fog servers in a highly dynamic manner.

1.7 Conclusion

Fog computing has appeared as a paradigm that extends Cloud computing and offers interesting and promising possibilities to overcome the limitations of the traditional environment. However, fog computing still faces a challenge with respect to mobility when the tasks come from ubiquitous mobile devices or applications in which the data sources are in constant movement. Therefore, this chapter pointins a spotlight on the development and advancement in the MFC and its related models. In this process, it reflects the areas where MFC is needed and has a very positive impact on enhancing the existing systems and opening new directions for evolving the environment, such as in the infrastructures, equipment and devices, land and marine vehicles, autonomous vehicles, etc. In addition, it investigates the communication technologies that have emerged or upgraded based on MFC with an emphasis on the added value that facilitated the process of solving some of the big challenges existing in the traditional design. Furthermore, the chapter also presents, in a comprehensive manner, the basics of nonfunctional requirement needed to achieve the basic QoS principles. Finally, it addresses the important open challenges that still need to be addressed in the new environment designed under the MFC framework.

Acknowledgment

The work is supported by the Estonian Centre of Excellence in IT (EXCITE), funded by the European Regional Development Fund.

References

1 Chang, C., Srirama, S.N., and Buyya, R. (2019). *Internet of Things (IoT) and New Computing Paradigms*, ch. 1, p. 1. Wiley.

2 Alam, K.M., Saini, M., and El Saddik, A. (2015). Toward social iInternet of vehicles: concept, architecture, and applications. *IEEE Access* 3: 343–357.

3 Gharibi, M., Boutaba, R., and Waslander, S.L. (2016). Internet of drones. *IEEE Access* 4: 1148–1162.

4 Al-Zaidi, R., Woods, J.C., Al-Khalidi, M., and Hu, H. (2018). Building novel VHF-based wireless sensor networks for the Internet of marine things. *IEEE Sensors Journal* 18 (5): 2131–2144.

5 Santagati, G.E. and Melodia, T. (2017). An implantable low-power ultrasonic platform for the Internet of medical things. In: *INFOCOM 2017-IEEE Conference on Computer Communications, IEEE*, –1, 9. IEEE.

6 Chang, C., Hadachi, A., Srirama, S.N., and Min, M. (2018). *Mobile Big Data: Foundations, State of the Art, and Future Directions*, 1–12. Cham: Springer International Publishing.

7 Chang, C., Srirama, S.N., and Mass, J. (2015). A middleware for discovering proximity-based service-oriented industrial Internet of Things. In: *2015 IEEE International Conference on Services Computing (SCC)*, 130–137. IEEE.

8 Satyanarayanan, M., Bahl, P., Cáceres, R., and Davies, N. (2009). The case for VM-based cloudlets in mobile computing. *IEEE Pervasive Computing* 4: 14–23.

9 ETSI (2017). *ETSI and OpenFog Consortium collaborate on fog and edge applications*. Berlin, Germany: MEC Congress https://www.etsi.org/newsroom/news/1216-2017-09-news-etsi-and-openfog-consortium-collaborate-on-fog-and-edge-applications.

10 M. Iorga, L. Feldman, R. Barton et al., Fog Computing Conceptual Model: Recommendations of the National Institute of Standards and Technology, Special Publication NIST SP 500-325, National Institute of Standards and Technology, March 2018.

11 Satyanarayanan, M. (2017). The emergence of edge computing. *Computer* 50 (1): 30–39.

12 Fahmy, H.M., El Ghany, M.A., and Baumann, G. (2018). Vehicle risk assessment and control for lane-keeping and collision avoidance at low-speed and high-speed scenarios. *IEEE Transactions on Vehicular Technology* 57 (6): 4805–4818.

13 Hafner, M.R., Cunningham, D., Caminiti, L., and Del Vecchio, D. (2013). Cooperative collision avoidance at intersections: algorithms and experiments. *IEEE Transactions on Intelligent Transportation Systems* 14 (3): 1162–1175.

14 Nguyen, V., Kim, O.T.T., Pham, C. et al. (2018). A survey on adaptive multichannel mac protocols in vanets using markov models. *IEEE Access* 6: 16493–16514.

15 Vinel, A., Lyamin, N., and Isachenkov, P. (2018). Modeling of V2V communications for C-ITS safety applications: a CPS perspective. *IEEE Communications Letters* 22 (8): 1600–1603.

16 Kalatzis, N., Avgeris, M., Dechouniotis, D. et al. (2018). Edge computing in IoT ecosystems for UAV-enabled early fire detection. In: *2018 IEEE International Conference on Smart Computing (SMARTCOMP)*, 106–114. IEEE.

17 Chang, C., Srirama, S.N., and Buyya, R. (2017). Indie fog: an efficient fog-computing infrastructure for the Internet of Things. *Computer* 50 (9): 92–98.

18 Soo, S., Chang, C., Loke, S.W., and Srirama, S.N. (2017). Proactive mobile fog computing using work stealing: data processing at the edge. *International Journal of Mobile Computing and Multimedia Communications (IJMCMC)* 8 (4): 1–19.

19 Soto, V., De Grande, R.E., and Boukerche, A. (2017). REPRO: time-constrained data retrieval for edge offloading in vehicular clouds. In: *Proceedings of the 14th ACM Symposium on Performance Evaluation of Wireless Ad Hoc, Sensor, & Ubiquitous Networks*, 47–54. ACM.

20 Zhu, C., Tao, J., Pastor, G. et al. (2018). Folo: latency and quality optimized task allocation in vehicular fog computing. *IEEE Internet of Things Journal* 6 (3): 4150–4161.

21 Yao, H., Bai, C., Zeng, D. et al. (2015). Migrate or not? Exploring virtual machine migration in roadside cloudlet-based vehicular cloud. *Concurrency and Computation: Practice and Experience* 27 (18): 5780–5792.

22 Cordeschi, N., Amendola, D., and Baccarelli, E. (2015). Reliable adaptive resource management for cognitive cloud vehicular networks. *IEEE Transactions on Vehicular Technology* 64 (6): 2528–2537.

23 Chen, Y.-A., Walters, J.P., and Crago, S.P. (2017). Load balancing for minimizing deadline misses and total runtime for connected car systems in fog computing. In: *2017 IEEE International Symposium on Parallel and Distributed*

Processing with Applications and 2017 IEEE International Conference on Ubiquitous Computing and Communications (ISPA/IUCC), 683–690. IEEE.

24 Fan, X., He, X., Puthal, D. et al. (2018). CTOM: collaborative task offloading mechanism for mobile cloudlet networks. In: *2018 IEEE International Conference on Communications (ICC)*, 1–6. IEEE.

25 Marshall, W.E. and Dumbaugh, E. (2018). Revisiting the relationship between traffic congestion and the economy: a longitudinal examination of us metropolitan areas. *Transportation*: 1–40.

26 Wang, C., Li, Y., Jin, D., and Chen, S. (2016). On the serviceability of mobile vehicular cloudlets in a large-scale urban environment. *IEEE Transactions on Intelligent Transportation Systems* 17 (10): 2960–2970.

27 Wang, Z., Zhong, Z., Zhao, D., and Ni, M. (2018). Vehicle-based cloudlet relaying for mobile computation offloading. *IEEE Transactions on Vehicular Technology* 67 (11): 11181–11191.

28 Yang, T., Cui, Z., Wang, R. et al. (2018). A multivessels cooperation scheduling for networked maritime fog-ran architecture leveraging SDN. *Peer-to-Peer Networking and Applications* 11 (4): 808–820.

29 R. Sosa, R. Sucasas, A. Queralt et al., Towards an open, secure, decentralized and coordinated fog-to-cloud management ecosystem, D5.1 mF2C reference architecture (integration IT-1), mF2C Consortium, 2018.

30 Xu, G., Shen, W., and Wang, X. (2014). Applications of wireless sensor networks in marine environment monitoring: a survey. *Sensors* 14 (9): 16932–16954.

31 Mohamed, N., Al-Jaroodi, J., Jawhar, I. et al. (2017). UAV fog: a UAV-based fog computing for Internet of Things. In: *2017 IEEE SmartWorld, Ubiquitous Intelligence & Computing, Advanced & Trusted Computed, Scalable Computing & Communications, Cloud & Big Data Computing, Internet of People and Smart City Innovation (SmartWorld/SCALCOM/UIC/ATC/CBDCom/IOP/SCI)*, 1–8. IEEE.

32 Radu, D., Cretu, A., Parrein, B. et al. (2018). Flying ad hoc network for emergency applications connected to a fog system. In: *International Conference on Emerging Internetworking, Data & Web Technologies*, 675–686. Springer.

33 451 Research, "Size and impact of fog computing market," OpenFog Consortium, 2017.

34 Puliafito, C., Mingozzi, E., and Anastasi, G. (2017). Fog computing for the Internet of mobile things: issues and challenges. In: *2017 IEEE International Conference on Smart Computing (SMARTCOMP)*, 1–6. IEEE.

35 Silva, P.M.P., Rodrigues, J., Silva, J. et al. (2017). Using edge-clouds to reduce load on traditional Wi-Fi infrastructures and improve quality of experience. In: *2017 IEEE 1st International Conference on Fog and Edge Computing (ICFEC)*, 61–67. IEEE.

36 Chang, C. and Srirama, S.N. (2018). Providing context as a service using service-oriented mobile indie fog and opportunistic computing. In: *European Conference on Software Architecture*, –219, 235. Springer.

37 Siddiqui, F., Zeadally, S., and Salah, K. (2015). Gigabit wireless networking with ieee 802.11 ac: technical overview and challenges. *Journal of Networks* 10 (3): 164.

38 Rejiba, Z., Masip-Bruin, X., Jurnet, A. et al. (2018). F2C-aware: enabling discovery in Wi-Fi-powered fog-to-cloud (F2C) systems. In: *2018 6th IEEE International Conference on Mobile Cloud Computing, Services, and Engineering (MobileCloud)*, 113–116. IEEE.

39 Chowdhury, M., Steinbach, E., Kellerer, W., and Maier, M. (2018). Context-aware task migration for HART-centric collaboration over FiWi based tactile internet infrastructures. *IEEE Transactions on Parallel and Distributed Systems* 29 (6): 1231–1246.

40 Enayet, A., Razzaque, M.A., Hassan, M.M. et al. (2018). A mobility-aware optimal resource allocation architecture for big data task execution on mobile cloud in smart cities. *IEEE Communications Magazine* 56 (2): 110–117.

41 Taleb, T., Dutta, S., Ksentini, A. et al. (2017). Mobile edge computing potential in making cities smarter. *IEEE Communications Magazine* 55: 38–43.

42 Akter, M., Zohra, F.T., and Das, A.K. (2017). Q-MAC: QoS and mobility aware optimal resource allocation for dynamic application offloading in mobile cloud computing. In: *International Conference on Electrical, Computer and Communication Engineering (ECCE)*, 803–808. IEEE.

43 Lei, L. (2016). Stochastic modeling of device-to-device communications for intelligent transportation systems. In: *2016 23rd International Conference on Telecommunications (ICT)*, 1–5. IEEE.

44 ITUR, Requirements related to technical performance for IMT-advanced radio interface(s), Report M.2134, International Telecommunications Union, 2008.

45 ITUR, Minimum requirements related to technical performance for IMT 2020 radio interface(s), Report M.2410, International Telecommunications Union, 2017 (22nd February Draft).

46 IMT vision – framework and overall objectives of the future development of IMT for 2020 and beyond, Recommendation ITU-R M.2083-0, pp. 2083–2090, International Telecommunications Union, 2015.

47 Dong, P., Zheng, T., Yu, S. et al. (2017). Enhancing vehicular communication using 5g-enabled smart collaborative networking. *IEEE Wireless Communications* 24: 72–79.

48 Yu, F., Chen, H., and Xu, J. (2018). DMPO: dynamic mobility-aware partial offloading in mobile edge computing. *Future Generation Computer Systems* 89: 722–735.

49 Wang, Z., Zhao, Z., Min, G. et al. (2018). User mobility aware task assignment for mobile edge computing. *Future Generation Computer Systems* 85: 1–8.

50 Nasrin, W. and Xie, J. (2018). SharedMEC: sharing clouds to support user mobility in mobile edge computing. In: *2018 IEEE International Conference on Communications (ICC)*, 1–6. IEEE.

51 Yang, J., Wen, J., Jiang, B. et al. (2018). Marine depth mapping algorithm based on the edge computing in Internet of Things. *Journal of Parallel and Distributed Computing* 114: 95–103.

52 Jeong, S., Simeone, O., and Kang, J. (2018). Mobile edge computing via a UAV-mounted cloudlet: optimization of bit allocation and path planning. *IEEE Transactions on Vehicular Technology* 67 (3): 2049–2063.

53 Salem, A., Salonidis, T., Desai, N., and Nadeem, T. (2017). Kinaara: distributed discovery and allocation of mobile edge resources. In: *2017 IEEE 14th International Conference on Mobile Ad Hoc and Sensor Systems (MASS)*, 153–161. IEEE.

54 Liyanage, M., Chang, C., and Srirama, S.N. (2018). Adaptive mobile web server framework for mist computing in the Internet of Things. *International Journal of Pervasive Computing and Communications.*

55 Sucipto, K., Chatzopoulos, D., Kosta±, S., and Hui, P. (2017). Keep your nice friends close, but your rich friends closer — computation offloading using nfc. In: *IEEE INFOCOM 2017 IEEE Conference on Computer Communications*, 1–9.

56 Characteristics of VHF radio systems and equipment for the exchange of data and electronic mail in the maritime mobile service RR appendix 18 channels, Recommendation M.1842-1, International Telecommunication Union, 2008.

57 Al-Zaidi, R., Woods, J., Al-Khalidi, M. et al. (2017). Next generation marine data networks in an iot environment. In: *2017 Second International Conference on Fog and Mobile Edge Computing (FMEC)*, 50–55. IEEE.

58 Manoufali, M., Alshaer, H., Kong, P.-Y., and Jimaa, S. (2013). Technologies and networks supporting maritime wireless mesh communications. In: *Wireless and Mobile Networking Conference (WMNC), 2013 6th Joint IFIP*, 1–8. IEEE.

59 Bardram, A.V.T., Larsen, M.D., Malarski, K.M. et al. (2018). Lorawan capacity simulation and field test in a harbour environment. In: *2018 Third International Conference on Fog and Mobile Edge Computing (FMEC)*, 193–198.

60 Raza, U., Kulkarni, P., and Sooriyabandara, M. (2017). Low power wide area networks: an overview. *IEEE Communication Surveys and Tutorials* 19: 855–873.

61 Martin, B.A., Michaud, F., Banks, D. et al. (2017). Openfog security requirements and approaches. In: *2017 IEEE Fog World Congress (FWC)*, 1–6.

62 Sui, Y., Wang, X., Pengt, M., and An, N. (2017). Optimizing mobility and energy charging for mobile cloudlet. In: *2017 IEEE International Conference on Communications (ICC)*, 1–6. IEEE.

63 Tang, F., Fadlullah, Z.M., Mao, B. et al. (2018). On a novel adaptive UAV-mounted cloudlet-aided recommendation system for LBSNs. *IEEE Transactions on Emerging Topics in Computing* 7 (4): 565–577.

64 Truong-Huu, T., Tham, C.-K., and Niyato, D. (2014). To offload or to wait: An opportunistic offloading algorithm for parallel tasks in a mobile cloud. In: *2014 IEEE 6th International Conference on Cloud Computing Technology and Science (CloudCom)*, 182–189. IEEE.

65 Chang, C., Srirama, S.N., and Buyya, R. (2016). Mobile cloud business process management system for the Internet of Things: a survey. *ACM Computing Surveys* 49, pp. 70:1–70:42.

66 Li, M., Yu, F.R., Si, P. et al. (2018). Software-defined vehicular networks with caching and computing for delay-tolerant data traffic. In: *2018 IEEE International Conference on Communications (ICC)*, 1–6. IEEE.

67 Kumar, N., Rodrigues, J.J.P.C., Guizani, M. et al. (2018). Achieving energy efficiency and sustainability in edge/fog deployment. *IEEE Communications Magazine* 56: 20–21.

68 Ruan, Y., Durresi, A., and Uslu, S. (2018). Trust assessment for Internet of Things in multiaccess edge computing. In: *2018 IEEE 32nd International Conference on Advanced Information Networking and Applications (AINA)*, –1155, 1161. IEEE.

69 Wang, S., Xu, J., Zhang, N., and Liu, Y. (2018). A survey on service migration in mobile edge computing. *IEEE Access* 6: 23511–23528.

70 Zhang, P., Zhou, M., and Fortino, G. (2018). Security and trust issues in fog computing: a survey. *Future Generation Computer Systems* 88: 16–27.

71 Lipp, M., Schwarz, M., Gruss, D. et al. (2018). Meltdown: reading kernel memory from user space. In: *27th Security Symposium*, 973–990.

72 Hussain, R., Son, J., Eun, H. et al. (2012). Rethinking vehicular communications: merging vanet with cloud computing. In: *2012 IEEE 4th International Conference on Cloud Computing Technology and Science (CloudCom)*, 606–609. IEEE.

73 Nobre, J.C., de Souza, A.M., Rosário, D. et al. (2019). Vehicular software-defined networking and fog computing: integration and design principles. *Ad Hoc Networks* 82: 172–181.

2

Edge and Fog: A Survey, Use Cases, and Future Challenges

Cosmin Avasalcai, Ilir Murturi, and Schahram Dustdar

Distributed Systems Group, TU Wien, Vienna, Austria

2.1 Introduction

In the past couple of years, the cloud computing paradigm was at the center of the Internet of Things' (IoT) ever-growing network, where companies can move their control and computing capabilities, and store collected data in a medium with almost unlimited resources [1]. It was and continues to be the best solution to deploy demanding computational applications with the main focus on processing vast amounts of data. Data are generated from geo-distributed IoT devices, such as sensors, smartphones, laptops, and vehicles, just to name a few. However, today this paradigm is facing growing challenges in meeting the demanding constraints of new IoT applications.

With the rapid adoption of IoT devices, new use cases have emerged to improve our daily lives. Some of these new use cases are the smart city, smart home, smart grid, and smart manufacturing with the power of changing industries (i.e. healthcare, oil and gas, automotive, etc.) by improving the working environment and optimizing workflow. Since most of the use cases consist of multiple applications that require fast response time (i.e. real-time or near real-time) and improved privacy, most of the time the cloud fails to fulfill these requirements (i.e. network congestion and ensuring privacy).

To overcome these shortcomings, researchers have proposed two new paradigms, fog computing and edge computing, to enable more computational resources (i.e. storage, networking, and processing) closer to the edge of the network. Fog computing (FC) extends cloud capabilities closer to the end devices, such that a cloud-to-things continuum is obtained that decreases latency and network congestion while enforcing privacy by processing the data near the user [2]. On the same note, the edge computing vision is to migrate some computational

Fog Computing: Theory and Practice, First Edition.
Edited by Assad Abbas, Samee U. Khan, and Albert Y. Zomaya.

resources from the cloud to the heterogeneous devices placed at the edge of the network [3].

Embracing the vision of these paradigms and focusing on the deployment of multiple applications in close proximity of users, researchers have suggested new fog/edge devices. Among these devices, the most notable are mini servers, such as cloudlets [4], portable edge computers [5], and edge-cloud [6], which enable an application to work in harsh environments; mobile edge computing (MEC) [7] and mobile cloud computing [8] improve user experience and enable higher computational applications to be deployed on smartphones by offloading parts of the application on the device locally.

Many surveys are found in the literature that describe each paradigm in detail and its challenges [3, 9, 10]. However, there is no paper that compares the two; most of the time the terms *fog* and *edge* are both used to describe the same IoT network. Generally speaking, the visions of the two paradigms overlap, aiming to make available more computational resources at the edge of the network. Hence, the most significant difference is given by the naming convention used to describe them. The aim of this chapter is to offer a detailed description of the two aforementioned paradigms, discussing their differences and similarities. Furthermore, we discuss their future challenges and argue if the different naming convention is still required.

The remainder of the chapter is structured as follows: Section 2.2 defines the edge computing paradigm by describing its architectural features. Next, Section 2.3 presents in detail the fog computing paradigm and describes two use cases by emphasizing the key features of this architecture. Section 2.4 describes several illustrative use cases for both edge and fog computing. Section 2.5 discusses the challenges that these paradigms must conquer to be fully adopted in our society. Finally, Section 2.6 presents our final remarks on the comparison between fog and edge computing.

2.2 Edge Computing

As we explore new IoT applications and use cases, the consideration of proximity between edge nodes and the end-users is becoming increasingly obvious. The physical distance between the edge and the user affects highly end-to-end latency, privacy, network, and availability. Recently, this leads to a new paradigm allowing computation to be performed in close proximity of user and IoT devices (i.e. sensors and actuators). Edge computing [11] is a new paradigm aiming to provide storage and computing resources and act as an additional layer, composed of edge devices, between the end-user IoT device and the cloud layer. In edge computing, we define "edge" as any computing and network resources along the path

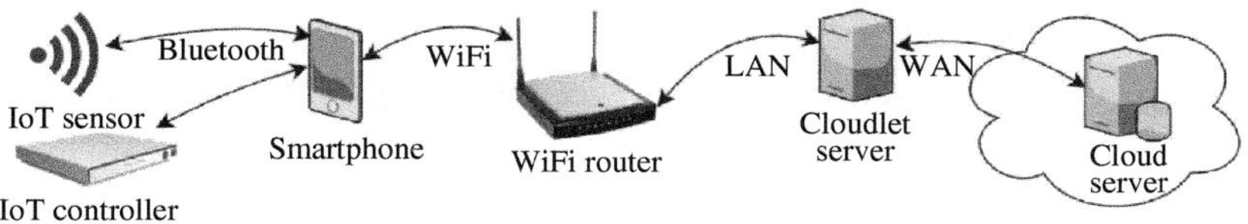

Figure 2.1 Edge computing solution using an IoT and edge devices [12].

between the initial source of data and destination storage of data (fog nodes, cloud data centers).

Edge computing is ever stronger when converging with IoT, offering novel techniques for IoT systems. Multiple definitions of edge computing are found in the literature; however, the most relevant is presented in [3]. Authors define edge computing as a paradigm that enables technologies allowing computation to be performed at the edge of the network, on downstream data on behalf of cloud services and upstream data on behalf of IoT services. The proposed paradigm is a relatively new concept and due to the same nature, the term "edge computing" in literature may refer to all other architectures, such as MEC, cloudlet computing (CC), or fog computing (FC). However, we vision edge computing as a bridge between IoT things and the nearest physical edge device that aims to facilitate the deployment of the new emerging IoT applications in users' devices, such as mobile devices (see Figure 2.1).

Authors [12] in Figure 2.1, present the main idea of the edge computing paradigm by adding another device in the form of an edge device. Such a device can be referred to as a personal computer, laptop, tablet, smartphone, or another device capable of locally processing the data generated by IoT devices. Furthermore, depending on device capabilities, it may offer different functionalities, such as the capability of storing data for a limited time. In addition, an edge device can react to emergency events by communicating with the IoT devices and can aid other devices like cloudlet, MEC server, and cloud data center, by preprocessing and filtering the raw data generated by the sensors. In such scenarios, the edge computing paradigm offers processing near to the source of data and reduces the amount of transmitted data. Instead of transmitting data to the cloud or fog node, the edge device, as the nearest device to the source of the data, will do computation and response to the user device without moving data to the fog or cloud.

Edge computing is considered a key enabler for scenarios where centralized cloud-based platforms are considered impractical. Processing data near to the logical extremes of a network – at the edge of the network – reduces significantly the latency and bandwidth cost. By shortening the distance that data has to travel, this paradigm could address concerns also in energy consumption, security, and privacy [13]. However, the rapid adoption of IoT devices, resulting in millions of interconnected devices, are challenges that Edge Computing must overcome.

2.2.1 Edge Computing Architecture

The details of edge computing architecture could be different in different contexts. For example, Chen et al. [14] propose an edge computing architecture for IoT-based manufacturing. The role of edge computing is analyzed with respect to edge equipment, network communication, information fusion, and cooperative mechanism with cloud computing. Zhang et al. [15] proposes the edge-based architecture and implementation for a smart home. The system consists of three distinct layers: the sensor layer, the edge layer, and the cloud layer.

Generally speaking, Wei Yu et al. divides the structure of edge computing into three aspects: the front-end, near-end, and far-end (see Figure 2.2). A detailed description is given below:

- The **front end** consists of heterogeneous end devices (e.g. smartphones, sensors, actuators), which are deployed at the front end of the edge computing structure. This layer provides real-time responsiveness and local authority for the end-users. Nevertheless, the front-end environment provides more interaction by keeping the heaviest traffic and processing closest to the end-user devices. However, due to the limited resource capabilities provided by end devices, it is clear that not all requirements can be met by this layer. Thus, in such situations, the end devices must forward the resource requirements to the more powerful devices, such as fog node or cloud computing data centers.
- The **near end** will support most of the traffic flows in the networks. This layer provides more powerful devices, which means that most of the data processing and storage will be migrated to the near-end environment. Additionally, many tasks like caching, device management, and privacy protection can be deployed at this layer. By increasing the distance between the source of data and its processing destination (e.g. fog node) it also increases the latency due to

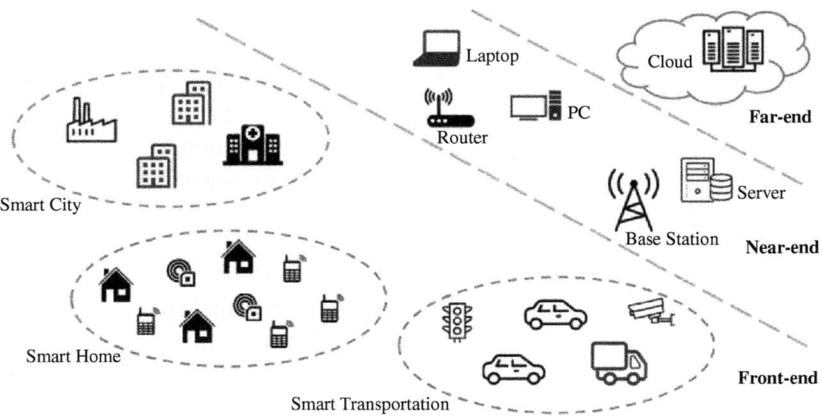

Figure 2.2 An overview of edge computing architecture [16]. (*See color plate section for the color representation of this figure*)

the round-trip journey. However, the latency is very low since the devices are one hop away from the source where the data is produced and consumed.

- The **far end** environment is cloud servers that are deployed farther away from the end devices. This layer provides devices with high processing capabilities and more data storage. Additionally, it can be configured to provide levels of performance, security, and control for both users and service providers. Nevertheless, the far-end layer enables any user to access unimaginable computing power where thousands of servers may be orchestrated to focus on the task, such as in [17, 18]. However, one must note that the transmission latency is increased in the networks by increasing the distance that data has to travel.

2.3 Fog Computing

Fog computing is a platform introduced by Cisco [10] with the purpose of extending the cloud capabilities closer to the edge of the network. Multiple definitions of fog computing are found in the literature; however, the most relevant is presented in [19]. According to them, fog computing is a geographically distributed computing architecture connected to multiple heterogeneous devices at the edge of the network, but at the same time not exclusively seamlessly backed by cloud services. Hence, we envision fog as a bridge between the cloud and the edge of the network that aims to facilitate the deployment of the newly emerging IoT applications (see Figure 2.3).

A fog device is a highly virtualized IoT node that provides computing, storage, and network services between edge devices and cloud [2]. Such characteristics are found in the cloud as well; thus, a fog device can be characterized as a mini-cloud

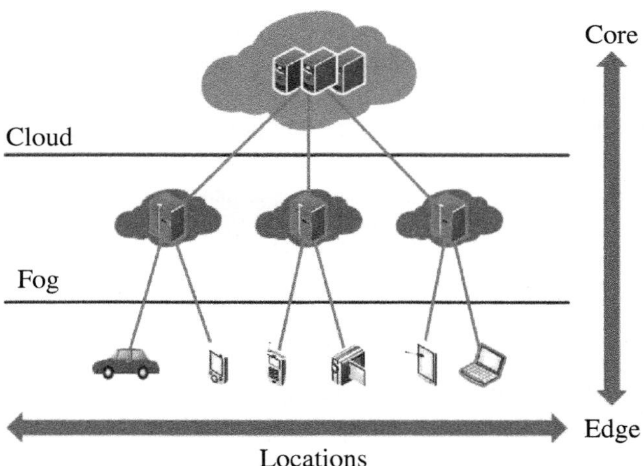

Figure 2.3 Fog computing a bridge between cloud and edge [20].

which utilizes its own resources in combination with data collected from the edge devices and the vast computational resources that cloud offers.

By migrating computational resources closer to the end devices, fog offers the possibility to customers to develop and deploy new latency-sensitive applications directly on devices like routers, switches, small data centers, and access points. Depending on the application, this can impose stringent requirements, which refers to fast response time and predictable latency (i.e. smart connected vehicles, augmented reality), location awareness (e.g., sensor networks to monitor the environment) and large-scale distributed systems (smart traffic light, smart grids). As a result, the cloud cannot fulfill all of these demands by itself.

To fill the technological gap in the current state of the art where the cloud computing paradigm is at the center, the fog collaborates with the cloud to form a more scalable and stable system across all edge devices suitable for IoT applications. From this union, the developer benefits the most since he can decide where is the most beneficial, fog or cloud, to deploy a function of his application. For example, taking advantage of the fog node capabilities, we can process and filter streams of collected data coming from heterogeneous devices, located in different areas, taking real-time decisions and lowering the communication network to the cloud. Thus, fog computing introduces effective ways of overcoming many limitations that cloud is facing [1]. These limitations are:

1. **Latency constraints.** The fog shares the same fundamental characteristics as the cloud in being able to perform different computational tasks closer to the end-user, making it ideal for latency-sensitive applications for which their requirements are too stringent for deployment in the cloud.
2. **Network bandwidth constraints.** Since the fog offers the possibility of performing data processing tasks closer to the edge of the network, lowering in the process the amount of raw data sent to the cloud, it is the perfect device to apply data analytics to obtain fast responses and send to the cloud for storage purposes only filtered data.
3. **Resource constrained devices.** Fog computing can perform computational tasks for constrained edge devices like smartphones and sensors. By offloading parts of the application from such constrained devices to nearby fog nodes, the energy consumption and life-cycle cost are decreased.
4. **Increased availability.** Another important aspect of fog computing represents the possibility of operating autonomously without reliable network connectivity to the cloud, increasing the availability of an application.
5. **Better security and privacy.** A fog device can process the private data locally without sending it to the cloud for further processing, ensuring better privacy and offering total control of collected data to the user. Furthermore, such devices can increase security as well, being able to perform a wide range of security functions, manage and update the security credential of constrained devices and monitor the security status of nearby devices.

The previous section introduces the edge computing paradigm as a solution to the cloud computing inefficiency for data processing when the data is produced and consumed at the edge of the network. Fog computing focuses more on the infrastructure side by providing more powerful fog devices (i.e. fog node may be high computation device, access points, etc.) while edge computing focuses more toward the "things" side (i.e. smartphone, smartwatch, gateway, etc.). The key difference between the edge computing and fog computing is where the computation is placed. While fog computing pushes processing into the lowest level of the network, edge computing pushes computation into devices, such as smartphones or devices with computation capabilities.

2.3.1 Fog Computing Architecture

Fog computing architecture is composed of highly dispersed heterogeneous devices with the intent of enabling deployment of IoT applications that require storage, computation, and networking resources distributed at different geographical locations [21]. Multiple high-level fog architectures have been proposed in the literature [22–24] that describe a three-layer architecture containing (1) the smart devices and sensor layer which collects data and send it forward to layer two for further processing, (2) the fog layer applies computational resources to analyze the received data and prepares it for the cloud, and (3) the cloud layer, which performs high intensive analysis tasks.

Bonomi et al. [10] present a fog software architecture (see Figure 2.4) consisting of the following key objectives:

- *Heterogeneous physical resources*. Fog nodes are heterogeneous devices deployed on different components, such as edge routers, access points, and high-end servers. Each component has a different set of characteristics (i.e. RAM and storage) that enables a new set of functionalities. This platform can run on multiple OSes and software applications, deriving a wide range of hardware and software capabilities.
- *Fog abstraction layer*. The fog abstraction layer consists of multiple generic application programming interfaces (APIs) enabling monitoring and controlling available physical resources like CPU, memory, energy, and network. This layer has the role of making accessible the uniform and programmable interface for seamless resource management and control. Furthermore, using generic APIs, it supports virtualization by monitoring and managing multiple hypervisors and OSes on a single machine with the purpose of improving resource utilization. Using virtualization enables the possibility of having multitenancy by supporting security, privacy, and isolation policies to ensure the isolation of different tenants on the same machine.
- *Fog service orchestration layer*. The fog service orchestration layer has a distributed functionality and provides dynamic and policy-based management

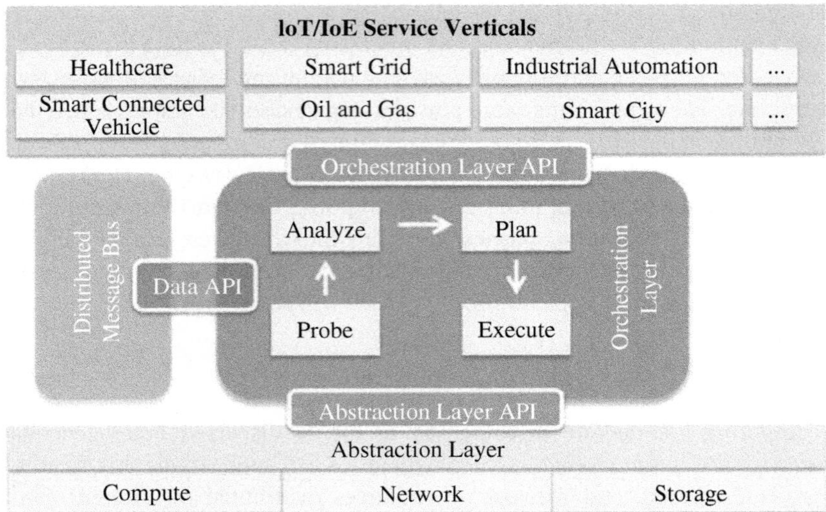

Figure 2.4 Fog computing architecture [10]. (*See color plate section for the color representation of this figure*)

of fog services. This layer has to manage a diverse number of fog nodes capabilities; thus, a set of new technologies and components are introduced to aid this process. One of these components is a software agent called foglet capable of performing the orchestration functionality by analyzing the deployed services on the current fog node and its physical health. Other components are a distributed database that stores policies and resource metadata, a scalable communication bus to send control messages for resource management, and a distributed policy engine that has a single global view and can perform local changes on each fog node.

2.4 Fog and Edge Illustrative Use Cases

In this section, we present two illustrative use cases for both the fog and edge paradigms with the purpose of showing key features, helping us to further understand the concept and applicability in real-world applications.

2.4.1 Edge Computing Use Cases

The rapid growth of big data frameworks and applications such as smart cities, smart vehicles, healthcare, and manufacturing has pushed edge computing among the major topics in academia and industry. With the increasing demand for high availability in such systems, system requirements tend to increase over time. The stringent requirements of IoT systems have recently suggested the architectural placement of a computing entity closer to the network edge. This architectural shifting has many benefits, such as process optimization and interaction speed. For example, if a wearable ECG sensor were to use the cloud instead of the edge it will consistently send all data up to the centralized cloud. As a consequence, in such a scenario it will cause high communication latency and unreliable availability between the sensor and the centralized cloud. In real-time, safety-critical IoT use-cases, devices must comply to stringent constraints to avoid any fatal events. In this scenario, the latency delay introduced by sending the data to the cloud and back is inadmissible. Thus, in case of a critical event detected by the sensor, a local decision must be taken by the edge device, rather than sending data to the cloud.

Edge computing is well suited for IoT deployments where both storing and processing data can be leveraged locally. For example, consider a smart home where the sensory information is stored on the edge device. Simply by doing encryption and storing sensory information locally, edge computing shifts many security concerns from the centralized cloud to its edge devices. In addition, IoT applications consume less bandwidth, and they work even when the connection to the cloud is affected. Furthermore, edge devices may assist in scheduling the operational time of each appliance, minimizing the electricity cost in a smart home [25]. This strategy considers user preferences on each appliance determined from appliance-level energy consumption. All such examples can benefit from the edge computing paradigm and to demonstrate the role of this paradigm in different scenarios, we describe in this section two possible use cases in healthcare [12] and a smart home [3, 15].

2.4.1.1 A Wearable ECG Sensor

This scenario consists of a wearable ECG sensor attached to the human body through a smartwatch and a smartphone that acts as an edge device, as presented in Figure 2.5. As for the communication, Bluetooth is used to connect the ECG

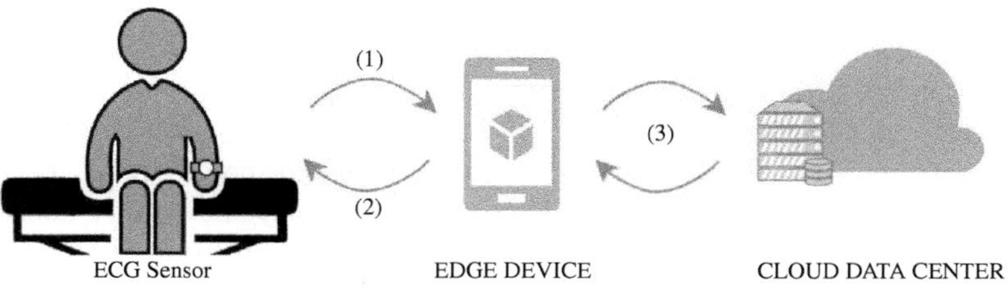

Figure 2.5 A wearable ECG sensor.

sensors with the edge device, while WiFi is used to connect the smartphone to fog devices and cloud.

Generally, users prefer smartwatch devices that provide monitoring heart functions while they continue normal physical activities. Due to the limited battery life and storage capacity of the smartwatch, we assume that the data produced by this device is around 1 KB per second and it is stored in the smartphone. Based on this assumption, daily produced data by a wearable device is around 86 MB per day and 2.6 GB monthly. One must note that smartphones have limited battery life and storage capacity. Hence, the smartphone at some point has to transfer the gathered data to another device that provides more storage capacity.

Referring to Figure 2.5, one can witness that data streaming is realized between the wearable device and the smartphone. Both devices remain connected to each other during the operation time. In case of any critical event, the wearable device interacts with the edge device and notifies the user for any situations. The process (1) start with getting real-time values from a wearable device to the smartphone. The smartphone application checks (2) periodically the wearable device to see if the connection between them is active. In addition, the smartphone may run out of free disk space and one can configure the application for daily synchronization (3) with another storage capability device, or with a central cloud storage or even with a fog node.

Since the wearable device and the edge device has limited resource capabilities, one must consider the energy consumption of both devices. In such system architecture, the first recommended approach is to decide what data to transmit to the cloud, what to store locally, and the last is to develop better monitoring algorithms. In the other words, when designing such systems, the critical point is to consider the energy consumption, which is affected by three main functions that are realized between devices, such as (1) communication, (2) storage, and (3) processing requirements. Hence, developers have to code software with highly efficient streaming algorithms, storing essential monitoring information, and avoiding continuously data transfers with the central cloud.

2.4.1.2 Smart Home

The smart home or smart apartment is an intelligent home network capable of sensing the home's occupants actions, and assisting them by providing appropriate services. In the following scenario, we will describe an example to illustrate a situation under development at the smart home, where an edge device is considered as a mediator between the IoT things deployed in a home environment. The smart home provides resources to the residents that are deployed in rooms. Each room has smart doors, smart windows, sensors (i.e. temperature, humidity, proximity, fire alarm, smoke detector, etc.), radio-frequency identification (RFID) tags, and readers.

The traditional implementation of the presented use case situation requires collecting data to a back-end cloud-based where system stores, processes, and replies to both real-time user queries. The configuration must be done on the cloud server, and each device sends the information to a central server for further processing of the data. Each of the devices contains a unique identification number and a lot of information saved in the cloud. Even if two devices reside near each other, the retrieval of the data must be done through communication with the cloud. A similar implementation of a system for monitoring environmental conditions by using a wireless sensor network (WSN) is given in [26]. However, just adding a WiFi connection to the current network-enabled devices and connecting it to the cloud is not enough for a smart home. In addition, in a smart home environment, besides the connected device, it must support communication with non-IP based resources, cheap wireless sensors, and controllers deployed in rooms, pipes, and even floor and walls.

Deploying a huge number of things in a smart home environment results in an impressive amount of produced data. One must consider that the data produced has to be transported to the processing units, assuring privacy and providing high availability. Since personal data must be consumed in the home, an architecture based only on the cloud computing paradigm is not suited for a smart home. In contrast, edge computing is perfect for building a smart home where data reside on an edge device running edge operating system (edgeOS). As a result, all deployed edge devices can be connected and managed easily and data can be processed locally by an edge device.

Figure 2.6 shows the structure of a variant of edgeOS in the smart home environment. EdgeOS provides a communication layer that supports multiple communication methods, such as WiFi, Bluetooth, ZigBee, or a cellular network. By using one of the methods, edgeOS collects data from the deployed things and mobile devices. In a smart home, most of the things will send data periodically to the edge device, respectively to the edgeOS. Collected data from different things need to be fused and massaged in the data abstraction layer. It is desirable that human interaction with edge devices is minimized. Hence, the edge device should consume/process all the data and interact with users in a proactive fashion. Additionally, the data abstraction layer will serve as a public interface for all things connected to edge devices where it enables the applicability of operations on the things.

Finally, on top of the data abstraction layer is the service management layer. This layer is responsible to guarantee a reliable system including differentiation (i.e. critical services must have higher priority compared to a normal service), extensibility (i.e. new things can be added dynamically), and isolation (i.e. if something crashes or is not responding, the user should be able to control things without crashing the whole edgeOS).

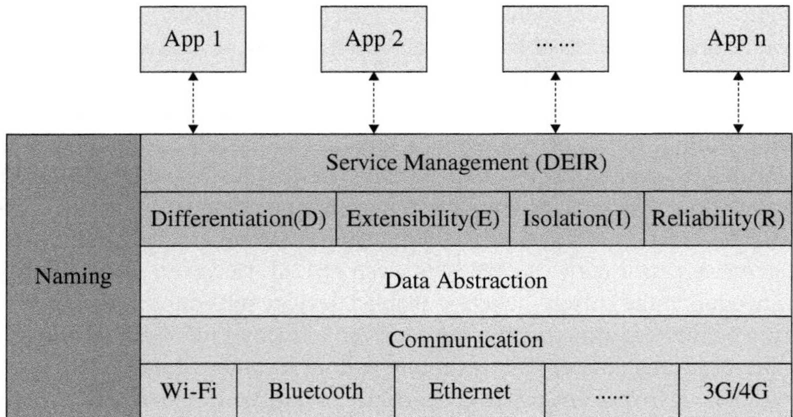

Figure 2.6 Structure of edgeOS in the smart home environment [3].

2.4.2 Fog Computing Use Cases

The new fog computing provides an improved quality of service (QoS), low latency and ensures that specific latency-sensitive applications meet their requirements. There are many areas like the healthcare, oil and gas, automotive, and gaming industries that can benefit from adopting this new paradigm. For example, by performing predictive maintenance the downtime of manufacturing machines can be reduced, optimizing the workflow in a manufacturing plant, or fog computing can simply monitor the structural integrity of buildings, ensuring the safety of workers and clients. However, by implementing such architecture not only businesses can profit. At the same time, life in the city as we know it today can be improved. Multiple day-to-day activities can be optimized to yield better living comfort. For example, consider the following scenario: we can improve congestion on the highway by using smart traffic congestion systems, optimize energy by creating smart grids, and lower the fuel consumption and waiting time in traffic by using a smart traffic light system. All such examples can benefit from this paradigm and, to demonstrate the role of fog in different scenarios, we describe in this section two possible use cases in a smart city, i.e. a smart traffic light system [10] and a smart pipeline monitoring system [27].

2.4.2.1 Smart Traffic Light System

In a smart traffic light system scenario, the objective is to lower the congestion in the city and optimize traffic flow. The immediate outcome of adopting this approach is the protection of the environment by lowering CO_2 emissions and reducing fuel consumption. Enabling an optimization like this requires the implementation of a hierarchical approach that enables real-time and near real-time operations, as well as analysis of data over long periods of time.

Each intersection in the city represents a component of our system where a smart traffic light application is deployed. The application is in charge of analyzing the collected data from local sensors and CCTV cameras and performs three major tasks: (1) compute the distance of every approaching vehicle in all directions and adapt the traffic light accordingly; (2) monitor pedestrians and cyclists to prevent any accidents, and (3) collect relevant data to help improve the overall system performance. Note that these functionalities require fast response time in case of (1) and (2), the exception being the last functionality (3), which only sends data to a higher layer for further investigation, without waiting for a response.

Another important component of our use case is the global node that creates a control function for each intersection. The key role for a global node is to collect all data from each smart traffic light and determine different commands, such that a steady flow of traffic is maintained. Notice that compared with the time requirements for the tasks deployed at an intersection, the functionality here requires a near real-time response.

The aforementioned hierarchical architecture of our traffic light system benefits from the advantages introduced by the fog computing paradigm. An immediate advantage over the cloud computing paradigm is its capabilities of orchestrating a wide range of distributed devices placed at each intersection. At the same time, it enables devices capable of analyzing data and performing fast response-time actions. Our system can be designed as a four-layer architecture, composed by the sensor layer, a fog device layer present locally at each intersection, another fog layer composed of the global node and the cloud layer. An overview of this architecture is presented in Figure 2.7.

2.4.2.2 Smart Pipeline Monitoring System

The smart pipeline monitoring system is an application deployed in the concept of smart cities, with the scope of monitoring the integrity of pipelines and preventing any serious economic and ecologic consequences. As an illustration, consider the case in which a pipeline that transports extracted oil from an offshore platform has failed, and the repercussion of failure has a big impact on the environment.

A pipeline system has an important role in our lives, being an essential infrastructure used to transport gas and liquids. It spreads throughout the entire city and provides us with basic needs like drinkable water. However, the integrity of a pipeline diminishes due to aging and sudden environmental changes. In the end, the risk of failure rises as corrosion and leaks appear.

To prevent such threats, a pipeline monitoring system has the capabilities of detecting any serious threats, reducing the overall failure by predicting three types of emergencies: (1) local disturbances (leakage, corrosion), (2) significant perturbations (damaged pipelines or approaching fire), and (3) city emergency situations. Since the infrastructure covers an entire city, our use case requires an architecture that supports geo-distribution of devices and low/predictable latency.

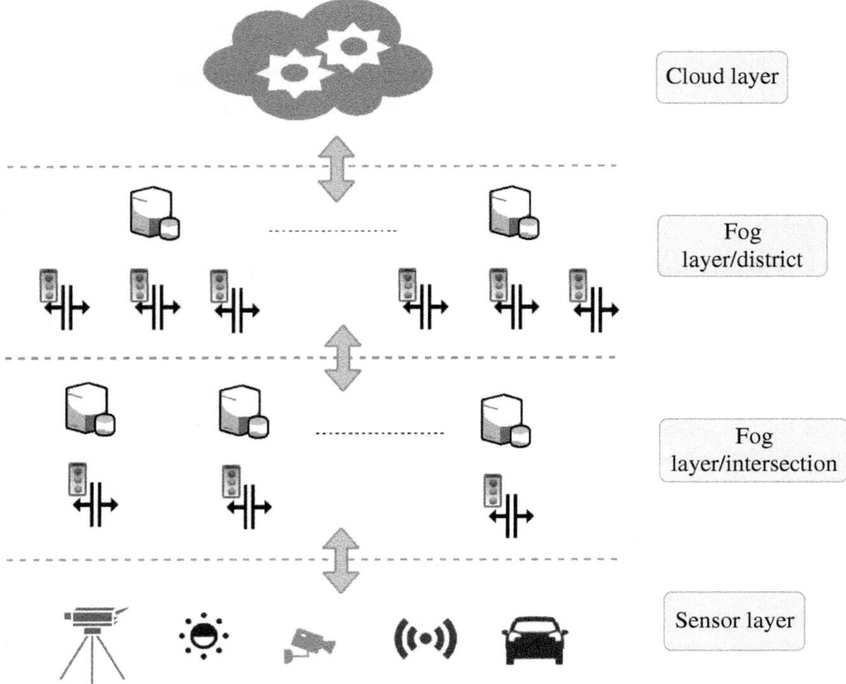

Figure 2.7 Smart Traffic light system.

An immediate solution is a four-layer fog computing architecture, described in Figure 2.8.

As in the case of a smart traffic light system, the fog architecture requires distinct layers to enable different time-scale responses. Depending on how wide the monitored area is, four or more layers can be used. The first layer represents the cloud and has the purpose of offering a global perspective of the entire system. In this layer, hard computational analysis is performed to prevent and respond to citywide disasters. Since the layer performs analysis on historical data, it does not require real-time or near real-time responses. Following, the second layer represents fog devices that are responsible to prevent any failure, at a smaller scale than cloud, in every neighborhood. The key purpose behind this layer is to identify potential hazardous events, based on the collected measurements from multiple devices. In this situation, near real-time responses are required for better prediction. Next, the third layer is composed of local fog nodes that identify potential threats based on the data received from the sensors and act by outputting control signals when it is necessary for fast recovery from a found threat. Furthermore, these devices

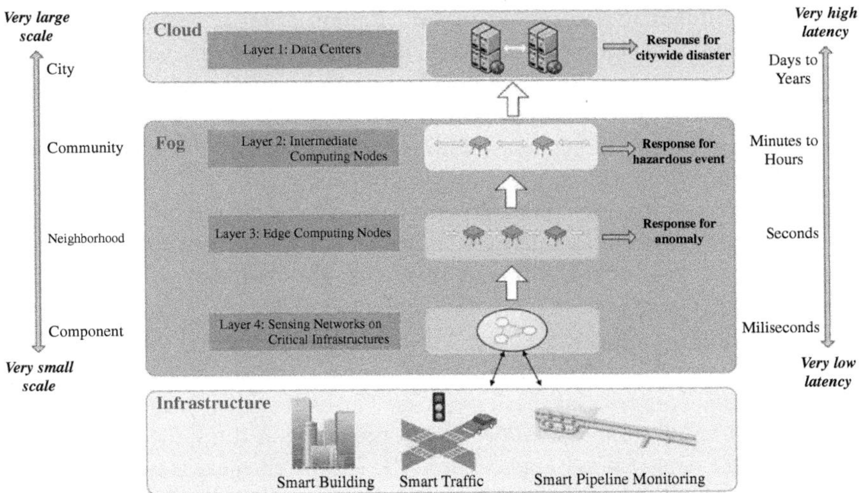

Figure 2.8 Smart pipeline monitoring system architecture.

process the raw data and prepare it for the layer above. Finally, the last layer is represented by sensors that generate measurements for the aforementioned layers in the architecture.

2.5 Future Challenges

Fog and edge computing vision introduce multiple advantages by migrating some computational resources at the edge of the network. The underlining of these paradigms is to create an IoT network environment covered with a vast amount of interconnected distributed heterogeneous devices having the purpose to deploy and manage demanding applications closer to the user. Yet, it is a nontrivial task to design platforms where all these required characteristics are met.

In this section, we identify and discuss the challenges that these paradigms must conquer in order to fulfill their full potential. We group these challenges in three main areas, i.e. resource management, security, and privacy, and network management.

2.5.1 Resource Management

Moving computational resources from the cloud closer to the end nodes stand in the center of the fog and edge paradigm. Therefore, novel resource management to fully utilize the available resources and process applications in close proximity of the user is imperative to the successful adoption of these systems. Since IoT devices are resource-constrained devices, applying resource management techniques at the edge will allow edge nodes to optimize their resource utilization (e.g. energy-aware smart devices that increase their battery levels by of loading computation to other nearby nodes), improve data privacy, and enable devices to collaborate and share resources to process IoT applications.

A taxonomy of resource management at the edge, based on the current state-of-the-art research in this area, is presented in [28]. According to this classification, a total of five different categories are identified considering the objective of the technique.

The first category refers to resource estimation and represents one of the fundamental requirements in resource management, i.e. the capability of estimating how many resources a certain task requires. This is important for handling the uncertainties found in an IoT network and providing at the same time a satisfactory QoS for deployed IoT applications. The second category is represented by resource discovery and aims to aid the user to discover available resources already deployed at the edge. Resource discovery complements resource estimation by keeping the pool of available computational resources updated.

Once the system can estimate and discover resources, a third category appears having the purpose of allocating IoT applications in close proximity to the users. This technique, called resource allocation, utilizes the knowledge of available resources to map parts of the applications at different edge devices such that its requirements are met. There are two different perspectives of the allocation: (1) it represents the initial deployment to the edge of the network, deciding where to map the application; and (2) it serves as a migration technique by self-adapting when a node has failed. Moreover, one challenge arises when sharing resources between distributed edge devices, i.e. a close collaboration between nodes enforced by security and privacy is required. Solving this challenge creates the fourth category, i.e. resource sharing.

Finally, the last technique is called resource optimization and is obtained by combining the aforementioned resource management approaches. The main objective is to optimize the usage of available resources at the edge according to the IoT application constraints. Usually, the developer creates the QoS requirement of his application before deploying it to the edge.

2.5.2 Security and Privacy

Adopting the vision of fog and edge computing, more applications that today reside in the cloud are moved to the edge of the network. By deploying and connecting IoT devices, we can transform our homes in a more digitalized environment that adapts automatically, based on our behavior. However, with such benefits arise a set of privacy and security issues that we must address. For example, one can easily study the behavior of a family by simply accessing the generated data from sensors deployed in the house. Hence, ensuring data privacy and security remains a crucial factor in the evolution of edge and fog paradigms.

To evaluate the security and privacy enforced in systems based on fog and edge devices, the designer can use the confidentiality, integrity, and availability (CIA) triad model, representing the most critical characteristics of a system [29]. While any breach of the confidentiality and integrity components yields a data privacy issue, the availability component refers to the property of the nodes to share their resources when required. Since fog and edge represents an extension of the cloud, such systems inherit not only the computational resources but also the security and privacy challenges. Besides these challenges, due to the deployment of devices at the edge of the network more security challenges appear. Yi et al. identify the most important security issues of fog computing as authentication, access control, intrusion attack, and privacy [9].

Considering the dynamic structure of an IoT network, authentication is an important key feature of fog and edge computing and was identified. as the main security issue in fog computing [20]. The authentication serves as the connectivity mechanism that allows to securely accept new nodes into the IoT network. By providing means to identify each device and establish its credentials, a trust is created between the new added node and network. The current security solutions proposed for cloud computing may have to be updated for fog/edge computing to account for threats that do not exist in its controlled environment [21]. One solution to securely authenticate edge devices is presented in [30].

A comprehensive study of security threats for edge paradigms (i.e. fog and edge computing, and MEC, among others) was presented in [31], where the importance of security is motivated for the overall system and each individual component. An edge ecosystem consists of different edge nodes and communication components, ranging from wireless to sensors and Internet-connected mobile devices, distributed in a multilayer fog architecture. While each individual component has its own security issues, new different security challenges appear by combining and creating an edge ecosystem. By reviewing the scope and nature of potential security attacks, the authors propose a threat model that analyzes possible security risks (see Table 2.1).

For this model, the authors in [31] discover all important components of edge paradigms and describe all attacks that can occur against them. As depicted from Table 2.1., we can observe that five different targets i.e. network infrastructure, service infrastructure composed of edge data center and core infrastructure, virtualization infrastructure and user devices [31] are identified. The network infrastructure represents the various communication networks that connect edge devices which an adversary can attack using one of the following: denial

Table 2.1 Threat model for fog and edge computing [21].

Fog components **Security issues**	**Network infrastructure**	**Service infrastructure (edge data center)**	**Service infrastructure (core infrastructure)**	**Virtualization infrastructure**	**User devices**
DoS	✓			✓	
Man-in-the-middle	✓				
Rogue component (i.e. data center, gateway, or infrastructure)	✓	✓	✓		
Physical damage		✓			
Privacy leakage		✓	✓	✓	
Privilege escalation		✓		✓	
Service or VM manipulation		✓	✓	✓	✓
Misuse of resources				✓	
Injection of information					✓

of service (DoS), man-in-the-middle attacks, and rogue datacenter. An example of a man-in-the-middle attack on an IoT network is presented in [32]. On the one hand, an adversary could attack the service infrastructure, at the edge of the network, by using physical damage, rogue component privacy leakage, privilege escalation, and service or virtual machine (VM) manipulation. On the other hand, the core infrastructure is more secure being prone to attacks like rouge component, privacy leakage, and VM manipulation [31]. Finally, the virtualization infrastructure is exposed to attacks, such as DoS, privacy leakage, privilege escalation, service or VM migration, and misuse of resources; while user devices are susceptible to attacks like VM manipulation and injection of information.

Privacy, defined as the protection of private data, ensures that a malicious adversary cannot obtain sensitive information while data is in transit [33]. At the moment, privacy is most vulnerable since the data of end users is sent directly to the cloud. From this point of view, edge and fog paradigms enforce privacy by moving the computation closer to the user. In doing so, data can be processed locally and the user can control what third parties are accessing his private data based on a defined role-based access control policy. However, some privacy challenges remain open, such as (i) the awareness of privacy in the community where, for example, almost 80% of WiFi user still use their default passwords for their routers and (ii) the lack of efficient tools for security and privacy for constrained devices [3].

2.5.3 Network Management

The network management plays the most important role in both edge and fog paradigms since it represents the means of connecting all smart devices at the edge and ultimately providing available resources by deploying more nodes. Since the nature of an IoT network consists of heterogeneous devices, which are highly dispersed across large areas, an engaging task is to manage and maintain connectivity. Newly emerging technologies like software-defined networks (SDNs) and network function virtualization (NFV) are seen as a possible solution that may have a significant impact in implementing and maintaining the network increasing the scalability and reducing cost [19].

Considering the volatile nature of the network, providing a seamless connectivity mechanism is critical since both mobile and stationary devices coexist in the network. Therefore, another aspect of network management is related to connectivity. This mechanism must be able to provide the possibility of connecting/disconnecting easily from the network such that the uncertainty introduced by mobile devices is accommodated. Moreover, providing this encourages an increased deployment of smart devices by users and manufacturers alike, without extra cost or expert knowledge.

An effort in this direction is made by the I3: the intelligent IoT integrator, developed by USC [34], having the purpose of creating a marketplace where users can share their private data with application developers and receive incentives for it. There are two main advantages of designing the marketplace like this: first, the users are encouraged to deploy more edge devices, which in return extends the IoT network with more resources that app developers can use; and second, there is a pool of data that developers can utilize to improve their IoT applications.

2.6 Conclusion

The never-ending increase in interconnected IoT devices and the stringent requirements of new IoT applications has posed severe challenges to the current cloud computing state-of-the-art architecture, such as network congestion and privacy of data. As a result, researchers have proposed a new solution to tackle these challenges by migrating some computational resources closer to the user. The approach taken in this solution made the cloud more efficient by extending its computational capabilities at the end of the network, solving its challenges in the process.

Continuing to improve this solution, multiple paradigms appeared, having as their underlying vision the same goal of deploying more resources at the edge of the network. Besides their common vision, some paradigms were influenced by their considered use case, e.g. MEC paradigm enables constrained devices like smartphones to offload parts of the applications to save resources. However, two of the most popular paradigms (i.e. fog and edge computing) are widely used in research today.

These two paradigms were designed to enable processing IoT applications at the endpoints of the network, sharing more similarities than others. Other than the naming convention, the difference at the beginning for the two, i.e. fog computing extends the cloud creating a cloud-to-things continuum and edge computing places the application directly on the edge devices, was represented by the location where computations are performed. Since in the past couple of years there were tremendous advances for edge devices, this difference between the two has disappeared, both fog and edge aiming to deploy applications as close as possible to the edge of the network. Considering the similarities they share, we argue that there is no difference between their purpose of them.

Acknowledgment

The research leading to these results has received funding from the European Union's Horizon 2020 research and innovation programme under the Marie Skłodowska-Curie grant agreement No. 764785, FORA (Fog Computing for Robotics and Industrial Automation). This publication was partially supported by the TUW Research Cluster Smart CT.

References

1 Chiang, M. and Zhang, T. (2016). Fog and IoT: an overview of research opportunities. *IEEE Internet of Things Journal* 3 (6): 854–864.

2 Bonomi, F., Milito, R., Zhu, J., and Addepali, S. (2012). *Fog computing and its role in the Internet of Things, 1st ACM Mobile Cloud Computing Workshop*, 13–15.

3 Shi, W., Cao, J., Zhang, Q. et al. (2016). Edge computing: vision and challenges. *IEEE Internet of Things Journal* 3 (5): 637–646.

4 Satyanarayanan, M., Bahl, P., Caceres, R., and Davies, N. (2009). The case for VM-based cloudlets in mobile computing. *IEEE Pervasive Computing* 8 (4): 14–23. [Online]. Available: http://dx.doi.org/10.1109/MPRV.2009.82 http://http://ieeexplore.ieee.org/document/5280678.

5 Rausch, T., Avasalcai, C., and Dustdar, S. (2018). Portable energy-aware cluster-based edge computers. In: *2018 IEEE/ACM Symposium on Edge Computing (SEC)*, 260–272.

6 Elias, A.R., Golubovic, N., Krintz, C., and Wolski, R. (2017). Where's the bear? Automating wildlife image processing using IoT and edge cloud systems. In: *2017 IEEE/ACM Second International Conference on Internet-of-Things Design and Implementation (IoTDI)*, 247–258.

7 M. T. Beck, M. Werner, S. Feld, and S. Schimper, Mobile edge computing: a taxonomy. Citeseer.

8 Fernando, N., Loke, S.W., and Rahayu, W. (2013). Mobile cloud computing: a survey. *Future Generation Computer Systems* 29 (1): 84–106, including Special section: AIRCC-NetCoM 2009 and Special section: Clouds and Service-Oriented Architectures. [Online]. Available: http://www.sciencedirect.com/science/article/pii/S0167739X12001318.

9 Yi, S., Li, C., and Li, Q. (2015). A survey of fog computing: concepts, applications and issues. In: *Proceedings of the 2015 Workshop on Mobile Big Data*, 37–42. ACM.

10 Bonomi, F., Milito, R., Natarajan, P., and Zhu, J. (2014). *Fog Computing: A Platform for Internet of Things and Analytics*, 169–186. Cham: Springer International Publishing [Online]. Available: https://doi.org/10.1007/978-3-319-05029-4 7.

11 Shi, W. and Dustdar, S. (2016). The promise of edge computing. *Computer* 49 (5): 78–81.

12 Gusev, M. and Dustdar, S. (2018). Going back to the roots|the evolution of edge computing, an IoT perspective. *IEEE Internet Computing* 22 (2): 5–15.

13 Pate, J. and Adegbija, T. (2018). Amelia: an application of the Internet of Things for aviation safety, in 15th. In: *IEEE Annual on Consumer Communications & Networking Conference (CCNC), 2018*, 1–6. IEEE.

14 Chen, B., Wan, J., Celesti, A. et al. (2018). Edge computing in IoT-based manufacturing. *IEEE Communications Magazine* 56 (9): 103–109.

15 Zhang, S., Li, W., Wu, Y. et al. Enabling edge intelligence for activity recognition in smart homes. In: *2018 IEEE 15th International Conference on Mobile Ad Hoc and Sensor Systems (MASS)*, vol. 2018, 228–236. IEEE.

16 Yu, W., Liang, F., He, X. et al. (2018). A survey on the edge computing for the Internet of Things. *IEEE Access* 6: 6900–6919.

17 Chen, Z., Xu, G., Mahalingam, V. et al. (2016). A cloud computing based network monitoring and threat detection system for critical infrastructures. *Big Data Research* 3: 10–23.

18 Xu, X., Sheng, Q.Z., Zhang, L.-J. et al. (2015). From big data to big service. *Computer* 48 (7): 80–83.

19 Yi, S., Hao, Z., Qin, Z., and Li, Q. (2015). Fog computing: platform and applications. In: *2015 Third IEEE Workshop on Hot Topics in Web Systems and Technologies (HotWeb) (HOTWEB)*, vol. 00, 73–78. [Online]. Available: http://doi.ieeecomputersociety.org/10.1109/HotWeb.2015.22.

20 Stojmenovic, I. and Wen, S. (2014). The fog computing paradigm: scenarios and security issues. In: *2014 Federated Conference on Computer Science and Information Systems*, 1–8.

21 Osanaiye, O., Chen, S., Yan, Z. et al. (2017). From cloud to fog computing: a review and a conceptual live VM migration framework. *IEEE Access* 5: 8284–8300.

22 Dastjerdi, A.V. and Buyya, R. (2016). Fog computing: helping the Internet of Things realize its potential. *Computer* 49 (8): 112–116.

23 Sarkar, S., Chatterjee, S., and Misra, S. (2018). Assessment of the suitability of fog computing in the context of Internet of Things. *IEEE Transactions on Cloud Computing* 6 (1): 46–59.

24 Shi, Y., Ding, G., Wang, H. et al. (2015). The fog computing service for healthcare. In: *2015 2nd International Symposium on Future Information and Communication Technologies for Ubiquitous HealthCare (Ubi-HealthTech)*, 1–5.

25 Xia, C., Li, W., Chang, X. et al. (2018). Edge-based energy management for smart homes. In: *2018 IEEE 16th International Conference on Dependable, Autonomic and Secure Computing, 16th Intl Conf on Pervasive Intelligence and Computing, 4th International Conference on Big Data Intelligence and Computing and Cyber Science and Technology Congress (DASC/PiCom/DataCom/CyberSciTech)*, 849–856. IEEE.

26 Bajrami, X. and Murturi, I. (2018). An efficient approach to monitoring environmental conditions using a wireless sensor network and nodemcu. *e & i Elektrotechnik und Informationstechnik* 135 (3): 294–301. [Online]. Available: https://doi.org/10.1007/s00502-018-0612-9.

27 Tang, B., Chen, Z., Hefferman, G. et al. (2017). Incorporating intelligence in fog computing for big data analysis in smart cities. *IEEE Transactions on Industrial Informatics* 13 (5): 2140–2150.

28 Tocze, K. and Nadjm-Tehrani, S. (2018). A taxonomy for management and optimization of multiple resources in edge computing. *Wireless Communications and Mobile Computing* 2018, Art. No: 7476203: 1–23.

29 Farooq, M.U., Waseem, M., Khairi, A., and Mazhar, S. (2015). A critical analysis on the security concerns of Internet of Things (IoT). *International Journal of Computer Applications* 111 (7).

30 Puthal, D., Obaidat, M.S., Nanda, P. et al. (2018). Secure and sustainable load balancing of edge data centers in fog computing. *IEEE Communications Magazine* 56 (5): 60–65.

31 Roman, R., Lopez, J., and Mambo, M. (2018). Mobile edge computing, fog et al.: a survey and analysis of security threats and challenges. *Future Generation Computer Systems* 78: 680–698. [Online]. Available: http://www.sciencedirect.com/science/article/pii/S0167739X16305635.

32 Wang, Y., Uehara, T., and Sasaki, R. (2015). Fog computing: issues and challenges in security and forensics. In: *2015 IEEE 39th Annual Computer Software and Applications Conference*, vol. 3, 53–59.

33 Zhou, M., Zhang, R., Xie, W. et al. (2010). Security and privacy in cloud computing: a survey. In: *2010 Sixth International Conference on Semantics, Knowledge and Grids*, 105–112.

34 University of Southern California, I3: The intelligent IoT integrator (i3), https://i3.usc.edu.

3

Deep Learning in the Era of Edge Computing: Challenges and Opportunities

Mi Zhang[1], Faen Zhang[2], Nicholas D. Lane[3], Yuanchao Shu[4], Xiao Zeng[1], Biyi Fang[1], Shen Yan[1], and Hui Xu[2]

[1]*Department of Electrical and Computer Engineering, Michigan State University, East Lansing, MI, USA, 48824*
[2]*AInnovation, Beijing, China, 100080*
[3]*Department of Computer Science, Oxford University, Oxford, United Kingdom, OX1 3PR*
[4]*Microsoft Research, Redmond, WA, USA, 98052*

3.1 Introduction

Of all the technology trends that are taking place right now, perhaps the biggest one is edge computing [1, 2]. It is the one that is going to bring the most disruption and the most opportunity over the next decade. Broadly speaking, edge computing is a new computing paradigm that aims to leverage devices that are deployed at the Internet's edge to collect information from individuals and the physical world as well as to process the collected information in a distributed manner [3]. These devices, referred to as edge devices, are physical devices equipped with sensing, computing, and communication capabilities. Today, we are already surrounded by a variety of such edge devices: our mobile phones and wearables are edge devices; home intelligence devices such as Google Nest and Amazon Echo are edge devices; autonomous systems such as drones, self-driving vehicles, and robots that vacuum the carpet are also edge devices. These edge devices continuously collect a variety of data, including images, videos, audios, texts, user logs, and many others with the ultimate goal to provide a wide range of services to improve the quality of people's everyday lives.

Although the Internet is the backbone of edge computing, the true value of edge computing lies at the intersection of gathering data from sensors and extracting meaningful information from the collected sensor data. Over the past few years, deep learning (i.e. deep neural networks [DNNs]) [4] has become the dominant data analytics approach due to its capability to achieve impressively high

Fog Computing: Theory and Practice, First Edition.
Edited by Assad Abbas, Samee U. Khan, and Albert Y. Zomaya.

accuracies on a variety of important computing tasks, such as speech recognition [5], machine translation [6], object recognition [7], face detection [8], sign language translation [9], and scene understanding [10]. Driven by deep learning's splendid capability, companies such as Google, Facebook, Microsoft, and Amazon are embracing this technological breakthrough and using deep learning as the core technique to power many of their services.

Deep learning models are known to be expensive in terms of computation, memory, and power consumption [11, 12]. As such, given the resource constraints of edge devices, the status quo approach is based on the cloud computing paradigm in which the collected sensor data are directly uploaded to the cloud; and the data processing tasks are performed on the cloud servers, where abundant computing and storage resources are available to execute the deep learning models. Unfortunately, cloud computing suffers from three key drawbacks that make it less favorable to applications and services enabled by edge devices. First, data transmission to the cloud becomes impossible if the Internet connection is unstable or even lost. Second, data collected at edge devices may contain very sensitive and private information about individuals. Directly uploading those raw data onto the cloud constitutes a great danger to individuals' privacy. Most important, as the number of edge devices continues to grow exponentially, the bandwidth of the Internet becomes the bottleneck of cloud computing, making it no longer feasible or cost-effective to transmit the gigantic amount of data collected by those devices to the cloud.

In this book chapter, we aim to provide our insights for answering the following question: can edge computing leverage the amazing capability of deep learning? As computing resources in edge devices become increasingly powerful, especially with the emergence of artificial intelligence (AI) chipsets, we envision that in the near future, the majority of the edge devices will be equipped with machine intelligence powered by deep learning. The realization of this vision requires considerable innovation at the intersection of computer systems, networking, and machine learning. In the following, we describe eight research challenges followed by opportunities that have high promise to address those challenges. We hope this book chapter act as an enabler of inspiring new research that will eventually lead to the realization of the envisioned intelligent edge.

3.2 Challenges and Opportunities

3.2.1 Memory and Computational Expensiveness of DNN Models

Memory and computational abilities are expensive for DNN models that achieve state-of-the-art performance. To illustrate this, Table 3.1 lists the details of some of

Table 3.1 Memory and computational expensiveness of some of the most commonly used DNN models.

DNN	Top-5 error (%)	Latency (ms)	Layers	FLOPs (billion)	Parameters (million)
AlexNet	19.8	14.56	8	0.7	61
GoogleNet	10.07	39.14	22	1.6	6.9
VGG-16	8.8	128.62	16	15.3	138
ResNet-50	7.02	103.58	50	3.8	25.6
ResNet-152	6.16	217.91	152	11.3	60.2

the most commonly used DNN models. As shown, these models normally contain millions of model parameters and consume billions of floating-point operations (FLOPs). This is because these DNN models are designed for achieving high accuracy without taking resources consumption into consideration. Although computing resources in edge devices are expected to become increasingly powerful, their resources are way more constrained than cloud servers. Therefore, filling the gap between high computational demand of DNN models and the limited computing resources of edge devices represents a significant challenge.

To address this challenge, the opportunities lie at exploiting the redundancy of DNN models in terms of parameter representation and network architecture. In terms of parameter representation redundancy, to achieve the highest accuracy, state-of-the-art DNN models routinely use 32 or 64 bits to represent model parameters. However, for many tasks like object classification and speech recognition, such high-precision representations are not necessary and thus exhibit considerable redundancy. Such redundancy can be effectively reduced by applying parameter quantization techniques that use 16, 8, or even fewer bits to represent model parameters. In terms of network architecture redundancy, state-of-the-art DNN models use overparameterized network architectures, and thus many of their parameters are redundant. To reduce such redundancy, the most effective technique is model compression. In general, DNN model compression techniques can be grouped into two categories. The first category focuses on compressing large DNN models that are pretrained into smaller ones. For example, [13] proposed a model compression technique that prunes out unimportant model parameters whose values are lower than a threshold. However, although this parameter pruning approach is effective at reducing model sizes, it does not necessarily reduce the number of operations involved in the DNN model. To overcome this issue, [14] proposed a model compression technique that prunes out unimportant filters which effectively reduces the computational cost of DNN models. The second category focuses on designing efficient small DNN models directly. For example, [15] proposed the use of depth-wise separable convolutions that are small and computationally efficient to replace conventional convolutions that are large and computationally expensive, which reduces not only model size but also computational cost. Being an orthogonal approach, [16] proposed a technique referred to as knowledge distillation to directly extract useful knowledge from large DNN models and pass it to a smaller model that achieves similar prediction performance as the large models, but with fewer model parameters and lower computational cost.

3.2.2 Data Discrepancy in Real-world Settings

The performance of a DNN model is heavily dependent on its training data, which is supposed to share the same or a similar distribution with the potential test data. Unfortunately, in real-world settings, there can be a considerable discrepancy between the training data and the test data. Such discrepancy can be caused by variation in sensor hardware of edge devices as well as various noisy factors in the real world that degrade the quality of the test data. For example, the quality of images taken in real-world settings can be degraded by factors such as illumination, shading, blurriness, and undistinguishable background [17] (see Figure 3.1 as an example). Speech data sampled in noisy places such as busy restaurants can be contaminated by voices from surround people. The discrepancy between training and test data could degrade the performance of DNN models, which becomes a challenging problem.

To address this challenge, we envision that the opportunities lie at exploring data augmentation techniques as well as designing noise-robust loss functions. Specifically, to ensure the robustness of DNN models in real-world settings, a large volume of training data that contain significant variations is needed. Unfortunately, collecting such a large volume of diverse data that cover all types of variations and noise factors is extremely time consuming. One effective technique to overcome this dilemma is data augmentation. Data augmentation techniques generate variations that mimic the variations occurred in the real-world settings. By using the large amount of newly generated augmented data as part of the training data, the discrepancy between training and test data is minimized. As a result, the trained DNN models become more robust to the various noisy factors in the real world. A technique that complements data augmentation is to design loss functions that are robust to discrepancy between the training data and the test data. Examples of such noise-robust loss functions include triplet loss [18] and variational autoencoder [19]. These noise-robust loss functions are able to enforce a DNN model to learn features that are invariant to various noises that degrade the quality of test data even if the training data and test data do not share a similar distribution.

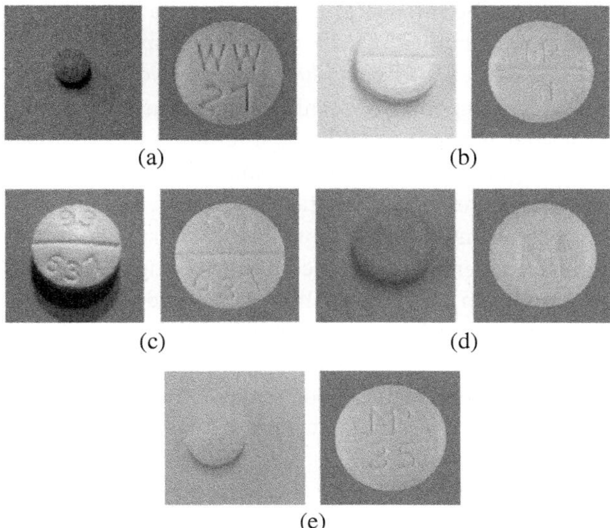

Figure 3.1 Illustration of differences between training and test images of the same pills under five different scenarios [17]. For each scenario, the image on the left is the training image; and the image on the right is the test image of the same pill. Due to the deterioration caused by a variety of real-world noisiness such as shading, blur, illumination, and background, training image and test image of the same pill look very different. (a) Size variation, (b) Illumination, (c) Shading, (d) Blur, (e) Undistinguishable background.

3.2.3 Constrained Battery Life of Edge Devices

For edge devices that are powered by batteries, reducing energy consumption is critical to extending devices' battery lives. However, some sensors that edge devices heavily count on to collect data from individuals and the physical world such as cameras are designed to capture high-quality data, which are power hungry. For example, video cameras incorporated in smartphones today have increasingly high resolutions to meet people's photographic demands. As such, the quality of images taken by smartphone cameras is comparable to images that are taken by professional cameras, and image sensors inside smartphones are consuming more energy than ever before, making energy consumption reduction a significant challenge.

To address this challenge, we envision that the opportunities lie in exploring smart data subsampling techniques, matching data resolution to DNN models, and redesigning sensor hardware to make it low-power. First, to reduce energy consumption, one commonly used approach is to turn on the sensors when needed. However, there are streaming applications that require sensors to be always on.

As such, it requires DNN models to be run over the streaming data in a continuous manner. To reduce energy consumption in such a scenario, opportunities lie at subsampling the streaming data and processing those informative subsampled data points only while discarding data points that contain redundant information.

Second, while sensor data such as raw images are high resolution, DNN models are designed to process images at a much lower resolution. The mismatch between high-resolution raw images and low-resolution DNN models incurs considerable unnecessary energy consumption, including energy consumed to capture high-resolution raw images and energy consumed to convert high-resolution raw images to low-resolution ones to fit the DNN models. To address the mismatch, one opportunity is to adopt a dual-mode mechanism. The first mode is a traditional sensing mode for photographic purposes that captures high-resolution images. The second mode is a DNN processing mode that is optimized for deep learning tasks. Under this model, the resolutions of collected images are enforced to match the input requirement of DNN models.

Lastly, to further reduce energy consumption, another opportunity lies at redesigning sensor hardware to reduce the energy consumption related to sensing. When collecting data from onboard sensors, a large portion of the energy is consumed by the analog-to-digital converter (ADC). There are early works that explored the feasibility of removing ADC and directly using analog sensor signals as inputs for DNN models [20]. Their promising results demonstrate the significant potential of this research direction.

3.2.4 Heterogeneity in Sensor Data

Many edge devices are equipped with more than one onboard sensor. For example, a smartphone has a global positioning system (GPS) sensor to track geographical locations, an accelerometer to capture physical movements, a light sensor to measure ambient light levels, a touchscreen sensor to monitor users' interactions with their phones, a microphone to collect audio information, and a camera to capture images and videos. Data obtained by these sensors are by nature heterogeneous and are diverse in format, dimensions, sampling rates, and scales. How to take the data heterogeneity into consideration to build DNN models and to effectively integrate the heterogeneous sensor data as inputs for DNN models represents a significant challenge.

To address this challenge, one opportunity lies at building a multimodal deep learning model that takes data from different sensing modalities as its inputs. For example, [21] proposed a multimodal DNN model that uses restricted Boltzmann machine (RBM) for activity recognition. Similarly, [22] also proposed a multimodal DNN model for smartwatch-based activity recognition. Besides building multimodal DNN models, another opportunity lies in combining information from heterogeneous sensor data extracted at different dimensions and scales. As an example, [23] proposed a multiresolution deep embedding approach for processing heterogeneous data at different dimensions. [24] proposed an integrated convolutional and recurrent neural networks (RNNs) for processing heterogeneous data at different scales.

3.2.5 Heterogeneity in Computing Units

Besides data heterogeneity, edge devices are also confronted with heterogeneity in on-device computing units. As computing hardware becomes more and more specialized, an edge device could have a diverse set of onboard computing units including traditional processors such as central processing units (CPUs), digital signal processing (DSP) units, graphics processing units (GPUs), and field-programmable gate arrays (FPGAs), as well as emerging domain-specific processors such as Google's Tensor Processing Units (TPUs). Given the increasing heterogeneity in onboard computing units, mapping deep learning tasks and DNN models to the diverse set of onboard computing units is challenging.

To address this challenge, the opportunity lies at mapping operations involved in DNN model executions to the computing unit that is optimized for them. State-of-the-art DNN models incorporate a diverse set of operations but can be generally grouped into two categories: parallel operations and sequential operations. For example, the convolution operations involved in convolutional neural networks (CNNs) are matrix multiplications that can be efficiently executed in parallel on GPUs that have the optimized architecture for executing parallel operations. In contrast, the operations involved in RNNs have strong sequential dependencies, and better-fit CPUs that are optimized for executing sequential operations where operator dependencies exist. The diversity of operations suggests the importance of building an architecture-aware compiler that is able to decompose a DNN models at the operation level and then allocate the right type of computing unit to execute the operations that fit its architecture characteristics. Such an architecture-aware compiler would maximize the hardware resource utilization and significantly improve the DNN model execution efficiency.

3.2.6 Multitenancy of Deep Learning Tasks

The complexity of real-world applications requires edge devices to concurrently execute multiple DNN models that target different deep learning tasks [25]. For example, a service robot that needs to interact with customers needs to not only track faces of individuals it interacts with, but also recognize their facial emotions at the same time. These tasks all share the same data inputs and the limited resources on the edge device. How to effectively share the data inputs across concurrent deep learning tasks and efficiently utilize the shared resources to maximize the overall performance of all the concurrent deep learning tasks is challenging.

In terms of input data sharing, currently, data acquisition for concurrently running deep-learning tasks on edge devices is exclusive. In other words, at runtime, only one single deep learning task is able to access the sensor data inputs at one time. As a consequence, when there are multiple deep learning tasks running concurrently on edge devices, each deep learning task has to explicitly invoke system Application Programming Interfaces (APIs) to obtain its own data copy and maintain it in its own process space. This mechanism causes considerable system overhead as the number of concurrently running deep learning tasks increases. To address this input data sharing challenge, one opportunity lies at creating a data provider that is transparent to deep learning tasks and sits between them and the operating system as shown in Figure 3.2. The data provider creates a single copy of the sensor data inputs such that deep learning tasks that need to acquire data all access to this single copy for data acquisition. As such, a deep learning task is able to acquire data without interfering other tasks. More important, it provides a solution that scales in terms of the number of concurrently running deep learning tasks.

In terms of resource sharing, in common practice, DNN models are designed for individual deep learning tasks. However, existing works in deep learning show that DNN models exhibit layer-wise semantics where bottom layers extract basic structures and low-level features while layers at upper levels extract complex structures and high-level features. This key finding aligns with a subfield in machine learning named multitask learning [26]. In multitask learning, a single model is trained to perform multiple tasks by sharing low-level features while high-level features differ for different tasks. For example, a DNN model can be trained for scene understanding as well as object classification [27]. Multitask learning provides a perfect opportunity for improving the resource utilization for resource-limited edge devices when concurrently executing multiple deep learning tasks. By sharing the low-level layers of the DNN model across different

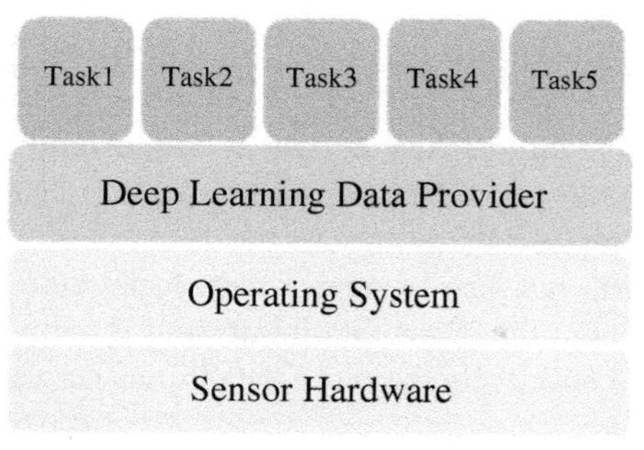

Figure 3.2 Illustration of data sharing mechanism.

deep learning tasks, redundancy across deep learning tasks can be maximally reduced. In doing so, edge devices can efficiently utilize the shared resources to maximize the overall performance of all the concurrent deep learning tasks.

3.2.7 Offloading to Nearby Edges

For edge devices that have extremely limited resources such as low-end Internet of Things (IoT) devices, they may still not be able to afford executing the most memory and computation-efficient DNN models locally. In such a scenario, instead of running the DNN models locally, it is necessary to offload the execution of DNN models. As mentioned in the introduction section, offloading to the cloud has a number of drawbacks, including leaking user privacy and suffering from unpredictable end-to-end network latency that could affect user experience, especially when real-time feedback is needed. Considering those drawbacks, a better option is to offload to nearby edge devices that have ample resources to execute the DNN models.

To realize edge offloading, the key is to come up with a model partition and allocation scheme that determines which part of model should be executed locally and which part of model should be offloading. To answer this question, the first aspect that needs to take into account is the size of intermediate results of executing a DNN model. A DNN model adopts a layered architecture. The sizes of intermediate results generated out of each layer have a pyramid shape (Figure 3.3), decreasing from lower layers to higher layers. As a result, partitioning at lower layers would generate larger sizes of intermediate results, which could increase the transmission latency. The second aspect that needs to take into account is the amount of information to be transmitted. For a DNN model, the amount of information generated out of each layer decreases from lower layers to higher layers. Partitioning at lower layers would prevent more information from being transmitted, thus

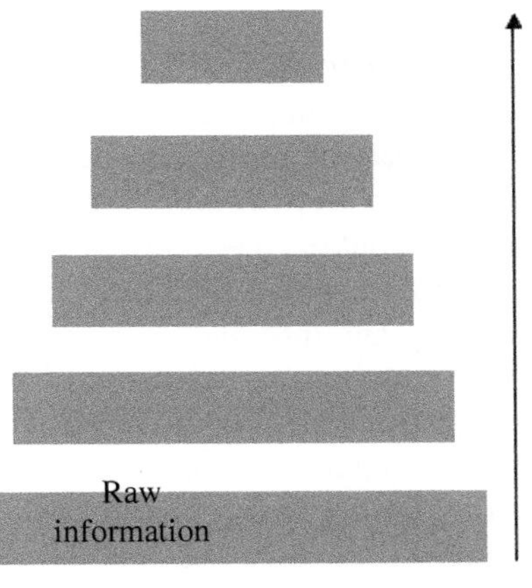

Figure 3.3 Illustration of intermediate results of a DNN model. The size of intermediate results generated out of each layer decreases from lower layers to higher layers. The amount of information generated out of each layer also decreases from lower layers to higher layers.

preserving more privacy. As such, the edge offloading scheme creates a trade-off between computation workload, transmission latency, and privacy preservation.

3.2.8 On-device Training

In common practice, DNN models are trained on high-end workstations equipped with powerful GPUs where training data are also located. This is the approach that giant AI companies such as Google, Facebook, and Amazon have adopted. These companies have been collecting a gigantic amount of data from users and use those data to train their DNN models. This approach, however, is privacy-intrusive, especially for mobile phone users because mobile phones may contain the users' privacy-sensitive data. Protecting users' privacy while still obtaining well-trained DNN models becomes a challenge.

To address this challenge, we envision that the opportunity lies in on-device training. As computer resources in edge devices become increasingly powerful, especially with the emergence of AI chipsets, in the near future, it becomes feasible to train a DNN model locally on edge devices. By keeping all the personal data that may contain private information on edge devices, on-device training provides a privacy-preserving mechanism that leverages the compute resources inside edge devices to train DNN models without sending the privacy-sensitive personal data to the giant AI companies. Moreover, today, gigantic amounts of data are generated by edge devices such as mobile phones on a daily basis. These data contain valuable information about users and their personal preferences. With such personal information, on-device training is enabling training personalized DNN models that deliver personalized services to maximally enhance user experiences.

3.3 Concluding Remarks

Edge computing is revolutionizing the way we live, work, and interact with the world. With the recent breakthrough in deep learning, it is expected that in the foreseeable future, majority of the edge devices will be equipped with machine intelligence powered by deep learning. To realize the full promise of deep learning in the era of edge computing, there are daunting challenges to address.

In this chapter, we presented eight challenges at the intersection of computer systems, networking, and machine learning. These challenges are driven by the gap between high computational demand of DNN models and the limited battery lives of edge devices, the data discrepancy in real-world settings, the need to process heterogeneous sensor data and concurrent deep learning tasks on heterogeneous computing units, and the opportunities for offloading to nearby edges and on-device training. We also proposed opportunities that have potential to address these challenges. We hope our discussion could inspire new research that turns the envisioned intelligent edge into reality.

References

1 Shi, W., Cao, J., Zhang, Q. et al. (2016). Edge computing: vision and challenges. *IEEE Internet of Things Journal* 3 (5): 637–646.

2 Shi, W. and Dustdar, S. (2016). The promise of edge computing. *Computer* 49 (5): 78–81.

3 Satyanarayanan, M. (2017). The emergence of edge computing. *Computer* 50 (1): 30–39.

4 LeCun, Y., Bengio, Y., and Hinton, G. (2015). Deep learning. *Nature* 521 (7553): 436–444.

5 Hinton, G., Deng, L., Yu, D. et al. (2012). Deep neural networks for acoustic modeling in speech recognition: the shared views of four research groups. *IEEE Signal Processing Magazine* 29 (6): 82–97.

6 Dzmitry Bahdanau, Kyunghyun Cho, and Yoshua Bengio, Neural machine translation by jointly learning to align and translate, ICLR, 2014.

7 Krizhevsky, A., Sutskever, I., and Hinton, G.E. (2012). Imagenet classification with deep convolutional neural networks. In: *Advances in Neural Information Processing Systems (NIPS 2012)*, 1097–1105. Curran Associates: New York.

8 Taigman, Y., Yang, M., Ranzato, M.'.A., and Wolf, L. (2014). Deepface: closing the gap to human-level performance in face verification. In: *Proceedings of the IEEE Conference on Computer Vision and Pattern Recognition*, 1701–1708. *IEEE.*

9 Biyi Fang, Jillian Co, and Mi Zhang, DeepASL: enabling ubiquitous and non-intrusive word and sentence-level sign language translation, in *Proceedings of the 15th ACM Conference on Embedded Networked Sensor Systems (SenSys)*, Delft, The Netherlands, 2017.

10 Zhou, B., Lapedriza, A., Xiao, J. et al. (2014). Learning deep features for scene recognition using places database. In: *Advances in Neural Information Processing Systems (NIPS 2014)*, 487–495. Curran Associates: New York.

11 He, K., Zhang, X., Ren, S., and Sun, J. (2016). Deep residual learning for image recognition. In: *Proceedings of the IEEE Conference on Computer Vision and Pattern Recognition*, 770–778. *IEEE.*

12 Karen Simonyan and Andrew Zisserman, Very deep convolutional networks for large-scale image recognition, in International Conference on Learning Representations (ICLR), 2015.

13 Song Han, Huizi Mao, and William J Dally, Deep compression: Compressing deep neural networks with pruning, trained quantization and huffman coding, in International Conference on Learning Representations (ICLR), 2016.

14 Hao Li, Asim Kadav, Igor Durdanovic et al., Pruning filters for efficient ConvNets, in International Conference on Learning Representations (ICLR), 2016.

15 Andrew G. Howard, Menglong Zhu, Bo Chen et al., Mobilenets: Efficient convolutional neural networks for mobile vision applications, arXiv preprint arXiv:1704.04861, 2017.

16 Geoffrey Hinton, Oriol Vinyals, and Jeff Dean, Distilling the knowledge in a neural network. NIPS Deep Learning Workshop, 2014.

17 Zeng, X., Cao, K., and Mi, Z. (2017). MobileDeepPill: a small-footprint Mobile deep learning system for recognizing unconstrained pill images. In: *Proceedings of the 15th ACM International Conference on Mobile Systems, Applications, and Services (MobiSys)*, 56–67. Niagara Falls, NY, USA. ACM: New York.

18 Florian Schroff, Dmitry Kalenichenko, and James Philbin, Facenet: a unified embedding for face recognition and clustering, IEEE Computer Society Conference on Computer Vision and Pattern Recognition (CVPR), 2015.

19 Diederik P. Kingma and Max Welling, "Auto-encoding variational bayes." International Conference on Learning Representations (ICLR), 2014.

20 LiKamWa, R., Hou, Y., Gao, J. et al. (2016). RedEye: analog ConvNet image sensor architecture for continuous mobile vision. In: *2016 ACM/IEEE 43rd International Symposium on Computer Architecture (ISCA)*, 255–266. IEEE Press.

21 Radu, V., Lane, N.D., Bhattacharya, S. et al. (2016). Towards multimodal deep learning for activity recognition on mobile devices. In: *Proceedings of the 2016 ACM International Joint Conference on Pervasive and Ubiquitous Computing: Adjunct*, 185–188. ACM.

22 Bhattacharya, S. and Lane, N.D. (2016). From smart to deep: robust activity recognition on smartwatches using deep learning. In: *Pervasive Computing and Communication Workshops (PerCom Workshops), 2016 IEEE International Conference, pp. 1–6. IEEE.*

23 Chang, S., Han, W., Tang, J. et al. (2015). Heterogeneous network embedding via deep architectures. In: *Proceedings of the 21th ACM SIGKDD International Conference on Knowledge Discovery and Data Mining*, 119–128. ACM.

24 Shuochao Yao, Shaohan Hu, Yiran Zhao et al., Deepsense: A unified deep learning framework for time-series mobile sensing data processing. The World Wide Web Conference (WWW), 2017.

25 Fang, B., Zeng, X., and Mi, Z. (2018). NestDNN: Resource-aware multi-tenant on-device deep learning for continuous mobile vision. In: *Proceedings of the 24th Annual International Conference on Mobile Computing and Networking (MobiCom)*, 115–127. New Delhi: India. ACM: New York.

26 Caruana, R. (1997). Multitask learning. *Machine Learning* 28 (1): 41–75.

27 Bolei Zhou, Aditya Khosla, Agata Lapedriza et al., Object detectors emerge in deep scene CNNs. International Conference on Learning Representations (ICLR), 2015.

4

Caching, Security, and Mobility in Content-centric Networking

Osman Khalid[1], Imran Ali Khan[1], Rao Naveed Bin Rais[2], and Assad Abbas[1]

[1] *Department of Computer Sciences, COMSATS University Islamabad, Abbottabad Campus, Pakistan, 22010*
[2] *College of Engineering and Information Technology, Ajman University, Ajman, UAE*

4.1 Introduction

Over the years, our world has seen significant growth in the number of mobile devices, such as personal laptops, smart phones, and tablets with sensing capabilities and multiple wireless technologies. With the evolution of 4G/5G technologies, previously resource-constrained devices like sensors and smart objects can easily connect to the outer world on a large scale. Generally, a large number of devices involved in wireless networking are mobile in nature, which tends to modify network topology with the passage of time. Traditional end-to-end Internet communication protocols have limitations in coping with such mobility and dynamic network conditions. A user requesting an item needs to have the IP address of the information provider to reach the content. This makes the traditional Internet protocols host centric, i.e. they need to reach the content host. However, the end user is usually not interested in knowing the location of a content item; rather, the user is interested only in the content, no matter from where the content is retrieved for the user. Therefore, there is a need for a communication mechanism that is more oriented to be data centric, rather than host centric. Keeping in view these issues, new architectures and paradigms are required to be proposed. For this purpose, content-centric networking (CCN) has been developed as an emerging candidate for upcoming data-centric Internet architecture.

CCN has gained considerable popularity as it appears to be a resourceful architecture for the future Internet. Several related projects are active worldwide [1]. Basically, CCN is an emerging technology in which content is accessed using its designated name, rather than an IP address of the host storing the content. Since the content is stored by name rather than IP address, both communication cost and

Fog Computing: Theory and Practice, First Edition.
Edited by Assad Abbas, Samee U. Khan, and Albert Y. Zomaya.

bandwidth usage are reduced. Kim et al. proposed a technology that provides an essential decrease of congestion produced by network traffic, rapid distribution of information, and contents security [2]. Mobile devices have been transformed into multimedia computers with embedded features (like messaging, browsing, cameras, and media players) [3]. The devices are made capable to create and use their own content easily and rapidly. Usually in CCN, media servers and web portals are used to host contents. To retrieve content, an end-to-end connection is required to be established. For that purpose, CCN is a self-organized approach that enables to push only the relevant data when desired.

CCN differs from traditional TCP/IP network in many aspects. CCN is an extension of the information-centric network (ICN) research projects comprising four main entities: the producers, the consumers, the publishers, and the routers. Currently CCN 1.0 is the latest protocol design for these networks. The forwarding information base (FIB), pending interest table (PIT), and content store (CS) are elements of a router in a CCN, as shown in Figure 4.1. The FIB replaces the traditional routing tables. The PIT is a cache for the requests. The CS is an optional element and acts as an extra cache where requests (also known as interests) might be stored. Manifests, interests, and content objects are types of messages dealt with in a CCN, which provides a system with data names that are independent of their location. Content search and retrieval is based on these names directly, i.e. content is identified and accessed solely by their names. Communication begins with the request of the consumer. An interest packet is requested, which is basically a

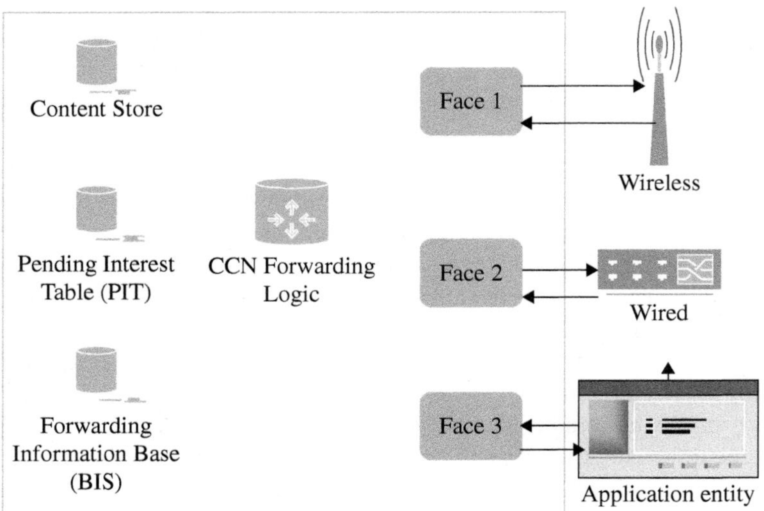

Figure 4.1 CCN router components.

content "by name." The node that contains the requested data can either be the owner or any temporary holder, and responds to the request with the data and additional authentication, ownership, and integrity information. With all aforementioned advantages of CCN, there are still challenges that need to be addressed such as naming, mobility management, and security of CCN.

The major research directions within CCN include: architecture, naming, mobility, security, caching, and routing [4]. These areas have been addressed by some literature surveys on content-centric networks [4]. However, the authors have considered a general discussion on the aforementioned research directions not going into enough details of any specific field. This chapter purely focuses on three major areas of research in CCN, which are caching, security, and mobility management. The discussion encompasses content management, i.e. where to place content (caching), how to secure it from threats (security), and how network mobility affects cached content (mobility).

The rest of the chapter is as follows. Section 4.2 discusses caching and fog computing. Section 4.3 explains the comprehensive analysis of various recent approaches proposed for mobility management. Section 4.4 describes the security aspects and approaches proposed for security management of the network while caching concept of CCN is discussed in Section 4.5, and finally Section 4.6 concludes the chapter.

4.2 Caching and Fog Computing

Recent years have seen fog computing as an emerging technology to address the latency-related issues in the traditional cloud computing paradigms [5]. Fog computing aims to bring the computing and storage capabilities from cloud and deploy them closer to the end user. The fog servers can be placed over multiple layers between network edge and cloud. Having the resources placed near the end users helps save bandwidth consumption at the network core, and reduces the latency, as a user's request can now be processed locally, instead of forwarding to remote cloud servers. This helps in meeting the demands of applications that require real-time response, e.g. augmented reality, real-time health monitoring, and localized traffic management apps, to name a few.

Data caching plays a pivotal role in fog computing—the efficient design of caching on a fog node exploits temporal locality of data and broader coverage of requests to save network bandwidth and reduce delay [5]. The CCNs can be integrated with fog networks, along with efficient caching mechanisms to improve the data content availability for end users. The caching in such scenarios can be classified as reactive/transparent or proactive. In transparent caching, the end user is not fully aware of the caching at the fog server, and the caching is made transparent to the end user. In proactive caching, the data is precached on the nodes before it is actually requested by the end user devices. For example, the software updates can be cached at fog nodes, before they are actually requested by the users. This saves the network from high bandwidth utilization. Another example is caching proximity-related data: geo-social networks (GSNs) like Google Latitude and Yelp store region-related content. Mobile users often use these services to request information about geographically nearby locations and places (restaurants, etc.). Proactive caching is highly related to content distribution networks and is expected to lead to further improvements in terms of bandwidth reduction for the core net.

4.3 Mobility Management in CCN

With the growing number of mobile devices, global mobile data traffic increased up to 18-fold from 2011 to 2016, and especially, mobile-connected devices using mobile video services have increased up to 1.6 billion. Current architecture of the Internet has been designed for a fixed network environment. In order to support mobile nodes over the network, existing topology of the Internet requires more operations like multi-staged address resolution, frequent location updates, and so on. CCN, as an alternative to IP-based content access, has enhanced the mobility features. IP addresses of sender and receiver are used for the delivery of contents over a pre-established network. In such networks, movement of mobile users results in disconnections. Moreover, because content is accessed through its name instead of IP, an end-to-end connection is no longer required to retrieve content. Consequently, a mobile user can achieve a seamless hand-off while continuously accessing content from a new and/or different location. CCN mobility can be classified as a client mobility and content mobility. *Client mobility* is the type of mobility that deals with client movements during requests for data objects. Likewise *content mobility* refers to the handling of location changes made by an object or set of objects. Network mobility is another kind of mobility in which a complete network changes location (e.g. a body area network or a train network). In CCNs, when a complete collection of an object set is changing its location, the new path needs to be broadcasted while keeping the old routing records safe. Besides, movements of some chunks of objects belonging to a collection of objects to a new location (e.g. a company employee takes a laptop on a trip) also needs to be taken care of. Moving an entire network results in too many routing updates—for instance, a network comprising a heterogeneous set of publishers. Too many updates result in bottlenecks unless they allow for relative route broadcasting.

4.3.1 Classification of CCN Contents and their Mobility

Keeping in view the stability and mobility concerns in CCN, there are two types of contents named as data contents and user contents. Data content is significantly termed as access to the content itself. Persistent content providers are used to serve the data contents and disseminate their replication for the ease of access to content. So, locations are comparatively stable for applications like web pages, streaming services, and files. While user-contents, on the other hand, contain data having real-time characteristics, e.g. Internet phone, chatting, and e-gaming constitutes user-content. Creation and management of the content is performed by individual mobile devices. Content service components such as rendezvous points and control messages are used for serving data contents and user contents. A rendezvous point is basically a location manager used to keep track of locality information of all providers containing location information of peers. Moreover, a rendezvous point is used as mobility handler by intervening in the relay of all interest and response packets. It is also known as indirection point. While control messages address the issues related to changing of content provider's routable prefixes. However, a sender-driven control message is needed for source mobility.

4.3.2 User Mobility

Mobility for the buffered contents in CCN on the user side is supported by using re-transmission mechanism of request and response packets that first come to Internet service provider (ISP). On the caller's hand-off, it receives a sequence number of the contents from a caller. During communication, lost chunks of information data can be recovered by simple process known as retransmission of data packets. User-side mobility deals with persistent interests. Kim et al. sectioned the CCN services as three types: real-time documents, channels, and on-demand documents based on service characteristics [6]. There is difference in persistent interest and the regular interest. In case of persistent interest, after hand-off, the route remains persistent to the old location and responses are multi-casted to the old location. This concept of multicasting results in waste of network resources. To address this, the signaling message can be used if the frame router could not successfully deliver content corresponding to persistent interest. When a threshold is reached, the edge router generates "host unreachable" with the specific name, for example, (persistent interest name)/Host unreachable, and so on. This signaling message is then delivered to the content server. The gateway used to communicate that signaling message can abolish the related obsolete persistent request and responses from their PIT. While the hand-off occurs due to the mobile client, persistent interests are retransmitted to the new path. Process of retransmission includes a dedicated bit that eliminates the outdated PIT entry. If the interest received on content server, it sets it to "on" in the next responses. It requires additional clients to send interests in addition to preserving the valid path to the clients.

4.3.3 Server-side Mobility

For receiving interests smoothly after hand-off, the name of the mobile is required to be updated in the routing table for each individual content. For this purpose, binding service between changing names is required at both old and new locations. To change the content name, the application server, such as a rendezvous point or a proxy agent binding service, are useful to provide binding service. Rendezvous-based approaches are used for changing names and binding services; each is discussed in subsequent sections.

4.3.4 Direct Exchange for Location Update

The hand-off during direct exchange of location updates indicates data source mobility. Callee is then informed about the new classified pathname of the caller for smooth running of ongoing service. Callee can continue getting replies even during hand-off. Recovery of lost data portion by simple retransmission is impossible. In order to recover the lost data, application regenerates the interests with the new hierarchical pathname. With the factors of simplicity and effectiveness, the issues arise when hand-offs are performed simultaneously by the peers.

4.3.5 Query to the Rendezvous for Location Update

On initiation of caller service with the mobile callee, caller queries a rendezvous server for obtaining callee's location record. As location information of all clients is maintained with rendezvous server, so on occurrence of hand-off new path-name of the client is communicated to the rendezvous server. Therefore, using the rendezvous services, a peer can obtain the routable prefix even after simultaneous hand-offs.

4.3.6 Mobility with Indirection Point

Another technique for reducing the hand-off delay is the indirection scheme. To initiate a request, caller issues a request and response packet with the name (callee). On successful receiving of packet, it issues another interest to the callee with its new routable name. On callee's response, the indirection server issues a corresponding response packet that is referred toward the caller. During a caller hand-off, it notifies occurrence of the hand-off event toward the indirection server. However, on receiving interest for the caller, indirection server buffers that interest and delays generating another corresponding interest toward the respective caller. When caller conveys new hierarchical path name to the server, the indirection server continues to generate corresponding interests and its service toward the caller.

4.3.7 Interest Forwarding

In order to cope with the drawbacks in previous schemes, layered approach known as "interest forwarding" is proposed. Unlike the previous schemes, in this approach a new name is not required for seamless service. It requires the modifiable routers, as incoming interests are buffered in CCN routers toward a hand-off client. Overhead caused by addition of new entry routers on the way between the access routers to corresponding hand-offs is somewhat reduced. In addition to this, hand-off process does not require a new routable prefix, while hand-off latency is diminished as well.

4.3.8 Proxy-based Mobility Management

This scheme is based on the concept that user proxies with CCN environment are configured over overall architecture of IP network. Contents are available to users via proxy node when required. User will send query packet to proxy node instead of making connection with all other devices having required contents. CCN on receiving the query packet (interest) will try to get content, thus reducing the overhead of network configuration. Said scheme is composed of the following steps:

Handover detection. By using the information of physical link or router advertisement, mobile node detects the change network status.

Handover indication. While the handover detection event is in progress, hold request message will be sent to proxy node by the mobile node before detection of actual handover. On receiving the hold message, proxy node will stop the process of sending content data to MS's old location and stores content data only to its local repository for future retransmissions.

Handover completion. In case of acquiring new IP address, mobile node notifies the new IP to proxy node by using handover message piggybacking. CCN proxy node transmits the stored data to new location and does not transmit the repeated interest packets. On receiving content packet at old location, the mobile node can ask for next content data segment using normal CCN interest packet.

4.3.9 Tunnel-based Redirection (TBR)

This scheme performs the redirection of incoming interest packets between content router of MSC's home domain and the content router in the moved domain. Tunneling is used for this redirection. To guarantee the presence of a router, it periodically broadcasts the prefix names. Point of attachment in CCN is located by the comparison of this broadcasted information. This information is also used to know whenever a network initiates a prefix update (PU) round to the home domain. Home content router announces the prefix of the name of the MSC initially to inform the other cognitive radio (CR). The PU message about its current movement is broadcasted by MSC and its active state is also announced to the local routers of the domain. Then, the CR of the new domain sends the PU message to the home domain. The redirection path can be extracted by the exchange of the prefix update and its acknowledgment messages (P-ACK). The data packets and interests are then exchanged among the MSC and home CR through this name prefix. This scheme works in the following three steps Figure 4.2:

Step 1. Indicates movement as the first step that identifies whether MSC would change the status of networks by the network address information or through the underlying physical link information. Normally, data is provided by wireless access point (AP) to MSC containing all CRs in the access domain. When the change of domain occurs in network location, depending on new domain router's prefix, then initially MSC forms a tentative name-prefix in order to tunnel the interest packets. The tentative name prefix composed of MSC's original name prefixes and moved domain name. In order to check validity of tentative name prefix, MSC performs duplicate name prefix detection (DND). Interest packets are broadcasted by MSC to the all in-range peripherals that are on a physical link residing in CR's domain that have been moved. This broadcast checks whether its tentative name prefixes is used by someone else. If it is used by anyone else, then the MSC repeats the DND process for creating new tentative name. Thus, repeated DND is just to guarantee unique tentative name-prefix to adjacent neighbors. After a successful DND process MSC forwards the PU message indicating its new location to its home-based domain CR. Tentative name prefixes and MSC's signature in the PU is used for checking its validity.

Step 2. This step deals with the configuration of the redirect path. The CR that has received a PU message compares name prefix domain of the PU message with that of its own. CR sends the PU message to the content router based on FIB reference. This information is based on difference of its own domain prefix from that of the PU message. Otherwise, the PU message is not forwarded, instead a PACK message is sent to the MSC. When a PU message is received, each content router pulls out the name prefix information immediately from the router to find the message's home domain router. Then, PIT entry is formed with the name-prefix of the MSC's home domain CR. On receiving PU, a home domain CR checks whether the PU message is valid or not. If the signature validity is proved, the CR sends an acknowledgment message to indicate a successful PU. At that point, the home switch's entrance for that MSC is changed to the new one which the PU message has set. While getting the PACK message, middle of the route CRs look into the direction table once more. The CRs consume the significant PIT section and afterwards forward the PACK message.

Step 3. This is the last step of the scheme and deals with interest redirection. The consumer requests are directed toward the MSC. Home CR looks up in the routing table to send the interest packet. The newly created interest packet is encapsulated by the home router and forwarded to the tentative name prefix that was extracted from the routing table. Intermediate CRs between the home content router and the MSC deliver this message through regular CCN forward techniques. The MSC receives and decrypts the encapsulated interest packet and transmit the data response.

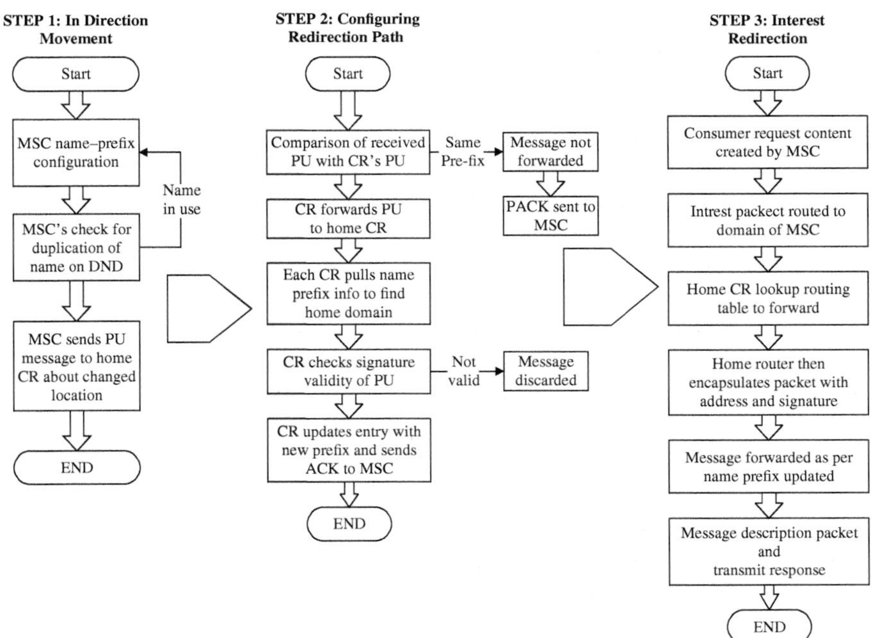

Figure 4.2 Tunnel-based redirection (TBR) scheme.

4.4 Security in Content-centric Networks

In this section, we highlight the architectural weaknesses of generic content centric networks. This will not include implementation weaknesses due to faulty cryptography schemes and algorithm selection. The major issues discussed will be the system inherent loopholes [7]. One of the major areas regarding research in security of CCNs is the naming convention. Most of the currently proposed naming schemes provide integrity of content using public/private key pairs. The real-world objects are linked to their keys using public-key infrastructure (PKI). Security risks relevant to naming schemes and PKIs are still under-researched and need more focus [8]. Signatures are also included in packets to provide authentication [4]. Certification chains are hierarchical and require an inheritance-based system for its application. The security of CCN is majorly linked to the naming schemes and architectures [9]. For human-readable names to be implemented, we need to trust a third party for the verification of data and name. The third party can be replaced if the network develops a trusted relationship with the system performing name resolutions. Although flat names support self-certification, which is extremely useful, they are not human readable. In this situation, the third party becomes mandatory and needs to be trusted since we require flat names to be mapped on by the human-readable names. The name of the data being requested by a user is made available to the nodes. This includes all those CCN nodes that will take part in processing the request [10].

Name resolution involves two steps. Content name resolution to a single IP address or a set of IP addresses is the first step in this process. These location-defining addresses are called locators. The routing of the request after its locators are marked is the second step. The message request is routed to one of the locators previously configured and is carried out by ISIS, OSPF, or any other shortest path-routing mechanism. In name-based routing, the content directly propagates back to the requesting party. This is ensured since names are used instead of addresses for routing and route-tracking information is maintained throughout the route.

Naming schemes are broadly classified as flat names and hierarchical or human-friendly names. Using flat names hides the original meaning and context of the object since they are hashed. The drawback of this mechanism is that they are difficult to remember. On the other hand, the hierarchical naming schemes are human friendly because of the fact that their structure is more like that of a URL, describing the content and context of the requested object. Human friendliness makes it easier for users to find the objects they are interested in but also poses a few challenges like authenticity, security binding, and ensuring the uniqueness of objects globally [11]. A few naming architectures have been designed recently to overcome security issues caused by naming schemes, as described below.

CCNs are host independent and data dependent. Because of this, we need to associate the security protocols with either the naming schemes or the data rather than securing the channel or the machine. In [1], the authors present such a naming scheme that enables verification of data integrity, owner authentication, and identification along with naming the content. This paper analyzes information-centric network architecture and highlights its requirement based on naming policies and schemes. The scheme proposed in this paper embeds these requirements into an overall architecture. The scheme provides a combination of name persistence, self-certification, owner authentication, and owner identification, all at once. Name persistence is ensured even if there is a change in content specifications, e.g. owner or location. Self-certification makes it easier for the user to verify the data integrity without requiring trust in any third party or the producer of the content. It is usually provided by joining the message/content digest with the original message. Owner authentication and owner identification binds an ID to an entity/content, which enables its verification at the consumer end. This is provided by keeping the encryption key pairs of the owner separate from the self-certification key pairs (which authenticate the content of the data).

In the NetInf—the proposed naming scheme [1]—any entity/content is represented by a globally unique ID. Together with the entity's own data and metadata, the resultant object provides the features mentioned above. Metadata contains information needed for the security functions of the NetInf naming scheme, e.g. public keys, content hashes, certificates, and a data signature authenticating the content. It also includes non-security-related information, i.e. any attributes associated with the object. Metadata is a mandatory part of a generic content-centric network data object according to this naming scheme and can be stored independently. The prototype of the naming scheme is built in Java and is tested in Windows, Linux and Android platforms. It has proven easy to implement because the security mechanism used for hashing, certificates, and encryption are library based and easily available. The naming scheme can be easily implemented in various web browsers and clients as a plug-in.

Wong and Nikander propose a secure naming scheme that can locate resources in information-centric networks, specifically generic-content-centric networks [12]. It is an extension of the existing URL-naming scheme. Allowing secure content retrieval from various unknown and untrusted sources is among the primary goals of this algorithm. The features of this scheme include backward compatibility with URL naming schemes and permission for content identification independently. The storage methods and routing and forwarding mechanisms do not play any role in identifying the content that is transferred. The source and the rules defining a location of URL/URI in its authority fields are separated to provide the independence level in the solution. As mentioned above, backward compatibility is the main feature of this scheme. Besides these, it provides secure content retrieval from varying sources, content authentication and validation with reference to the source, and uninterrupted mobility over the network.

A name-based trust and security approach is proposed in [13]. It is constructed on top of the identity-based cryptography, which is based on allotting tokens of identification to the users and only the authorized users can access data. The identity token derived from any feature of the user's or owner's identity, name of the content, or its prefix is also used as the public. Since the signed identity creates a direct link between the name (identity) and the content, it can be easily accessed by a malicious user and an attack can be launched [14]. An attacker can be anyone from the ISP to the end-users of the system. The technique presented in [14] is an enhanced version of a previous scheme [13], which has replaced identity-based cryptography with hierarchical identity-based cryptography to overcome identified drawbacks.

Besides the naming-related security issues, we briefly discuss the effect of caching on the security and privacy of generic content-centric networks, and denial of service (DOS) attacks.

4.4.1 Risks Due to Caching

The feature of caching in generic content-centric networks makes them different from traditional TCP/IP networks as it stores all the data in the router caches. Since data leaves its traces even after it is removed or forwarded, and can be retrieved by an attacker by probing and other such techniques. Hence caching puts the system security at risk in spite of being a defining feature of the CCNs. Attacks due to caching of data can be reduced by keeping data strictly up to date and restricting the replay of any old data.

4.4.2 DOS Attack Risk

DOS attacks, although difficult to carry out in these networks, are still possible. They are carried out through interests. Since data packets of request and response always follow the same path, they can be exploited and a destination can be flooded with irrelevant data packets. The complications arise due to the fact that the origin of the packet with respect to its location cannot be identified. Due to the architecture of the Internet and the generic CCNs, DOS attackers can easily cover their tracks. In order to cope with these attacks, a few security patches (to be added to the existing system) and trust mechanisms like firewalls and spam filtering have been developed. New security protocols that complement the existing networking protocols are also introduced to solve this problem [4].

4.4.3 Security Model

Establishing security of data in such a system, which is largely content-oriented, is mandatory before it can be implemented on a large scale. This has been a relatively unexplored area as compared to the caching or mobility of generic CCNs and needs further investigation to determine a better security model.

4.5 Caching

A key feature of CCN is in-network caching of content [15]. The CCN idea is based upon showing of interest in content and in turn serving of actual content against the interest. This idea is achieved by allocating a globally unique identifier for each content in the network. Such identifier can be comprehended by all network nodes including routers. Hence routers can cache data packets and, therefore, serve future content interest requests via its own cache instead of forwarding interests to server. Due to the increase in content number and variety, caching has become the leading variable for efficient utilization of CCNs. The problems of caching in CCN include: (a) which content is to be cached, (b) when is appropriate timing for caching, (c) how content would be cached (storing and eviction), and (d) on which network path a content be cached. Furthermore, there is a problem of cache allocation to network routers that states how much space must be allocated to each router. There are two approaches: (1) homogeneous and (2) heterogeneous [15]. The homogeneous approach allocates equal cache size to each router while heterogeneous allocates higher space to some routers and lower to others, based on some optimal criteria. Some recent research projects on the aforementioned caching problems are analyzed and compared in following subsections.

4.5.1 Cache Allocation Approaches

The authors in [15] presented an algorithm to compute optimal cache allocation among routers. The algorithm focuses on maximization of aggregate benefit, where benefit means reduction in hops taken by an interest packet from client to server. The interest packet is assumed to follow shortest path from client to server. Therefore, a shortest path tree rooted at the server that has the desired data for clients is formed. The shortest path tree structure leads the problem of cache allocation to be divided in two subproblems: (1) maximizing benefit of a data packet that has highest probability of being requested in a shortest path tree and (2) maximizing overall benefit of all data packets in whole network that will be sum of all maximizations in previous step. The former problem is being solved with help of k-means clustering algorithm while a greedy approach is adopted for the latter. The proposed greedy algorithm produces a binary matrix having nodes in rows and content packets in columns, depicting content allocations on nodes. The complexity of proposed optimization algorithm is max ($O(sn^3)$,$O(c_{total}$ log$N$)) where s is the number of servers, n is total number of nodes, N is number of contents, and c_{total} is the total cache space of the network. Experimental evaluation showed that cache allocation should be heterogeneous and related to topology and content popularity.

In [16], the authors modeled network topology as an arbitrary graph and computed several centrality metrics of the graph (betweenness, closeness, stress, graph, eccentricity, and degree centralities). Total amount of network's cache is distributed heterogeneously based on centrality values among nodes of the graph. One homogeneous and 12 heterogeneous cache allocations were simulated using ccnSim, an open-source simulator. Simulation results depicted that cache allocation based on degree centrality is better among all centrality metrics. Moreover, heterogeneous cache allocation had very little (2.5%) performance gain over homogeneous cache allocation.

The cache allocation problem is described as a part of content-aware network planning problem in a budget constrained scenario in [17]. The content-aware network planning derives an optimal strategy to migrate from an IP network to content-aware network. The proposed algorithm for that purpose applies linear programming approach. The algorithm finds optimal placement of content on network routers based on minimization of flow and price per unit traffic. Probability of caching each content on each router is calculated. If cost of storage addition on a router for caching of a content with certain probability is less than budget, then cache is allocated to that router. Content popularity distribution has a major impact on the proposed approach for cache allocation in [17]. For highly skewed content popularity, there is about 87% less storage needed than the one deployed for uniformly distributed popularity content.

In [18], a centralized global cooperative caching placement strategy named generalized dominating set-based caching (GDSC) has been proposed for content-centric ad hoc networks. GDSC minimizes the network power consumption by selection of proper cache nodes and also achieves a better trade-off between round trip nodes and cost of caching. The selection of nodes depends upon nodes degree and number of hops, and the node with high degree will cache the content. Changing the number of hops selection results in different nodes being selected for caching. The number of hops that results in reduced power consumption for the whole network is selected.

The problem of trade-off between in-network storage provisioning cost and network performance has been addressed in [19]. The authors initially proposed a holistic model for intra-domain networks to portray network performance of routing contents to clients and network cost sustained by globally coordinating in-network storage capability. Furthermore, an optimal strategy for provisioning of storage capability has been derived that results in optimization of overall network performance and cost. The authors also demonstrated that the optimal strategy can result in significant improvement on both the load reduction at origin servers and routing performance. Moreover, provided an optimal coordination level, a routing-aware content placement algorithm has been designed in [19] that runs on a centralized server. The algorithm computes and assigns contents to every router for caching, which leads to minimization of overall routing cost in terms of transmission delay or hop counts.

4.5.2 Data Allocation Approaches

Cache management and content request routing policies have been proposed in [20]. Cache management policy decides whether a piece of data that passes through a router must be cached or not. If so, then the policy determines which portion of data residing in the cache must be removed to make room for data to be cached. In-caching and removal process of data, link congestion along the path of data retrieval, and data popularity are considered. A utility function is used in the decision of caching a piece of content on a node. The utility function takes minimum bandwidth of content retrieval path and content popularity as parameters. If a piece of content has higher utility value than lowest utility value of any content at a node, the content is cached in place of lowest utility value content. The main idea of cache management policy is that content forwarded over congested links is retained by caches while evicting the content forwarded over uncongested links. The content request routing policy utilizes a scoped-flooding protocol. The protocol locates requested cached content with minimum download delay as there are multiple caches having a piece of content. The *scope* is boundary for the protocol to stop searching for cached content, and is calculated in terms of number of hops that a packet traverses. An issue with such caching policy is that there might be no interested node for a content after some time that is residing with a high utility value in a cache. Such a content will never be removed from a cache due to higher utility value.

Caching approach presented in [21] focuses on fair sharing of the available cache capacity on a path among various content flows. Several clients connected to a server with different hop counts may share a network path. A probability factor called *ProbCache* is proposed in to resolve which content must be cached among all the contents of several users passing through a router. ProbCache is a product of two factors called *TimesIn* and *CacheWeight*. TimesIn determines the number of times a path can afford to cache a packet. TimesIn takes into consideration the remaining hop count to client and cache size of the path. CacheWeight resolves contention for caching among contents of several users on a router. CacheWeight decides where to cache the number of copies indicated by TimesIn. A content with highest CacheWeight on a router gets cached on that router. Only that client's content will have higher CacheWeight for which the client is nearest to the cache. CacheWeight is calculated by the ratio of number of hops from cache to server to the number of hops from client to server. The ultimate objective of ProbCache is to utilize caches efficiently to reduce caching redundancy and, in turn, network traffic redundancy. The caching approach achieved 20% fewer server hits and 8% hop count reduction in simulation results. However, only binary tree topology was being considered in evaluation.

In [22], the authors proposed a latency-aware cache management mechanism for CCN. The mechanism is based upon the principal that whenever a content is retrieved, it is stored into cache with a probability proportional to its recently observed retrieval latency. There is a trade-off between cache size and delivery time as reducing delivery time of every content requires a cache of size appropriate to store every content at every node [22]. The proposed mechanism minimizes this trade-off by assigning caching priority to contents that are more distant. It is a fully distributed mechanism since the network caches do not exchange caching information. Two main benefits of the proposed mechanisms are: (1) faster delivery time as compared to latency insensitive approaches, and (2) fast convergence to optimal caching situation. However, dynamic network conditions have not been considered while calculating optimal data caching.

A betweenness centrality-based caching strategy has been presented in [23]. The authors argued that a ubiquitous caching strategy results in data redundancy. Furthermore, a random caching strategy is useful where a high number of overlapping paths exists in network. The proposed caching strategy assumes that each node's betweenness centrality has already been calculated. A content request message stores every node's centrality value on its path to server. The content message generated by a server against a request message gets cached on a node with highest centrality value on the path to client. There is no collaboration between network caches, therefore it is a distributive caching strategy. The proposed caching strategy also considers dynamic network environments where topology tends to change with time. An ego network betweenness measure has been calculated instead of betweenness centrality in dynamic networks. An ego network consists of only immediate neighbors of a node.

In [24], the authors propose an age-based cache replacement policy that specifies which content objects to be cached in routers. An age is associated with each content object. The age decides lifetime of a content copy in a router. Age of a duplicate content object is calculated when it is added to a cache. The age value depends on two factors: (1) distance in terms of node count of the duplicate content object from server, and (2) popularity of the duplicate content object. The age is directly proportional to distance of duplicate content object from server and popularity of duplicate content object. This scheme allows for the content chunks to be pushed to network edges and release redundant content object storage at intermediate nodes. Moreover, this scheme assigns popular contents greater priority for getting cached by routers. The duplicate content object is removed when its age expires. Cooperation exists between routers to modify the age, hence it is a collaborative approach. The approach makes assumptions that content's popularity and network topology are already known.

It has been revealed in [25] that by jointly considering forwarding and meta-caching decisions, the performance gains are achieved in terms of content average distance traveled. The potential gains offered by smart forwarding policies such as ideal nearest replica routing (iNRR) are enabled by meta-caching policies such as leave a copy down (LCD). Cache pollution dynamics completely counterpoise these potential gains otherwise. The authors in [26] argued that iNRR with LCD acquires significant performance improvement over iNRR with leave a copy everywhere (LCE), provided best-case CCN topology. LCD has been designed for hierarchical topologies, so in general case (alternative meta-caching policy), least recently used (LRU) may be desirable.

In [26], the authors argued against the necessity of an indiscriminate in-path caching strategy in ICN and investigated the possibility to achieve higher performance gain by caching less. It was proved that a simple random caching strategy can overtake the current pervasive caching paradigm under certain network topology conditions. The authors proposed a caching policy on the basis of betweenness centrality that caches the content at the nodes having the highest probability of getting a cache hit along the content delivery path. The main goal of designing this policy is to ensure that content always spreads toward users and hence it reduces the content access latency. An approximation of the caching policy was also proposed by the authors for scalable and distributed realization in dynamic network environments where full topology is not known a priori. This approximation is based on betweenness centrality of ego networks. The research in [26] is an extension to work in [23], discussed previously. The authors realized that effectiveness of betweenness is dependent on the betweenness distribution. It was concluded that topologies that exhibit power law distribution such as WWW networks ensure effectiveness of betweenness policy.

The authors have proposed a collaborative caching policy unlike any caching policy mentioned before in [27]. The nodes collaborate by serving each other's request instead of independently determining if a specific item should be cached or which item must be eliminated to make room for new item. This way the policy allows nodes to eliminate redundancy of their cached items. The policy is based on a simple rule that an originally cached item should either be kept in place or be found at one of the neighbor nodes. This way, same performance of distributed caching policy can be preserved while increasing the available free caching slots.

In [28], a low complexity content placement scheme combined with dynamic request routing in CCN infrastructure has been proposed. The objective of the scheme is to reduce caching redundancy and make more efficient use of available cache resources. That way the scheme increases overall cache utilization and potentially increase user-perceived quality. The authors in [28] designed a self-assembly caching (SAC) scheme of collaborative content placement along a path of caches that is based on an analysis of the feasible cache size distribution. The hot content is being pulled by the SAC scheme toward the edge of the network while the cold content is pushed back to the core of the network based on popularity. Therefore, a low caching redundancy and low download delay is achieved because SAC could automatically distribute the content to the proper cache location according to its popularity. A trail having content cache locations is created at some special routers along the content delivery path. The trail is updated during the content chunk caching or eviction, leading to enabling of dynamic and absolute direction of following request search. The trail information makes the cached chunk detectable locally and so improves the in-network caching utilization.

In [29], a cross-layer cooperative caching strategy for CCN is proposed. Three parameters that are user preference, betweenness centrality, and cache replacement rate are introduced as application layer, network layer, and physical layer metrics. A caching probability function is derived based on Grey relational analysis (GRA) among all nodes along the content delivery path. Multiple requests for the same content are aggregated in PIT, which keeps track of interest packets forwarded upstream so that returned data packets can follow the path back to end-point of content requester. Some nodes aggregate requests and termed as aggregated nodes. The aggregate node computes the caching probabilities and adds them in the PIT. Caching probability fields of data packet is updated by the caching probability in PIT when the data packet arrives at the aggregated node.

A novel caching method to deliver content over the CCN has been proposed in [30]. The authors analyze caching content distribution and related interest distribution by considering the mechanism of content store and pending interest table of CCNs. Based on the analysis, a caching algorithm has been proposed to efficiently use the capacity of content store and pending interest table. The caching is done based on a cost model derived in the paper. The cost model is based on total delay to retrieve a piece of content. The content store will calculate the cost to remove a cached content for caching a newly requested piece of content. If the cost is below a threshold, the new content is cached. The proposed scheme does not account for the dynamic network conditions and is a distributed caching mechanism.

In [31], the authors studied the user-behavior-driven CCN caching and jointly investigated video popularity and video drop ratio. It has been said that this is the first paper taking the video drop ratio into consideration in CCN caching policy design. The video drop ratio means that video consumer does not always watch the video till the end. An intra-domain caching in CCN system has been considered that is composed of servers/repositories, routers, and users. Content store (CS) of each router acts as a buffer memory. The routers are divided in levels hierarchy and videos are ranked according to popularity. The main idea of proposed scheme is that ith chunk is cached in jth router if, and only if, the higher ranking chunks than ith chunk can be stored in lower level routers than jth router, and if the lower level routers cannot cache all the chunks having higher ranking than ith chunk. The proposed algorithm achieves optimal values in terms of average transmission hops needed and server hit rate, given a cascade network topology.

A caching policy similar to the one mentioned in [31] has been proposed in [32], where the similarity is used for classification of routers into levels. The proposed strategy maximizes cache utilization and improves content diversity in networks by content popularity and node level matching-based caching probability. Three parameters including hop count to the requester, betweenness centrality, and cache space replacement rate are considered for evaluation of nodes' caching property along the delivery path. The proposed strategy performs classification of nodes into levels on the basis of their caching property. High caching property nodes are defined as first-level nodes; while less capable nodes are termed as second-level nodes. Moreover, in this approach, the popularity of content is matched to the affiliated node using probabilistic methods. Content is cached in nodes based on their popularity in an increasing order, starting by placing the most popular content in the first level, the lesser popular in the second level, and so on. This decreases the replication of less popular content and access delay. The proposed policy improved up to 23% cache hit ratio, reduced up to 13% content access delay, and accommodated up to 14% more contents, compared with leave copy everywhere (LCE) policy.

The research work conducted in [33] emphasizes the need of an algorithm for selection of a subset of nodes among those available to serve new incoming requests for longer time. The authors proposed a cost-effective caching (CEC) algorithm that selects eligible node based on node's remaining capacity to cache more content. In CEC, a node with highest space left in content store will be designated to cache the content. The term *cost* in CEC is the cost of computation carried on each node in network for the process of sending interest packet and receiving data packet. CEC tries to minimize this cost. The motivation behind CEC is that if a new content replaces already cached content then the cost increases. The cost is minimized by caching a content on some other node that has sufficient free space to store that content without replacement. That way CEC maximizes cache hit and minimizes both cache miss and content replacement.

Table 4.1 Caching schemes comparison.

Reference	Topology used	Caching objective	Cache allocation type (in case of cache allocation)	Cache allocation criteria (in case of cache allocation)	Data allocation criteria (in case of data allocation)	Performance metric	Performance metric used
Wang et al. [15]	Barbasi-Albert, Watts-Strogatz	Cache allocation	Heterogeneous and homogeneous	Topology and interest dependent (heterogeneous in BA while homogeneous in WS)	—	Network-centric	Remaining traffic
Rossi and Rossini [16]	Arbitrary topologies	Cache allocation	Heterogeneous	Nodes centrality dependent	—	Network-centric and user-centric	Cache hit probability and path stretch
Badov et al. [20]	Grid, scale-free, Rocketfuel, hybrid	Data allocation	—	—	Links congestion and interest dependent	User-centric	Content download delay
Psaras et al. [21]	Binary tree	Data allocation	—	—	Hop count from client to cache	Network-centric	Server hits and hop reduction
Mangili et al. [17]	Netrail, Abilene, Claranet, Airtel, Geant	Data and cache allocation	Heterogeneous	Data caching probability and budget	Traffic cost and flow minimization	Network and user-centric	Budget and traffic flow
Carofiglio et al. [22]	Line, binary tree	Data allocation	—	—	Content popularity and latency	Network-centric	Data retrieval latency
Chai et al. [23]	k-ary tree, scale-free	Data allocation	—	—	Betweenness centrality of node	Network-centric	Hop reduction ratio, server hits reduction ratio
Ming et al. [24]	CERNET2	Data allocation	—	—	Content popularity	User-centric	Network delay
Rossini and Rossi [25]	10 × 10 grid, 6-level binary tree	Data allocation	—	—	Ubiquitous	Network- centric	Content traveling average distance
Chai et al. [26]	k-ary tree, scale-free	Data allocation	—	—	Betweenness centrality of node	Network and user-centric	Latency and congestion

Wang et al. [27]	AS 1755, AS 3967, Brite1, Brite2	Data allocation	—	—	Benefit of eviction of a content	Network-centric	Content redundancy and cache hit rate
Li et al. [28]	5-level binary tree,	Data allocation	—	—	Content popularity	Network and user-centric	Content redundancy and content download delay
Wu et al. [29]	Scale-free	Data allocation	—	—	User preference, betweenness centrality, cache replacement rate	Network-centric	Cache hit ratio, average hops
Zhou et al. [30]	Unspecified	Data allocation	—	—	Network delay cost	User-centric	Network delay
Liu et al. [31]	Cascade topology	Data allocation	—	—	Video popularity and drop ratio	Network and user-centric	Average transmission hops and server hit rate
Li et al. [32]	Scale-free	Data allocation	—	—	Probability based on nodes level	Network and user-centric	Cache hit ration, content number, and content access delay
Mishra and Dave [33]	Unspecified	Data allocation	—	—	Cache free space	Network and user-centric	Cache hit, cache miss, and total number of replacements
Zhou et al. [18]	Arbitrary	Cache allocation	Heterogeneous	Power consumption	—	Network and user-centric	Average round trip hops, power consumption, number of caching nodes
Li et al. [19]	Abilene, CERNET, GEANT, US-A	Cache allocation	Heterogeneous	Network performance and cache provisioning cost	—	Network and user-centric	Hop count, latency

Due to dynamic change in bandwidth in VANETs, dynamic adaptive streaming (DAS) technology is used for delivery of video content with different bit rates according to available bandwidth. DAS requires several versions of a same video content, resulting in reduced cache utilization. Hence a cache management scheme for adaptive scalable video streaming in Vehicular CCNs has been proposed. The scheme aims to provide high quality of experience (QoE) video-streaming services through caching of appropriate bit-rate content near consumers. A video content is divided into layers where each layer is the video content with specific bit rate. Chunk of layers are pushed to neighbors according to their available bandwidth through broadcast hop-by-hop.

Table 4.1 compares the CCN caching schemes mentioned so far on various critical parameters. Table 4.2 depicts the other objectives (if any) of aforementioned caching schemes coupled with caching.

Table 4.2 Objectives-based comparison.

References	Caching	Routing	Forwarding
Wang et al. [15]	✓	×	
Rossi and Rossini [16]	✓	×	×
Badov et al. [20]	✓	✓	×
Psaras et al. [21]	✓	×	×
Mangili et al. [17]	✓	×	×
Li et al. [19]	✓	✓	×
Carofiglio et al. [22]	✓	×	×
Chai et al. [23]	✓	×	×
Ming et al. [24]	✓	✓	×
Rossini and Rossi [25]	✓	×	✓
Chai et al. [26]	✓	×	×
Wang et al. [27]	✓	×	×
Li et al. [28]	✓	✓	×
Wu et al. [29]	✓	×	×
Zhou et al. [30]	✓	×	×
Liu et al. [31]	✓	×	×
Li et al. [32]	✓	×	×
Mishra and Dave [33]	✓	×	×
Zhou et al. [18]	✓	×	×

4.6 Conclusions

This chapter has explored the features of ICN, looking at how the architecture could support mobility, caching, and security elements. Different management techniques of each of the challenges have been discussed and analyzed in detail. Due to the caching of contents at local level, replication is challenging for CCN architectures. There are two issues of caching in CCN: (1) data allocation, and (2) cache allocation. Data allocation issue has been addressed by simpler techniques like centrality based or by complex probabilistic techniques. Moreover, routing and forwarding techniques are being coupled with caching schemes by some of the studied approaches to enhance network performance. The cache allocation problem is further classified into homogeneous and heterogeneous cache allocation to the routers. Most of the researchers prefer heterogeneous while some argue that heterogeneous policies have a very lesser gain than homogeneous allocations. Based on caching schemes analysis, we can conclude that there are a lot of parameters such as network topology that affect caching scheme performance. Every scheme achieves some gains based on the prior assumptions on those parameters. Hybrid caching schemes could be designed, but in terms of complexity trade-off on network routers. According to all three aspects, security is the main focus of the survey that has to be explored for future of CCN.

References

1 C. Dannewitz, J. Golic, B. Ohlman, and B. Ahlgren, Secure naming for a network of information, INFOCOM IEEE Conference on Computer Communications Workshops, 2010.

2 Kim, D., Kim, J., Kim, Y. et al. (2012). Mobility support in content-centric networks. In: *Proceedings of the Second Edition of the ICN Workshop on Information-centric Networking*. ACM.

3 Lee, J., Cho, S., and Kim, D. (2012). Device mobility management in content-centric networking. *IEEE Communications Magazine* 50 (12): 28–34.

4 Xylomenos, G., Ververidis, C.N., and Siris, V.A. (2014). A survey of information-centric networking research. *IEEE Communications Surveys & Tutorials* 16 (2): 1024–1049.

5 Bilal, K., Khalid, O., Erbad, E., and Khan, S.U. (2018). Potentials, trends, and prospects in edge technologies: fog, cloudlets, mobile edge, and micro data centers. *Computer Networks* 130: 94–120.

6 Kim, D., Kim, J., Kim, Y. et al. (2015). End-to-end mobility support in content-centric networks. *International Journal of Communication Systems* 28 (6): 1151–1167.

7 Kuriharay, J., Uzun, E., and Wood, C.A. (2015). An encryption-based access control framework for content-centric networking. In: *IFIP Networking Conference (IFIP Networking)*. IEEE.

8 Ellison, C. and Schneier, B. (2000). Ten risks of PKI: what You're not being told about public key infrastructure, Computer. *Security Journal* 16 (1).

9 A. Ghodsi, T. Koponen, J. Rajahalme et al., Naming in content-oriented architectures, ACM Workshop on Information-Centric Networking (ICN), 2011.

10 A. Ghodsi, S. Shenker, T. Koponen et al., Information-centric networking: seeing the forest for the trees, ACM Workshop on Hot Topics in Networks (HotNets), 2011.

11 A. Ghodsi, Naming in content-oriented architectures, ACM SIGCOMM Workshop on Information Centric Networking, 2011.

12 W. Wong and P. Nikander, Secure naming in information-centric networks, *Proceedings of the Re-Architecting the Internet Workshop,* Philadelphia, Pennsylvania, November 30–31, 2010.

13 Hamdane, B., Serhrouchni, A., Fadlallah, A., and El Fatmi, S.G. (2012). Named-data security scheme for named data networking. In: *Network of the Future (NOF)*. IEEE.

14 B. Hamdane, R. Boussada, M. E. Elhdhili, and S. G. El Fatmi, Towards a secure access to content in named data networking, IEEE 26th International Conference on Enabling Technologies: Infrastructure for Collaborative Enterprises (WETICE), Poznan, Poland, 2017.

15 Y. Wang, Z. Li, G. Tyson, S. Uhlig, and G. Xie, "Optimal cache allocation for content-centric networking," 21st IEEE International Conference on Network Protocols (ICNP), October 2013.

16 D. Rossi and G. Rossini, On sizing CCN content stores by exploiting topological information, IEEE Conference on Computer Communications Workshops (INFOCOM WKSHPS), March 2012.

17 M. Mangili, F. Martignon, A. Capone, and F. Malucelli, "Content-aware planning models for information-centric networking," Global Telecommunications (GLOBECOM), IEEE Conference and Exhibition, 2014, DOI: 10.1109/GLOCOM.2014.7037078.

18 L. Zhou, T. Zhang, X. Xu et al., Generalized dominating set-based cooperative caching for content-centric ad hoc etworks, IEEE/CIC ICCC Symposium on Next Generation Networking, 2015.

19 Li, Y., Xie, H., Wen, Y. et al. (2015). How much to coordinate? *IEEE Transactions on Network and Service Management* 12 (3): 420–434.

20 M. Badov, A. Seetharam, J. Kurose, and A. Firoiu, "Congestion-aware caching and search in information-centric networks," ACM-ICN '14: Proceedings of the 1st ACM Conference on Information-Centric Networking, September 2014, pp. 37–46, https://doi.org/10.1145/2660129.2660145.

21 I. Psaras, W. K. Chai, and G. Pavlou, "Probabilistic In-Network Caching for Information-Centric Networks," In Proceedings of ACM Sigcomm Conference, pp. 50–60, 2012.

22 G. Carofiglio, L. Mekinda, and L. Muscariello, "LAC: Introducing latency-aware caching in Information-Centric Networks," 2015 IEEE 40th Conference on Local Computer Networks (LCN), DOI: 10.1109/LCN.2015.7366343, 2015.

23 Chai, W.K., He, D., Psaras, I., and Pavlou, G. (2012). Cache "less for more" in information-centric networks. In: *Part I, LNCS 7289* (ed. R. Bestak), 27–40. IFIP International Federation for Information Processing.

24 Z. Ming, M. Xu, and Dan Wang, "Age-based cooperative caching in information-centric networking," 23rd International Conference on Computer Communication and Networks (ICCCN), September 2014.

25 G. Rossini and D. Rossi, Coupling caching and forwarding: benefits, analysis, and implementation, ICN'14, September 2014.

26 G. Rossini and D. Rossi, Proceedings of the 1st ACM Conference on Information-Centric Networking, September 2014, pp. 127–136, https://doi.org/10.1145/2660129.2660153.

27 J. M. Wang, J. Zhang, and B. Bensaou, Self assembly caching with dynamic request routing for information-centric networking, ICN'13, August 2013, pp. 61–66.

28 Li, Y. Xu, T. Lin, G. Zhang, Y. Liu, and S. Ci, "Self assembly caching with dynamic request routing for Information-Centric Networking," 2013 IEEE Global Communications Conference (GLOBECOM), Atlanta, GA, USA, DOI: 10.1109/GLOCOM.2013.6831394, (2013).

29 Wu, L., Zhang, T., Xu, X. et al. (2015). Grey relational analysis based cross-layer caching for content-centric networking. In: *Symposium on Next Generation Networking IEEE/CIC*. ICCC.

30 Zhou, S., Dongfeng, F., and Bo, H. (2015). Caching algorithm with a novel cost model to deliver content and its interest over content centric networks. *China Communications* 12 (7): 23–30.

31 Z. Liu, Y. Ji, X. Jiang, and Y. Tanaka, "User-behavior Driven Video Caching in Content Centric Network," ICN'16 September 26–28, 2016, Kyoto, Japan.

32 Y. Li, T. Zhang, X. Xu et al., Content popularity and node level matched–based probability caching for content-centric networks, IEEE/CIC International Conference on Communications in China (CCC), July 2016.

33 G. P. Mishra and M. Dave, "Cost Effective Caching in Content-Centric Networking," 2015 1st International Conference on Next Generation Computing Technologies (NGCT), DOI: 10.1109/NGCT.2015.7375111, (2015).

5

Security and Privacy Issues in Fog Computing

*Ahmad Ali, Mansoor Ahmed, Muhammad Imran, and Hasan Ali Khattak**

Department of Computer Science, COMSATS University Islamabad, Islamabad, 44000, Pakistan

5.1 Introduction

Smart devices provide the building blocks for the Internet of Things (IoT). IoT is populating the research and development landscape exponentially in different dimensions like home automation, industrial automation, and manufacturing, as well as healthcare. Fog computing is the process that helps to extend the data storage, communication power and computing speed near to end users. In fog computing networks, the edge devices are connected and integrated closely, which helps to enhance the performance and system efficiency. To some extent, fog computing is also called edge computing, because both of them are used to increase the efficiency of the system; however, their processing systems are different.

We can move the data to the optimum place for processing; and decisions can be based on how quickly a result is needed. For example, time-sensitive decisions should be made closer to the things producing and acting on the data. In contrast, big data analytics on historical data needs the computing and storage resources of the cloud. While in fog computing, they work locally on the data that needs analysis instead of sending it to the server or cloud computing servers, as shown in Figure 5.1. In a fog computing system, there are nodes used for connection and working of this system, which are protected by the policy and procedure of the regular IT environment.

Security, privacy, and accountability of service providers are the core areas when it comes to the implementation and integration of smart devices that are connected to the fog-enabled infrastructures, as shown in Figure 5.2. It is obvious that "trust" would be included in a collective concept of security and privacy, as shown in Figure 5.3. Furthermore, IoT Security may be explored as "things identification," "authentication," and "authorization." Unique identification

Fog Computing: Theory and Practice, First Edition.
Edited by Assad Abbas, Samee U. Khan, and Albert Y. Zomaya.

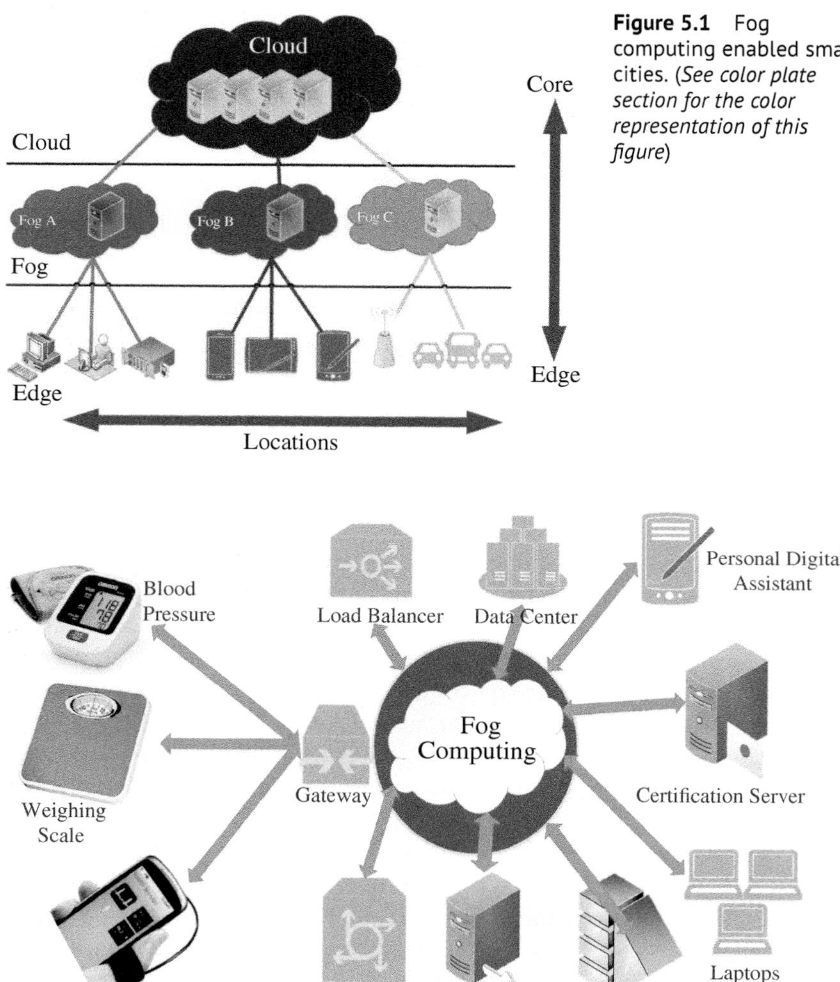

Figure 5.1 Fog computing enabled smart cities. (*See color plate section for the color representation of this figure*)

Figure 5.2 A generic fog enabled IoT environment. (*See color plate section for the color representation of this figure*)

and authentication provide the legitimacy of the thing. Access control is carried out by authorization methods, which define available roles and functions for a specific thing. However, with this state-of-the-art technology, several issues arise among which security and privacy are considered as the most significant challenges in fog computing. The security and privacy measures that were being

Figure 5.3 Internet of Things security phenomenon.

applied in cloud computing cannot be applied in fog computing due to its different structure and functionalities like large-scale geo-distribution.

This chapter is organized as follows: initially, we start by describing the basics of trust and its important components. Then we move on to explore security, in which we present related work in sections of authentication and authorization. Privacy being one of the most important components discusses related work and basic requirements to ensure the trust of end users in the IoT and fog by proposing privacy-preserving techniques. The section on web semantics and trust management discusses the importance of web semantics and the possible problems that it can help us eliminate. Finally, the discussion section concludes our chapter by using a healthcare scenario and presenting the possible solutions that one may encounter while deploying an IoT-based solution.

5.2 Trust in IoT

Trust is a vital component in today's connected world. The users should believe that the resource providers could comprehensively fulfill the requested tasks as they committed, published, and provide security and privacy to confidential information [1]. The availability of claimed services and applications is the foremost thing in trust establishment. Security aspects in the IoT can be achieved by providing secure and stringent authentication and authorization procedures. There are multiple methods used in practice to provide user authentication. Multifactor authentication with mutual consent is required to be enforced. Privacy is "The right of the individual to decide what information about himself should be communicated to others and under what circumstances" [2]. Privacy should be applicable on both users' data as well as users' activities.

IoT provides a layered architecture for Trust management as shown in Figure 5.4. Information security is required at each layer. Security to information

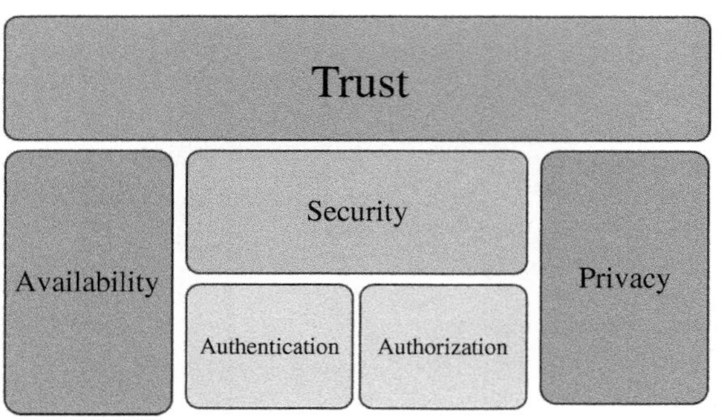

Figure 5.4 Layered depiction of components of trust.

can be provided by providing basic information security measures [1, 3–5]. Provision of controlled and meticulous access to its users is the basic principle of Information Security in Electronic Data Processing (EDP). After qualifying physical access parameters, the next step for implementing Information Security is electronic authentication. Electronic access control covers the legitimacy of a user, i.e. user identification and then authorization of access on predefined resources to the legitimate users. A user authentication mechanism initially requires user identification by the principle of something you know (user ID) or something you have (physical access) or something you are (biometrics). The user may use any of the above discussed methods for his identification, but for assuring the legitimacy of a user, the system has to validate his identity by using different other parameters like primary PIN and secondary PIN. Smart cards and USB mass storage devices are in use as physical components, such as keys of locks for legitimate authentication.

After a comprehensive arrangement of availability of services and data in an IoT, the purposes of information security encompass the key features, e.g. confidentiality (avoiding unauthorized disclosure of the resources), integrity (data health and accuracy), authentication (concurrence of sources and data) and non-repudiation (denial of any activity).

Trust can only be achieved by providing information security and privacy measures, i.e. authentication [6–10], authorization [11], and privacy [5, 12–16]. We can achieve trust by applying measures for AAP (authentication, authorization, and privacy) at each level and for each component in the environment. It will provide a trustworthy environment whether you are in a local area network (LAN), distributed computing, IoT, Web of Things (WoT), or even in cloud computing environment.

Identification is highly coupled with security and privacy in IoT management. Different and dynamic methods of identification are key components in multiple layers of IoT, some of them are embedded in the end devices whereas some of those use message routing and discovery. Each type of identification (numbering, addressing, and naming) has a set of important influencing factors that create divergence and it is also important to appreciate that these differences are often necessary and sometimes advantageous. As IoT exploits established elements and applications there is a legacy environment that cannot be ignored, and which must be addressed in some part or in its totality. There are various ways to achieve this, but each technique ultimately has an effect upon IoT's scope.

Distributed environments are challenging, even those that are closed, bounded by similar functional and interoperable technologies, and supported by a clear governance structure.

5.3 Authentication

Authentication plays a key role in the security of any system. With the advancement in technology and an increased number of threats, systems need secure authentication. The authentication of assets/resources depends on their prominence and significance.

There are different authentication methods required at different layers of IoT. Device, application, and user authentications are the three major layers of authentication. Uniqueness of the devices in IoT can be achieved by enforcing architectures and policies to avoid cloning the devices [1]. Physical unclonable functions (PUF) is the concept of using intrinsic physical characteristics of the devices for identification. This concept provides hardware level physical characteristics to restrict and save the device from cloning problems. Dongle-based coupled devices is also another method for device authentication. Some devices are bounded with a detachable dongle for device authentication [17]. Both symmetric as well as asymmetric encryption techniques are widely used for secure authentication. The use of encryption for authentication credentials as well as public-key infrastructure (PKI) digital certificates are recommended for secure device authentication [18–22]. Application authentication is carried out using digital certificates, which are the key solutions to enforce application authentication [22].

The legitimacy of a user is carried out by his/her authentication using different authentication methods. Threats and vulnerabilities, an authentication procedure, and different authentication schemes are covered in the next section.

5.3.1 Related Work

Developments in electronic authentication are revolutionary and have reached their current state through a number of improvements to provide security to resources. With the advent of technology, initial text-based single-factor authentication was unable to meet the information security requirements. This resulted in the development of a two-factor authentication and further paved the way for multifactor authentication along with encryption and coupled devices (Table 5.1).

Recently, a comprehensive criticism on two-factor authentication is carried out by Wang et al. [7]. They highlighted weaknesses of two-factor authentications with the practical implementation of adversary's tools. After the implementation of two-factor authentication, usage of coupled devices is introduced. Van der Haar et al. surveyed the crucial implementation of recursive usage of IoT. IoT needs authentication for security, facilities, and benefits of using wearable sensors for authentication. There is the recursive implementation of IoT; e.g. IoT needs authentication, as well as these, are used for authentication in terms of wearable sensors also [39]. Munch-Ellingsen et al. presented their view point of two-factor authentication while using coupled/hardware-based authentication. CIPURSE contactless cards were originally designed to meet the requirements in the transportation sector and the first version of the specification mirrored and followed Open Standard for Public Transportation (OSPT) Alliance. Smartphones are very much equipped with Bluetooth. Bluetooth devices are proposed for controlling smartphones as coupled devices [17]. Host card emulation (HCE) and the near-field communication (NFC) are the two basic features of smartphones that are used for authentication. These provide single factor authentication in IoT. Authors highlighted vulnerabilities and misuse of these features. To counter these vulnerabilities, the authors proposed an additional approach of short message service (SMS) service as a second factor authentication [37].

Security and privacy risks of a user are highlighted in home automation systems by Jacobsson et al. explored the area of smart-homes [38]. A risk analysis of a smart home automation system resulted in the identification of human behavior is the main cause of these severe risks and related to the software components. The whole research concluded on the importance of integration of security and privacy in the design phase of any new development [38]. Impersonation, replay, and similar attacks are very common to OAuth. To overcome these types, work was done to introduce another concept of security manager [9]. This security manager improves the security and performance using a database having token expiration time along with other useful information to reduce multiple registrations and multiple logins for IoT network.

Arno et al. came up with the idea of using smart devices for securing assets and as an example introduced a smart lock for a bicycle using smartphones. The main logic of this lock is authentication using an accelerometer. The sample data is generated using an NFC, GPS, and Bluetooth modules and the concept is implemented using Android-based smartphones [48]. User authentication should be imparted and required when cloud and IoT service providers need periodic access to IoT/smart devices for firmware updates and other maintenance tasks [49]. Henze et al. collected research and response on privacy in which

Table 5.1 State of the art research work timeline.

2004–2010	2011–2014	2015–2018
Roaming agreements for handover procedures in mobiles [23] Proposed two-factor authentication [24] OS-based authentication [25] MFA with handheld device [26] Unique identification scheme [27] Algorithm for two-factor authentication [28] Biometrics and smart cards [8] Interoperability and standards issues Regulatory, legal, and rights issues [5] touch pattern threats (smudge attacks) [29] Threats to RFID-based authentication [16] Encrypted dynamic IDs and passwords [20]	Visible security (two-factor authentication) [30] Need of OS for IoT as well as WSNs [31] Authentication in homogeneity as well as heterogeneity of IoT Redundant authentication [32] Electroencephalography in authentication [33] Two finger patterns [6] Three-factor authentication [34] (including USB mass storage) Authentication by secure identification [35] Mutual authentication by key management [36] Issues in two-factor authentication [7] Ciphering and physical authentication [18] Security and privacy concerns in IoT [1]	SMS-based authentication for NFC-like communication [37] Security and privacy by design [38] A central security manager for OAuth [9] Multifactor authentication using biometrics [10] Continuous authentication [21] Verification of authentication [39] Delegation-based authentication [40] Another idea of dynamic IDs [41] IoT communication in 5G [42] Bluetooth as a coupled device [17] Identification using fuzzy [19] Asymmetric cryptography [22] Node-based mutual authentication [43] IPSec along with RFID [15] EAKA protocol [44] IoT literature [45] Multilevel privacy [12] Security issues in IoT [4] Private mutual authentication [14] Two-factor zero knowledge authentication [11] Markov chain devices authentication in trust management [46] Security and privacy issues in IoT and cloud [47]

authors highlighted the issues in making users' decisions about privacy. Decisions about an individual's privacy settings should not be done by the developers or service providers. This paper presented a scheme for customizable user privacy. This scheme introduced multilevel privacy scheme and provides users with a transparent and adaptable interface for configuring privacy in IoT.

The idea of utilizing dynamic ID is very much active now and research is ongoing for securing IoT using dynamic IDs [41]. Node-based mutual authentication scheme for IoT considering the login ID, password hash and mandatory access control (MAC) address along with a DBMS (database management system) for the management and logging of authorized and unauthorized access controls is recommended [43]. Ruan et al. raised the issues of identity misuse. Impersonation attack is very common in the misuse of identity. A random oracle model is suggested to counter impersonation attack by extending the two-party setting to the "n" parties setting and developed an efficient two-party-based EAKA protocol as covered in the standard model [44].

Delegation-based authentication in IP-based IoT is carried out [40] to cope with the security and privacy matters in IoT. Private mutual authentication scheme to cope with privacy and discovery issues using new protocols are introduced using public-key infrastructures encryption schemes. Authors implemented this scheme using open source Vanadium framework [14]. Authors proposed identity-based cryptography (IBC) and elliptic curve cryptography (ECC) for end-to-end authentication. Asymmetric cryptography for end-to-end encryption [22]. Integration of Ciphering and other physical authentication approaches are suggested for further security and third factor authentication [18]. It is also highlighted that desirable security goals can be achieved by providing Dynamic IDs'-based authentication. An advanced framework for sharing multisite knowledge with Ciphered Dynamic credentials is also demonstrated in [20]. A protocol for Continuous Authentication in IoT where smart devices communicates limited data/messages frequently with short intervals is proposed by Bamasag et al. The protocol is based on Shamir's secret sharing scheme, with the advancement of mutual authentication. Claimer identity is verified using tokens issued for a function of time [21].

Development of new protocols and incorporate new features of IPV6 and 5G communication in IoT [42]. Mao et al. presented the idea of encryption in IoT using Fuzzy. This identity-based fuzzy encryption scheme is a lightweight, fully semantically secure and does not rely on random oracle models along with very short public parameters and quiet efficient because of its lightweight features [18]. A new term, *threat index*, is introduced to calculate vulnerabilities in IoT and suggested the development of new methods for security at each layer of IoT [4]. An introductory but comparative article on IoT in the area of IoT security, privacy, and trust [1]. Need for enforcement (i.e. governance in IoT) is the main crux of the

Table 5.2 Authentication grid.

Constraints	Parameters		
	Something you know	Something you have	Something you are
Only me	ID/Password	RFID	Biometric
Only now		Session key	
Only here		Geo location bounded	

paper. A comparative study was carried out where the latest European security projects in IoT analyzed and found "enforcement" to be missing [1].

In the use of IoT, the privacy of an individual always remains at stake. With the advent of smart devices, it is very crucial to take care of user privacy. Mostly radio-frequency identification (RFIDs) are used for identification purposes in IoT. Authors suggested the use of IPSec along with RFID to maintain user privacy. In this approach, "Need to Know"-based rule is introduced [15]. Similarly, Díaz et al. introduced the implementation of Zero-Knowledge Authentication Protocol along with some other authentication factors in IoT authentication. One-time password (OTP) and SMS are some other factors that can be used for authentication [11, 47].

Different types of authentications are implemented to strengthen the procedure. This can be explored using the grid given in Table 5.2, which discusses parameters for each constraint.

5.4 Authorization

The authorization of the system information in IoT is a fundamental management responsibility. Practically all applications that deal with security, confidentiality, or defense include some form of access controls. An important objective of access control is to protect available system resources against unwanted and inappropriate access. Sometimes access is granted after successful user authentication; however, most of the systems require more complex and sophisticated access control. Access control can be viewed as how authorization is structured. Most of the time authorization is designed according to the organization structure, while in some cases it is based on sensitivity and clearance of documents. Some commonly used access control mechanisms are access control list (ACL), role-based access control (RBAC), capability-based access control (CBAC) and extensible access control markup language (XACML). These mechanisms are well executed in a centralized information system. Though, in such distributed environments authorization becomes a formidable challenge [4, 50].

Table 5.3 Authorization requirements in the Internet of Things.

Constraints	Parameters			
Authorization management	User managed access	Usage access management	Authorization policy management	Fine-grained access privileges
Third-party authorization	Authentication, authorization, and accountability	Data confidentiality	Multiaccess	End-to-end communication Least privileges
Semantic authorization	Contextualization	Interoperability	Heterogeneous	Integrity

An access control system should deliberate three notions: access control mechanisms, models, and policies. Access control policies are the high-level abstraction that specifies what type of access privileges are allowed, to whom, under what circumstances, such as location or time [51]? Access control polices enforce the access control mechanism (ACL) that can be used to protect an automated system resources by identifying and preventing unauthorized access. However, an access control model (i.e. discretionary access control model) represents a formal demonstration of security policies and thus acts as a bridge between policy and mechanism. Table 5.3 demonstrates secure authorization in IoT. Authorization management [52] is required for multiaccess. Accordingly, semantic authorization requires contextualization, heterogeneous, integrated, and interoperable environment [53]. Thus, it is easy to control the access for a single ontology store. Similarly, third-party authorization is needed whenever the third party wants to access a person's specific data.

5.4.1 Related Work

Discretionary access control (DAC), one of the most frequently used access control models, is a user centric approach that is based on user directions. However, ACL is a relatively simple example of DAC, on the contrary in MAC individual owner has no right to change the access policies. IoT deals with real-time data streaming and processing of smart objects [54] over the Internet. Thus, to support the IoT ecosystem it requires a deep revision and adaptation of existing access control mechanisms. A lot of existing work will deal with these aspects.

Attribute-based access control (ABAC) is a policy-based mechanism, in which certain pre-agreed policies are used to grant users access. Su et al. [55] present attribute-based signature scheme with standard Diffie–Hellman assumption to reduce computational cost. Ye et al. [56] define ABAC policies by using a light-weighted encryption mechanism called elliptic curve cryptography in a distributed ecosystem. However, this extended proposed mechanism lacks confidentiality, which is mandatory for the distributed environment. In [57], Ning et al. suggested aggregated-proof-based hierarchical authentication (APHA) to provide hierarchal attribute-based access control. This proposed scheme uses homomorphism function for Chebyshev chaotic maps and directed path descriptors to achieve data integrity and data confidentiality. However, in APHA the following assumptions are made, such as that path descriptor is fresh and a trusted third party has authority on the entitled values, so any malicious activity can cause the loss of user's personal data. Misra et al. [58] proposed a theoretical next-generation IoT architecture using distributive policy-based access control. Although ABAC used to assign permissions dynamically but nested groups are the major limitations of ABAC.

Role-based access control (RBAC) is more powerful and generalized than MAC and DAC because it grants access rights by assigning roles. The main advantage of using RBAC and ABAC in term of IoT is that it can be easy to modify the access rights dynamically by simply modifying the role and attribute assignments. However, there is a need to introduce a new form of RBAC [59–62] and ABAC [56, 58, 63, 64] styles in IoT because users are allowed to access particular data in real-time environment. Gusmeroli et al. [65] affirm that several authorization frameworks like ABAC [66] and RBAC do not provide an effective and manageable mechanisms to support IoT, because such frameworks cannot provide least privileges. For this, Adda et al. [59] proposed a CollABAC and CollRBAC models to evaluate IoT Collaborative requirements. Lee et al. [66] proposed a dynamic model called location-temporal access control (LTAC). This model combines time and location with security level for access control for stable communication. Access is granted only if a given node is located within an appropriated time interval and location with respect to another node.

The author in [67] introduces the concept of capabilities (called capability-based access control CapBAC) to achieve least privileges for access control and suggested the identity authentication and capability-based access control (IACAC) scheme to grant access on the local network. In CapBAC, user's authorization capabilities have to be presented to the service provider. Whereas in traditional ACL it is the responsibility of service provider to check if the user is indirectly or directly authorized for the requested resource. Mahalleet et al. [68] represent capability-based context aware access control for federated IoT networks. This scheme is flexible, integrated, and scalable but still lacks granularity, reliability, and trust. A distributed capabilities-based access control model (DCapBAC) [69] uses a distributed approach where smart objects are enabled with access control logic with IoT devices. Hernández-Ramos et al. [70] presented an efficient, scalable, integrated, usable, and fully DCapBAC for certification and authorization.

However, this proposed model does not provide trust mechanism, so access control decisions are taken according to trustworthiness values. Bernabe et al. [71] proposed a trust-aware access control mechanism based on DCapBAC, which provides a flexible, efficient, and end-to-end security access control mechanism. However, it does not provide any privacy-aware feature that allows smart objects to create capability tokens anonymously. Hernandez-Ramos et al. [72] proposed enhanced privacy-aware DCapBAC mechanism for the IoT, by using anonymous authorization tokens in order to deal with privacy concerns with the integration of idemix. But there is a need that these capabilities be usable and easy to understand by the user. It is also important that least privileges are granted by default.

From the previously discussed work, the major challenges that are emerging in IoT scenario are the following.

- Which access control mechanism is required that not only grants access to the users but also smart objects that could be authorized to interact with the information systems?
- To deal with flexible and scalable IoT ecosystem, is it useful to exploit the distributed, centralized, or semi-distributed approach?
- How can the identification of smart objects (Things) be supported?

To deal with the identity issue, Li et al. [73] designed a practical identity-based framework for wireless sensor networks (WSN) in the context of IoT by using heterogeneous signcryption (HSC) scheme. Integrity, nonrepudiation, authentication, and confidentiality are simultaneously attained. However, two types of adversaries have an effect on the security of certificate-less systems. The first type of adversarial model refers to a trusted third party (TTP) who can easily fake the user's public key to the user's random secret value. In the second type of adversarial model, a key generating center (KGC) can be compromised and act maliciously to access the users' private and partial public keys. Furthermore:

- How is access control managed? How do we deal with the issuance of certificates and registration of smart objects and users?
- How do we define specific types of roles and functions in IoT systems?

To answer these issues, new solutions have been recently proposed that use the OAuth protocol to secure the network protocol. They leverage the idea of RBAC and reduce the numbers of rules by using the key-value attribute system. Customized fine-grained access policies can be used so that things can interact in an IoT environment. This demonstrates a further step in the direction of the management of registering things, users, and their certificates, but still, an extra effort is required to establish standards that will be globally accepted.

The Information Technology Laboratory (ITL) at the National Institute of Standards and Technology (NIST) includes practical, physical, and management standards of information systems for security and privacy. Almost all applications that deal with security, confidentiality, or defense include some form of access controls. Some commonly used access control mechanisms and services are role-based access control [74], capability-based access control, ACL, and XACML. These well-known access control models and mechanisms are implemented in a centralized information system. However, in a distributed environment, authorization becomes a formidable challenge when it is distributed to multiple systems [50].

5.5 Privacy

Privacy and security are undoubtedly the most important issues as they are at the core of trust, relationship building, and exchange [75]. In IoT and fog, the trust is developed based on security and privacy, and they lie under the umbrella of trust as shown in Figure 5.4. So, with reference to security, three essential components – confidentiality, integration, and availability (CIA) – are needed in addition to the authentication and authorization. Therefore, in privacy data protection and personal information of user have to be certified, whereas devices carry a lot of sensitive data.

In IoT, privacy is a major concern to keep the information of individuals from the coverage of the other devices in IoT environment. Privacy concerns in the IoT are improved by the way they expand the feasibility and spread the observation and tracking of devices. The motive behind this research is to explore the importance of privacy and how personal data is collected, analyzed, used, and protected in the IoT environment.

Table 5.4 Privacy requirements in the light of Internet of Things.

Constraints	Parameters			
Device privacy	Authorization	Reliability	Accessibility	Identification and tracking
Communication privacy	Accuracy	Transferability	Encryption	Timeouts
Storage privacy	Data minimization and portability	Data management, access control	Life cycle transition privacy and inventory attacks	Anonymity, linkage
Processing privacy	Transparent disclosure	Right to be forgotten, inevitability	Profiling, event detection,	Freedom of choice

As individuals have their daily activities and performances measured, recorded, and analyzed, there is an insistent need for inventors to keep in mind who collects which kind of personal information, how it is then stored and used, and to whom and for what purposes it is disclosed. Our main concern in IoT and fog is about the privacy concerns that are covered under trust and how they should be treated.

5.5.1 Requirements of Privacy in IoT

In IoT the privacy is defined as "the right of an entity, acting in its own behalf, to determine the degree to which it will interact with its environment, including the degree to which the entity is willing to share information about itself with others" [76]. The privacy of a person using a device [77] may be revealed by the smartphone and it contains information about the user. Table 5.4 shows our findings that are related to protecting privacy in the IoT. There are four stages of privacy that we discuss in our grid, and every stage has some constraints.

5.5.1.1 Device Privacy

Device privacy is quite important in the context of IoT. It helps to preserve the secrecy of sensitive device data. It includes authorization, reliability, accessibility, identification, and tracking of malicious activity performed on data that should be recovered.

5.5.1.2 Communication Privacy

In communication privacy, encryption is the most important factor to guarantee data confidentiality during transmission, timeouts/session, accuracy, and transferability.

5.5.1.3 Storage Privacy

When it comes to trust management in IoT, storage privacy is the core of privacy. The least of personal information needs, to some extent, to be stored. Personal information would be reserved for a certain time period. "Need-to-know" should always be the basis of revealing information. Storage privacy includes data management, access control, data minimization and portability, life-cycle transition privacy, inventory attacks, and anonymity linkage.

5.5.1.4 Processing Privacy

Parameters that are part of communication privacy are transparent disclosure, right to be forgotten, and freedom of choice. Both privacy preservation and primary principles would include freedom of choice parameters. Profiling accumulates information records around entities in order to conclude securities by incorporating profiles and data. Inevitability is the ability for users to intervene: the right to access, change, correct, block, revoke consent, and delete their personal data.

Users can manage the process themselves and control their environment and communicate data with other humans entities. To achieve privacy in IoT, various privacy models from the literature are studied and summarized to determine whether they can be extended in the context of IoT. Protecting device privacy becomes increasingly difficult as the IoT becomes more prevalent. In [63] the attribute-based encryption is used to analyze the results of time, data overhead, energy consumption, and CPU/memory usage in PC-class and Android-based smartphones. In [78] the cooperative distributed systems (CDS) model of IoT is proposed for the prevention of the unauthorized acts of execution. In [79], the conditional privacy-preserving authentication with access linkability (CPAL) model proposed for roaming service provides multilevel privacy preservation for networking.

Communication privacy is one of the important factors that cannot be neglected in IoT. In [80], authors proposed a scheme called privacy preserved data mining (PPDM) in sensor data to evaluate the privacy risk in IoT. Similarly, [81] purposed an ecosystem within broad smartphone ecosystem that will be the platform responsible for the distribution of apps for smart home and IoT devices. In [82], authors purposed a key-policy attribute-based encryption (KP-ABE) scheme supporting any monotonic access structure with constant size ciphertext and verified that the proposed scheme is semantically secure.

IoT devices can collect data about people in one zone and transmit that data to another zone for data storage. A decentralized anonymous credential system is introduced and implemented by [83] within the privacy-preserving self-adaptive model of IoT community targets. In [72], certificate basic foundations under public key cryptography and anonymous credential system is proposed for privacy. The authors in [84] proposed a delay-free method of anonymization for privacy preservation in electronic health data stream, which can also be used in IoT.

Privacy during information processing in the IoT is also a very important factor. Authors in [85] presented a novel token-based approach to attain distributed privacy-preserving access control in single-owner multiuser sensor networks. Another approach proposes a communication protocol for the measurement of the overall security and privacy architecture of smart meters by counting friendly distribution [86]. Authors of the data streams usually contain confidential data are help to reveal other useful information for K-anonymity [87]. Figure 5.2 shows the taxonomy of privacy, where privacy preserving and privacy management schemes are further divided into categories.

5.6 Web Semantics and Trust Management for Fog Computing

Current Internet technologies facilitate the integration of information from a syntactical point of view, but future proposals are working toward a semantic approach. The initial goal of the semantic web was to achieve an intelligent and sophisticated web that will alleviate the human factor [88]. They presented the semantic Web services that leverage ontologies to generate and parse machine readable web pages. Along with the demand in a fully integrated web, the critical need for information management and its security becomes paramount. Various models and systems have been proposed as future proposals [89] to protect the data and information from unauthorized use as well as corruption since the early days of the web as shown through the detailed taxonomy in Figure 5.5

In this era where the emergence of web and information technology systems are excessively growing day by day, data will be an important information resource in many business areas. Essentially the semantic web is a collection of layers that take advantage of underlying technologies, the most basic layer is the protocol layer that is not included in the literature while discussing semantic web technologies as described in Figure 5.6. The next is eXtensible Markup Language (XML) layer, a document representation language, which lays the ground for a resource description framework (RDF) layer. This RDF provides a framework for specifying ontology languages, such as web ontology language (OWL), it not only addresses the inadequacy of RDF but also provides the power of reasoning about various policies using Semantic Web Rules Language (SWRL) [90] and Rule Markup Language (RuleML) [91]. On top of these, the literature proposes the Trust layer [92], which not only is important but a much-needed part of the whole Semantic web stack.

5.6.1 Trust Through Web Semantics

Trust is achieved through different access control models, such as MAC [93], DAC, RBAC [94], location-based access control (LBAC), temporal access control, and ABAC. MAC usually enforces access control by attaching security labels or tags to users and objects while DAC enforces permissions on objects based on the object owner's configuration. These two models can be simulated by RBAC, which is a loosely coupled form of a mixture of components. A role is a collection of permissions for performing a certain task, a permission is an access mode that can be exercised on an object, and, finally, a session that relates the user to the object for a certain role. Most of the modern access control methods not only rely on but also support web semantics for decision making when it comes to ensuring trust management.

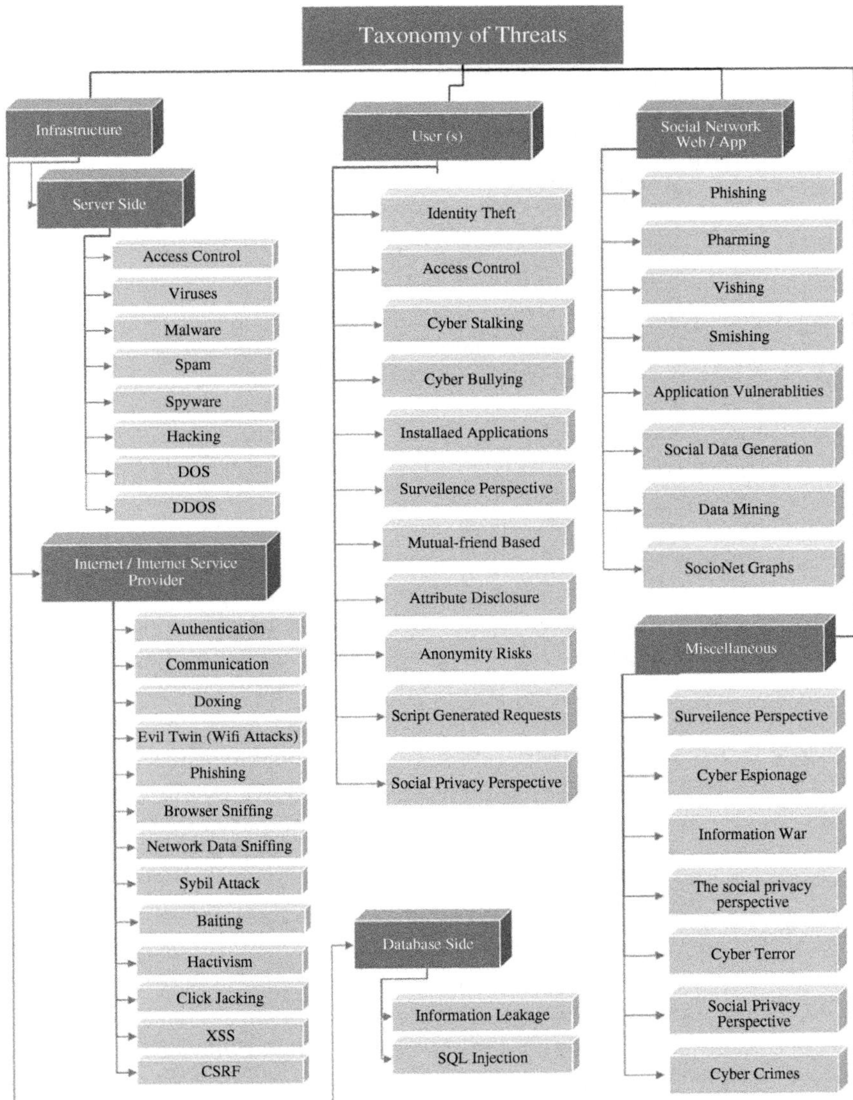

Figure 5.5 Detailed taxonomy of threats.

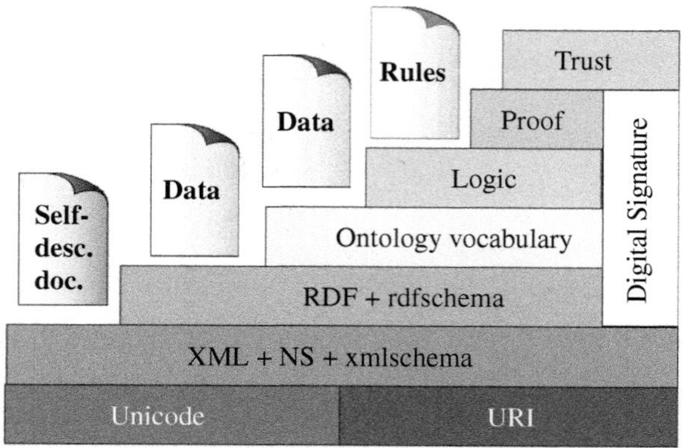

Figure 5.6 Semantic web technology stack by Tim Berners-Lee 2000 (http://w3c.org).

Traditionally, security (authentication and authorization) and semantics are discussed in different contexts, mainly because security is considered as a separate layer in federated IoT ecosystem. To deploy IoT-based cost-effective solutions, a tremendous amount of work has been done to develop standards and protocols. The most challenging aspect of IoT standards is the availability of too many standards, which generates an interoperability issue. Interoperability becomes a worse problem in IoT because different architectures are used for communication in different ecosystems. For example, in web-based applications for healthcare, a RESTful protocol stack, such as CoAP [95] is used. However, in the device-to-device communication in IoT and fog, several other stacks and protocols can be used, such as data distribution service (DDS) and message queuing telemetry transport (MQTT). Because of using multiple and a variety of protocols, there is another issue called integrity that arises. To overcome this issue there exist several access control models and policies to control unauthorized access [96]. If various security access models break the interoperable communication, it is still a problem because it creates compatibility issues between multiple secure access models. In these situations, a secure trusted environment is needed that allows entities to share and interact their information easily. A semantic-based approach allows controlling the access using a single ontology store [97].

Accordingly, semantic authorization requires to support contextualized, heterogeneous, integrated, and interoperable environment [53]. In a secure semantic web environment, the privacy-preserving methods are deliberated in access control model and trust modeling techniques. The access control models focus on subject-oriented requirements of information about resource access. Moreover, it also finds strong semantic relations for data models in knowledge. Proof of valid semantic web information is also considered under trust management. The privacy in semantic web regards keeping user-sensitive information secret throughout semantic web interaction. Privacy should be preserved, and confidential data should not be disclosed. Moreover, in information systems, both security and privacy must be provided protection of individual and nonmalicious users. In the semantic web, environment user should be protected against misuse and exposition of sensitive data where data is accessed, shared, exchanged, and mined. The semantic web would be facing a challenge regarding "how to collect and use personal information."

Privacy is considered to be a nonfunctional requirement in the semantic web; thus, it should be present in all the layers of the sematic web. The main focus of the semantic web regarding privacy is to prevent information resources from unauthorized access, modification of data, and unwanted use of resources [98]. Consequently, implementation of the access control model and trust models' mechanisms should assure the execution of the privacy requirements. Moreover, privacy apprehensions relating to these mechanisms have an obligatory role for considering privacy-preserving methods in the semantic web.

5.7 Discussion

We demonstrate the findings of our work in trust for the IoT using a use-case scenario depicting the interactions as well as the components of an IoT-based patient monitoring system. The system is connected to different parts of the patient and continuously monitoring her. Trusting a new system is a difficult task on its own, and when the system involves a total machine-to-machine communication ideology, then it becomes even more difficult. Though this can be made possible with the enforcement of Identification, Authentication, Authorization, and Privacy mechanisms.

Alice is an elderly patient with cardiovascular complications and hearing impairment and has received a pacemaker, which can be connected to a prepaired Bluetooth control device while the hearing aid is installed to her ear, as shown in Figure 5.7. She also has healthcare insurance subscription, which entitles her to emergency as well as specialized healthcare treatment. The said subscription can be verified through a smart card that stores her information, not only related to her health record, but also her social and economic situation, to ensure the best possible healthcare policy from the insurance company. This RFID card is microchip-enabled, which provides necessary services to ensure her privacy.

On a certain day, Alice feels she needs to have a doctor's visit for some pain due to the hearing aid and thus wants to visit the hospital that covers her medical visits according to her insurance policy. On her arrival she passes through the information desk and upon swiping the smart card the hospital's automatic machine prints her a visitor's slip containing the booking with the ENT specialist. This scenario will tackle the situation in which Alice's smart card is used to authenticate

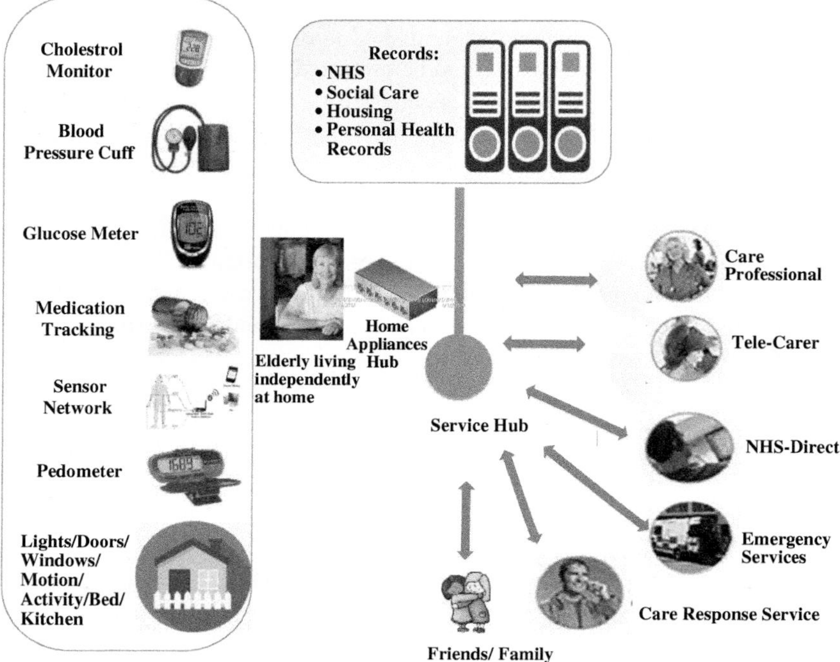

Figure 5.7 With the appropriate sensors and wireless technology, several wireless sensors enabled services can be provided to patients, such as the elderly. Source: Freescale Semiconductor.

her for accessing different services on the hospital premises. The sensors present in the hospital would be authorized to access her medical record upon providing legitimate signatures while the communication is done through public key cryptography to ensure her privacy.

5.7.1 Authentication

General parameters for electronic authentication are unique identification and temporal variables. Identification variables may be used singularly as well as in a set of two or three to identify a legitimate user. Temporal variables are additional parameters to secure the authentication. Some known identification variables are (1) something you know, (2) something you have, and (3) something you are. Similarly, temporal variables are intended (1) only for an identified user, (2) only for a specified time, and (3) only at specified geo location to restrict individual, duration, and location.

Multiple methods are used in practice to provide user authentication; for example, one factor uses user ID and password. For more security, the second factor is used; for example, authentication verification via SMS, use of biometric devices, passcode through e-mail, or even using a phone call. Use of encryption techniques in transferring credentials provides the third factor for secure authentication. Another approach for secure authentication is three-dimensional (3D) authentication used in credit/debit card transaction validations. Smart cards also play an important role in the provision of secure authentication.

There are different authentication schemes covered in a comprehensive taxonomy presented in Figure 5.8:

- Text-based (using textual input like user id, password, SMS, email, etc.)
- Three-dimensional (third party guaranty required for the successful transaction on cards)
- Biometric (fingerprints, retina scans, voice signatures, and wearable sensors)
- Electronic gadgets (coupled devices like dongles, etc.)

We have to focus not only on identification but also for the authentication of the activity and transactions in terms of financial services. IoT operating system is another idea for central authentication for enabling control at the distributed level. With the development, availability and application of such operating systems will significantly increase users trust in IoT. The WoT operating system will also enhance user's trust in higher scale in WoT, i.e. scalability of IoT. Such operating systems will also be of great interest of success and confidence for local as well as universal control of IoT and WoT. Integration of Open Source Architecture with IoT and WoT in terms of hardware, as well as the software will enhance users trust in these technologies. Open Source Development of hardware and software for IoT and WoT by multiple vendors and further certification of things used in IoT and WoT will also enhance user's trust in IoT as well as WoT.

5.7.2 Authorization

Authorization management in the proposed grid provides fine-grained access policy management. Constrained devices in IoT may be attacked physically or through Denial of Service (DOS) attack. Protection from physical attacks is not in the scope of this document, but it should be kept in mind by developers of authorization solutions. There are different authorization frameworks and models, which are shown in a comprehensive taxonomy presented in Figure 5.9.

Sensors implanted in the elderly patient's body, whom we will refer to as Alice, are connected to the Internet using wireless technology. These devices send alerts and the current readings/values along with Alice's identity and geolocation to the

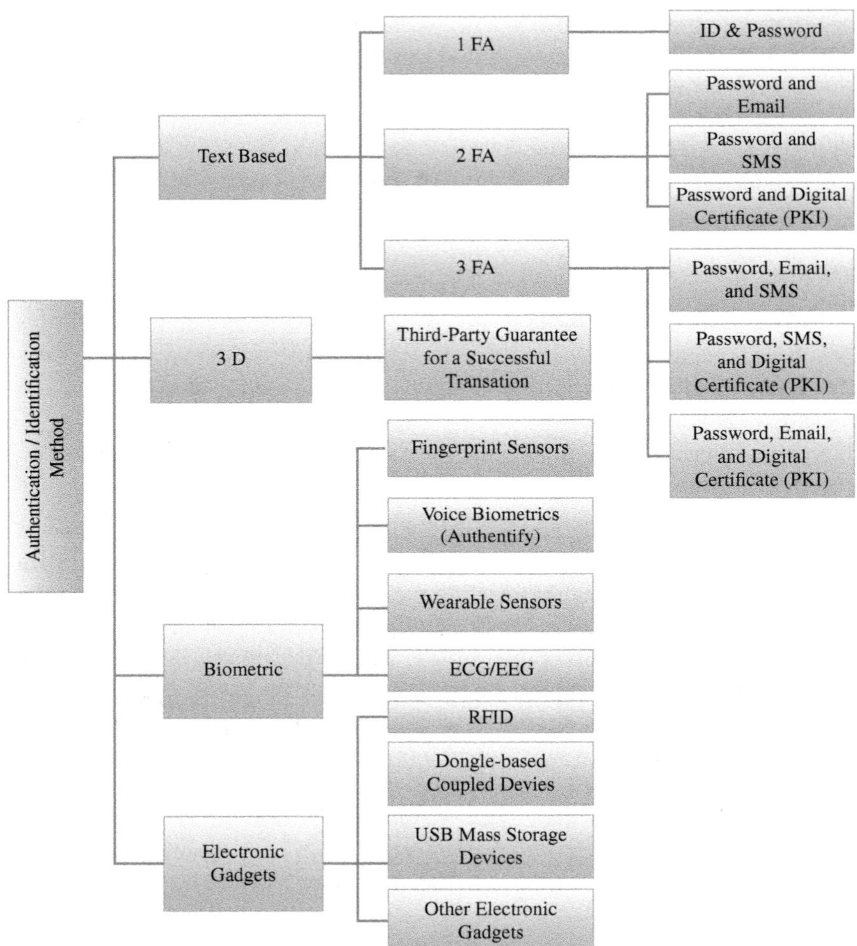

Figure 5.8 Authentication methods taxonomy.

duty doctor available in the hospital. Alice's medical data contains very sensitive information and therefore good protection is needed against unauthorized access. A frequent, contradictory requirement is needed as the capability for emergency access. In the meantime, typically users are not trained for security, thus secure default settings are used with an easy-to-use interface. Moreover, smart devices have short battery life, so changing a battery frequently is unacceptable.

In the case of emergency, such as if Alice patient suffers a cardiac arrest, the gateway at her home will automatically send alerts and the current readings/values along with identity and geolocation information to the duty doctor available in the

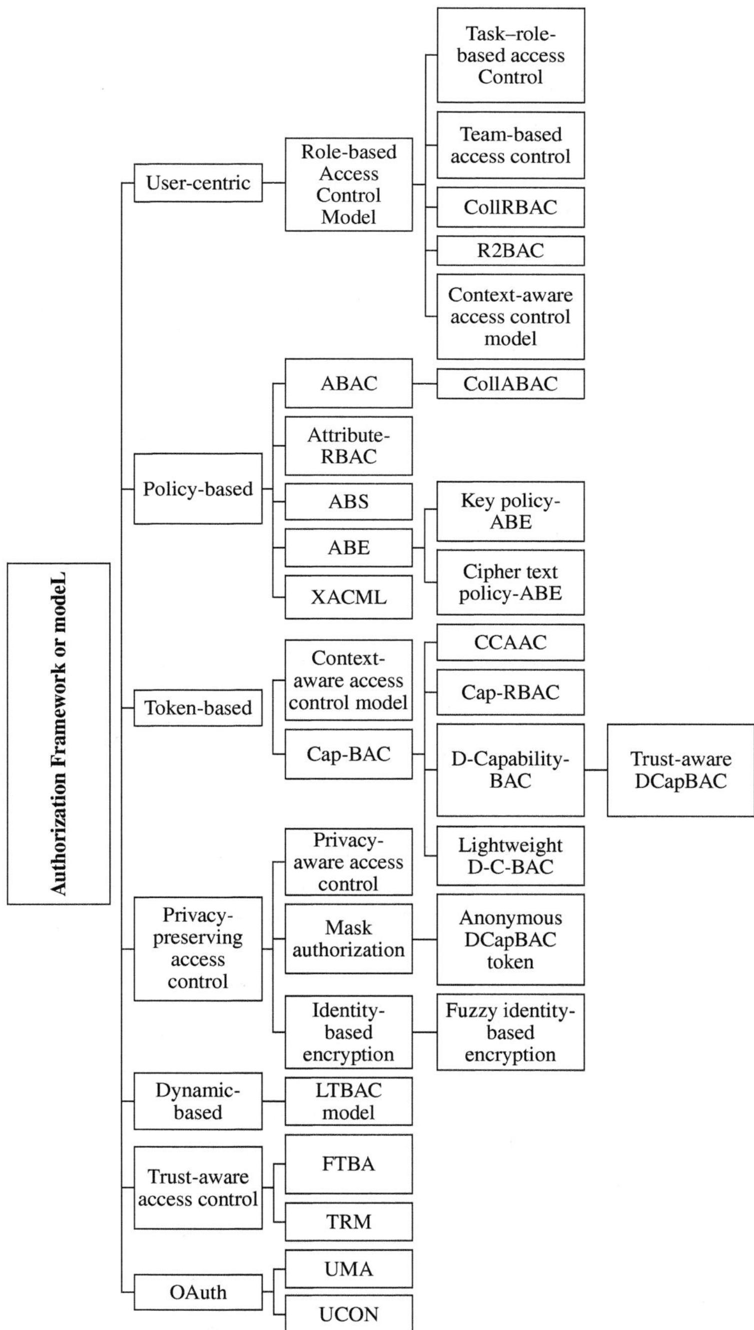

Figure 5.9 Authorization frameworks and models.

hospital. This device uses some intermediate nearby device, such as a smart card, to transmit such an alarm (P1). Alice configures a list of people (such as a close relative and doctor) to be notified in case of emergency (P2) (P3). Moreover, the smart card stored Alice's heartbeat data, which can be accessed later by a medical specialist.

Alice may have some privacy concerns, so she does not want John to monitor her location when there is not an emergency (P4). Sensors should generate alert only when the pulse rate is very high or very low. Finally, Alice is comfortable with this latest technology and easily uses the device because she is not trained in device security. If Alice cannot understand the meaning of device settings, then she can assume that manufacturers (third party) have initialized the devices to secure settings (P5) (in case the user does not have enough knowledge about their attached device. Their security may be breached, or malicious activity may occur and give unambiguous data).

In this scenario, a remote authorization grid is very useful. In the case of pre-configured access rights, there is a need for authorization management in which fine-grained access privileges and policies are granted, according to the user and resource usage. After configuring the authorization management, device user's security can be managed using third-party authorization that allows multiple accesses and ensures data confidentiality and integrity. Access rights must be dynamically changed and for this it is necessary that smart devices must have semantic authorization, which means the context of authorization. For example, in the above scenario, Alice requests that John monitor her location only in the context of an emergency. A taxonomy of the frameworks and models used for privacy is shown in Figure 5.10

The insurance companies may not get Alice's medical records; rather they receive only the information for a certain period when the device is embedded and it has access to the medical condition. Moreover, the National Health Service (NHS) does not allow Alice's data to be published, unless anonymity is granted to that data they can use or publish. In IoT the storage privacy is one of the most important factors, where devices are embedded, and they have to communicate with each other. Storage data privacy management is one of the major issues in communication between embedded devices. The mechanism to send alert system must be preserved by applying some encryption schemas like Shamir's Secret Sharing, fine-grained access control, or by applying attributes.

Alice's embedded device data must be stored at a remote location under an authorized person. The embedded device has not enough capacity to store data, if more issues arise, such as processing speed, storage capacity, battery time, etc. In case Alice's information may be revealed to an unauthorized party there must be transparent disclosure regarding who can reveal information. Conclusively, from the start to the end, in communication between the pacemaker to hospital alerts, system privacy applies at each step.

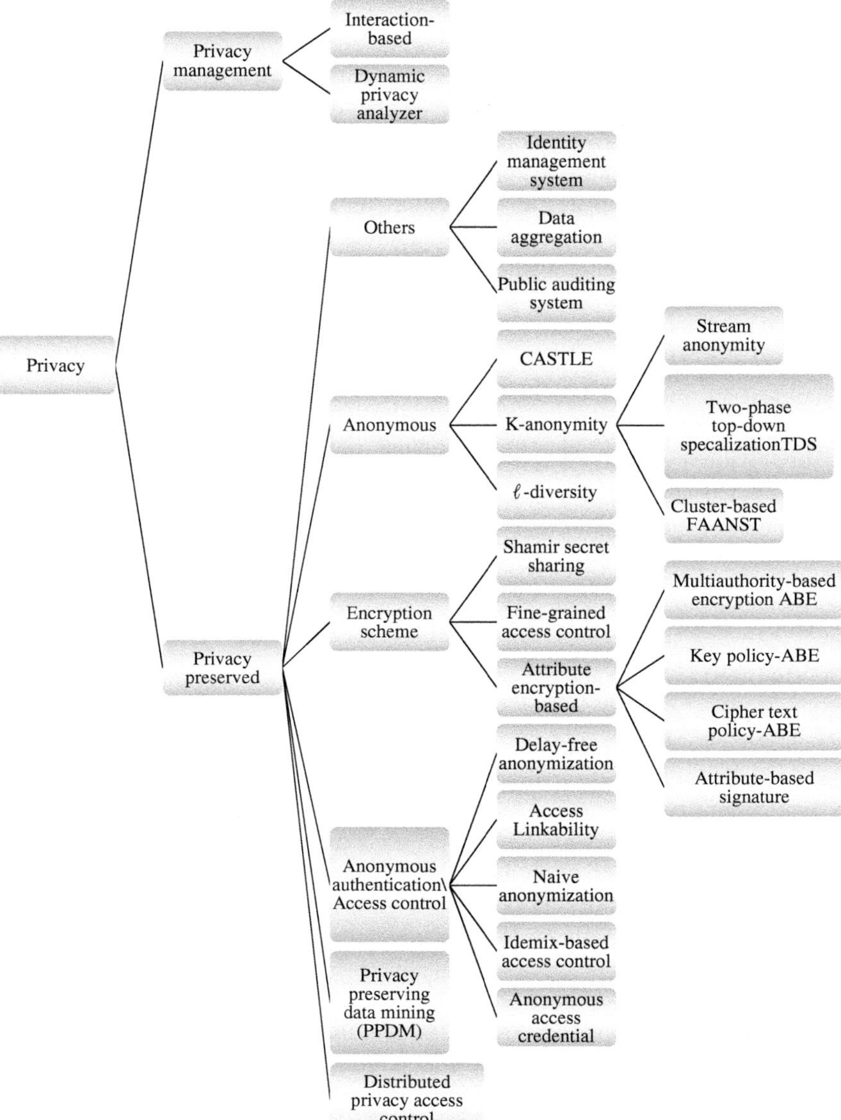

Figure 5.10 Taxonomy of the frameworks and models used for privacy.

5.8 Conclusion

IoT is a growing area of research. However, it still faces several challenges related to the application of access control frameworks due to the heterogenous nature of the connected devices. As access control is critical, so least privileges must be assigned to assure that during maintenance, the maintainer cannot use all authorized rights, as happens in RBAC and ABAC. In traditional access control frameworks, a thorough revision and adaptation is needed. For this, a capability-based access system is being available openly for access capabilities and revocation services. In recent years, because of the ubiquitous nature of IoT networks and devices, the protection of smart services and resources is addressed by centralized architecture. In such an environment, back-end servers are responsible for authentication and authorization tasks. The use of traditional access control mechanism prevents end-to-end security. Consequently, central architecture cannot provide scalability for smart objects in IoT.

In the scenario we discussed, a remote authorization grid is very useful. In the case of preconfigured access rights there is a need for authorization management in which fine-grained access privileges and policies are granted according to the user and resource usage. After configuring the authorization management, device user's security can be managed using third-party authorization that allows multiple accesses and ensures data confidentiality and integrity. A mechanism to dynamically change access rights can be ensured by embedding smart devices to have a policy for semantic authorization. The device-to-device privacy must be ensured if cardiologist needs hearing device access to monitor the patient's overall health. Only the desired data should be provided to the cardiologist, with the agreement of the patient, Alice. Appropriately, to deliver privacy in IoT, we have discussed the privacy properties that affect privacy connections, categorized them into four classifications, and shown that complete privacy preserving would need attention for parts of these categories in diverse frameworks and designed for diverse use cases. A general IoT privacy scheme model needs to satisfy some of the suggested parameters that are based on privacy requirement objectives for complete IoT privacy. As part of the future work, we will implement the proposed privacy grid to achieve privacy in IoT and present a comparative study with the help of simulations and practical implementations.

References

1 Sicari, S., Rizzardi, A., Grieco, L.A., and Coen-Porisini, A. (2015). Security, privacy and trust in internet of things: the road ahead. *Computer Networks* 76: 146–164.

2 Westin, A.F. (1970). Privacy and freedom. *Science and Society* 34: 360–363.

3 Farooq, M.U., Waseem, M., Khairi, A., and Mazhar, S. (2015). A critical analysis on the security concerns of internet of things (IoT). *International Journal of Computers and Applications* 111 (7): 1–6.

4 Kumar, S.A., Vealey, T., and Srivastava, H. (2016). Security in Internet of Things: challenges, solutions and future directions. In: *2016 49th Hawaii International Conference on System Sciences (HICSS)*, 5772–5781. IEEE Computer Society.

5 Medaglia, C.C.M. and Serbanati, A. (2010). *An Overview of Privacy and Security Issues in the Internet of Things*. Springer.

6 Sun, J., Zhang, R., Zhang, J., and Zhang, Y. (2014). Touching: sightless two-factor authentication on multi-touch mobile devices. *IEEE Conference on Communications and Network Security (CNS)* 2014: 436–444.

7 Wang, D., He, D., Wang, P., and Chu, C.-H. (2015). Anonymous two-factor authentication in distributed systems: certain goals are beyond attainment. *IEEE Transactions on Dependable and Secure Computing* 12 (4): 428–442.

8 Li, C.-T. and Hwang, M.-S. (2010). An efficient biometrics-based remote user authentication scheme using smart cards. *Journal of Network and Computer Applications* 33 (1): 1–5.

9 Emerson, S., Choi, Y.-K., Hwang, D.-Y. et al. An OAuth-based authentication mechanism for IoT networks. In: *2015 International Conference on Information and Communication Technology Convergence (ICTC)*, vol. 2015, 1072–1074. IEEE.

10 U. Gupta, Application of multifactor authentication in Internet of Things domain, arXiv Prepr. arXiv1506.03753, 2015.

11 Rafidha Rehiman, K.A., Veni, S., Rehiman, K.A.R., and Veni, S. (2016). A secure authentication infrastructure for IoT enabled smart mobile devices – an initial prototype. *Indian Journal of Science and Technology* 9 (9): 520–523.

12 Henze, M., Hermerschmidt, L., Kerpen, D. et al. (2016). A comprehensive approach to privacy in the cloud-based internet of things. *Future Generation Computer Systems* 56: 701–718.

13 Roman, R., Zhou, J., and Lopez, X. (2013). On the features and challenges of security and privacy in distributed internet of things. *Computer Networks* 57 (10): 2266–2279.

14 D. J. Wu, A. Taly, A. Shankar, and D. Boneh, Privacy, discovery, and authentication for the Internet of Things," arXiv Prepr. arXiv1604.06959, 2016.

15 Gross, H., Hölbl, M., Slamanig, D. et al. (2015). Privacy-aware authentication in the internet of things. In: *Cryptology and Network Security*, 32–39.

16 Feng, H. and Fu, W. (2010). Study of recent development about privacy and security of the internet of things. *International Journal of Computer Applications* 2 (7): 91–95.

17 Jeong, H.-D.J., Lee, W., Lim, J., and Hyun, W. (2015). Utilizing a Bluetooth remote lock system for a smartphone. *Pervasive and Mobile Computing* 24: 150–165.

18 Crossman, M.A. and Liu, H. (2015). Study of authentication with IoT testbed. In: *2015 IEEE International Symposium on Technologies for Homeland Security (HST)*, 1–7. IEEE.

19 Mao, Y., Li, J., Chen, M.-R.R. et al. (2016). Fully secure fuzzy identity-based encryption for secure IoT communications. *Computer Standards & Interfaces* 44 (4): 117–121.

20 Ilyas, M., Ali, A., and Kueng, J. (2010). WebSeA: a secure framework for multi-site knowledge representation in software engineering. In: *International Conference on Bio-Inspired Models of Network, Information, and Computing Systems, Springer*, 682–686.

21 Bamasag, O.O. and Youcef-Toumi, K. (2015). Towards continuous authentication in internet of things based on secret sharing scheme. In: *Proceedings of the WESS'15: Workshop on Embedded Systems Security, ACM*, 1.

22 Markmann, T., Schmidt, T.C., and Wählisch, M. (2015). Federated end-to-end authentication for the constrained internet of things using IBC and ECC. *ACM SIGCOMM Computer Communication Review* 45 (4): 603–604.

23 Bargh, M.S., Hulsebosch, R.J., Eertink, E.H. et al. (2004). Fast authentication methods for handovers between IEEE 802.11 wireless LANs. In: *Proceedings of the Second ACM International Workshop on Wireless Mobile Applications and Services on WLAN Hotspots, WMASH 2004*, 51–60. ACM.

24 Schneier, B. (2005). Two-factor authentication: too little, too late. *Communications of the ACM* 48 (4): 136, ACM.

25 Farooq, M.O. and Kunz, T. (2011). Operating systems for wireless sensor networks: a survey. *Sensors (Basel)* 11 (6): 5900–5930.

26 Sabzevar, A.P. and Stavrou, A. (2008). Universal multi-factor authentication using graphical passwords. In: *IEEE International Conference on Signal Image Technology and Internet-based Systems, 2008. SITIS'08*, 625–632.

27 Sarma, A.C. and Girão, J. (2009). Identities in the future Internet of Things. *Wireless Personal Communications* 49 (3): 353–363.

28 Aloul, F.A., Zahidi, S., and El-Hajj, W. (2009). Two-factor authentication using mobile phones. *IEEE/ACS International Conference on Computer Systems and Applications*: 641–644.

29 Aviv, A.J., Gibson, K., Mossop, E. et al. (2010). Smudge attacks on smartphone touch screens. *USENIX Conference on Offensive Technologies* 10: 1–7.

30 Wimberly, H. and Liebrock, L.M. (2011). Using fingerprint authentication to reduce system security: an empirical study. In: *2011 IEEE Symposium on Security and Privacy*, 32–46.

31 Baccelli, E., Hahm, O., Gunes, M. et al. (2013). RIOT OS: towards an OS for the Internet of Things. *2013 IEEE Conference on Computer Communications Workshops (INFOCOM WKSHPS)*: 79–80.

32 Huang, X., Xiang, Y., Bertino, E. et al. (2014). Robust multi-factor authentication for fragile communications. *IEEE Transactions on Dependable and Secure Computing* 11 (6): 568–581.

33 Pham, T., Ma, W., Tran, D. et al. (2014). Multi-factor EEG-based user authentication. In: *2014 International Joint Conference on Neural Networks (IJCNN)*, 4029–4034. IEEE.

34 D. He, N. Kumar, J.-H. Lee, and R. S. Sherratt, Enhanced three-factor security protocol for consumer USB mass storage devices, *IEEE Transactions on Consumer Electronics*, Vol. 60, No. 1, pp. 30–37, IEEE, 2014.

35 Edwards, C. (2014). Ending identity theft and cyber crime. *Biometric Technology Today* 2014 (2): 9–11.

36 M. Turkanović, B. B. Brumen, and M. Hölbl, A novel user authentication and key agreement scheme for heterogeneous ad hoc wireless sensor networks, based on the internet of things notion, *Ad Hoc Networks*, Vol. 20, pp. 96–112, Elsevier, 2014.

37 Munch-Ellingsen, A., Karlsen, R., Andersen, A., and Akselsen, S. (2015). Two-factor authentication for android host card emulated contactless cards. In: *2015 First Conference on Mobile and Secure Services (MOBISECSERV)*, 1–6.

38 Jacobsson, A., Boldt, M., and Carlsson, B. (2015). A risk analysis of a smart home automation system. *Future Generation Computer Systems* 56: 719–733.

39 van der Haar, D. (2015). Canvis: a cardiac and neurological-based verification system that uses wearable sensors. In: *2015 Third International Conference on Digital Information, Networking, and Wireless Communications (DINWC)*, 99–104.

40 T. Borgohain, A. Borgohain, U. Kumar, and S. Sanyal, Authentication systems in Internet of Things," arXiv Prepr. arXiv1502.00870, 2015.

41 Zhai, J., Cao, T., Chen, X., and Huang, S. (2015). Security on dynamic ID-based authentication schemes. *International Journal of Security and Its Applications* 9 (1): 387–396.

42 Mahmoud, R., Yousuf, T., Aloul, F., and Zualkernan, I. (2015). Internet of Things (IoT) security: current status, challenges and prospective measures. In: *2015 10th International Conference for Internet Technology and Secured Transactions (ICITST)*, 336–341. IEEE.

43 Devi, G.U., Balan, E.V., Priyan, M.K., and Gokulnath, C. (2015). Mutual authentication scheme for IoT application. *Indian Journal of Science and Technology* 8 (26): 15–19.

44 Ruan, O., Kumar, N., He, D., and Lee, J.-H.H. (2015). Efficient provably secure password-based explicit authenticated key agreement. *Pervasive and Mobile Computing* 24: 50–60.

45 Rose, K., Eldridge, S., and Chapin, L. (2015). The Internet of Things: an overview. *Internet Society* 80: 1–50.

46 Kang, D., Jung, J., Mun, J. et al. (2016). Efficient and robust user authentication scheme that achieve user anonymity with a Markov chain. *Security Communication Networks* 9: 1032–1035.

47 Díaz, M. et al. (2016). State-of-the-art, challenges, and open issues in the integration of internet of things and cloud computing. *Journal of Network and Computer Applications* 67 (1): 99–117.

48 Arno, A., Toyoda, K., and Sasase, I. (2015). Accelerometer assisted authentication scheme for smart bicycle lock. In: *2015 IEEE 2nd World Forum on Internet of Things (WF-IoT)*, 520–523. IEEE.

49 L. Barreto, A. Celesti, M. Villari et al., An authentication model for IoT clouds, in *Proceedings of the 2015 IEEE/ACM International Conference on Advances in Social Networks Analysis and Mining 2015, IEEE*, 2015, pp. 1032–1035.

50 Ferraiolo, D.F., Hu, V.C., and Kuhn, D.R. (2007). Assessment of access control systems, Interagency Report 7316. *National Institute of Standards and Technology*.

51 Keoh, S.L., Kumar, S.S., and Tschofenig, H. (2014). Securing the Internet of Things: a standardization perspective. *IEEE Internet of Things Journal* 1 (3): 265–275.

52 Seitz, L., Selander, G., and Gehrmann, C. (2013). Authorization framework for the Internet-of-Things. In: *2013 IEEE 14th International Symposium and Workshops on a World of Wireless, Mobile and Multimedia Networks (WoWMoM)*, 1–6. IEEE.

53 Alam, S., Chowdhury, M.M.R., and Noll, J. (2011). Interoperability of security-enabled Internet of Things. *Wireless Personal Communications* 61 (3): 567–586.

54 Ho, G., Leung, D., Mishra, P. et al. (2016). Smart locks: Lessons for securing commodity Internet of Things, Devices. In: *Proceedings of the 11th ACM on Asia Conference on Computer and Communications Security, Xi'an, China*, 461–472.

55 Su, J., Cao, D., Zhao, B. et al. (2014). ePASS: an expressive attribute-based signature scheme with privacy and an unforgeability guarantee for the Internet of Things. *Future Generation Computer Systems* 33: 11–18.

56 Ye, N., Zhu, Y., Wang, R.-C. et al. (2014). An efficient authentication and access control scheme for perception layer of Internet of Things. *Applied Mathematics & Information Sciences* 8 (4): 1617.

57 Ning, H., Liu, H., and Yang, L.T. (2015). Aggregated-proof-based hierarchical authentication scheme for the internet of things. *IEEE Transactions on Parallel and Distributed Systems* 26 (3): 657–667.
58 P. Misra, Y. Simmhan, and J. Warrior, Towards a practical architecture for the next generation Internet of Things, arXiv Prepr. arXiv1502.00797, 2015.
59 Adda, M., Abdelaziz, J., Mcheick, H., and Saad, R. (2015). Toward an access control model for IOTCollab. *Procedia Computer Science* 52: 428–435.
60 Hummen, R., Shafagh, H., Raza, S. et al. (2014). Delegation-based authentication and authorization for the IP-based internet of things. In: *2014 Eleventh Annual IEEE International Conference on Sensing, Communication, and Networking (SECON)*, 284–292. IEEE.
61 Chen, D., Chang, G., Sun, D. et al. (2012). Modeling access control for cyber-physical systems using reputation. *Computers and Electrical Engineering* 38 (5): 1088–1101.
62 Thomas, R.K. (1997). Team-based access control (TMAC): a primitive for applying role-based access controls in collaborative environments. In: *Proceedings of the Second ACM Workshop on Role-Based Access Control*, 13–19. ACM.
63 Wang, X., Zhang, J., Schooler, E.M., and Ion, M. (2014). Performance evaluation of attribute-based encryption: toward data privacy in the IoT. In: *2014 IEEE International Conference on Communications (ICC)*, 725–730. IEEE.
64 Bao, F. and Chen, I.-R. (2012). Dynamic trust management for Internet of Things applications. In: *Proceedings of the 2012 International Workshop on Self-Aware Internet of Things*, 1–6. IEEE.
65 Gusmeroli, S., Piccione, S., and Rotondi, D. (2013). A capability-based security approach to manage access control in the internet of things. *Mathematical and Computer Modelling* 58 (5): 1189–1205.
66 Lee, C., Guo, Y., and Yin, L. (2013). A location temporal-based access control model for IoTs. *AASRI Procedia* 5: 15–20.
67 Mahalle, P.N., Anggorojati, B., Prasad, N.R., and Prasad, R. (2013). Identity authentication and capability-based access control (iacac) for the internet of things. *Journal of Cyber Security and Mobility* 1 (4): 309–348.
68 Anggorojati, B., Mahalle, P.N., Prasad, N.R., and Prasad, R. (2012). Capability-based access control delegation model on the federated IoT network. In: *2012 15th International Symposium on Wireless Personal Multimedia Communications (WPMC)*, 604–608. IEEE.
69 Skarmeta, A.F., Hernandez-Ramos, J.L., and Moreno, M. (2014). A decentralized approach for security and privacy challenges in the Internet of Things. In: *2014 IEEE World Forum on Internet of Things (WF-IoT)*, 67–72. IEEE.

70 Hernández-Ramos, J.L., Jara, A.J., Marın, L., and Skarmeta, A.F. (2013). Distributed capability-based access control for the Internet of Things. *Journal of Internet Services and Information Security* 3 (3/4): 1–16.

71 Bernabe, J.B., Ramos, J.L.H., and Gomez, A.F.S. (2015). TACIoT: multidimensional trust-aware access control system for the Internet of Things. *Soft Computing* 20: 1–17.

72 Hernndez-Ramos, J.L. et al. (2015). Preserving smart objects privacy through anonymous and accountable access control for a M2M-enabled internet of things. *Sensors (Basel, Switzerland)* 15 (7): 15611–15639.

73 Li, F., Han, Y., and Jin, C. (2016). Practical access control for sensor networks in the context of the Internet of Things. *Computer Communications* 89–90: 154–164.

74 Ross, R. and Oren, J.C. (2014). Systems security engineering. *NIST Special Publication* 800: 160.

75 Weinberg, B.D., Milne, G.R., Andonova, Y.G., and Hajjat, F.M. (2015). Internet of Things: convenience vs privacy and secrecy. *Business Horizons* 58 (6): 615–624.

76 Schiller, J.I. (2002). *Strong Security Requirements for Internet Engineering Task Force Standard Protocols*, vol. 1, 1–8. ACM.

77 Cheng, Y., Naslund, M., Selander, G., and Fogelström, E. (2012). Privacy in machine-to-machine communications a state-of-the-art survey. In: *2012 IEEE Internationa. Conference on Communicaton Systems ICCS 2012*, 75–79. IEEE.

78 Samani, A., Ghenniwa, H.H., and Wahaishi, A. (2015). Privacy in Internet of Things: a model and protection framework. *Procedia Computer Science* 52 (1): 606–613.

79 Lai, C., Li, H., Liang, X. et al. (2014). CPAL: a conditional privacy-preserving authentication with access linkability for roaming service. *IEEE Internet of Things Journal* 1 (1): 46–57.

80 Ukil, A., Bandyopadhyay, S., and Pal, A. (2014). IoT-privacy: to be private or not to be private. In: *Proceedings-IEEE INFOCOM*, 123–124. IEEE.

81 Ahmad, W., Sunshine, J., and Wynne, A. (2015). Enforcing fine-grained security and privacy policies in an ecosystem within an ecosystem. In: *Proceedings of the 3rd International Workshop on Mobile Development Lifecycle*, 28–34. ACM.

82 Wang, C. and Luo, J. (2013). An efficient key-policy attribute-based encryption scheme with constant ciphertext length. *Mathematical Problems in Engineering* 2013, 810969-810976. https://doi.org/10.1155/2013/810969.

83 Alcaide, A., Palomar, E., Montero-Castillo, J., and Ribagorda, A. (2013). Anonymous authentication for privacy-preserving IoT target-driven applications. *Computers & Security* 37: 111–123.

84 Kim, S., Sung, M.K., and Chung, Y.D. (2014). A framework to preserve the privacy of electronic health data streams. *Journal of Biomedical Informatics* 50: 95–106.

85 Zhang, R., Zhang, Y., and Ren, K. (2012). Distributed privacy-preserving access control in sensor networks. *IEEE Transactions on Parallel and Distributed Systems* 23 (8): 1427–1438.

86 Rottondi, C., Verticale, G., and Krauß, C. (2013). Distributed privacy-preserving aggregation of metering data in smart grids. *IEEE Journal on Selected Areas in Communications* 31 (7): 1342–1354.

87 Zakerzadeh, H. and Osborn, S.L. (2011). FANST: fast anonymizing algorithm for numerical streaming DaTaA. In: *Data Privacy Management and Autonomous Spontaneous Security*, 36–50. Springer.

88 Berners-Lee, T., Hendler, J., Lassila, O. et al. (2001). The semantic web. *Scientific American* 284 (5): 34–43.

89 H. A. Khattak, Internet of Things in the future Internet, 1st Workshop on the State of the Art and Challenges of Research Efforts, Politecnico di Bari, 2014.

90 Horrocks, I., Patel-Schneider, P.F., Boley, H. et al. (2004). SWRL: a semantic web rule language combining OWL and RuleML. *W3C Member Submission*: 1–20.

91 Boley, H., Tabet, S., and Wagner, G. (2001). Design rationale of RuleML: a markup language for semantic web rules. In: *First International Conference on Semantic Web Working Symposium*, vol. 1, 381–401. CEUR-WS.

92 Aghaei, S., Nematbakhsh, M.A., and Farsani, H.K. (2012). Evolution of the world wide web: from WEB 1.0 to WEB 4.0. *International Journal of Web & Semantic Technology* 3 (1): 1.

93 Nyanchama, M. and Osborn, S. (1995). Modeling mandatory access control in role-based security systems. *Database Security* 51: 129–144.

94 Ferraiolo, D., Cugini, J., and Kuhn, D.R. (1995). Role-based access control (RBAC): features and motivations. In: *Proceedings of the 11th Annual Computer Security Application Conference*, 241–248. NIST.

95 Khattak, H.A., Ruta, M., Eugenio, E., and Sciascio, D. (2014). CoAP-based healthcare sensor networks: a survey. In: *Proceedings of 2014 11th International Bhurban Conference on Applied Sciences and Technology, IBCAST 2014*, 499–503. IEEE.

96 Darshak Thakore, C. (2016). IoT Security in the context of Semantic Interoperability. In: *Workshop on Internet of Things (IoT) Semantic Interoperability (IOTSI), Santa Claria, CA, March*, 17–18.

97 Alamri, A., Bertok, P., Thom, J.A., and Fahad, A. (2016). The mediator authorization-security model for heterogeneous semantic knowledge bases. *Future Generation Computer Systems* 55: 227–237.

98 Nematzadeh, A. and Pournajaf, L. (2008). Privacy concerns of semantic web. In: *Fifth International Conference on Information Technology: New Generations, 2008*. ITNG 2008, 1272–1273. Elsevier.

6

How Fog Computing Can Support Latency/Reliability-sensitive IoT Applications: An Overview and a Taxonomy of State-of-the-art Solutions

Paolo Bellavista[1], *Javier Berrocal*[2], *Antonio Corradi*[1], *Sajal K. Das*[3], *Luca Foschini*[1], *Isam Mashhour Al Jawarneh*[1], *and Alessandro Zanni*[1]

[1] *Department of Computer Science and Engineering, University of Bologna, 40136 Bologna, Italy*
[2] *Department of Computer and Telematics Systems Engineering, University of Extremadura, 10003, Cáceres, Spain*
[3] *Department of Computer Science, Missouri University of Science and Technology, Rolla, MO 65409, USA*

6.1 Introduction

The widespread ubiquitous adoption of resource-constrained Internet of Things (IoT) devices has led to the collection of massive amounts of heterogeneous data on a continual basis, that when coupled with data coming from highly trafficked websites forms a challenge that exceeds the capacities of today's most powerful computational resources. Those avalanches of data coming from sensor-enabled and alike devices hides a great value that normally incorporates actionable insights if extracted in a timely fashion.

IoT is loosely defined as any network of connected devices spanning from home electronic appliances, connected vehicles, and sensor-enabled devices and actuators that interact and exchange data in a nonstationary fashion. It has been predicted that by 2020 the number of IoT devices could reach 50 billion connected to the Internet, and twice that number are anticipated within the next decade.

IoT devices are normally resource-constrained, limiting their ability to contribute toward advanced data analytics. However, nowadays with cloud infrastructures, the computations required to perform costly operations are hardly ever an issue. Cloud computing environments have gained an unprecedented spread and adoption in the last decade or so, aiming at deriving (near) real-time actionable insights from fully loaded ingestion pipelines. This is in part attributed to the fact that those environments are normally elastic in their offering for dynamic provisioning of data management resources on a per-need basis. Amazon Web Services (AWS) remains the most widely accepted competitor in the market and looks set

Fog Computing: Theory and Practice, First Edition.
Edited by Assad Abbas, Samee U. Khan, and Albert Y. Zomaya.

to remain that way for some time. For those aims to be achieved, production-grade continuous applications must be launched, which typically raises several obstacles. Most important, however, is the ability to guarantee the end-to-end reliability of the overall structure, which is achieved by being resilient to failures such as those most common in upstream components, which ensures delivering highly dependable and available actionable insights in real time. Furthermore, the architecture should guarantee the correctness in handling late and out-of-order data, which is a fact in real-life scenarios.

Cloud computing environments offer a great ability to process highly trafficked loads of data. However, IoT applications have binding prerequisites. The vast majority of them are time-sensitive and require practically instant responsiveness while, in the meantime, quality of service (QoS), security, privacy, and location-awareness are well preserved [1]. With an expansive number of IoT devices continuously sending huge amounts of data, two-tier architectures fall short in meeting desired requirements [2].

There are several challenges that render solo-cloud deployments insufficient. From those, we focus on the case where there is an oscillation in data arrival rates, which is sometimes characterized as ephemeral, whereas in other circumstances it can be persistent and severe. Current cloud-based solutions highly depend either on elastic provisioning of resources on-the-fly or aggressively trading-off accuracy for latency by early discarding of some data or (worse) some processing stages. While those ad-hoc and glue-code solutions constitute conceptually appealing approaches in specific scenarios, they are undesirable in systems that are resistant to approximations and anticipate exact answers that, if not provided, can cause the system to become untrustworthy, thus requiring extra care when analysis of such data is essential. This does not detract from the values obtained by the two-layered cloud-edge architectures but rather complementing them in a manner that ensures reliability for scenarios that seek exact results on-the-fly. Scenarios where this applies are innumerable, e.g. smart traffic lights (STL), smart connected vehicles (SCV), and smart buildings to mention just a few. However, throughout our discussions we place due importance on both cloud-only and cloud-fog architectures. All that said, we next present an alternative in this chapter.

In the relevant literature, several works are proposed for facing up to the challenges introduced by the two-tiered architecture. Some of them merely depend on integrating lightweight sensors with the cloud, to counter common cloud issues such as latency, the capacity to support intermittent recurrent events, and the absence of flexibility when various remote sensors transmit information at the same time [3, 4]. Another solution focuses on increasing the number of layers in the architecture to push part of the processing load uphill to intermediate layers [5], subsequently reducing information traffic, reaction time, and the response location-awareness.

There is significant overload caused by spikes in network traffic, when data arrival rates grow at an unprecedented and mostly unpredictable pace. To meet this challenge, fog computing comes into play, which is simply treated as a programming and communication paradigm that conveys the cloud assets closer to the IoT devices. Stated another way, it fills in as an interface that associates cloud with IoT devices in a manner that improves their cooperation fundamentally by keeping the advantages of the two universes by broadening the application field of cloud computing and expanding the asset accessibility in IoT settings.

Fog importance stems from the fact that there are situations in which a little computation is performed just-near the edge, thus lightening the burden on the shoulders of network hops and the cloud. Fog computing is somewhat novel but is receiving elevated attention among researches and becoming a widely discussed topic. For example, fog nodes can handle some nonsubstantial computing loads to act as an intermediate caching system that stores some intermediate computational results (e.g. preserving the state of an online aggregation framework), thus preventing cold-start scenarios (those scenarios that mandates recomputation in case of state loss in stateful operations), so that a stream processing engine can resume from where it left off before the (non-)intentional continuous query restarts. However, one hindering challenge could be the fact that fog is still in its infancy. Nevertheless, there are a considerable number of works that focus on different aspects for fog optimization, ranging from communication requirements to security to privacy and to responsiveness, to mention just a few. Current efforts are mostly following layer stack-up trends. Older systems of the technology are maintained while a person or group is trying to adapt to new technologies. This serves as a new jumping-off point for incorporating newer approaches. As such, we here posit the importance of incorporating fog computing as a core player in the current two-tiered architecture in order to reap the benefits of fog. To better corroborate our conclusions, we justify the incorporation of six service layers that span all three worlds (fog, cloud, and edge). With this setting in mind, some may be misled by intuitively concluding that sudden data arrival spikes are no longer a problem when seeking reliable answers. More rigorously, by this division we aim at breaking the architecture into its granular constituent parts to simplify their comprehension.

In this chapter, we are flipping the switch on significant architectural changes that, most importantly, incorporate fog as a central player in a novel three-tiered architecture. We demystify the explanation of the three-tiered architecture that we dub as cloud-fog-edge by dividing into major sections the pattern defined herein for the stacked-up architecture. First, we establish some basic definitions before getting under the hood. Thereafter, we start detailing mechanisms required for fully building a convenient self-organized fog between cloud and edge devices that hides management details so to relieve the user from reasoning about the underlying technical details and focuses on analytics instead of the heavy lifting of resource management. Most this chapter is devoted to summarizing the underlying components that constitute our three-layered architecture and how each contributes toward a fast fault-tolerant low-latency IoT application's operation. We herein point out that several alternative designs and proposals for related architectural models are found in the relevant literature that differ significantly in the way they present the interplay between the three worlds. We specifically refer the interested reader to a recent survey by [6]. The downside, however, is that the authors focus mainly on comparing existing models without introducing a novel counterpart. A similar trend also follows in [7].

This chapter is organized as follows. We first start by defining major aspects of core players in the architecture, such as fog, IoT, and cloud. In what follows, we discuss the challenging requirements of fog when applied to IoT application domains. In a later section, we draw a taxonomy of fog computing for IoT. In the last section, we discuss challenges and recommended research frontiers. Finally, we close the chapter with some concluding remarks.

6.2 Fog Computing for IoT: Definition and Requirements

In this section, first, we give a depiction of what fog computing is, what it performs, and the upgrades it conceivably presents in IoT application areas and deployment environments. Then, we clarify the inspirations that lead to the presentation of the fog computing layer atop the stack and the inappropriateness of a two-tier architecture composed only by cloud computing and IoT. Finally, we propose an original reference architecture model with the aim of clarifying its structure and the interactions among all the elements, providing also a description of its components.

6.2.1 Definitions

Fog computing alludes to a distributed computing paradigm that off-loads calculation, for some parts, close to the edge nodes of the network system with the purpose of lightening their computational burden and thus speed up their responses. We stress the fact that fog is becoming increasingly focal in enriching the responsiveness of computations by intermediating the stacked-up architecture between the two IoT and cloud universes, bringing substantial computing power near the edge and helping time-intensive applications in their front-stage loads that may need instant reactions, which normally cannot wait for the whole cycle, from sending data upstream to cloud to getting results back. Fog can be considered as a significant extension of the cloud computing concept, capable of providing virtualized computation and storage resources and services with the essential difference of the distance from utilizing end-points. While cloud exploits virtualization to provide a global view of all available resources and consists of mostly homogeneous physical resources, far from users and devices, fog tends to exploit heterogeneous resources that are geographically distributed (often with the addition of mobility support) and situated in proximity of data sources and targeted devices.

Moreover, fog is based on a large-scale sensor network to monitor the environment and is composed of extremely heterogeneous nodes that can be deployed in different locations and must cooperate and combine services across domains. In fact, fog communicates, at the same time, with a wide range of nodes at different level of the stack, from constrained devices (with very restricted resources) to the cloud (which has virtually infinite resources). Many applications require, at the same time, both fog localization and cloud centralization: fog to perform real-time processes and actions, and cloud to store long-haul information, and thereby perform long-haul analytics.

A primary idea emerging from existing fog solutions in the literature is to deploy a common platform supporting a wide range of different applications, and the same support platform with multitenancy features, can be used also by a multiplicity of client organizations that anyway should perceive their resources as dedicated, without mutual interference [8]. Figure 6.1 shows a high-level architecture

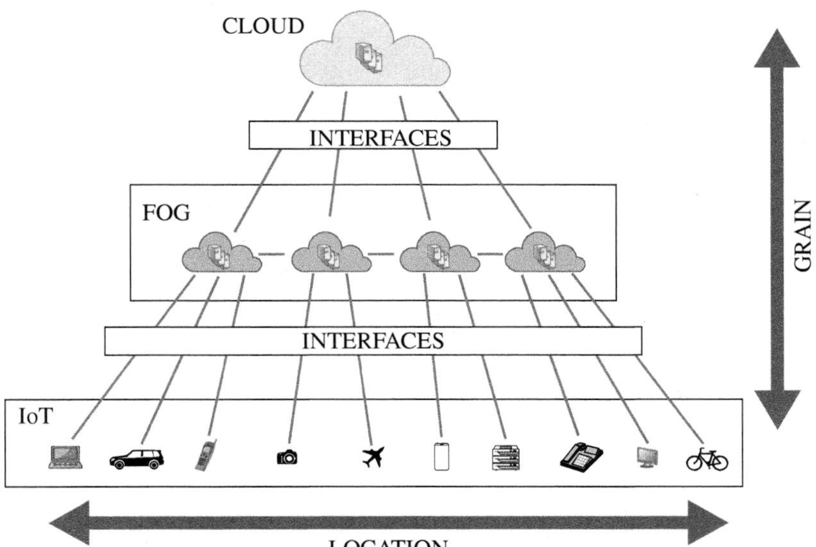

Figure 6.1 Cloud-fog-IoT architecture. (*See color plate section for the color representation of this figure*)

that summarizes the above vision by positioning the IoT, cloud, and fog computing layers.

Fog interacts with the other layers through interfaces with different communication specifications, simplifying the connection among the different technologies. Due to cloud technology maturity and well-defined standardizations, cloud-side interfaces are more defined, and it is currently easier to make cloud service platforms interact with outside users. Cloud interfaces allow connecting the cloud with any device, anywhere, as a virtually unique huge component, and independently of where cloud services are located. On the contrary, IoT-side interfaces and, even more, fog ones are more various and heterogeneous, and much work should be done to homogenize the different approaches and implementations that are emerging.

6.2.2 Motivations

The incorporation of cloud in IoT applications is twofold, bringing significant preferences to the suppliers and end clients on one hand, yet bringing new unsuitableness in the integration with ubiquitous services on the other hand. In spite of its generous offering when it comes to the resource-rich assets that cloud may bring to an IoT setting, excessive misuse of cloud assets by ubiquitous IoT devices may present a few technical difficulties, such as network latency, traffic, and communication overhead, to mention a few. In particular, "dumbly" connecting a bunch of sensors legitimately to the cloud computing framework is resource-demanding, which are not designed, implemented, and deployed for high-frequency remote interactions, e.g. in the extreme case of one cloud invocation per each sensor duty cycle. Ubiquitous devices gather enormous quantity of data during normal operations, a condition that is worsened in crowded places during peak-load conditions or in future applications, because the purpose of IoT systems is to sense as much as possible and, thus, increasingly collecting more data, with the end result of exceeding the bandwidth capacity of the networks. In addition, IoT sensors usually use a high sampling rate, in particular for critical applications, to better monitor and act instantly, which typically generates a huge amount of data that need to be managed.

Interfacing a horde of sensors straightforwardly with cloud is incredibly demanding and potentially challenges the capacity of cloud resources. The result is a continuous iteration of cloud, which stays occupied per every sensor duty cycle, rendering cloud's "scale per sensor" property inefficient. To this end, we argue that an architecture that is based on a direct communication between cloud and IoT devices is infeasible. A direct rationale for this judgment is that a network's bandwidth is no longer able to support excessive data loads. In addition, the planned optimizations of networking and intercommunication capabilities are not promising for keeping up with the unprecedented avalanches of data, which are growing at a continuous rate and coming downstream at a fast pace. Such huge amounts of real-time data challenges the processing capabilities of any cloud deployment and potentially causes performance to take a severe dive in case the data are fed directly to the cloud layer, thus rendering the whole processing power a single point of failure and counteracting the benefits of parallelization.

Future web applications, which are arising from the advancement of IoT, are large-scale, latency-sensitive, and are never again meant to work in seclusion, but instead will potentially share foundation and inter-networking assets. These applications have stringent new requirements, such as mobility support, large-scale geographic distribution, location awareness, and low [9]. As a general idea, it begins to be broadly perceived that an architectural model that is just founded based on a direct interconnection between IoT devices and the cloud is unseemly for some IoT application situations [8, 9]. Cloud environments are too globally available and "far" from IoTs devices, which hinders meeting the IoT application's requirements and critical issues.

A distributed intelligent intermediate layer that adds extra functionalities to the system is required by, e.g. pushing some processing workloads into the data sources themselves, thus off-loading some heavy lifting that otherwise may cause congestion if sent abroad to the cloud. For this to happen, a support infrastructure is needed for the system to work properly and efficiently, thus providing QoS and capitalizing on the great potential of cloud. For a comprehensive survey that covers relevant literature methods that can be employed for off-loading time-sensitive tasks just-near the edge, in fog, we refer the interested reader to [10]. Authors basically discuss various methods and algorithms for off-loading, mostly by taking a utilitarian perspective that depicts the relevance of each method in achieving a streamlined off-loading conduct.

In the relevant literature, some works propose moving part of resources toward network edge to overcome limitations of cloud computing. In this chapter, we recapitulate main fog-related research directions including cloudlet, edge computing, and follow-me cloud (FMC).

The term *cloudlet* was first coined by [11] and describes small clouds. In simple terms, it is composed of a cluster of multicore computers with gigabit internal connectivity near endpoints that aim at bringing the computing power of cloud data centers closer to end devices in order to satisfy real-time and location-awareness requirements. An important distinction with what existed in traditional data centers is that a cloudlet has only soft state. In simpler terms, this means that management burden is kept considerably low and, once configured, a cloudlet can dynamically self-provision from remote data center [12]. Satyanarayanan et al. [13] highlights that usually cloudlet depends on a three-level stacking (mobile devices, cloudlet, and cloud) and is totally transparent under typical conditions, giving portable clients the deception that they are directly communicating with the cloud. Cloudlet stands as a possible realization of the resource-rich nodes, while those components deployed in the cloudlet are able to respond to requests coming from the resource-poor nodes in a timely manner [11].

Edge computing aims to move applications, data and services from cloud toward the edge of the network. Firdhous et al. [14] has summarized different advantages of edge computing, including a significant reduction in data movement across the network resulting in reduced congestion, cost and latency, elimination of bottlenecks, improved security of encrypted data (as it stays closer to the end-user, reducing exposure to hostile elements) and improved scalability arising from virtualized systems. Davy et al. [15] presents the idea of edge-as-a-service (EaaS), an idea that decouples the strict ownership relationship between network operators and their access network infrastructure that, through the development of a new novel network interface, allows virtual networks and functions to be requested on demand to support the delivery of more adaptive services.

FMC [16] is a technology developed to help novel mobile cloud processing applications, by giving both the capacity to move network end-points and adaptively relocating network services relying upon client's location, so as to ensure sufficient execution throughput and to have a fine-grained control over network resources. [16] analyses the scalability properties of an FMC-based system and proposes a role separation strategy based on distribution of control plane functions, which enable system's scale-out. [17] proposes a framework that aims at smoothing migration of an ongoing IP service between a data center and user equipment of a 3GPP mobile network to another optimal data center with no service disruption. [18] proposes and evaluates an implementation of FMC based on OpenFlow and underlines that services follow the user throughout his movements, always provided from data center locations that are optimal for the current locations of the users and the current conditions of the network. In a similar vein, [16] introduces an analytical model for FMC that provides the performance related to the user experience and to the cloud/mobile operator, stressing the importance of attention when triggering service migration.

In addition to these activities, some standardization efforts are geared toward improving interoperability and thus fostering fog and edge computing ecosystems. Recent initiatives include open edge computing [19], open fog computing [20], and mobile edge computing (MEC) [21].

The Open Edge Computing Consortium is a joint activity among industries and the scholarly community that calls for driving the advancement of an ecosystem around edge computing by giving open and internationally acknowledged standardized instruments, reference implementations, and live demonstrations. To that end, this community leverages cloudlets and provides a testbed environment for the deployment of cloudlet applications.

The OpenFog Computing Consortium aims mainly at defining standards in order to improve the interoperability of IoT applications. They indicate that cloud-only architectural approaches is not able to keep up with the fast data arrival rates and volume requirements of the IoT applications. Given this observation, efforts are geared toward a novel architectural view that emphasizes information processing and intelligence at logical edge. The upcoming reference architecture is a first step in creating standards for fog computing [22]. Being a multilayer architecture, where some are vertical while others are horizontal, it thus covers all aspects and requirements for achieving a multivendor interoperable fog computing ecosystem. The vertical layers, considering the cross-cutting properties, are as follows: (1) performance, including time critical computing, time-sensitive networking, network time protocols, etc.; (2) security, covering end-to-end security, and data integrity; (3) manageability, with remote access services (RAS), DevOps, Orchestration, etc.; (4) data analytics and control, containing machine learning, rules engines, and so on; and (5) IT business and cross fog applications, providing characteristic to properly operate applications at any level of the fog. The horizontal view, which aims at satisfying different stakeholder requirements is composed of (1) node view, including the protocol abstraction layer and sensors, actuators and control; (2) system view, providing support for the infrastructure and hardware virtualization; and (3) software view, providing services and supporting the deployment of applications.

MEC is a reference design and a standardization exertion by the European Telecommunication Standards Institute (ETSI). MEC gives a service environment and cloud-computing capabilities at the edge of the portable system, within the radio access network (RAN). Therefore, MEC environment is characterized by low latency, proximity, highly efficient network operation and service delivery, real-time insight into radio network information, and location awareness. Its key element is the MEC server, which is integrated at the RAN element and provides computing resources, storage capacity, connectivity, and access to user traffic and radio and network information. The MEC server's architecture comprises a facilitating framework and an application platform. The application platform gives the ability to facilitate applications and is composed of an application's virtualization administrator and application-platform services. MEC applications from third parties are deployed and executed within virtual machines (VMs) and managed by their related application manager. The application-platform services provide a set of middleware application services and infrastructure services to the hosted applications. Thus, ETSI is working on a standardized environment to enable the efficient and seamless integration of such applications across multivendor MEC platforms. This also guarantees ensure serving vast majority of the mobile operator's customers.

Finally, the fog vision was conceived to address applications and services that do not fit well the paradigm of the cloud [8]. Fog computing is pushed between IoT and cloud, leveraging the best from both worlds in enabling IoTs applications to become established as future enabling technologies. Along the same lines, [23] emphasizes that, as indicated by IoT developing paradigm, everything will be seamlessly associated to structure a virtual continuum of interconnected and addressable items in a global networking system. The outcome will be a strong hidden structure on which clients may create novel applications helpful for the whole community. Fog computing is considered a driver for enterprise/industrial-based IoTs that brings connections to the real world in an unprecedented way and aims at interfacing with new business models introduced by IoTs, rethinking about how to create and capture value.

6.2.3 Fog Computing Requirements When Applied to Challenging IoTs Application Domains

IoTs systems raise requirements that burden fog and cloud computing to accomplish the right activity and fulfill the client's expectations. The following subsections discuss these highlights and elucidate their definitions.

6.2.3.1 Scalability

Scalability is a core requirement, not only for big data management, but also a proper geo-distribution of devices. To state the obvious, [8] proposes to add geo-distributed property as a further data dimension in big data analysis, in order to manage the information distributed nature as a coherent whole. In this scenario, fog plays an important role, thanks to its proximity to the edge and, thus provides information's location-awareness.

By considering scalability referred to big data processes, we feature the characteristic of the framework to scale, depending on the amount of data and, if necessary, being able to manage great amount of data. On the other side, regarding "scalability" of device geo-distribution, we underline the ability of fog computing to manage a large number of nodes in a highly distributed system.

Big data scalability is a fundamental necessity for IoT applications, where a developing number of devices must be interconnected. Geo-distributed scalability is a demand that underlines the paradigm of fog computing to have a capacity of overseeing distributed services and applications, even profoundly distributed frameworks, which falls in stark contrast with the more centralized cloud settings. In exceptionally conveyed systems, fog is dealing with a huge number of nodes across the board in geographic zones, and nodes can likewise be spread out with different degrees of density on the ground. Fog computing is thus dealing with various sorts of topology and distributed configurations systems and have the capacity to scale and adjust so as to meet the demands for every scenario.

6.2.3.2 Interoperability

The IoT is a very heterogeneous setting that is normally found in real-life situations, in light of a wide scope of various devices that gather heterogeneous data from the encompassing geography. Sensors differs when it comes to range coverage, from short- to long-distance coverage. Bonomi et al. [8] lists various heterogeneity inside the fog: (1) fog nodes extend from high-end servers, edge switches, access points, set-top boxes and, even, end devices such as vehicles and cell phones; (2) the different hardware platform has varying levels of RAM memory, secondary storage, and real estate to support new functionalities; (3) the platforms run various kinds of operating systems and software applications resulting in a wide variety of hardware and software capabilities; (4) the fog network infrastructure is also heterogeneous, ranging from high-speed links, connecting enterprise data centers and the core, to multiple wireless access technologies, toward the edge. In addition, inside fog computing, services must be unified on the grounds that they require the participation of various suppliers [9]. Fog computing is a very virtualized setting that needs heterogeneous devices and their running services to be unified under one umbrella in a homogeneous way.

In complex settings, heterogeneity can influence technical interoperability, in addition to semantic interoperability. Technical interoperability concerns communication norms, components executions, or parts interfaces with various information formats or diverse media sorts of data streams. Whereas semantic interoperability is more concerned with the information inside data interleaved and the likelihood that two components comprehend and share similar data in an unexpected way. A standard method to portray and exchange data, together with an abstraction layer that anonymizes physical diversities among components are thus required to make interoperability possible. Diallo et al. [24] explains under which conditions systems are interoperable, proving definitions and classifications and many approaches to address interoperability at different levels.

Fog computing is there for empowering interoperability, so as to make an exceptional information stream to be later handled by sensing and information analytics parts or to host conventional application programming interfaces (APIs) that can be utilized by various applications, without the costly need to move calculations to the cloud layer.

6.2.3.3 Real-Time Responsiveness

Real-time responsiveness is a principle empowering agent for IoT applications and their organization in real-life situations. Fog computing is vital to accomplish low-latency prerequisite in cases where cloud-IoT collaborations are not able to achieve the target latency for several reasons including the distance. (1) IoT and cloud are, in practice, geometrically distant and information requires considerable time in the loop, arriving as an input to cloud and thereafter returning as results (final or intermediate) to IoT devices. Fog is a promising field that alleviates the overhead costs caused by long-distance traveling of information; instead it is clear that performing computations near the edge costs less than sending data all the way downstream to cloud nodes. For some jobs, which do not require high computational power or are less demanding, fog promotes instant computation. (2) Real-time interactions loosely mean processing the unbounded streams of fresh data that arrive continuously in the cloud. Moments where data arrival rates exceed the processing capacities of cloud resources, in addition to Internetwork communication overheads, are not unheard of. Such conditions normally challenge the capacities of fully resource-loaded cloud deployments, where neither reactive nor proactive solutions make a difference. The promising (near) real-time operation of cloud environments is hindered by such facts. To top that off, in a highly dynamic and real-time scenario, such as those in Industry 4.0 (I4.0) or smart cities, data from IoT is fed very quickly to cloud, and because different chunks follow different networking routes, so that it arrives, sometimes, out of order, thus negatively affecting the overall accuracy. Fog diminishes this by sensing information, processing it, and acting in real-time using data that reflect instantly the situation. (3) Sensors accumulate a tremendous amount of continuously arriving data that if sent to the cloud potentially causes system congestion and consequently causing system's performance to hit a wall. On the contrary, fog acts as a front-stage that locates data traffic in a defined space surrounding sensors and preprocesses information before uploading data to the cloud, with lower focused loads and reduced core network load.

6.2.3.4 Data Quality

Data quality is a pertinent demand in real-life scenarios, making it essential for high-performing systems' operations. Also, this element essentially performs initial data filtering so as to early discard unnecessary duplicate and noisy loads, thus improving significantly the overall system quality and performance.

Data quality support is provided by the fog layer with the aim of early discarding useless data, aiming at relieving the burden on subsequent computational stages and consequently decreasing network data traffic by confining traffic near the edge, and reducing the amount of data pushed toward the cloud. Data quality depends on the association of various strategies. The mix of data filtering, aggregation, standardization and analytics, big data and small data analysis is fundamental to understand the surrounding environment, thus performing proactive maintenance and anomaly detection in real time, among a huge amount of data gathered from IoT sensors. In a ubiquitous environment, finding noisy data is challenging, especially in nonstationary settings where data is arriving fast and may thus challenge a front-stage system's resources.

6.2.3.5 Security and Privacy

An essential challenge in ubiquitous settings and fog computing is to harmonize security and reliability of systems, citizens' privacy concerns and personal data control, with the possibility to access data to provide better services. Particularly, with multitenancy support, fog offers policies to specify security, isolation, and privacy management that are required for different applications.

Fog computing is utilized in real applications that appear in critical settings, so reliability and safety are basic requirements. Moreover, it must consider that actuator's operations may be irreversible. Subsequently, the presence of unforeseen conduct, even due to bugs in applications, must be minimized with precautionary measures. Security is a key issue that must be solved to help industrial organizations and it concerns the entire system's architecture, from IoT devices reaching down to cloud. We need to provide important features such as (1) confidentiality, ensuring that the data arrives to the target spot, thus counteracting divulgence of data to unapproved objects with access limitations; (2) integrity, detecting, and preventing unauthorized alteration of the system by steps that control the preserving of consistency, accuracy, and trustworthiness of data over its entire life-cycle; (3) availability, ensuring that services are available when requested by authorized users, and performing repairs, if necessary, to maintain a correct functioning of the system.

A rich arrangement of security features that empowers essential security for every condition for the entire framework is thus required to avoid implementing security mechanisms on a node-by-node basis. Many controls are needed at all levels, including network and communications, from the physical and computational points of view. In fact, intelligence, data processing, analysis, and other computing workloads move toward the edge but, on the downside, many devices will be located in low-security locations, thus protecting devices, and their data becomes a big challenge.

Distributed and internetworked security arrangements are required to ensure complete and superior intelligence and for responsiveness reactions with automated decisions based on M2M communications and M2M security control without human interventions.

What is more, privacy is an undeniably critical issue that is growing in importance with ubiquitous and pervasive settings, where clients are aware concerning the privacy of their private information. Storing encrypted sensitive data in traditional clouds for privacy is not a suitable option as it causes many processing problems when applications have to access these data. In fog computing, personal information is kept in the system for better protection of privacy. It is imperative to characterize the responsibility for information inside the fog, since applications must utilize o information that they have access to [1]. In particular, we must consider the geographic diversity of information and certain data. For example, sensitive military or government data cannot be sent outside of certain geometrical areas. Fog can anonymize and aggregate user data and is thus useful for localizing intelligence and preventing the discovery of protected data. Anyway, it is necessary to introduce additional ways to protect data privacy and thereby to incentive the utilization of fog computing in privacy-critical contexts.

6.2.3.6 Location-Awareness

In dynamic real-world scenario, such as IoT applications, the location-awareness is the property of fog computing, due to its proximity to the edge, to own a widespread knowledge of its subnetwork and to comprehend the outer setting in which it is submerged. Fog improves adaptability because of its behavior adjustability in response to different events, where it adapts itself to better suit certain circumstances, assisted by the awareness of the context.

6.2.3.7 Mobility

Extending the concept of availability and trying to satisfy novel IoT application requirements, fog adapts itself in accordance with geographical distribution of its devices, thus providing mobility support. MIoT (Mobile Internet of Things) is proving itself to be a challenge for distributed supports [25, 26]. The ubiquity of mobile devices raises the need to present mobility support in fog computing, which allows sensing information and reacting while they are moving around the environment. Fog computing, in order to be effective, even with systems that have mobility as a peculiarity, must adjust to oversee high mobility devices. In addition, fog computing supports the likelihood that devices can move between fog nodes without causing issues that may bring the system into halts.

6.2.4 IoT Case Studies

At present, there are diverse IoT scenarios that are broadly promoted in the relevant literature to inspire the distinctive prerequisites of these frameworks and fog computing settings. Likewise, numerous researches are dealing with those situations so as to propose novel arrangements hybridizing the three worlds – IoT, fog and cloud – to address the demands [8, 27]. We herein provide a brief description of most important case studies. Those scenarios are used throughout the rest of this chapter to present how different approaches and solutions are applied in fog-enabled environments.

Smart traffic light (STL). STLs focus on better handling of traffic congestion in metropolitan cities. These frameworks depend on camcorders and sensors distributed along the streets to detect vehicles and components in streets, distinguishing the nearness of bikers, vehicles, or ambulances. To decrease errors, these frameworks can be fine-tuned to sense when traffic lights must be turned green or red depending on vehicles congestion in a single direction. Similarly, when an ambulance with blazing lights is distinguished, the STLs change road lights to open paths for the emergency vehicle. In this situation, traffic lights are fog devices.

Smart connected vehicle (SCV). SCVs are systems located inside a vehicle controlling every sensor and actuator in the vehicle, such as tires pressure, temperature inside the car, and the street lane in which the car is located. All the information collected by different sensors and sent to the closest fog node, which is usually located in the vehicle, so that all information can be quickly processed, and a real-time response can be given to any dangerous situation, e.g. stopping the car if a puncture is detected in a tire. In addition, different information can be exchanged between vehicles (vehicle-to-vehicle [V2V] communication), with the road infrastructure (through the roadside units [RSUs]) or with the Internet at large (through Wi-Fi, 3G, etc.).

Wind farm. These are systems that aim at improving the capture of wind power as well as preserving the wind tower structure under adverse condition. Diverse sensors to distinguish the turbine speed, the produced power, or the weather conditions are essential. These data can be given to a nearby fog node situated in every turbine to tune it in order to build the effectiveness and to decrease the probabilities of harm because of wind conditions. Furthermore, wind farms may comprise several individual turbines that must be facilitated to achieve the highest possible efficiency. The optimization of a single turbine can likewise lessen the effectiveness of different turbines in a row at the back of a farm.

Smart grid frameworks are promoted to counteract the waste of electrical energy. Those frameworks analyze energy requests and evaluate the accessibility and the cost to adaptively switch to green powers such as sun and wind. For this to occur, distinctive fog nodes are deployed on system edge devices hosting software responsible for balancing the equations.

Smart building systems are one of the most demanding IoT applications. In this scenario, different sensors are deployed throughout a house or a building to get information about different parameters, such as temperature, humidity, light, or levels of various gases. In addition, a fog node could be deployed in-house for collecting and combining all that information in order to react to different situations (e.g. turning the air conditioning on if the temperature is too high or activating a fan depending on gas level). With fog computing applied in those systems, they can better control the waste of energy and water in order to execute actions to better conserve them.

For an overwhelming review that better sheds light on more case studies, we refer the interested reader to a recent survey by [28], which focuses mainly on smart city scenarios and compares more than 30 related research efforts. For another six more scenarios, we point to a recent survey by [29].

6.3 Fog Computing: Architectural Model

This section explains an architectural framework dubbed as cloud-fog-IoT for interwoven application scenarios (see Figure 6.2). It depicts a high-level view of the composing essential elements and their associations. These elements form the ground for our taxonomy of IoT fog computing that we present hereafter. They have been identified considering both the different solutions and approaches surveyed, and the requirements presented in Section 6.2.3. First, the architecture is divided into three areas: cloud, fog, and IoT. These areas mirror the diverse types of nodes that could normally execute activities and tasks of components comprised within those areas. A component can be completely performed by a specific kind of nodes or by diverse nodes relying on the granularity of tasks (for instance, the IoT layer may contain some activities done by fog nodes and others performed by IoT devices). Second, the architecture consists of six layers that span the three worlds (fog, cloud, and IoT). For example, the communication layer must act on improving the interconnection between IoT devices and fog nodes, among fog nodes themselves, and between fog nodes and cloud environments. We expand the explanation of those layers as follows.

6.3.1 Communication

The communication layer is in charge of the communication among the constituent nodes of the network. It contains different techniques for a proper communication between those nodes, including standardization mechanisms for facilitating the exchange of information between different nodes of the network or between different subsystems of the IoT application. These techniques directly address the infrastructure interoperability requirement (see Section 6.2.3.2 for more details). Furthermore, IoT applications are typically described by a high mobility of a portion of its IoT devices. To this end, this layer should contain methods permitting the relocation of a device from a subnetwork to a different one without corrupting a system's normal operation. Meanwhile, this layer is

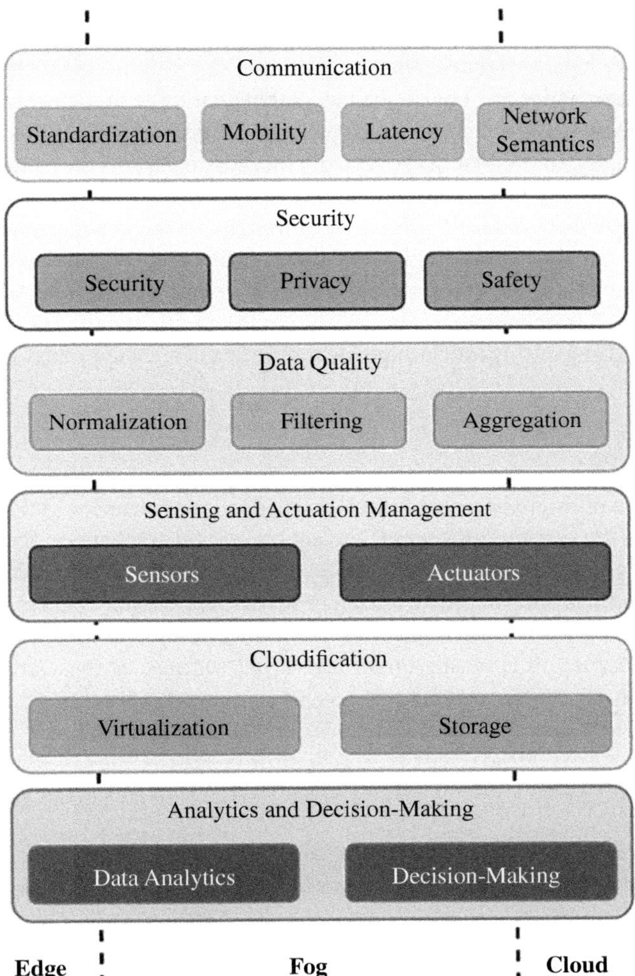

Figure 6.2 Our proposed architecture for cloud-fog-IoT integration.

significant for accomplishing real-time responsiveness, which can be obstructed by the inefficiency of communication protocols. In this manner, it additionally incorporates diverse procedures for decreasing communications latency. As a final perspective to consider, this layer needs to guarantee the dependability of communications, ensuring that data will not be lost in the system and that each node or subsystem expecting a particular information ingestion is getting it properly, subsequently improving information quality.

6.3.2 Security and Privacy

The security and privacy layer influences the entire design, since all interconnections, information, and activities must be done in ways that guarantee the safety of the system and its clients. This layer achieves security on three unique dimensions: security, privacy, and safety. In the first place, security focuses on various methods to guarantee the dependability, confidentiality and integrity of the interconnections between diverse nodes of the setting. Second, unreliable privacy-awareness strategies normally render the whole framework trustworthy. Hence, this layer incorporates access control component to submit data to just-approved clients. As a final consideration, IoT frameworks act in critical environments where safety matters a lot. Fog computing settings encourage the promotion of such approaches just near the edge.

6.3.3 Internet of Things

IoT applications consist of interconnected objects, embedded with sensors, gathering information from the surrounding world, and actuators, and acting upon the environment. The IoT layer is included by each one of those devices responsible for sensing the surroundings and adaptively acting in specific circumstances. This is implanted in IoT and fog worlds. Sensing straightforwardly influences the quality of generated data. Actuation is mostly important in IoT settings, as these are naturally required to interactively respond to flaws so as to avoid disasters. Fog computing can improve the actuation with convenient responses to data.

6.3.4 Data Quality

The data quality layer oversees the processing of all sensed and gathered data so as to build their quality and to diminish the size of data to be stored in the fog nodes or to be transmitted to the cloud environments. This layer comprises three different phases that are successively executed: data normalization, filtering, and aggregation. To start with, data normalization techniques get raw data sensed by heterogeneous devices in order to unify it in a common homogeneous language. Thereafter, since many of the data collected are useless (in the sense that they are not contributing to the final result) and only a part of them are valuable, different data-filtering techniques are employed to get just the contributing subsets, relevantly discarding worthless information, aiming at better exploiting scarce computational resources during subsequent steps. Finally, data aggregation is a process through which fog nodes take filtered data to construct a unique information stream and thereby improve its analysis. Fog should be able to follow aggregation rules in order to identify homogeneous information and, thus, produces a uniform data flow. Architectural components have long been utilized by various settings; they are rather essential for connecting heterogeneous sensors' data with computational assets spread out within the premises of the architecture remainder, thus improving significantly the quality, the scalability, and the general responsiveness of the framework.

6.3.5 Cloudification

The cloudification layer forms tiny cloud-analogy inside a fog. It helps to close the gap by bringing a defined set of cloud services just-near the edge of the targeted deployment environment or temporarily storing a portion of data and recurrently uploading them to a remote cloud setting, thus diminishing overwhelmed remote communications with the cloud. In order to achieve distributed clouds in the fog nodes, virtualization techniques must be provided so that different applications can be deployed in the same node. In addition, diverse services deployed on fog nodes should be composed and coordinated with the target of orchestrating challenging services for supporting higher-level business processes. Finally, the storage subcomponent is in charge of orchestrating distributed data, which are normally stored in diverse fog nodes, thus managing their processing when applications have to access these data and controlling their privacy. Different and non-irrelevant advantages are normal with that framework in light of the fact that each task is normally achieved in a location-aware setting with better analytics, responsiveness, and results. Moreover, it is not unheard of to confine traffic near IoT devices, without overwhelming traffic load on the shoulders of the network interconnections, improving also privacy, since users have their data in-proximity and can control them. Therefore, this layer provides important benefits in the parlance of responsiveness and user experience quality.

6.3.6 Analytics and Decision-Making

The analytics and decision-making segment are accountable for getting insights from the stored information to create distinctive analytics and to identify various situations. These analytics can lead to making specific decisions that systems should perform. In analogy with previously explained layers, this layer spans the three components of the architecture. In ubiquitous settings, where an immense number of sensors always accumulate data and forward it to fog, the mix of short and long-haul analytics, in fog and cloud respectively, normally accomplishes reactive and proactive decision-making, in a row.

The multifaceted nature of IoT environments prompts the need of a precise introductory conduct of the surroundings to characterize a substantial model to use in system. In fog computing, the prediction of input/output is favored by the proximity to end-users that allows a greater location-awareness of the environments where they execute. This enables further processes, performed by the system, to the external context and, thus improve every future task.

While short-lived analytics and lightweight processing with constrained data amounts are recommended to be handled by fog, cloud are normally performing long-haul and overwhelming asset-demanding activities in Big Data as underpinned in [30] for, specifically, health-care applications, these capabilities transfer over to any similar IoT scenario. Often within a Big Data analysis and processing cycle, significant resource assets are utilized for supporting data-intensive resource-demanding operations, which mostly cannot be satisfied by constrained devices with few available resources near IoT environment. However, big data processing algorithms that perform efficiently even with constrained-resources devices is not a situation that is unheard of. This specifically rationale our decision to expand this layer so that it covers the edge component. Ultimately, in case of resource incapability near the edge, cloud computing is taking the lead and offering great amount of computing power, relieving scalability, cost and performance issues. Moreover, those long-lived analysis cycles are normally utilized for performing orchestrated and proactive decisions processes.

6.4 Fog Computing for IoT: A Taxonomy

In this section, we propose a novel taxonomy for elucidating the principal constituent parts found normally in fog for IoT computing application scenarios. We specifically aim at explaining our categorization, which flows naturally from our previously sketched architectural model. Aiming at facilitating a comprehension of our taxonomy, we introduce a breed of over-the-shelf proposals that support specific elements or traits similar to those appearing in our architectural model. Thus, the proposed classification is populated with different approaches.

Instead of analyzing a complete list of solutions proposed for each characteristic of the taxonomy, we also detail different approaches, applying parts of them to different scenarios, as introduced in Section 6.2.4. We argue that this conceptualization serves as a comprehensive guideline, assisting IoT application designers in deploying effective cloud-fog-edge computing environments.

The presentation order of our taxonomy follows the same pattern as proposed in our logical architecture. Therefore, six parts of our taxonomy are drawn as one for each architectural element in a row. Section 6.4.1 presents the various interconnection aspects that should be considered for the interaction between IoT devices and fog and cloud nodes. Section 6.4.2 elucidates the various security and privacy measures required by IoT scenarios. Section 6.4.3 details interactions between fog nodes and IoT devices. Section 6.4.4 classifies how gathered data is processed. Section 6.4.5 spots the light on various aspects that need to be managed to enable off-loading some cloud loads to fog nodes. Finally, Section 6.4.6 sums up by explaining the taxonomy for data analytics and decision-making perspectives.

6.4.1 Communication

The communication layer offers four distinctive parts, providing help to the diverse attributes and prerequisites of IoT applications with respect to the communication between edge devices and fog and cloud nodes (see Figure 6.3). First, the standardization component, Section 6.4.1.1, explains various protocols that are used in standardizing the intercommunication aspects among the IoT devices ands fog and cloud nodes. Second, not only the communication between network elements should be standardized; rather, in critical IoT applications, the reliability of the transmitted information is essential for correct operation of the system. Section 6.4.1.2 details some of the main techniques to achieve the communication reliability. Likewise, the latency of transmitted information must also be considered in those applications requiring real-time communication. Section 6.4.1.3 analyzes some protocols and techniques focused on reducing communication latency. Finally, the mobility component is discussed in Section 6.4.1.4, which reviews some of the most important mechanism to reduce the mobility issues in IoT applications. For simplicity, Figure 6.3 sums up the proposed taxonomy elements. Some communication protocols actualize diverse

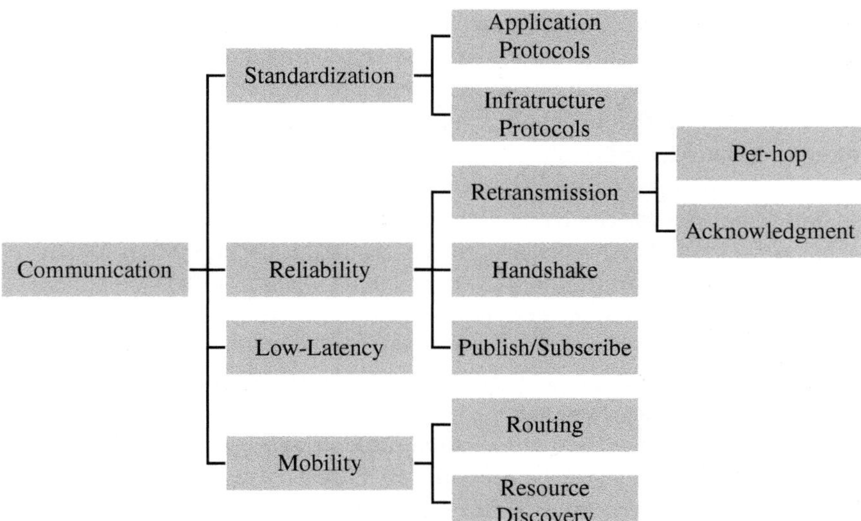

Figure 6.3 Taxonomy for the classification of the communication layer.

methods to support the above parts. However, each method has its pros and cons, making it appropriate for various conditions.

6.4.1.1 Standardization

A standout amongst the most basic focuses for the right coordination and intercommunication between IoT devices and IoT applications is the networking protocol utilized. Those protocols standardize the communication among sensors, actuators, and fog and cloud nodes, allowing programmers to accomplish proper levels of infrastructure interoperability in IoT settings. Various authors [31, 32] partition the infrastructure interoperability into two diverse arrangements of protocols: *application protocols* and *infrastructure protocols.* Application protocols are utilized to guarantee communicating messages among applications and their devices (Constrained Application Protocol [CoAP] [33], Message Queuing Telemetry Transport [MQTT] [34], Advanced Message Queuing Protocol (AMQP) [35], HTTP, Data Distribution Service [DDS] [36], ZigBee [37], Universal Plug and Play (UPnP) [38]). The latter are required to launch the underlying interconnection among various networks (RPL [39], 6LoWPAN [IPv6 low-power wireless personal area network] [40], IEEE 802.15.4, BLE [Bluetooth Low Energy] [41], LTE-A [Long-Term Evolution – Advanced] [42], Locator/ID Separation Protocol [LISP] [43]). Each system is to utilize a layered-up stack of protocols relevant to a set of requirements and the sensate traits of every application.

For instance, IEEE 802.15.4 offers a generally very secure wireless personal area network (WPAN), which focuses on low-cost, low-power usage, low-speed ubiquitous communication among devices, and with support to facilitate very large-scale management with a dense number of fixed endpoints [44]. On the top of the IEEE 802.15.4 standard, the ZigBee protocol takes advantage of the lower layers in order to facilitate the interoperability between wireless low-power devices, the optimization for time-critical environment, and the discoverability support for mobile devices. In addition, it is possible to extend the IEEE 802.15.4 standard, with 6LoWPAN network protocol that creates an adaptation layer that fits IPv6 packets, improving header compression with up to 80% compression rate, packet fragmentation, and thus direct end-to-end Internet integration [44].

The smart grid is composed of frameworks made out of an enormous number of distributed and heterogeneous devices spread out in various networks. Wang and Lu [45] represents Smart Grid communication network onto a hierarchical network composed by backbone network and millions of different local-area networks with ad-hoc nodes. Infrastructure interoperability is needed to allow devices and networks to cooperate in order to create a unique vision of the system state or to execute a common task. In smart building applications, in addition to the devices, the integration and communication among buildings allow managers to share the infrastructure and management costs, thus reducing the capital and operational expenses [46]. ZigBee is specifically utilized in smart building and smart grid applications because of its short-range and robustness under noise traits [47]. In [48], ZigBee is utilized in smart grid applications for associating sensors with smart meters, considering the overarching characteristic of its low bandwidth prerequisites and minimal-effort deployment.

ZigBee can also be used in vehicular applications, especially to perform short intra-vehicles communications among all the devices inside a vehicle and, for certain specific applications, to communicate outside the vehicle, as shown in [49], where it is used to improve the safety short-range system requirements. SCV acts in a heterogeneous scenario, with the involvement of a swarm of in-built sensors inside the vehicle that communicate, as well as many types of vehicles, externally seen as macro-endpoints, and access point stations that must cooperate among themselves. Several wireless technologies are used to communicate in the network environment [50]. Hence, infrastructure interoperability must be provided in order to allow V2V communications, access-point-to-vehicle and access-point to access-point.

Another well-known kind of WPAN connection is that of Bluetooth tech, characterized by an exceptionally low transmission range and a poor transmission rate, but on the upside with low power consumption. In IoT situations, BLE has gained momentum. This version extends the Bluetooth technology to face and support connections among constrained devices, optimizing lightweight coverage (about 100 m), latency (about 15 times lower than that of a classic Bluetooth), and energy requirements, with a transmission power range of 0.01–10 mW. BLE provides a good trade-off between energy requirements, communication range and flexibility, and a lower bit rate, combined with a low latency and reduced transmitting power, allows developers to transmit beacons beyond 100 m. For these reasons, BLE has several advantages that make it suitable for V2V applications, and thus could be successfully adopted in SCV systems [51].

Moreover, in heterogeneous environments, OSGi framework is widely adopted as a lightweight application container that defines dynamic managements of software components, allowing deployment of bundles that expose services' discoverability and accessibility. As examples, [52, 53] present platforms for SCV where the middleware is based on OSGi services and modules in order to improve the interaction between devices. In [54, 55] the proposal architecture, utilized in Smart Grid and Smart Building scenarios, capitalizes OSGi as a component platform, which enables to activate, deactivate, or deploy system modules easily, to provide a platform on which new components can be integrated in a plug-and-play fashion, keeping the design as flexible and technology independent as possible.

6.4.1.2 Reliability

Another imperative property of the communication protocols is that of the intercommunication reliability. This property guarantees the gathering of the information transmitted by diverse nodes of the setting. Presently, various systems can be utilized to guarantee the quality of the communications, e.g.: *retransmission, handshake,* and *multicasting*.

First, *retransmission* requires the acknowledgment of every packet and thereby retransmitting each lost packet. Several application protocols, like CoAP, MQTT, AMQP, DDS, concentrate on communication's dependability and are founded on retransmission schemes that are designed to handle the packet loss in lower layers. For example, the scheme per-hop retransmission (often called automatic repeat request [ARQ] at the mandatory access control [MAC] layer) tries to retransmit a packet several times before reaching a defined level before the packet is declared lost. Losses are on-the-spot realized and corrected, and even a few per-link retransmissions can substantially enhance end-to-end dependability [56, 57]. CoAP is based on UDP, an unreliable transport layer protocol, but it promotes the utilization of confirmable messages, that require an acknowledgment message, and non-confirmable messages, that does not need an acknowledgment [58].

Second, the *handshake* mechanism is designed so that two nodes or devices attempting to communicate can agree on parameters of the interconnection before data transmission. MQ Telemetry Transport (MQTT) and Advanced Message Queuing Protocol (AMQP) support three different layers of reliability that are used based on the domain-specific needs: (1) level 0, where a message is delivered at *most* once, with no acknowledgment of reception; (2) level 1, where every message is delivered at *least* once with a confirmation message; and (3) level 2, where the message is delivered *exactly* once and uses a four-way handshake mechanism.

Third, the *publish/subscribe* technique permits a device to publish some specific information. Other devices or nodes can be subscribed to that information. To state the obvious, each time the publisher posts new information, it is forwarded to the subscribers. Multiple nodes can subscribe to the same information; therefore, the information would be multicast to all of them. DDS uses multicasting for bringing excellent QoS and high reliability to its applications, with the support of various QoS approaches in connection with a wide scope of adjustable communication paradigms [59]: network scheduling policies (e.g. end-to-end network latency), timeliness policies (e.g. time-based filters to control data delivery rate), temporal aspects for specifying a rate at which recurrent data is refreshed (stated another way, timelines between data sampling), network transport priority policies, and other policies that influence how data is processed alongside the communication in relative to its reliability, urgency, importance, and durability [60].

In SVC and STL case studies, with huge increase in the number of connected vehicles, the number of sensors they incorporate, and with their unprecedented mobility, the support for low latency and unobstructed communication among sensors and fog nodes is crucial to guarantee a correct flow of applications [61]. DDS has been used as basis for building an intra-vehicle communication network. To that end, the vehicle was divided into six modules constituting the intra-vehicle network; the vehicle controller, inverter/motor controller, transmission, battery, brakes, and the driver interface and control panel. Fifty-three signals were shared among the different modules, some of them periodic and some sporadic. The test showed that this protocol improves the reliability and the QoS [60].

6.4.1.3 Low-Latency

As fog computing is implemented at the edge of the network, it is easier to provide low latency response, but it is also necessary to use the right protocol. Distinctive protocols can be utilized to improve the interplay between fog or cloud nodes or among devices and nodes. For instance, [59] utilizes MQTT publish/subscribe protocol to acknowledge continuous and low-latency streams of data, in a real-time setting dependent on fog computing capabilities toward cloud and IoT, utilizing fog layer in the meantime as broker and message interpreter: MQTT conveys data stream among fog and cloud and MQTT-SN, while the lightweight variant transports information from edge devices to the fog layer. CoAP [33] is yet another application protocol specifically utilized in IoT scenarios to provide low-latency cycles. In addition, [58] explains performance differences between MQTT and CoAP, focusing on the response delay variation in association with reliability and the QoS provided for communication: lower packet loss or bigger message size implies that MQTT outperforms CoAP; the opposite holds, reversing the conditions. So, deciding the protocol to use is essential depending on the type of application. In other terms, this means employing MQTT for reliable communications or for communication of large packets, and CoAP elsewhere for decreasing latency and thereby increasing the system's performance accordingly.

Finally, DDS [36] is a brokerless publish/subscribe protocol, recursively used for real-time M2M communication scenarios among resource-constrained devices [59]. Amel et al. [60] consider DDS as a favorable solution for real-time distributed industrial deployments and applies the protocol for improving the performance of vehicular application scenarios, evaluating the performance with tests that encapsulate hard real-time applications benchmark. Hakiri et al. [62] suggests utilizing DDS for enterprise-distributed real-time scenarios and embedded systems, those common in smart grid and smart building applications, for the efficient and predictable circulation of time-critical data.

In smart grid and smart building scenarios, most control capacities have strict latency prerequisites and need instant responses. Low-latency activities are fundamental to improve the framework's adaptability on the two sides of the electricity market. In smart grid scenarios, electricity markets expect to utilize demand-response instance pricing and charge clients for time-fluctuating costs that mirror the time-changing savings for power obtained at the discount level [63]. Wang and Lu [45] highlights the importance of low-latency actions in order to collect correlated data samples from local-area systems to enable a global power signal quality at a particular time instant. All samples must be collected by the phasor measurement unit (PMU) in a timely fashion to estimate the power signal quality for a certain instant and, depending on applications, the frequency of synchronization is usually 15–60 Hz, leading to delay requirements of tens of milliseconds for PMU data delivery. For modern power distribution automation, the IEDs (intelligent electronic devices) that are implanted in substations are sending their measurements to data collectors within 4 ms ranges in row, while intercommunications between data collectors and utility control centers need a network latency that falls roughly within the 8–12 ms range [32]; and for the standard communication protocol IEC 61850, maximum acceptable delay requirements vary from 3 ms; for fault isolation and protection purposes messages, to 500 ms; for less time-critical information exchange, such as monitoring and readings [45].

6.4.1.4 Mobility

One of the characteristics of IoT applications is the high mobility of some of their devices [9]. Currently, there are different protocols applying different techniques to support such mobility, such as *routing* and *resource discovery mechanism*.

Routing mechanisms oversee constructing and maintaining paths among remote nodes. Few protocols are responsible for building such routes, despite the need of some nodes for mobility peculiarities. For example, the LISP [43] indicates a design for decoupling host identity from its locational data in the present address scheme. This division is accomplished by supplanting the addresses utilized in the Internet with two separate name spaces: endpoint identifiers (EIDs), and routing locators (RLOCs). Isolating the host identity from its locations allows good improvements to its mobility, by enabling the applications to tie to a perpetual address, which is dubbed as the host's EID. Host location changes commonly amid an active connection. RPL is yet another routing protocol for constrained communications, utilizing insignificant routing necessities through structuring a strong topology over light connections and supporting straightforward and complex traffic models such as multipoint-to-point, point-to-multipoint, and point-to-point [31].

In STL and SCV case studies, mobility and routing support are essential needs for creating dependable applications ready for high rates of mobility of vehicles edges. Vehicles must be managed as macro-endpoints, allowing them to switch from one fog subnetwork to another. SCV must incorporate a routing mechanism for supporting vehicles' mobility externally but it does not require these mechanisms internally, since the swarm of sensors normally act statically inside the vehicle. RPL is adaptable for a multi-hop-to-infrastructure design, as a protocol that enables huge area coverage in real geometries, hosting connected vehicles with minimal deployment of infrastructure [64]. This protocol has practical applications in SCV and STL systems, and it is emerging as the reference Internet-related routing protocol for advanced metering infrastructure applications, since it can meet the requirements of a wide range of monitoring and control applications, such as building automation, smart grid, and environmental monitoring [65].

Resource discovery techniques strategies concentrate on recognizing adjacent nodes when a device relocates, so as to build up new communications. For instance, CoAP is able to discover nodes resources in a subnetwork, through URI that host a rundown of assets given by the server. On the other hand, MQTT does not offer out-of-the-box support resources discovery, and thus clients must understand the message design and associated topics to enable the communication. UPnP is a discovery protocol widely used in many application contexts that enables automatic devices' discovery in distributed environments. Fog solutions use UPnP+ [66], which is complementary to IoT applications. This version encapsulates light-tailed protocols and architectural parts (e.g. REST interface, JSON data format instead of XML) aimed at enhancing communication levels with resource-constrained devices. Moreover, Kim et al. [54] proposes an architectural view encompassing both smart building and smart grid utilizing UPnP for further detecting and adding new devices dynamically with no user intervention, unless the system wants additional information about the user's environment. Likewise, Seo et al. [67] propose using UPnP in vehicular applications, to allow external smart devices to communicate with an in-vehicle network, sharing data over a single network with the services provided by each device.

In other scenarios, like wind farm, smart grid, and smart building, the mobility of devices is not a priority in building the application, due to the static nature of the systems.

6.4.2 Security and Privacy Layer

The Security and Privacy layer encompasses three essential elements constituting parts of the IoT settings: those are safety, security, and privacy (see Figure 6.4). In the shown use cases, security and privacy essentially span the whole cycle from computational all the way down to physical aspects. First, few IoT systems are providing minimal safety policies for their users. Section 6.4.2.1 analyses few of the most significant safety mechanisms used in different fog computing–IoT applications. Second, security is a key element for every industrial development. Section 6.4.2.2 details the most important techniques and approaches in IoT systems. Finally, Section 6.4.2.3 presents the most important mechanism to preserve the privacy of the data. As detailed above, this is a vertical layer in the architecture because different kind security and privacy policies can be implemented throughout the data life cycle, from the gathering of the data to their storage in the fog nodes or the cloud environments. For the sake of clarity, Figure 6.4 sums up our taxonomy with all alternatives of modular constituting components.

6.4.2.1 Safety

Safety is fundamental for critical IoT systems. Safety is most often found in a lawful society and the corresponding business logic of IoT systems. That is, these systems must be designed to maximize the safety of any element, entity, good, or user of the system.

The most widely adopted safety practices are *activity coordination* for orchestrating actions with a concentration on maximizing the users' safety or even those of goods; *activity monitoring* for ensuring a streamlined and correct execution of actions; and, *action planning* for controlling the actions required in hazardous situations by either one of deterministic or stochastic designs [68].

Evaluating the application of these techniques in the different use cases, for instance, in STL, different applications can execute *coordinated activities* to construct green waves for assisting emergency vehicles in avoiding traffic jams or to reduce noise and fuel consumption [69]. SCV systems also rely on *action control* techniques for monitoring each and every operation, through acquisition of images or vehicles' movement patterns. More often, all users' actions are traced utilizing targeted surveillances.

As a final consideration, in wind farm scenarios, the framework must face fluctuating weather conditions, compare them with a predefined set of thresholds, and apply a lot of arranged activities, e.g. stopping the turbines in the event of a strong wind, for safety reasons. In [70], the authors survey diverse ways to deal with location vulnerability in wind control generation in the unit commitment issue, with fascinating results that demonstrate the presence of models that can adequately adjust expenses, revenue, and safety in a balanced fashion. What's more, as it is demonstrated in [68], the use of stochastic models for unit commitment, instead of deterministic models, can expand the penetration of wind power without any trade-off of safety.

6.4.2.2 Security

Security is a fundamental perspective that must be confronted by industrial settings. The security of IoT systems is usually supported by at least four basic pillars: the *confidentiality* of the information, ensuring the arrival of data to safe locations, thus preventing their circulation among unauthorized parties. *Data encryption*

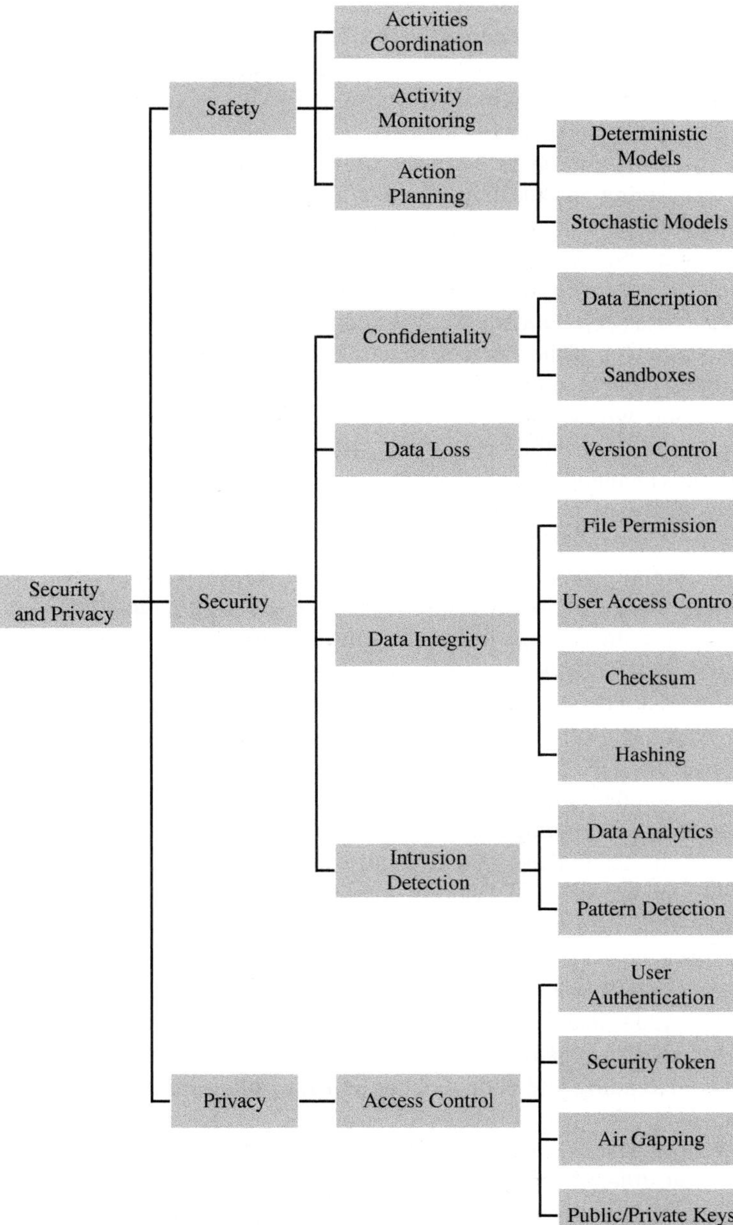

Figure 6.4 Taxonomy for the classification of the security and privacy layer.

and the use of *sandboxes* to isolate executions, data, and communications are standard methods of ensuring confidentiality [71, 72]. *Data loss*, can occur; information can be lost throughout circulation or in the different nodes of the fog environment. Normally, these situations are controlled by various protocols utilized for sending data to fog nodes or to cloud settings, through the channels of *version control,* and *configuration management* [73]. The *integrity* of the data must be ensured, discovering and disallowing nonauthorized dissemination of information throughout the whole cycle. Common ways to ensure integrity are *file permissions*, *user-access controls*, *checksum,* and *hashing* [74] methods. *Intrusion detection* allows a system to identify if an nonauthorized client is intending to access protected data [27]. *Data analytics* techniques are used to detect intrusions (observing the behavior of the system), to check for anomalies, and to discover fault recognitions. Finally, *pattern detection* techniques can be used to compare the users' behaviors with already known patterns, or with the prediction of the system's behavior. Notwithstanding, adapting intrusion detection mechanisms for each IoT setting is challenging as it requires domain-specific knowledge in addition to the technical aspects.

Some techniques are utilized for the shown case studies. For example, in smart grid and smart building, security levels needs are obviously high, because any potential breach could result in a blackout. In [45], the authors highlight that these attacks commonly come through intercommunication networks, from intruders with the motive and ability to perform malicious attacks, such as the following. (1) Denial-of-service (DoS) attacks can be targeted toward a variety of communication layers (applications, network/transport, MAC, physical) to degrade intercommunication performance and thus obstruct the normal operation of associated electronic devices. (2) Attacks can target integrity and confidentiality of data, trying to acquire/manipulate information. The great number of devices and providers connected to these systems require the implementation of different security policies to prevent and face any possible cyberattack. Those systems have to ensure that devices are protected against physical attacks, utilizing *user-access control* policies, and also that sensitive data will never be altered throughout their transmission life cycles, using *data encryption* and *sandboxes* techniques [32]. Additionally, fault-tolerant and integrity-check mechanisms are normally operated hand-in-hand with power systems for protecting data integrity and for thereby defending and anonymizing user's actions and their associated localizations.

In relation to protocols used in critical applications, the widely adopted solution is to use different protocols for different parts of the system. MQTT benefits in terms of security and privacy from encryption through *secure socket layer* (SSL)/*transport layer security* (TLS), but, at the same time, it has a weak authentication phase (e.g. short usernames and passwords) and inappropriate security/privacy policies (e.g. global namespaces that produce global topics).

Hence, MQTT in particularly is used to connect local elements in private networks, e.g. connections between a fog node and sensors in the subnetwork. AMQP extends the security and privacy of MQTT with *sandboxes* for the authentication phase and proxy servers to protect the servers with additional security levels (e.g. firewall protection). In addition, ADQP separates the message and the delivery information, providing metadata management and encrypted message. ADQP is normally used to connect public elements, e.g. connections between fog and cloud. Finally, CoAP enhances the security/privacy match using DTLS (datagram transport layer security) aimed at preventing eavesdropping, tampering, or message forgery toward integrity and confidentiality of exchanged messages.

6.4.2.3 Privacy

In fog computing, personal data are circulated throughout the fog's nervous system and not centralized in some components, which normally improves privacy. Controlling data access remains the top method for ensuring the privacy of personal data that must be protected from unauthorized parties. Few strategies such as *user authentication*, *security token,* or *air gapping* are normally utilized for increasing the privacy of data for privacy-sensitive applications [75]. Moreover, Stojmenovic and Wen [27] emphasize authentication at various stages yet it remains a security issue of high importance in fog settings. Every network device has an IP address and a hacker is able to tamper with a device and send false readings or imitate the IP addresses. To avoid this, a few authentication techniques, based on *public key infrastructures* or *key exchange*, are utilized.

In STL, e.g. privacy plays an important role and its concerns are related to the images acquisition of the vehicles approaching the intersection and the close-circuit cameras for vehicle presence detection and monitoring of traffic conditions, which also analyze patterns of vehicles' movements. Likewise, in SCV, the vehicles store information about the drivers' habits, their actions, and their driving patterns. Different user control techniques should be implemented to prevent unauthorized users to access that information.

In smart grid and smart building scenarios, privacy aspects mainly concern a mistaken spread of users' private data (e.g. monetary information and account balance) or information related to energy, from voltage/power readings to device running status to nonauthorized parties. This information is valuable and the system should seek to keep it safe and out of reach of malicious users. Ancillotti et al. [65] classifies the main smart grid vulnerabilities as the following: (1) device vulnerabilities, where attackers tamper with IEDs, which are normally utilized for monitoring electricity production/distribution, with probable not-to-be-underestimated consequences in the language of stealing data or destroying operations, or even aggravating by an unsafe WiFi communications between edge devices; (2) network vulnerabilities, where the adoption of architectures comprising open network is risky and open gates for routing modification attacks, DNS hacking, different DoS; (3) data vulnerabilities, where data is attacked aiming compromising the private data of customers, e.g. man-in-the-middle (MiM) attacks. Out-of-the-box smart grid applications normally ensure stringent access control with minimized levels of abilities performed by every node, which are usually resource-constrained nodes. As fog computing joined the play, smart grid nodes are now off-loading access control to the fog layer, which provides more computational resources and thus boosts more accurate and faster analysis near the edge.

6.4.3 Internet of Things

The IoT layer is of central relevance since the successful ability of the integrated system to be compliant with the possibly stringent requirements of the application domain, e.g. in terms of scalability and latency, highly depends on the appropriate behavior of IoT. This layer incorporates all components sensing data from the surrounding environments and launching actuators correspondingly to alter the environments in an appropriate reaction manner. It can be divided into two main modules. The first one, the sensors component, presented in Section 6.4.2.1, considers all these elements, gathering information from the environments. The second module, actuators, takes into account all those elements that can actuate or somehow change the environment. Figure 6.5 elucidates our taxonomy, highlighting the most significant alternatives for each of these elements.

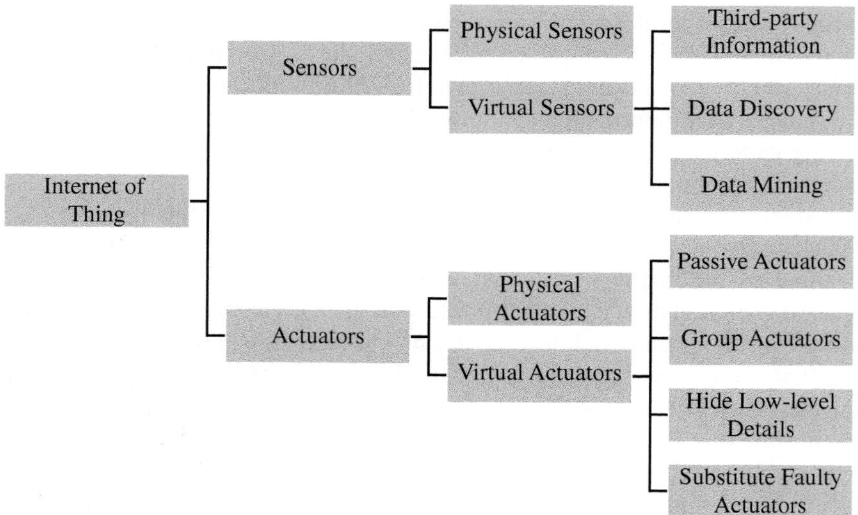

Figure 6.5 Taxonomy for the classification of the Internet of Things layer.

6.4.3.1 Sensors

Sensing is a capacity that normally IoT devices are performing. This element is concerned with two types of sensors: *physical sensors*, those that get data dynamically as directly connected and implanted within the environment utilizing specific hardware devices, and *virtual sensors*, those that indirectly acquire data through indirect channels (e.g. using a third-party system web service) [76, 77].

In STL and SCV, sensing is a main element that collects data for helping in accident prevention and maintaining a fluid flow of traffic. Sensing is important to understand the state of the road, and the internal state of the vehicle in SCV. For example, a smart vehicle may have *physical sensors* to identify wheel pressure, the distance to the nearby vehicles, or the lighting conditions. It can also have different *virtual sensors* to gather data on forecasting meteorology in its route or different traffic alerts obtained directly from the traffic authority [78]. Regarding wind farm systems, *physical sensors* gather real-time data related to fluctuations in weather, the wind speed, or the electrical power generated [8]. On the other hand, *virtual sensors* are fundamental for collecting relevant data and assisting in forecasting weather conditions [79]. Therefore, various reports are composed to confront weather conditions with generated power or even a deviation from forecasted loads.

In smart grid and smart building applications, *physical sensor* networks are required to sense the whole area pervasively and *virtual sensors* can provide information on networks incidents and system load. Thus, different statistics and reports are normally generated for comparing data from real-time operations in connection with data related to network incidents, loading, and connectivity. In smart building, sensors sense and share information among different rooms/floors of the building and obtain dynamic information about users' activities and energy supply conditions [46].

Few settings have only a sensing part, since they just watch the environment without concentrating on any actuation act. Business intelligence (BI) applications are a prominent case of this sort of frameworks dependent on sensing. BI applications utilizes procedures such as *data discovery* [80], *data mining* [81], business performance, analytics and processing, to convert sensed raw data into valuable insights for subsequent strategic decision-making or for solely visualizing results [82, 83]. In real-world settings, using BI in real-time scenarios in support of knowledge delivery is normally the act [84, 85], thus enhancing the strategic decision-making cycle and maximizing on the cost/profit cycles of the enterprise resources.

6.4.3.2 Actuators

Other systems provide relevant sensing and actuation phases; thus, they actively modify the environment, reacting to current events. Most real use cases are dependent on a sensing phase alongside a robust actuation phase. Henceforth, actuators are normally changing surrounding environments in an automatic or semi-automatic fashion.

This element comprises two types of actuators: *physical actuators*, which normally produce a physical modification that affects the surrounding entities via specific hardware; and *virtual actuators*, which are utilized for controlling a group of physical actuators [86], aiming at anonymizing the unnecessarily complicated peculiarities of low-level layers communicating with the physical actuators or otherwise substituting faulty actuators and preventing the system from thereby taking a severe dive [87, 88].

In STL and SCV, the actuation is fundamental in preventing accidents and maintain a fluid traffic flow. In STL, different reaction can be provided in relation of the circumstances. For instance, *physical actuators* may launch an alarm or alternate a traffic light so that it turns to red so as to slow down near vehicles. On the same side, *virtual actuators* are utilized for better controlling big areas and thus creating green waves of traffic lights to decrease pollution or for the passage of emergency vehicles. Likewise, SCV provides many virtual and physical reactions for sending warning messages to *actuate passively* on the driver showing visible or audible signals. Also, it can *physically activate* a warning or specific piece of hardware for recovering from anomalies as quickly as possible (e.g. slowing down the vehicle or turning some lights on) [89].

In wind farm scenarios, physical actuators are utilized for starting and possibly stopping turbines in response to a forecast and wind speed, thus preventing the system from breaking down [8]. In general we can resume some typical operational scenarios related to weather conditions: (1) in low wind speed, turbines switched off to avoid economic losses; (2) in normal wind speed, normal operation conditions, and blade optimized to maximized the electrical production; (3) in high wind speed, turbines, and power limited in order to avoid exceeding electrical and mechanical load limits; (4) in very high wind speed, turbines switched off to prevent electrical and mechanical breakdowns, due to the high possibility of exceeding allowable values. Also, in the recent literature, researchers of wind turbines replace faulty actuators through corresponding virtual actuators [87].

As a final consideration herein, in smart grid, operational planning and optimization actors employ simulation of network operations, put switching actions on schedule, dispatch repair crews, inform affected customers, and schedule the importing of power [90]. In smart building, the systems handle the energy consumption issue through a strong autonomous and continuous interaction sensing-actuation. Sensors sense and share information among different rooms/floors of the building and then make distributed decisions and activate physical actuators such as lowering the temperature, injecting fresh air, turning lights on and off, opening windows, or removing moisture from the air [91].

6.4.4 Data Quality

The *data quality* layer is in charge of processing the collected data so as to give a uniform specification for all information, on-the-fly discarding useless data elements, and thus reducing the amount of data sent to the fog or cloud nodes. The *data quality* layer comprises three essential elements for data processing: *data normalization*, *data filtering* and *data aggregation* (see Figure 6.6). First, *data normalization* is the process in which all the information is homogenized.

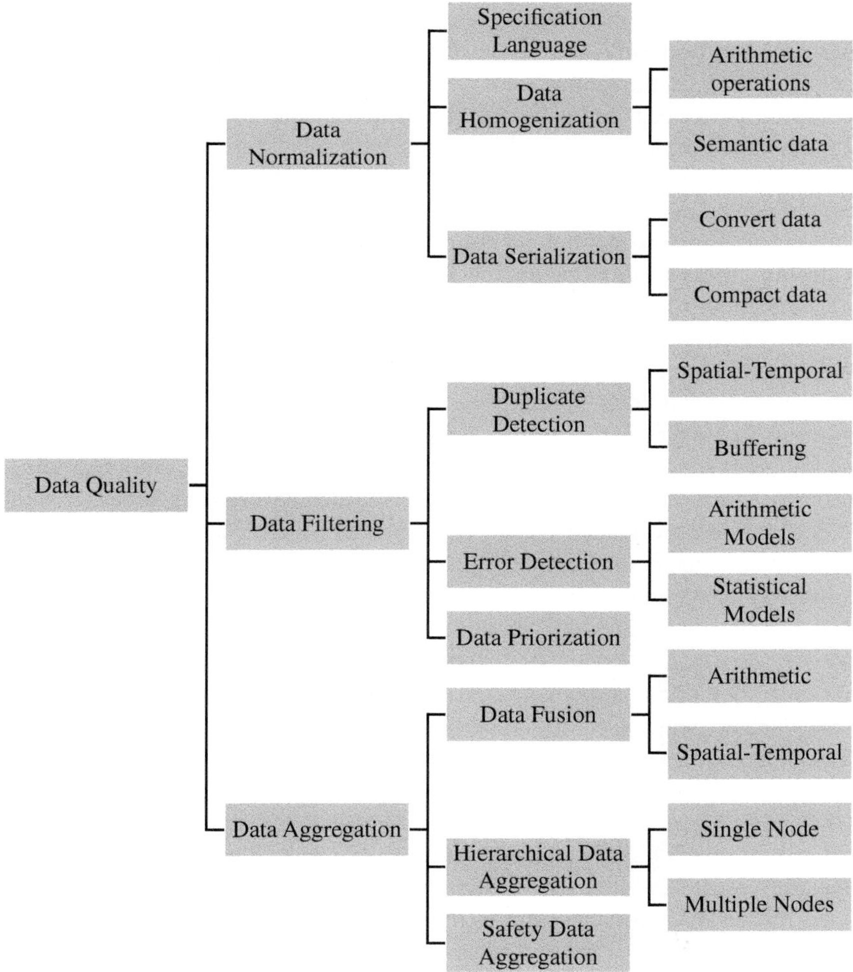

Figure 6.6 Taxonomy for the classification of the data quality layer.

Section 6.4.4.1 analyzes some of the most important approaches for the normalization of data. Second, *data filtering* focuses on eliminating any duplicate and erroneous data in order to reduce dataset size. Section 6.4.4.2 details the most important data-filtering solution. Finally, *data aggregation* is the process of merging large data set in a single flow. Section 6.4.4.3 presents some of the techniques used for aggregating collected information coming from heterogeneous sources.

6.4.4.1 Data Normalization

As detailed in Section 6.2.1, IoT is an extremely heterogeneous environment that is built upon a comprehensive range of devices that gather information, often heterogeneous, from surrounding environments. Sensors have normally different reachability ranges, fluctuating from weak to robust sensors, with significant differences in power consumption, data rate, and available resources. In the same vein, fog nodes are naturally heterogeneous, with the different levels of services they provide. Hence, all senses are normalized so as to facilitate data dissemination among various devices or between fog nodes. *Data normalization* is a main data preprocessing procedure for learning data forms [92].

Currently, there are different approaches to normalize all the exchanged data in order to improve the semantic interoperability between devices. In general, interoperability affects the system at different levels. Different approaches tackle this issue using different cross-applications, languages, and data homogenization and serialization mechanisms. *Specification languages* allow developers to transform all sensed data to a common format. *Data homogenization* mechanism unify data from different sensors and services. Even though semantic data and open standard middleware can homogenize data, the great variety of data sources leads to the need for new mechanisms. Each IoT application should adopt the technique that best fits their requirements [93]. *Data serialization* techniques convert and compact data to various formats to transmit them effectively.

Often these mechanics are combined to enhance data normalization. For instance, Zao et al. [59] employ two-level mechanisms for data normalization: first, they utilize a global *specification language* called Pigi [94] to gather data/metadata fields; secondly, they use the Google Protocol Buffers [95] to perform the *data serialization* for streaming and archiving the information with a compact binary data encoding.

STL and SCV are systems composed of heterogeneous components that should work appropriately together. Though multiple providers give different specifications, they must exhibit common interfaces and interoperability abilities. Data normalization is highly important in these environments. For example, in SCV, all sensed data are converted to a common language and serialized in order to be efficiently transmitted to a local fog node in the vehicle or to other vehicles or road infrastructures requiring that information.

Smart Grid systems usually comprise a huge mass of distributed and naturally heterogeneous devices, thus demanding the use of protocols so as to achieve the inter-communication between them. Greer et al. [90] indicates eight layers of the degrees of interoperation necessary to enable different transactions on the smart grid, divided among three drivers: (1) an informational driver, which focuses on what information is exchanged and its meaning (using specific languages); (2) an organizational driver, which emphasizes the pragmatic (business and policy) aspects of interoperation, especially those pertaining to the management of electricity; and (3) a technical driver, which involves technical aspects, such as mechanisms to establish physical and logical connections and to exchange messages between multiple systems across various networks. In addition, [90] promotes a few guidelines for identifying standards and protocol in support for interoperability throughout the smart grid, coupled with definitions of architectures for incorporating and supporting a wide spectrum of technologies, spanning from legacy to novel. Also, standard languages are common for performing interoperability between smart meters, smart devices, charging interfaces, and in exchanging information for all smart grid scenarios [96].

In smart building applications, *data normalization* is a critical activity due to the huge number of devices that must communicate [46]. In addition, these applications could be integrated with smart grid systems with the intention to manage consumption in response to supply conditions by selectively turning off appliances or reducing nonessential or non-time-critical services, with many benefits for suppliers (e.g. avoiding costly capital investments) and customers (e.g. sharing the savings resulting from the lower operational cost of energy production) [46]. In this situation, the language used to transmit the information between different systems is essential in order to achieve the required coordination. These iterations issues are exacerbated by the range of producers that build devices and smart building systems that generally use different type of implementations and with different policies. Hence, smart building applications must define and/or enforce standardized data normalization techniques and guidelines. In this scenario, Chen et al. [46] propose reference semantic models (RSMs) implementing the smart-building related industry standards to facilitate the exchange of information among different subsystems, through a variety of functions including measurements, planning/scheduling, and life cycle management.

Along these lines, a wind farm system is mainly responsible for sensing weather conditions, wind speed, and the turbine power and react interactively with a number of different actuators, thus, data normalization can be achieved. Nevertheless, the most appropriate techniques for data homogenization and serialization are selected so as to exchange information efficiently.

6.4.4.2 Data Filtering

IoT frameworks with several sensors continually sending data to the fog nodes and to the cloud could quickly clog the system and overwhelm the nodes' constrained assets. Data filtering is in part responsible for decreasing the quantity of data transmitted by taking out any information that is repetitive, mistaken, or broken [97]. Data filtering procedures ought to be actualized as close to the edge as possible to decrease information traffic. In spite of the fact that sensors may execute lightweight filtering to evacuate some noise at information accumulation stage, increasingly vigorous and complex data filtering is as yet required at the fog layer. The principal data-filtering techniques consists of *duplicate detection*, *errors detection*, and *data prioritization*. Normally, due to its location near to the edge, fog nodes do not have a widespread knowledge of the whole system, thus, they cannot perform advanced data-filtering operations, so they are left to subsequent stages.

Duplicate detection techniques focus on analyzing received data, from either a specific sensor or a set of sensors, so as to detect duplicated data that can be safely discarded. The Bloom algorithm [98], for instance, detects the redundant data via a buffer that stores the received data and periodically controls whether newly arrival data is present in the buffer. Different algorithms utilize the locality property of IoT systems. They indicate that there is a strong spatial-temporal interconnection between various sensors, with an elevated probability that they collect same data at times and that the same information will probably be gathered by those sensors soon.

Error detection mechanisms are used for identifying flaw data generated by incorrect measures obtained through sensors or noise. [99]. The knowledge of the system, the relation input-output, and the use of secondary data features of data gathered to compare several additional information are basic to detect faulty data. To detect these data, different techniques can be applied. For instance, *Statistical model* approaches can assist in creating models for predicting data distribution and thus detecting those data which do not fit the model (known as outliers).

Finally, *data prioritization techniques* filter time-critical data aiming at prioritizing and forwarding them to be processed early by the data analytics component. This is in stark contrast to the aim of reducing data size that is of the other filtering techniques mentioned earlier.

In the STL and SCV case studies, the variance and standard deviation for the vehicles observed speed compared to the average data received at same contexts, with same location, time, weather, and type of vehicle, metrics are normally utilized for identifying outliers and detecting an accurate speed of an approaching vehicle. Consequently, with the definition of data thresholds, it is possible to discard outliers and refine the information collected.

In addition, it is possible to introduce various management control system (e.g. Six Sigma, Lean, etc.) usually adopted in manufacturing or business processing, adapting them to be effective to face the specific problem of data quality improvements. These control systems extend variance/standard deviation measures and are based on a set of advanced statistical methods and techniques, trying to identify and remove all the causes of errors and waste in order to maximize the quality level in relation to the specific application where they act.

In smart building, there are works that focus on reducing the inaccurate sensed information by identifying outliers. The authors indicate that outlier's detection is important in application processing since these erroneous data can lead to abnormal behavior of an application, e.g. turning on the air conditioning when an erroneous measure of the building temperature has been identified. In [47], the authors use the Hodrick-Prescott and moving average techniques to identify the outliers. The Hodrick-Prescott filter is a mathematical tool used to remove the cyclical components of a time series from raw data. The moving average technique is a calculation to analyze data points by creating a series of averages of different subsets of the full data set [100]. The authors applied those techniques to minimize the fluctuation of the temperature/humidity parameters in smart buildings.

Finally, in smart grid, the signals are typically sampled and communicated at high rates – several tens or hundreds of times per second – to augment or even replace the conventional supervisory control and data acquisition system. In [101], the authors indicate that this situation creates some congestion problems, preventing the most critical information to be received at the right time at the nodes where decisions are made and hindering a real-time response. Therefore, they propose the technique of distributed execution with filtered data forwarding to assign more importance to time-critical applications compared to other, less time-critical monitoring applications, prioritizing data for applications having more stringent timeliness requirements compared to ones having lenient QoS.

6.4.4.3 Data Aggregation

The data aggregation element further focuses on minimizing collected information by means of aggregating them into groups based on some keys and constraints for some variables of interest. To such an end, it combines all the gathered information in order to form a unified picture and a unique flow of data. In addition, this component also improves the system interoperability, since it aggregates and merges data from various sensors, ensuring reliable data collected from sensors [102].

Data aggregation can be done using diverse integral perspectives focused on: merging large data sets, hierarchically aggregating data in different nodes and improving the degree of safety via data aggregation. *Data fusion* methods attempt to combine diverse sorts of information to diminish data size and acquire a novel data stream. Keeping that in mind, a distinctive number of arithmetic operations can be used to get increasingly steady and representative estimations from a big data sample. Spatial methods can be utilized to aggregate data relying upon the sample's geographical location. *Hierarchical data aggregation* promotes the idea of identifying whether the information can be aggregated on a single node or simultaneously on many nodes. In fog environments that incorporate middle nodes with diverse capacities, this technique normally allows the consequent application of aggregation techniques to enforce location-awareness property for every node. *Safety data aggregation* centers around gathering similar data from various sensors to have alternate points of view of a similar context with the objective of improving the security and safety of the IoT applications.

The application of these techniques is evaluated in the different cases of uses. For instance, both SCV and STL systems are highly distributed applications with numerous geo-distributed information gatherers that must impart and collect data so as to make productive traffic strategies to draw paths for vehicles.

In SCV, some approaches use data aggregation to monitor the roughness of road surfaces. In [14], the authors propose to collect and fuse different information from accelerometer and GPS sensors and progressively thin it at different levels through sampling and spatial/temporal aggregation techniques. Each collected point is mapped onto a map database and aggregated according to specific geometric constraints. This makes it possible to consistently map the sensed physical quantities of several adjacent points into a single aggregate data. Then, a temporal aggregation is performed. The stored points are kept updated with the last significant changes (incrementally down-weighting older points). Finally, by applying a linear predictive coding (LPC) algorithm to the collected samples, the roughness of the road surface upon which the vehicle travels are estimated as their arithmetic average. This estimate provides significant information on the quality of road surface, given the capability of the LPC algorithm of filtering out (to a certain degree) spurious components of the acceleration signals (engine vibrations, gravitation, inertial forces, etc.).

In STL, the control of the diverse states of the vehicles and the streets is basic to give the ideal safety to the drivers. Currently, different approaches rely on *safety data aggregation* methods to identify and trace vehicles via observation cameras and diverse sensors (for instance, BOLO Vehicle Tracking Algorithm [103]).

In smart grid, the data produced by the diverse elements of smart grid network is huge. Therefore, in [97], the author proposes the hierarchical aggregation of the data. Power consumption data from different meters are collected and aggregated at prespecified aggregator meters hierarchically. All collected data is aggregated at data aggregation points of power generation plants, distributed energy resource stations, substations, transmission and distribution grids, control centers, and so on. The aggregator meters perform the aggregation by using some arithmetic and temporal operations. The aggregator meter waits for the arrival of data packets associated with the same timestamp from lower meters. For time-aligning the packets from multiple nodes, it temporarily buffers the extracted data, and once all the data with the same timestamp arrive, they are aggregated to create a single data packet. Then, the data are transmitted to the next aggregator meter.

In wind farms, wind turbines transmit data to the base station using multiple hops. Intermediate fog nodes can operate as aggregators reducing the number of direct links from the wind turbines to the base station. In [79], authors apply different data fusion techniques to aggregate the sensed data in the intermediate nodes depending on the physical components, timestamp of the sensed information, sampling rate, and sensor type.

6.4.5 Cloudification

The cloudification layer permits the operation of various IoT settings within the layers of fog nodes, serving as a small-version cloud within the fog. This layer offers four fundamental components: *virtualization*, *composition, orchestration*, and *storage*. First, a virtualization component permits developers to incorporate IoT applications and deploy them in a fog node. Second, the composition component facilitates the mash-up of small services deployed in the fog nodes in order to provide higher-level and better adapted services. The orchestration module allows IoT systems manager to control the different IoT instances deployed in a fog node. Finally, a storage element supports the permanent storage of data that is sent or requested by users so as to increase system's responsiveness. Despite being geared toward fog components, cloudification by definition spans all the three components of our architecture, being a shrink volume of cloud deployed in fog or even in the edge itself.

In this section, we mainly focused in the virtualization and storage components because they are the most important ones and the areas in which a higher number of solutions and technologies have been proposed. Section 6.4.5.1 details the most important platforms, allowing virtualization and deployment of IoT settings in fog nodes. Section 6.4.5.2 presents some of the most important mechanisms providing support to the storage of data in the fog nodes. For the sake of clarity, Figure 6.7 summarizes the proposed taxonomy with the possible choices for any of the components.

6.4.5.1 Virtualization

Virtualization permits fog nodes to make VMs. Virtualization considers two principle attributes. Tho=ese are the technology utilized to epitomize the IoT framework and how the virtual images are relocated between fog nodes, supporting the clients and framework mobility needs. Currently, the main technologies for creating virtual images are hypervisor and container. *Hypervisor* is a virtualization

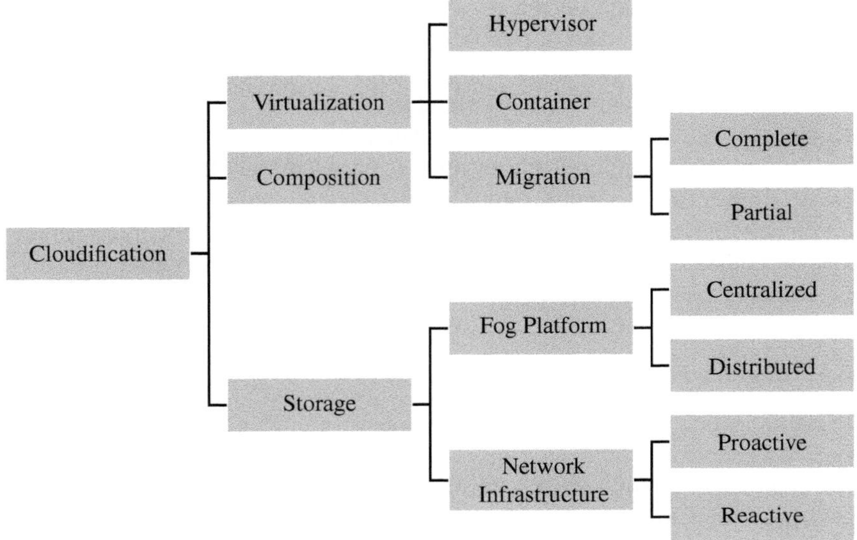

Figure 6.7 Taxonomy for the classification of the cloudification layer.

approach, where since a virtual image not only contains a final output but, rather, it delivers an operating system required for executing it. OpenStack and Open-Nebula [104] are some of the frameworks supporting this virtualization technique. Instead, *containers* is a lightweight solution for the deployment of isolated execution context in top of an already defined operating system (i.e. it contains the applications, but not the operating systems). Not supporting the emulation of various operating platforms, it improves the execution and the relocation of the containerization parts. LinuX Containers (LXC) [105] or Docker [106] are some representative examples of this technology.

Migration is yet another key characteristic in virtualization. VM migration is essential in fog computing to address the mobility requirement. When a user leaves the area covered by the current fog node, the VMs or the container may need to be migrated to another node covering the user's destination. This process should be fast enough for maintaining real-time and location-awareness needs of IoT settings. Two principal migration approaches are used; a *complete migration*, in which the entire virtual image is migrated from one node to another (using Internet Suspend/Resume [ISR] [107] or Xen live migration [108], for instance) or *partial migration*, in which only specific pieces of the virtual image are transferred (utilizing, for instance, alternatives for the previous methods [109]).

Currently, there are many approaches employing those virtualization methods in fog computing settings. Cloudlet was proposed before fog computing, but both share the same concept. Actually, as it is indicated in Section 6.2.2, the Open Edge Computing Consortium promotes its use. Cloudlet comprises a three-layered-up stack, with the base layer containing a working OS, such as Linux, in addition to the information cache from the cloud. The center layer incorporates a hypervisor to isolate the transient guest software environment from the cloudlet infrastructure. Concretely, this layer is based on OpenStack++ [110], which is a specific extension of OpenStack including a set of cloudlet-specific APIs. The third upper layer constitutes applications separated by various VMs. Finally, cloudlets again deploys a specific method for the subtotal relocation of VM instances, dubbed as dynamic VM synthesis. Each cloudlet node comprises a basic VM and every mobile device constitutes a tiny VM overlay. Hence, when a mobile device relocates, a source node stops the overlay and saves it in the mobile device. When the mobile device hits the destination, it transmits the VM overlay into the destination node, subsequently applying it to its base, and thereby beginning its running in a precise state where it left off.

In [71, 72], the authors characterize an exploratory fog processing platform. This stage utilizes a hypervisor virtualization method so as to give adaptability. Solidly, they utilize OpenStack together with the Glance module for the administration of VM images. Likewise, they additionally execute two diverse migration schemes to encourage the meander of VM images among fog nodes. In the main technique, they take a snapshot of the VM to be relocated, compress it, and afterward exchange the compacted data to the target fog. In the second strategy, the VM has a base snapshot entrenched on the two fogs with the goal that they just exchange the incremental part of the VM's snapshot. IOx [111] is the Cisco implementation of fog computing, providing uniform and consistent hosting capabilities for fog applications across its own network infrastructure. IOx works by hosting applications in a guest operating system running a hypervisor directly on the connected grid router. The platform also supports programmers in running applications embedded on Docker and LXC, packaged as a VM, compiled and run as Java or Python bytecodes.

In addition, different platforms have been developed specifically focused on supporting some of the case studies evaluated in this paper.

Truong et al. [50] proposes a platform using software-defined networking (SDN) and fog computing to reduce the latency and improve the responsiveness of SCV applications. Concretely, the authors deploy a fog infrastructure on the roadside-unit controller (RSUC) and on the cellular base station (BS), allowing the computation and storage of some information from the vehicles in these elements. For supporting the virtualization, they use the hypervisor technique. Then, different vehicular and traffic services can be hosted in the VMs allowing service migration and replication.

In smart building, [112] proposes the platform ParaDrop to exploit the underused resources of WiFi access point or cable set-top box, provided by the network operators, with the aim of making them smarter and reducing the information transmitted to the cloud. To that end, ParaDrop uses the LXC abstraction for providing resource isolation and allowing third-party developers to deploy their services using this container. The containers retain user state and can move with the users as the latter changes their points of attachment. This platform has been used to support the control of environmental sensors (humidity and temperature) and security cameras, improving the privacy and the latency and providing local networking context to the system, since all the information is computed in the local node and only the data requested by end-users are transmitted through Internet.

6.4.5.2 Storage

Given the huge amount of data generated by smartphones and IoT devices, data must be allocated as close as possible to users. Information is first stored on fog nodes so as to accelerate its processing, decreasing thereby the information exchange latency and increasing the systems QoS.

Various approaches include storing that data on fog nodes or on different components of the network.

From one viewpoint, the fog system is able to handle local storage for keeping data in a disk-residence fashion. Based on priority, data are stored locally in a compressed manner, thus improving their security and privacy. Typically, these OSs store information on a single node (similar to a semi-centralized mode) or on various nodes (thus similar to a distributed mode).

Every virtualization mechanism and solid framework used by the operating system can implement one or both models. As detailed above, one of the widely used Hypervisor frameworks is OpenStack. The basic implementation of OpenStack can complements different modules providing different storage capabilities. The most important modules are Cinder and Swift. Cinder gives persistent block storage to guest VMs. This module encourages the storage of information on a given fog node, utilizing a centralized mode. Cinder virtualizes the administration of block storage devices and gives end clients a self-managed API to ask for and expend those assets. Swift functions as a distributed, API-open storage system that can be incorporated straightforwardly into IoT applications or used for storing VM images.

The fog platform presented in [89] is based on OpenStack, including the module Cinder for data storage. Therefore, this platform allows the centralized data storage in a specific fog node. Cloudlet is built on OpenStack++, which is mainly focused on improving the cloudlet deployment. Nevertheless, as OpenStack++ is plugging on top of OpenStack, it supports the addition of Cinder or Swift components. In addition, in [113], the authors present CoSMiC, a cloudlet-based implementation of a hierarchical cloud storage system for mobile devices based on multiple I/O caching layers. The solution relies on Memcached as a cache system, preserving its powerful capacities, such as performance, scalability, and quick and portable deployment. The solution aims to be hierarchical by deploying memcached-based I/O cache servers across all the I/O infrastructure data paths.

Containers, such as Docker or LXC, again provide various functionalities for caching and storing data. For instance, in [114], the authors evaluate Dockers as an edge computing platform. Every Docker container is separated and comprises its network subsystem in addition to memory and file systems. For data storage, Docker utilizes a lightweight file system dubbed as UnionFS to provide the building blocks for containers, allowing the caching of data to decrease user local access and to improve overall application performance. Concretely, Flocker is a container data-volume manager. It allows state-full container, e.g. production database data, to be protected. Any container regardless of its location can use Flocker.

On the other hand, other relevant works consider caching users' information on network infrastructure. Suchinformation is typically stored in a reactive fashion, thus caching data as users request it; or proactively, analyzing users' requirements on data and precaching accordingly.

The content delivery network (CDN) [115] serves as a representative for caching networks. CDN acts as an Internet-based cache network by disseminating cache servers at Internet edge so as to reduce upstream lateness caused by contents arriving from remote sites.

Information centric network (ICN) [116] is a wireless cache infrastructure that provides content distribution services to mobile users with distributed cache servers. Different from the cache servers in ICN, the fog servers are intelligent computing units. The fog servers can be connected to the cloud and leverage the scalable computing power and big data tools for rich applications other than content distribution, such as IoT, vehicular communications, and smart grid applications [9, 117].

Bastug et al. [118] detail that the information demand patterns of mobile users are predictable to an extent and propose to proactively precache the desirable information before users request it. The social relations and device-to-device communications are leveraged. The proactive caching framework described by Bastug et al. can be applied in fog computing.

6.4.6 Analytics and Decision-Making Layer

This layer promotes two principal components for decision-making in IoT settings. Those are decision management and data analytics (see Figure 6.8). First, all gathered data must be analyzed so as to detect varying trends and contexts. Section 6.4.6.1 details some of the main data analytics approaches. Second, from

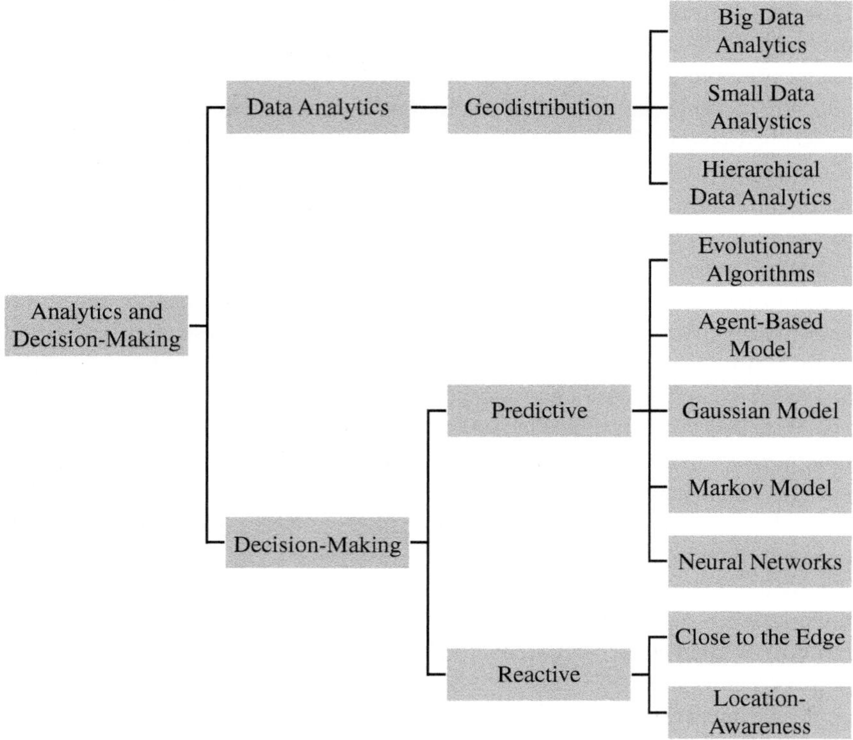

Figure 6.8 Taxonomy for the classification of the analytics and decision-making layer.

the data obtained by the sensors and the knowledge obtained processing them, different decisions are made in order to execute the appropriate business rules at the right time. Section 6.4.6.2 analyzes the main decision-making approaches for fog computing. For simplicity, Figure 6.8 depicts our taxonomy, elucidating all alternative components. Notice that this layer potentially spans the three components (fog, cloud, and edge) as it is a matter of fact that those kinds of advanced processing are nowadays being conducted at any of those layers with various degrees of load capacity. For example, some decisions are to be made in real-time near the edge nodes so as for the system to be able to feed actuators with appropriate decisions that guide their instant operation.

6.4.6.1 Data Analytics

Data analytics is an advanced analytics techniques application to data loads, aiming basically at gaining deep insights [61]. Concentrating basically on the location where data is analyzed, this component is normally divided into *big data analytics* and *small data analytics*. *Big data analytics* relies completely on cloud settings in order to perform complex operations for huge avalanches of fast-arriving big data loads, being able to identify complex associations, patterns, and trends [119]. On the contrary, *small data analytics* are performed near the edge, and thereby close to the IoT devices, suited to be handled by fog nodes. Small data analytics can handle only small amounts of fine-grained data that naturally provide beneficial information for a system, thus enabling real-time decisions near the edge in an as-fast-as-possible manner. The centralized model is perfectly suitable for orchestrating data generated by many applications with an appropriate response time. Likewise, a decentralized model provides real-time information for fast and short-term decision-making. Nevertheless, when many geographically distributed devices produce data requiring analysis alongside on-the-fly decisions, *hierarchical data analytics* models seem to respond better to different scenarios. Instead of sending all gathered data to the cloud, information can be stored, and small data analytics can be executed in the fog nodes. Then, relevant or complex data can be aggregated and posted to the cloud environment. In doing so, more complex and resource-demanding analytical techniques can be performed, and medium or long-term decisions can follow, depending on their outcomes [5].

Methods herein are employed for our aforementioned case studies. For instance, in STL and SCV systems, small data analytics are used for constructing a nearly exact picture of the surrounding current situation and thus assisting decision-maker components in reacting spontaneously. Both STL and SCV act in a critical context where the quickness of the decisions plays a key role in these systems than can even save many lives. These systems collect data regarding traffic density, vehicle-related data (for instance, speed, traces, internal condition), movements of different vehicles or bicyclists in lanes and emergency routing. The fog node should store this data and execute fast analysis to recognize certain events on the street, e.g. which vehicles are coming closer, crossing, or traveling along the road, and foresee where they will likely move. In this way, a basic setting for actuation must be able to prevent systems from crashing and improve safety. In this sense, Hong et al. [103] promoted the MCEP system for overseeing traffic via several patterns.

In the case of a wind farm, big data analytics are more often used, because not only are instant actions significant but also insights into future events are useful. A common issue in wind darm applications is to gather enough real-time data of good quality and, therefore, some ISO/RTO introduce mandatory data reporting requirements for wind power producers, with penalties for noncompliance [70]. As a general consideration, it is of great importance to ensure a specific exactness level when applying wind forecasting methods, especially those that are short-lived, so as to improve the quality of a wind power generator that the system offers to markets and to schedule appropriate levels of operating reserves needed to perform the different regulation tasks [68].

In smart grid and smart building use cases, layered-up data analytics are essential to guarantee that systems are working effectively and to accurately oversee dynamic end-client requests and distributed producing sources, supporting instant responses if an occurrence of unforeseen occasions should arise. Chen et al.'s [46] stacked-up data analytics are central to confront sustainable power source supply capriciousness that might be a important factor in connection to weather fluctuations, since each middle node serves as a functioning control unit, processing and responding to passing information [96]. In this scenario, many utility companies, with the intention to improve coordination and control techniques, have introduced smart meter readings to remotely meter-gathered data. As an example, consider consumption records, electricity production, and alarms [32].

6.4.6.2 Decision-Making

As the speed at which the collected information must be transmitted and processed in IoT scenarios increases, the readiness to settle on decisions to trigger explicit business procedures and standards in the correct minute is pivotal and it plainly influences the asset usage and consumer loyalty levels essentially [120]. Depending on how fast a decision is to be considered, the decision-making element is thus divided into *predictive* and *reactive*.

The *predictive model* stores all the data gathered by the sensors, the system's behaviors, its performance, etc., to get a profound learning of the surroundings and the framework so as to trigger the most suitable responses for every context and to surmise conceivable system advancement. These models focus on data calculation and gaining insights, thus constructing clear and predictive models from extensive volumes of data gathered so as to predict the relationship between settings' data. Despite what might be expected, *reactive models* react in the briefest conceivable time to various occasions that occur in the environment so as to endeavor to create corrections as early as possible. These models serve in a way that enables it to accomplish an ideal objective while interfacing with an environment.

More often, however, systems have diverse dimensions in the decision-making process consolidating at various degree the two models, yet our proposal depends on which part is progressively created and on which functionality the system is increasingly cantered around.

Few systems work in settings where predictions are vital, and they perform comprehensive data calculation and analytics.

Predictive Systems In numerous applications, the interaction among fog and cloud is straightforwardly related with the decision-making model desired.

Predictive systems depend more on cloud so as to gather a lot of information and perform long-haul analytics to distinguish the distinctive policies that ought to be executed.

Some of these IoT systems rely on *evolutionary or genetic algorithms* to optimize the system operations. In Wind Farm, for instance, prediction has been identified as an important tool to address the increasing variability and uncertainty, and to more efficiently operate power systems with large wind power penetration. They are based on the analysis of the atmospheric stability and wind patterns at different time scales in order to increase the efficiency of the overall system: (1) monthly yearly basis predictions are used to improve the deployment process of the system; (2) daily forecast to submit bids to the independent system operators ISO/RTO; (3) hourly forecast to adjust the commitments, responding to contingencies that may occur in the operating conditions (e.g. forced outages of generators, transmission lines, deviation from the forecast loads, and so on); (4) few-minutes forecast to dynamically optimize the wind farm operations [70]. Unit commitment components relay on evolutionary techniques [121, 122] to limit operating expenses while satisfying the total needs offered in the market. This is normally done by a controller that decides global or customized arrangements and enforces them for each subsystem. Obviously, they additionally have some reactive methodology so as to build productivity and to counteract harm, closing down the turbines if wind is excessively low or hitting hard. Other approaches apply *agent-based model*, *multivariate Gaussian model*, *hidden Markov model*, and *neural networks* as predictive strategies to foresee behaviors of the various parts of systems. For example, in smart grid and smart building systems, [123, 124] highlight that predictive strategies meet user demand better than typical baseline strategies, showing better energy-savings performance and even significantly better QoS than typical reactive strategies. Prediction-based methods are able to predict the indoor comfort level and the QoS of a specific user by learning the user's behaviors and comfort zone, and to satisfy a user's needs by managing the efficient use of available resources. In [125], Erickson et al. make use of an *agent-based* and *multivariate Gaussian* model to evaluate the inhabitancy in a huge multipurpose building and for foreseeing client mobility designs so as to effectively control energy use in a smart building. In [123], the authors employ *hidden Markov* to model the behavior of the individual in relation to the temporal nature of the occupancy changes, the interroom correlations, and occupant usage of the areas in order to generate probabilistic instructions, achieving significant energy savings while still maintaining service quality, or even improving upon it.

Response approaches in these environments usually need more time to achieve the target quality required by the user, since there is a delay from the occurrence of an event and the system stabilization and, thus, they require more energy. Reactive methods can be utilized at the same time with the predictive methods so as to refine the system if there should arise a need and get the best results for every circumstance. In this context, [124] proposes a procedure, in view of *neural networks* methods, that consolidates distributed production, storage, and demand-side load management techniques, accomplishing a superior coordination of demand and supply.

Reactive Systems Reactive systems exploit the capabilities of fog environments in order to decide as early as the data are gathered in order to produce (near) real-time responses. Real-time support is a key trait in fog that is particularly essential in those systems that require an instant action. In general, real-time response is directly related to the reactive part of the system. Nonetheless, in some systems there must also be a balance between fog and cloud in order to optimize real-time responses.

To obtain a sufficient response time, as a rule-of-thumb a fog node closer to the edge means a better response time. Therefore, these systems exploit the *close-to-the-edge* [117] and *location-awareness* [9] methods. For instance, in STL and SCV applications, low-latency actions are essential demands crucial for safety. In both systems, low-latency reactions must be provided to be able to prevent vehicle collisions or for timely emergency ambulance vehicles passage. Bonomi et al. estimates that reaction time must fall in the premises of minimal milliseconds and, maybe most notably significant, less than 10 ms to be considered effective and thus satisfying safety hard bounds. In such a context, fog's role is critical in sensing the situation, process data, and thereby identify the required actuation in significantly limited time bounds. As such, those systems should use the trait of *fog moving close to sensors/actuators* so as to diminish the latency on a scale. In SCV, the fog layer has a massive role in changing information among vehicles in order to make better, more dynamic decisions, and outside vehicles, to provide information to the RSUs (stationary infrastructure elements that are installed along the road [50], and can be envisaged as higher-level fog nodes in a multilevel architecture), reducing latency and communication overhead. In both contexts, there is a relevant different time scale in relation to the purpose of the action. Fog nodes must be sufficiently powerful to be able to allow low-latency communication, inside the vehicles and between decision makers and traffic lights, in order to perform real-time reactions. At the same time, elevated amounts of data are forwarded to the cloud for long-haul analytics to evaluate the impact on traffic jams and to monitor city pollution, to mention just a few factors. Finally, the *location-awareness* strategy enhances system's responsiveness, providing advance knowledge of the environment amid applications execution with a few advantages; e.g. an RSU can infer whether a vehicle is in danger (because it is approaching too fast to a bend), and react to the nearby traffic light cycle to alter the circumstance or notify the driver accordingly.

In addition, STL systems may cross boundaries of several authorities. Thus, many critical application APIs around the whole system are coordinating and organizing their policies across the whole system, creating the control policies to send to the individual traffic lights and, inversely, collecting data from traffic lights [8].

6.5 Comparisons of Surveyed Solutions

In this section, we contrast the studied methodologies with a stress on the principle benefits and the inadequacies of the conceivable arrangement that can be adopted for structuring successful fog computing environments for IoT applications. As in the previous section, we organize our comparison alongside our logical architecture (see Figure 6.2).

6.5.1 Communication

Table 6.1 shows approaches surveyed for communication support, and cross-matching them with the categories and subcategories of relevancy. Further details about each follows in subsequent sections.

Table 6.1 Comparison between surveyed communication approaches.

	Standardization		Reliability			Latency	Mobility	
	Application	Infrastructure	Retransmission	Handshake	Publish/subscribe	Latency	Routing	Resource Discovery
CoAP [33]	X		X			X		X
MQTT [34]	X		X	X		X		
Oasis [35]	X		X	X				
DDS [36]	X		X		X	X		
Zigbee [37]	X							
ISO [38]	X							X
Tim Winter et al. [39]		X					X	
IEFT WG [40]		X						
IEEE 802.15.4 [126]		X						
Bluetooth SIG [41]	X	X						
Qualcomm [42]		X						
Farinacci et al. [43]		X					X	

6.5.1.1 Standardization

The first step is selecting a protocol for communicating different system devices. This allow developers to standardize the communication between the different elements of the system and, at the same time, to choose the one that best suits the system's requirements. The selection of communication protocols is very dependent on the specific requirements of the application. Right now, various protocols can be chosen to improve the communication at network level, or between various devices or parts of the system. To state the obvious, numerous protocols at the application level depend on explicit protocols at the foundation level. Hence, certain attributes of the latter are acquired by the former. For instance, the ZigBee protocol depends on the IEEE 802.15.4 standard.

Specifically, by analyzing our case studies, we have noticed that ZigBee is especially recognized in IoT scenarios such as smart grid, smart building, and smart vehicle. This may be attributed in part to its short range and robustness with present noise conditions. In addition, especially in smart building and smart vehicle environments, the BLE has also gained momentum recently, in part due to its low-power energy consumption, short-range communication, and elasticity.

6.5.1.2 Reliability

The communication standardization is important for the correct exchange of information, but its reliability is also critical. This requires a system that confirms that appropriate data levels are being received instead of being lost, thus ensuring a correct flow of the system in most circumstances. Therefore, the protocols for communicating the different parts of the applications must contain techniques ensuring such reliability. As can be seen in Table 6.1, protocols such as CoAP, MQTT, AMQP, and DDS are based on retransmission techniques, allowing message retransmission a certain number of times until it is received. Algorithms such as MQTT and AMPQ also encompass handshake methods to establish the most appropriate parameters to ensure such communication. In addition, in order to further increase the reliability of the communications, other protocols implement multicasting techniques, so that different nodes can get the transmitted information.

Those techniques are commonly found in cases studies encompassing critical environments, such as smart vehicle and STL, since reliability is an essential requirement.

Nevertheless, as Table 6.1 shows, all the reliability methods found in the surveyed proposals are adopted by application protocols.

6.5.1.3 Low-latency Communication

A low-latency communication is important in IoT applications so that the system can respond adequately to certain situations. Low-latency must be accomplished among fog nodes, in addition to fog nodes and those in the cloud. CoAP, MQTT y DDS are protocols in support for low-latency of communication among various nodes. In addition, MQTT proposes the usage of various versions, a version for the communication between devices and fog nodes, and the other one is for the exchanging information between fog and cloud nodes. In the case studies evaluated, DDS is promoted in different situations associated with smart grid, smart building, and smart vehicle as it provides efficient and predictable distributions for time-critical data.

Similar to reliability, in the surveyed methods, low latency is focally accounted for by protocols of the applications (Table 6.1).

6.5.1.4 Mobility

The protocols figured beforehand normally encompass various techniques, such as routing and discovery mechanisms to improve and facilitate the mobility of the different nodes of the system. As indicated in Section 6.2, this is a fundamental requirement of IoT systems.

According to the analyzed solutions, mobility support is essential for infrastructure and application protocols in row. Two of the protocols that support this feature are RPL and LISP, making it easy for a device to migrate from one subnetwork to another. In addition, when a device changes from one context to another, it must facilitate the device in discovering services and resources disponible in that context. Normally, this characteristic is provided by application protocols. CoAP and UPnP are two of the analyzed protocols supporting this type of mobility.

Again, STL and SCV are the case studies where a correct implementation of the mobility techniques is crucial, since vehicles may be constantly moving from one fog node to another and the system operations should not be diminished by it. RTL is being adopted in other settings in addition to those described herein. In these same environments, the resource discovery techniques are used to identify and interact with new devices automatically.

6.5.2 Security and Privacy

Table 6.2 summarizes surveys works that fall in the security and privacy layer, with a matching to the correspondent categories.

6.5.2.1 Security

IoT systems are environments where data is normally scattered between different elements. Thus, the communication between them is constant. Therefore, techniques ensuring the security of such communications are essential.

This security is fundamentally given by the communication protocols. As can be seen in Table 6.2, some of them implement different techniques to encrypt the data transmitted between the different elements of the system. In addition, other protocols, such as CoAP, encapsulate techniques for improving integrity of data, ensuring that the exchanged messages have not been manipulated during the communication.

Table 6.2 Comparison between surveyed security and privacy approaches.

	Security				Privacy	Safety		
	Confidentiality	Data loss	Data integrity	Intrusion detection	Access control	Activities coordination	Activity monitoring	Action planning
Wang et al. [4]						X	X	
Botterud et al. [70]								X
Bouffard and Galiana [68]								X
Wang and Lu [45]	X		X					
Ancillotti et al. [32]	X		X		X			
MQTT [34]	X							
Oasis [35]	X							
CoAP [33]	X		X					

On the other hand, security has also to be provided by every system. As Table 6.2 shows, the methodologies and case studies assessed basically focus on privacy and integrity of the information to guarantee that all communications are made by approved people and to ensure that the exchanged data is not altered.

Nonetheless, distributed and interlinked security mechanisms are urgent to integrate supreme solutions with a superior intelligence for responsiveness reactions and automated decisions based on M2M communications and M2M security control, without human interventions.

6.5.2.2 Privacy

For all case studies, privacy is considered a focal job. They store and analyze privacy-sensitive data and any unapproved access involves an incredible hazard for the IoT frameworks and for their clients. Nevertheless, as can be seen in Table 6.2, few of the analyzed solutions include mechanisms for the data privacy control, concentrating mainly on applications security and safety.

6.5.2.3 Safety

Many IoT applications work in critical environments where safety techniques must exist for ensuring the correct operation in suspected situations.

While analyzing the reviewed methodologies, we have distinguished that those case studies including an incredible number of devices and requiring an instant actuation more often than not employ one of two methods: activity monitoring and coordination. The former is utilized to precisely know the condition of every component, while the previous is applied to trigger facilitated activities to act over the environment and meet the system's objectives.

But, on the case studies where the reaction ought not be continuous such as those cases found in wind farm, there is a more noteworthy emphasis on the utilization of planned action techniques, since this permits a fine planning of each activity so as to augment the system benefits and safety.

6.5.3 Internet of Things

Table 6.3 summarizes a comparison of the surveyed solutions related to the Internet of Things layer. In the following subsections, we elaborate each layer, explaining the behavior of related systems.

6.5.3.1 Sensors

Sensing is a capacity performed by all those case studies. They all gather sense the surroundings gathering that assist in obtaining credible results, analyze the

Table 6.3 Comparison of the surveyed solution related to the Internet of Things layer.

	Sensors		Actuators	
	Physical sensors	**Virtual sensors**	**Physical actuators**	**Virtual actuators**
Ronnie Burns [78]	X	X		
Bonomi et al. [8]	X	X	X	
Chen et al. [46]		X		
Duan and Xu [83]	X	X		
Baars et al. [82]	X	X		
Watson and Wixom [85]	X	X		
Azvine et al. [84]	X	X		
Ahmed and Kim [79]	X	X		
Rotondo et al. [88]	X	X	X	X
Blesa et al. [87]	X	X	X	X
Al-Sultan et al. [89]	X		X	
Greer et al. [90]	X		X	
Stojmenovic [91]	x		X	

circumstance and make decisions accordingly. Both virtual and physical sensors are used herein.

Few systems just require a sensing element, as an actuation phase is simply not there serving no purpose. More often than not, those systems acquire such data solely for analyzing the surrounding environment, or visualizing reports, or for conducting executive-level strategic decision making, such as establishing a company strategy for upcoming years.

6.5.3.2 Actuators

Real-time responsiveness requires more than sensing. Here is where actuators come into play. They work normally by changing the environments aiming at automatically or semi-automatically achieving desired goals. All case study systems clearly demand physical actuators. However, few of them use virtual actuators.

6.5.4 Data Quality

Table 6.4 shows a comparison of the surveyed solutions related to the Data Quality layer. We next explain the relationship between the surveyed solutions and each component of the layer.

Table 6.4 Comparison of the surveyed solution related to the data quality layer.

	Data normalization			Data filtering			Data aggregation		
	Language	Homogenization	Serialization	Duplicate	Erroneous	Prioritization	Fusion	Hierarchical	Safety
Zao et al. [59]	X		X						
Greer et al. [90]	X		X						
Ancillotti et al. [32]	X								
Chen et al. [46]	X								
Gupta et al. [47]				X	X				
Khandeparkar et al. [101]						X			
Freschi et al. [127]					X		X		
Hong et al. [103]							X		X
Ho et al. [97]							X	X	
Tyagi et al. [98]							X	X	
Ahmed and Kim [79]							X		

6.5.4.1 Data Normalization

Normalization collected is essential in unifying all heterogeneous data coming from different sources under one umbrella. In the case studies analyzed, we have recognized that present solutions are typically founded on proposing normal dialects improving the communication between various devices or nodes or some serialization instruments diminishing the asset consumption that is caused by exchanging data.

6.5.4.2 Data Filtering

Data filtering is a widely spread approach for reducing the sizes of data loads arriving in avalanches. Currently, many solutions employ methods for specifying duplicates and, detecting outliers. On the other part of the story, data prioritization is less famous and receives less attention specifically used in settings where different components of a system receive diverse priorities.

However, filtering critical data can be off-loaded to fog nodes, thus, prioritizing them on the fog environment level.

6.5.4.3 Data Aggregation

Data aggregation can normally be performed by fog nodes with certain computing and storage capacities. The vast majority of the techniques analyzed focus on fusing data, because it is the focal constituent part of this element. Various techniques also exploit a layered-up aggregation, on the price that this implies that common aggregation methods are deployed in each fog node or systems' coordinators have to perfectly know the resources for every fog node and its location so as to establish the aggregation method that is to be employed.

Also, data aggregation safety is yet another aspect widely accepted in critical scenarios. Nevertheless, they also require to know the different fog nodes in order to establish specific encryption and access control mechanisms.

6.5.5 Cloudification

Table 6.5 lists a comparison of the surveyed solution related to the Cloudification layer. In the next subsections, we explain how each component of the Cloudification layer is accommodated by the surveyed systems.

6.5.5.1 Virtualization

To create a fog platform capable of supporting different IoT applications, designers must analyze and clearly define the virtualization technology to use. As can be seen in Table 6.5, Hypervisor is the most extended technology, because is flexible and a wider range of IoT applications are exploiting it for their deployments, because the required OS is emerged in the VM. However, for settings where fog

Table 6.5 Comparison of the surveyed solution related to the cloudification layer.

	Virtualization			Storage	
	Hypervisor	Container	Migration	Fog platform	Network infrastructure
Al Ali et al. [11]	X		X	X	
S. Yi et al. [71, 72]	X		X	X	
Cisco [111]	X	X	X	X	
Truong et al. [50]	X		X	X	
Willis et al. [112]		X	X	X	
Peng [115]					X
Ahlgren et al. [116]					X
Bastug et al. [118]					X

nodes have lower computing resources, or the type of deployed applications, and their requirements, are known and more or less fixed, containers can provide additional advantages, because it is a lightweight solution. Specifically, analyzing the surveyed solutions, those approaches proposing general fog platforms usually implement the hypervisor technology or both. More specific approaches, such as ParaDrop, use containers.

Also, all platforms allow the relocation of VMs. Often, containerized applications are totally relocatable, while platforms depending on the hypervisor technology permit a complete move of the VM images or complete and partial relocation. This is in part due to a large size of hypervisor-based VM images. It worth noting that some approaches propose few efficient algorithms for effectively migrating VMs images using IoT devices for transmitting information without overburdening network shoulders.

6.5.5.2 Storage

When it comes to storage, all fog nodes are welcoming. Naturally, data storage respects a centralized model, storing data in the same fog node that receives it. However, different approaches also implement the distributed storage of information, thus enhancing users' mobility alongside network and capacity of storage but increasing network overload.

Moreover, other approaches are working on directly storing data on the network infrastructure. These solutions allow the deployment of a large number of servers to cache the information in a distributed way. In addition, some proposals are working on predicting the information that will be required by users in order to store it in the network infrastructure close to the users that will request it. These approaches usually do not implement any method for the deployment of IoT applications on these servers, but they can be used to store the information produced by the IoT applications deployed on near fog nodes or incorporated in the fog platform.

6.5.6 Analytics and Decision-Making Layer

Table 6.6 shows a comparison of the surveyed solutions related to the Analytics and Decision-Making layer. In the subsections that follow, we detail the way that related systems are designed to handle each component of the layer.

6.5.6.1 Data Analytics

IoT systems aim mainly at analyzing massive amounts of sensor-collected datasets from diverse contextual conditions, and consecutively extracting a knowledge-base that serves top tiers in stacked-up architectures mainly for visualization and reporting in support for strategic decision making. Adopted data deep analytical

Table 6.6 Comparison of the surveyed solution related to the analytics and decision-making layer.

	Data analytics	Decision-making	
	Geo-distribution	**Predictive**	**Reactive**
Hong et al. [103]	x		
Botterud et al. [70]	X	X	X
Bouffard and Galiana [68]	X		
Chen et al. [46]	X		
Allalouf et al. [96]	X		
Ancillotti et al. [32]	X		
Uyar and Türkay [122]		X	
Huang et al. [121]		X	
Erickson et al. [123]		X	
Molderink et al. [124]	X	X	X
Erickson et al. [125]		X	
Truong et al. [50]			X
Bonomi et al. [8]	X	X	X

techniques depend highly on the volume of the data in addition to its being time-sensitive, safety-sensitive or alike. Those techniques, often, consider the geo-locality of the arrived data so as to decide keeping it in fog or turning it on to a remote resource-rich cloud. Thus, the stringent the temporal constrains and the smaller the data set to analyze, the greater the number of techniques that should be relegated to the fog nodes. Instead, if the data set is very large, the analytics techniques are resource consuming, and the temporal requirements are not very strict, a higher number of these techniques should be executed in the cloud.

Referring to our case studies, we know intuitively that most of them require instant decision making, thus are relying highly on processing some tasks near the edge, in fog specifically. However, some systems, such as those of wind farms, which normally require long-haul forecasting analytics more than often delegate such compute-intensive tasks to a resourceful cloud. It worth mentioning however that hybridizing capacities of the both worlds, small data analytics and big data analytics, is capable of providing the system with a performance boost.

6.5.6.2 Decision-Making

Upon analyzing collected data, the resulting information is normally used for making different decisions. If the decision-making process is automated or semi-automated, it can be performed in both the cloud and the fog nodes. As mentioned earlier, this depends, for the most parts, on the target response time, in such a way that those that are time-sensitive can perform some parts of their jobs near the edge (in fog), whereas those with more relaxed time constraints can wait normally for the whole cycle to complete, from sending collected data remotely to cloud all the way back to visualization and reporting In this sense, the data analytics and the decision-making elements are mostly related, since their distribution between fog and cloud nodes is usually similar.

Wind Farm, smart building, and smart grid are those scenarios such that the temporal requirements are normally eased, so many of their decision-making functionalities can be relegated to the cloud.

On the other hand, in critical systems, such as STL and SCV, the low-latency reaction is a key necessity and is pivotal for ensuring safety. These environments are predominantly reactive.

Nevertheless, most systems comprise various levels constituting the decision-making process hybridizing both models at different degrees.

6.6 Challenges and Recommended Research Directions

Fog is a promising powerful computational model that is expected to highly boost IoT applications' performance by scales of magnitudes. However, many challenges remain still in the path for a production-grade fog setting. Prioritizing their research is essential for taking current initiatives way steps further. We recommend the following prioritization list in this domain, to be considered as potential future research frontiers that characterize significantly the way that this paradigm may spread out within the premises of the next decade:

- *Multilevel organization.* Depending on the target application and the associated scenarios where an IoT setting is deployed, one can define a multilevel structure that is constituting of groups of nodes and interwoven and their constituent subgroups. In such a staggered organization, every fog node serves a particular job and responsibility. More often than not, in real-case applications, fog nodes ought to be organized into a hierarchical structure with a potential elasticity and scalability.
- *Node specialization.* In hierarchical structures or mesh typologies, each fog node is specialized to perform a specific work and it is optimized to handle each specific issue or weakness, with significant performance improvements. In addition, each node can be composed differently, with specific and concentrate resources in order to enhance its ability to perform the tasks assigned. Every setting should design fog in a way that optimizes the overall system's performance. For example, in wide-scale vehicular applications, fog nodes may be really heterogeneous, with very strong mobility and geo-distributed abilities outside the vehicle because they normally administer quick mobility endpoints in widespread real geometries and, on the contrary, inside the car, nodes concentrate resources on strong sensing and actuation phases to take the right decisions, timely, among static sensors.
- *Actuation capacity.* Fog computing is characterized by the interplay with cloud computing and many use cases are strongly based on it to analyze the situation and take the best decisions. An overarching trait of fog computing is that of its ability to sense whether an instant actuation is urgent, which triggers an action directly by fog itself, or otherwise the data can be sent safely all the trip down to a cloud where its slowly sending back results for a noninstant decision making. Depending on the environment where a fog setting is implanted, a set of rules should be able to define instant actuations and other cloud-depending relatively slow reactions. Moreover, each application has to define the level of fog-cloud interwoven. To take a more utilitarian perspective, in a vehicular IoT setting, cloud analytical capacities are normally employed for evaluating traffic patterns, thus detecting favorable ways for reaching out a destination. On the other hand, fog is responsible for real-time analysis related to the vehicle insides.
- *Efficient fog-cloud communications.* Fog computing often upload data to the cloud and similarly the cloud can send data and actuation commands/strategies to the fog. We aim to transmit as much data as possible, without generating traffic congestion and performance degradation that can slow down the system or too many iterations that can raise cost. It would be interesting to detect algorithms that are able to optimize these communications and improve fog

computing components toward an efficient fog-cloud interface. Regarding the specific scenario, it is possible to choose the type of interactions that better suits the specific use case: in (1) latency-tolerant applications, the interplay with cloud communication can be reinforced and more frequent, with intensive batching communications that exchange heavy data and increase the focus on big data processing even to react to a current situation; in (2) latency-sensitive applications, the interplay between fog and cloud must be accurately planned and make the most of it with few interactions and raising small data processing importance for those systems. Moreover, independently to the specific application, the implementation of lightweight and efficient M2M protocols improving the communication and optimizing the resources used to exchange data are required.

- *Efficient distributed data processing*. Fog computing is responsible for managing the handling of massive amounts of fast arriving sensor real-time data, thus conforming to the hard requirements of real-time insights analytics. Sensing, data aggregation, data filtering, data analytics, fault detection techniques, Big Data and Small Data processing at different roles, must understand which data are useful and which can be wasted because are only "ground noise," process data and, if required, be able to store only a minimal quantity of them. All these phases should be improved to speed-up performance and systems accuracy, for instance, with balancing the load computations or, if necessary, delegating few computing and storage jobs for resource-rich nodes.
- *Interworking of different Fog localities*. Fog deployments are widespread into environment localities and each fog locally coordinates a subgroup of IoT devices. Demand are raised for ways that are able to coordinate fog efforts in ways that enables achieving unified global objectives via the interwoven of various networks that share different information among different multi-level fog nodes or nodes spread out in various locations that have the awareness of several application scenarios. In this way, it could be possible to aggregate data from sensors located in different networks and with the possibility to use actuators spread in different networks too, extending the sensing-actuation cycle within larger areas, thus offering services in a more distributed fashion, with more complete functionalities and, thus, creating more sophisticated analysis and processing information more effectively.

In conclusion, we believe fog computing is a promising concept that has the potentiality to be an enabler and a significant driver for IoT environments. Further research is needed, and many challenges must be solved to establish the fog role practically and to enable industrially deployed IoT, in particular in critical and dynamic real-world applications.

6.7 Concluding Remarks

Cloud computing provides an elastic on-demand resource provisioning that is capable of achieving high big data processing throughputs in a fashion that was impossible before. Acting in parallel, Cloud deployments compete with server-based beefed architectures, gathering colloquial power in a recipe that is driven by appropriate share of computing resources. Having said that, IoT have been driven by such an evolution, which motivates and incentivize gathering huge amounts of data in the front-ends, knowing that back-end powers are available to process them in timely fashion aiming at actionable real-time insights. However, the past decade or so has witnessed an unprecedented adoption of IoT devices in all aspects of life, presenting the current two-tier cloud-IoT architectures with new challenges that exceeds their capacities.

Consider real-life end-to-end scenarios that integrate storage back-ends with serving systems and batch jobs in a fault-tolerant and consistent fashion that aims at keeping decision makers in the know. Those pipelines normally run into multiple complications that are not affecting traditional database management systems (DBMSs) or beefed-up server-based architectures. Those are complications that fog sets out to solve. By pushing down some processing loads toward edge or fog nodes, so as for those to serve as quick-and-clean sieve that filters unnecessarily downstream-loaded data (those that do not contribute to a critical result). This normally comes at the cost of higher consolidation requirements that spans multiple layers of the envisioned cloud-fog-IoT architecture. Works in the relevant literature have focused mainly on discrete aspects of fog-enabled IoT deployments.

Beyond its theoretical impact, fog technologies are gaining momentum and regarded more appropriate for scenarios that seek exact results. Healthcare, city planning, industrial IoT and Industry 4.0 (I4.0) are just few examples.

Relevant literature has mainly geared efforts toward cloud-only architectures, which easily become a bottleneck in sudden data arrival spike scenarios, where it become brittle and prone to failures. A situation that is mostly solved by approximation, which does not play well with fault tolerance.

We provide a high-level perspective pictured in a multilayer architecture. However, functionalities are nowadays hidden in black boxes so that a user does not have to reason about low-level aspects of communication and near-edge light processing. To our knowledge, there no conclusive work that gathers technologies developed for IoT requirement's support in a systematic manner. Also, there is no work that colloquially define (or at best are ill-defined) the role that each participated component of each layer plays in the background of the interconnected components that constitute our architecture. To this end, this chapter aims at analyzing the main IoT applications needs. We have conceptualized a global model of a fog architectural model satisfying the analyzed needs and a taxonomy so as to compare and contrast various proposals and elucidate their applicability in IoT domains to specific domains.

Stated another way, we herein draw a map that constitutes a reference guideline that assists practitioners in designing multilayered fog-enabled deployments, thus fostering and incentivizing a faster adoption. Our proposed architecture is modular and seamlessly allows incorporating elements in a hot-swappable manner. We mainly aim at achieving an overarching objective of better exploiting fog computing in accomplishing a seamless transition from a loosely coordinated set of existing cloud-based solutions to a fit-for-purpose fog-enabled architectures, which granularly aims at filling existing in-situ cloud-based architectural gaps.

We detail an explanation for all enabling technologies in order for this vision to see the light. From communication hardware-assisted levels all the way downhill to complex analytics.

This serves as a conclusive study that demonstrates systemically ways of interleaving the three worlds into an entwined architecture such as the one we propose in this chapter. We further expand on an "alchemy" behind a successful hybridization among the three elements. The purpose of this chapter is unveiling a rollout of a new working architecture through which we analyze and discuss our approach. We demonstrated it in a systematic way that aims at enabling it to serve as a compass for research and practice in the field of fog computing.

References

1 Vaquero, L.M. and Rodero-Merino, L. (2014). Finding your way in the fog: towards a comprehensive definition of fog computing. *SIGCOMM Computer Communication Review* 44: 27–32. https://doi.org/10.1145/2677046.2677052.

2 Bellavista, P., Cinque, M., Cotroneo, D., and Foschini, L. (2005). Integrated support for handoff management and context awareness in heterogeneous wireless networks. In: *Proceedings of the 3rd International Workshop on Middleware for Pervasive and Ad-Hoc Computing, MPAC'05*, 1–8. New York, NY, USA: ACM https://doi.org/10.1145/1101480.1101495.

3 Podnar Zarko, L., Antonic, A., and Pripužic, K. (2013). Publish/subscribe middleware for energy-efficient Mobile Crowdsensing. In: *Proceedings of the 2013 ACM Conference on Pervasive and Ubiquitous Computing Adjunct Publication, UbiComp '13 Adjunct*, 1099–1110. New York, NY, USA: ACM https://doi.org/10.1145/2494091.2499577.

4 Wang, W., Lee, K., and Murray, D. (2012). Integrating sensors with the cloud using dynamic proxies. In: *2012 IEEE 23rd International Symposium on Personal, Indoor and Mobile Radio Communications – (PIMRC)*, 1466–1471. Sydney, NSW, Australia: IEEE https://doi.org/10.1109/PIMRC.2012.6362579.

5 Yannuzzi, M., Milito, R., Serral-Gracià, R. et al. (2014). Key ingredients in an IoT recipe: fog computing, cloud computing, and more fog computing. In: *2014 IEEE 19th International Workshop on Computer Aided Modeling and Design of Communication Links and Networks (CAMAD)*, 325–329. Athens, Greece: IEEE https://doi.org/10.1109/CAMAD.2014.7033259.

6 Li, C., Xue, Y., Wang, J. et al. (2018). Edge-oriented computing paradigms: a survey on architecture design and system management. *ACM Computing Surveys* 51, pp. 39:1–39:34 doi: 10.1145/3154815.

7 Ferrer, A.J., Marquès, J.M., and Jorba, J. (2019). Towards the decentralised cloud: survey on approaches and challenges for Mobile, ad hoc, and edge computing. *ACM Computing Surveys* 51, pp. 111:1–111:36 https://doi.org/10.1145/3243929.

8 Bonomi, F., Milito, R., Natarajan, P., and Zhu, J. (2014). Fog computing: a platform for Internet of things and analytics. In: *Big Data and Internet of Things: A Roadmap for Smart Environments, Studies in Computational Intelligence*, 169–186. Cham:Springer https://doi.org/10.1007/978-3-319-05029-4_7.

9 Bonomi, F., Milito, R., Zhu, J., and Addepalli, S. (2012). Fog computing and its role in the Internet of things. In: *Proceedings of the First Edition of the MCC Workshop on Mobile Cloud Computing, MCC '12*, 13–16. New York, NY, USA: ACM https://doi.org/10.1145/2342509.2342513.

10 Wang, J., Pan, J., Esposito, F. et al. (2019). Edge cloud off-loading algorithms: issues, methods, and perspectives. *ACM Computing Surveys* 52, pp. 2:1–2:23 doi: 10.1145/3284387.

11 Al Ali, R., Gerostathopoulos, I., Gonzalez-Herrera, I. et al. (2014). An architecture-based approach for compute-intensive pervasive systems in dynamic environments. In: *Proceedings of the 2nd International Workshop on Hot Topics in Cloud Service Scalability, HotTopiCS'14*. New York, NY, USA, pp. 3:1–3:6: ACM https://doi.org/10.1145/2649563.2649577.

12 Satyanarayanan, M., Bahl, P., Caceres, R., and Davies, N. (2009). The case for VM-based cloudlets in mobile computing. *IEEE Pervasive Computing* 8: 14–23. https://doi.org/10.1109/MPRV.2009.82.

13 Satyanarayanan, M., Lewis, G., Morris, E. et al. (2013). The role of cloudlets in hostile environments. *IEEE Pervasive Computing* 12: 40–49. https://doi.org/10.1109/MPRV.2013.77.

14 Firdhous, M., Ghazali, O., Hassan, S., and Publications, S.D.I.W.C. (2014). Fog computing: will it be the future of cloud computing? In: *The Third International Conference on Informatics & Applications (ICIA2014)*. Malaysia: Kuala Terengganu.

15 Davy, S., Famaey, J., Serrat, J. et al. (2014). Challenges to support edge-as-a-service. *IEEE Communications Magazine* 52: 132–139. https://doi.org/10.1109/MCOM.2014.6710075.

16 Bifulco, R., Brunner, M., Canonico, R. et al. (2012). Scalability of a mobile cloud management system. In: *Proceedings of the First Edition of the MCC Workshop on Mobile Cloud Computing, MCC '12*, 17–22. New York, NY, USA: ACM https://doi.org/10.1145/2342509.2342514.

17 Taleb, T. and Ksentini, A. (2013). Follow me cloud: interworking federated clouds and distributed mobile networks. *IEEE Network* 27: 12–19. https://doi.org/10.1109/MNET.2013.6616110.

18 Taleb, T., Hasselmeyer, P., and Mir, F.G. (2013). Follow-me cloud: An OpenFlow-based implementation. In: *2013 IEEE International Conference on Green Computing and Communications (GreenCom) and IEEE Internet of Things (iThings) and IEEE Cyber, Physical and Social Computing(CPSCom)*, 240–245. Beijing, China: IEEE https://doi.org/10.1109/GreenCom-iThings-CPSCom.2013.59.

19 OEC, Open edge computing [WWW document], 2019. http://openedgecomputing.org (accessed October 29, 2019).

20 OFC, OpenFog Consortium, 2019. https://www.openfogconsortium.org (accessed February 14, 2019).

21 S. Dahmen-Lhuissier, Multi-access edge computing [WWW document], 2019. ETSI. http://www.etsi.org/technologies-clusters/technologies/multi-access-edge-computing (accessed October 29, 2019).

22 OF Reference, OpenFog Reference Architecture for Fog Computing, OpenFog Consortium, 2019. https://www.openfogconsortium.org/wp-content/uploads/OpenFog_Reference_Architecture_2_09_17-FINAL.pdf (accessed September 5, 2019).

23 Borgia, E. (2014). The Internet of things vision: key features, applications and open issues. *Computer Communications* 54: 1–31. https://doi.org/10.1016/j.comcom.2014.09.008.

24 Diallo, S., Herencia-Zapana, H., Padilla, J.J., and Tolk, A. (2011). *Understanding interoperability, Proceedings of the 2011 Emerging M&S Applications in Industry and Academia Symposium, Boston, MA*, 84–91. New York, NY, USA: ACM.

25 Bellavista, P., Corradi, A., and Magistretti, E. (2005). REDMAN: an optimistic replication middleware for read-only resources in dense MANETs. *Pervasive and Mobile Computing* 1: 279–310. https://doi.org/10.1016/j.pmcj.2005.06.002.

26 Toninelli, A., Pantsar-Syväniemi, S., Bellavista, P., and Ovaska, E. (2009). Supporting context awareness in smart environments: a scalable approach to information interoperability. In: *Proceedings of the International Workshop on*

Middleware for Pervasive Mobile and Embedded Computing, M-PAC '09. New York, NY, USA, pp. 5:1–5:4: ACM https://doi.org/10.1145/1657127.1657134.

27 Stojmenovic, I. and Wen, S. (2014). The fog computing paradigm: scenarios and security issues. In: *2014 Federated Conference on Computer Science and Information Systems, FedCSIS 2014*, 1–8. Warsaw, Poland: IEEE https://doi.org/10.15439/2014F503.

28 Perera, C., Qin, Y., Estrella, J.C. et al. (2017). Fog computing for sustainable smart cities: a survey. *ACM Computing Surveys* 50, pp. 32:1–32:43 https://doi.org/10.1145/3057266.

29 Puliafito, C., Mingozzi, E., Longo, F. et al. (2019). Fog computing for the Internet of things: a survey. *ACM Transactions on Internet Technology* 19, pp. 18:1–18:41 https://doi.org/10.1145/3301443.

30 Ferrer-Roca, O., Tous, R., and Milito, R. (2014). Big and small data: the fog. In: *Presented at the 2014 International Conference on Identification, Information and Knowledge in the Internet of Things*, 260–261. Beijing, China: IEEE https://doi.org/10.1109/IIKI.2014.60.

31 Al-Fuqaha, A., Guizani, M., Mohammadi, M. et al. (2015). Internet of things: a survey on enabling technologies, protocols, and applications. *IEEE Communication Surveys and Tutorials* 17: 2347–2376. https://doi.org/10.1109/COMST.2015.2444095.

32 Ancillotti, B.R. and Conti, M. (2013). The role of communication systems in smart grids: architectures, technical solutions and research challenges. *Computer Communications* 36: 1665–1697. https://doi.org/10.1016/j.comcom.2013.09.004.

33 CoAP, CoAP – Constrained Application Protocol [WWW document], 2019. http://coap.technology (accessed October 24, 2019).

34 MQTT, MQTT – Message Queue Telemetry Transport, 2019. http://mqtt.org (accessed October 24, 2019).

35 Oasis, AMQP – Advanced Message Queuing Protocol [WWW document], 2019. https://www.amqp.org (accessed October 24, 2019).

36 DDS, DDS – Data Distribution Services [WWW document], 2019. http://portals.omg.org/dds (accessed October 24, 2019).

37 Zigbee, Zigbee, 2019. http://www.zigbee.org (accessed October 24, 2019).

38 ISO, UPnP – ISO/IEC 29341-1:2011 Device Architecture [WWW document], 2019. https://www.iso.org/standard/57195.html (accessed October 24, 2019).

39 T. Winter, P. Thubert, A. Brandt et al., RPL: IPv6 Routing Protocol for Low-Power and Lossy Networks [WWW document], 2019. https://tools.ietf.org/html/rfc6550 (accessed October 24, 2019).

40 IEFT WG, 6LoWPAN – IPv6 over Low-Power Wireless Area Networks [WWW document], 2019. http://6lowpan.tzi.org (accessed Octoer 24, 2019).

41 Bluetooth SIG, BLE – Bluetooth Low Energy [WWW document], 2019. www.bluetooth.com (accessed October 24, 2019).

42 Qualcomm, LTE Advanced [WWW document]. Qualcomm, 2014. https://www.qualcomm.com/invention/technologies/lte/advanced (accessed October 24, 2019).

43 D. Farinacci, D. Lewis, D. Meyer, and V. Fuller, LISP – The Locator/ID Separation Protocol [WWW document], 2019. https://tools.ietf.org/html/rfc6830 (accessed October 24, 2019).

44 Bartoli, A., Dohler, M., Hernández-Serrano, J. et al. (2011). Low-power low-rate goes long-range: the case for secure and cooperative machine-to-machine communications. In: *NETWORKING 2011 Workshops, Lecture Notes in Computer Science* (eds. V. Casares-Giner, P. Manzoni and A. Pont), 219–230. Springer Berlin Heidelberg.

45 Wang, W. and Lu, Z. (2013). Survey cyber security in the smart grid: survey and challenges. *Computer Networks* 57: 1344–1371. https://doi.org/10.1016/j.comnet.2012.12.017.

46 Chen, H., Chou, P., Duri, S. et al. (2009). The design and implementation of a smart building control system. In: *2009 IEEE International Conference on e-Business Engineering*, 255–262. Macau, China: IEEE https://doi.org/10.1109/ICEBE.2009.42.

47 Gupta, M., Krishnanand, K.R., Chinh, H.D., and Panda, S.K. (2015). Outlier detection and data filtering for wireless sensor and actuator networks in building environment. In: *2015 IEEE International Conference on Building Efficiency and Sustainable Technologies*, 95–100. Singapore: IEEE https://doi.org/10.1109/ICBEST.2015.7435872.

48 Gungor, V.C., Sahin, D., Kocak, T. et al. (2011). Smart grid technologies: communication technologies and standards. *IEEE Transactions on Industrial Informatics* 7: 529–539. https://doi.org/10.1109/TII.2011.2166794.

49 Selvarajah, K., Tully, A., and Blythe, P.T. (2008). ZigBee for intelligent transport system applications. In: *IET Road Transport Information and Control – RTIC 2008 and ITS United Kingdom Members' Conference*, 1–7. Manchester, UK: IEEE https://doi.org/10.1049/ic.2008.0814.

50 Truong, N.B., Lee, G.M., and Ghamri-Doudane, Y. (2015). Software-defined networking-based vehicular ad hoc network with fog computing. In: *2015 IFIP/IEEE International Symposium on Integrated Network Management (IM)*, 1202–1207. Ottawa, Canada: IEEE https://doi.org/10.1109/INM.2015.7140467.

51 Frank, R., Bronzi, W., Castignani, G., and Engel, T. (2014). Bluetooth low energy: an alternative technology for VANET applications. In: *2014 11th Annual Conference on Wireless on-Demand Network Systems and Services (WONS)*, 104–107. Obergurgl, Austria: IEEE https://doi.org/10.1109/WONS.2014.6814729.

52 Lee, B., An, S., and Shin, D. (2011). A remote control service for OSGi-based unmanned vehicle using SmartPhone in ubiquitous environment. In: *2011 Third International Conference on Computational Intelligence, Communication Systems and Networks*, 158–163. Bali, Indonesia: IEEE https://doi.org/10.1109/CICSyN.2011.43.

53 Park, P., Yim, H., Moon, H., and Jung, J. (2009). An OSGi based in-vehicle gateway platform architecture for improved sensor extensibility and interoperability. In: *2009 33rd Annual IEEE International Computer Software and Applications Conference*, 140–147. Seattle, WA: IEEE https://doi.org/10.1109/COMPSAC.2009.203.

54 Kim, J.E., Boulos, G., Yackovich, J. et al. (2012). Seamless integration of heterogeneous devices and access control in smart homes. In: *2012 Eighth International Conference on Intelligent Environments*, 206–213. Guanajuato, Mexico: IEEE https://doi.org/10.1109/IE.2012.57.

55 Koß, D., Bytschkow, D., Gupta, P.K. et al. (2012). Establishing a smart grid node architecture and demonstrator in an office environment using the SOA approach. In: *2012 First International Workshop on Software Engineering Challenges for the Smart Grid (SE-SmartGrids)*, 8–14. Switzerland: IEEE, Zurich https://doi.org/10.1109/SE4SG.2012.6225710.

56 Bharghavan, V., Demers, A., Shenker, S., and Zhang, L. (1994). MACAW: a media access protocol for wireless LAN's. In: *Proceedings of the Conference on Communications Architectures, Protocols and Applications, SIGCOMM'94*, 212–225. New York, NY, USA: ACM https://doi.org/10.1145/190314.190334.

57 Gnawali, O., Yarvis, M., Heidemann, J., and Govindan, R. Interaction of retransmission, blacklisting, and routing metrics for reliability in sensor network routing. In: *2004 First Annual IEEE Communications Society Conference on Sensor and Ad Hoc Communications and Networks, 2004*. IEEE SECON 2004, 34–43. Santa Clara, CA: IEEE https://doi.org/10.1109/SAHCN.2004.1381900.

58 Thangavel, D., Ma, X., Valera, A. et al. (2014). Performance evaluation of MQTT and CoAP via a common middleware. In: *2014 IEEE Ninth International Conference on Intelligent Sensors, Sensor Networks and Information Processing (ISSNIP)*, 1–6. Singapore: IEEE https://doi.org/10.1109/ISSNIP.2014.6827678.

59 Zao, J.K., Gan, T.T., You, C.K. et al. (2014). Augmented brain computer interaction based on fog computing and linked data. In: *2014 International Conference on Intelligent Environments*, 374–377. Shanghai, China: IEEE https://doi.org/10.1109/IE.2014.54.

60 Amel, B.N., Rim, B., Houda, J. et al. (2014). Flexray versus Ethernet for vehicular networks. In: *2014 IEEE International Electric Vehicle Conference (IEVC)*, 1–5. Florence, Italy: IEEE https://doi.org/10.1109/IEVC.2014.7056123.

61 Varun, Menon, G. (2017). Moving from vehicular cloud computing to vehicular fog computing:issues and challenges. *International Journal of Computational Science and Engineering* 9: 14–18.

62 Hakiri, A., Berthou, P., Gokhale, A. et al. (2014). Supporting SIP-based end-to-end data distribution service QoS in WANs. *Journal of Systems and Software* 95: 100–121. https://doi.org/10.1016/j.jss.2014.03.078.

63 Rusitschka, S., Eger, K., and Gerdes, C. (2010). Smart grid data cloud: a model for utilizing cloud computing in the smart grid domain. In: *2010 First IEEE International Conference on Smart Grid Communications, IEEE, Gaithersburg, MD, USA*, 483–488. https://doi.org/10.1109/SMARTGRID.2010.5622089.

64 Lee, K.C., Sudhaakar, R., Ning, J. et al. (2012, 2012). A comprehensive evaluation of RPL under mobility. *International Journal of Vehicular Technology*: 1–10. https://doi.org/10.1155/2012/904308.

65 Ancillotti, B.R. and Conti, M. (2013). The role of the RPL routing protocol for smart grid communications. *IEEE Communications Magazine* 51: 75–83. https://doi.org/10.1109/MCOM.2013.6400442.

66 UPNP Forum, Leveraging UPNP+. The next generation of universal interoperability, 2015. [online] Available at http://upnp.org/resources/whitepapers/UPnP_Plus_Whitepaper_2015.pdf.

67 Seo, H.S., Kim, B.C., Park, P.S. et al. (2013). Design and implementation of a UPnP-can gateway for automotive environments. *International Journal of Automotive Technology* 14: 91–99. https://doi.org/10.1007/s12239-013-0011-5.

68 Bouffard, F. and Galiana, F.D. (2008). Stochastic security for operations planning with significant wind power generation. *IEEE Transactions on Power Apparatus and Systems* 23: 306–316. https://doi.org/10.1109/TPWRS.2008.919318.

69 Gupta, H., Chakraborty, S., Ghosh, S.K., and Buyya, R. (2017). Fog computing in 5G networks: an application perspective. In: *Cloud and Fog Computing in 5G Mobile Networks* (eds. E. Markakis, G. Mastorakis, C.X. Mavromoustakis and E. Pallis), 23–56. London: The Institution of Engineering and Technology https://doi.org/10.1049/PBTE070E_ch2.

70 Botterud, A., Wang, J., Monteiro, C., and Mir, V. (2009). *Wind Power Forecasting and Electricity Market Operations*. Argonne National Laboratory.

71 Yi, S., Hao, Z., Qin, Z., and Li, Q. (2015). Fog computing: platform and applications. In: *Presented at the 2015 Third IEEE Workshop on Hot Topics in Web Systems and Technologies (HotWeb)*, 73–78. Washington, DC: IEEE https://doi.org/10.1109/HotWeb.2015.22.

72 Shanhe Yi, Z. and Qin, Q.L. (2015). Security and privacy issues of fog computing: a survey. In: *Wireless Algorithms, Systems, and Applications: 10th*

International Conference, WASA 2015, Qufu, China, August 10–12, 2015, Proceedings (eds. K. Xu and H. Zhu), 685–695. Cham: Springer International Publishing https://doi.org/10.1007/978-3-319-21837-3_67.

73 Kumari, M. and Nath, R. (2015). Security concerns and countermeasures in cloud computing paradigm. In: *Proceedings of 2015 Fifth International Conference on Advanced Computing & Communication Technologies (ACCT 2015)*, 534–540. IEEE https://doi.org/10.1109/ACCT.2015.80.

74 Sivathanu, G., Wright, C.P., and Zadok, E. (2005). Ensuring data integrity in storage: techniques and applications. In: *Proceedings of the 2005 ACM Workshop on Storage Security and Survivability, StorageSS '05*, 26–36. New York, NY, USA: ACM https://doi.org/10.1145/1103780.1103784.

75 Stojmenovic, I., Wen, S., Huang, X., and Luan, H. (2016). An overview of fog computing and its security issues. *Concurrency and Computation: Practice and Experience* 28: 2991–3005. https://doi.org/10.1002/cpe.3485.

76 Aazam, M. and Huh, E.N. (2015). Dynamic resource provisioning through fog micro datacenter. In: *2015 IEEE International Conference on Pervasive Computing and Communication Workshops (PerCom Workshops)*, 105–110. St. Louis, MO, USA: IEEE https://doi.org/10.1109/PERCOMW.2015.7134002.

77 Kabadayi, S., Pridgen, A., and Julien, C. (2006). Virtual sensors: abstracting data from physical sensors. In: *2006 International Symposium on a World of Wireless, Mobile and Multimedia Networks(WoWMoM'06)*, vol. 592, 6. Buffalo–Niagara Falls, NY, USA: IEEE https://doi.org/10.1109/WOWMOM .2006.115.

78 Ronnie Burns, Method and system for providing personalized traffic alerts,. US Patent 6590507 B2, 2003.

79 Ahmed, M.A. and Kim, Y.-C. (2016). Wireless communication architectures based on data aggregation for internal monitoring of large-scale wind turbines. *International Journal of Distributed Sensor Networks* 12 https://doi.org/10.1177/1550147716662776.

80 Di Liping (2007). Geospatial sensor web and self-adaptive earth predictive systems (SEPS). In: *Proceedings of the Earth Science Technology Office (ESTO)/Advance Information Systems Technology (AIST) Sensor Web Principal Investigator (PI). Presented at the NASA AIST PI Conference, San Diego, CA*, 1–4.

81 Srivastava, A.N., Oza, N.C., and Stroeve, J. (2005). Virtual sensors: using data mining techniques to efficiently estimate remote sensing spectra. *IEEE Transactions on Geoscience and Remote Sensing* 43: 590–600. https://doi.org/10.1109/TGRS.2004.842406.

82 Baars, H., Kemper, H.G., Lasi, H., and Siegel, M. (2008). Combining RFID technology and business intelligence for supply chain #x0A0: optimization scenarios for retail logistics. In: *Proceedings of the 41st Annual Hawaii International Conference on System Sciences (HICSS 2008)*, 73–73. Waikoloa, HI, USA: IEEE https://doi.org/10.1109/HICSS.2008.93.

83 Duan, L. and Xu, L.D. (2012). Business intelligence for enterprise systems: a survey. *IEEE Transactions on Industrial Informatics* 8: 679–687. https://doi.org/10.1109/TII.2012.2188804.

84 Azvine, B., Cui, Z., and Nauck, D.D. (2005). Towards real-time business intelligence. *BT Technology Journal* 23: 214–225. https://doi.org/10.1007/s10550-005-0043-0.

85 Watson, H.J. and Wixom, B.H. (2007). The current state of business intelligence. *Computer* 40: 96–99. https://doi.org/10.1109/MC.2007.331.

86 Azzara, A. and Mottola, L. (2015). Virtual resources for the Internet of Tthings. In: *2015 IEEE 2nd World Forum on Internet of Things (WF-IoT)*, 245–250. Milan, Italy: IEEE https://doi.org/10.1109/WF-IoT.2015.7389060.

87 Blesa, J., Rotondo, D., Puig, V., and Nejjari, F. (2014). FDI and FTC of wind turbines using the interval observer approach and virtual actuators/sensors. *Control Engineering Practice* 24: 138–155. https://doi.org/10.1016/j.conengprac.2013.11.018.

88 Rotondo, D., Nejjari, F., and Puig, V. (2014). A virtual actuator and sensor approach for fault tolerant control of LPV systems. *Journal of Process Control* 24: 203–222. https://doi.org/10.1016/j.jprocont.2013.12.016.

89 Al-Sultan, S., Al-Doori, M.M., Al-Bayatti, A.H., and Zedan, H. (2014). A comprehensive survey on vehicular ad hoc network. *Journal of Network and Computer Applications* 37: 380–392. https://doi.org/10.1016/j.jnca.2013.02.036.

90 C. Greer, D.A. Wollman, D.E. Prochaska et al., NIST framework and roadmap for smart grid interoperability standards, Release 3.0. Special Publication NIST SP 1108r3, National Institute of Standards and Technology, 2014. https://doi.org/916755

91 Stojmenovic, I. (2014). Fog computing: a cloud to the ground support for smart things and machine-to-machine networks. In: *2014 Australasian Telecommunication Networks and Applications Conference (ATNAC)*, 117–122. Southbank, VIC, Australia: IEEE https://doi.org/10.1109/ATNAC.2014.7020884.

92 Nayak, S., Misra, B.B., and Behera, H.S. (2013). Impact of data normalization on stock index forecasting. *International Journal of Computer Information Systems and Industrial Management Applications* 6: 257–269.

93 Díaz, M., Martín, C., and Rubio, B. (2016). State-of-the-art, challenges, and open issues in the integration of internet of things and cloud computing. *Journal of Network and Computer Applications* 67: 99–117. https://doi.org/10.1016/j.jnca.2016.01.010.

94 Piqi, Pigi – The Piqi Project, 2019. http://piqi.org (accessed October 25, 2019).

95 Google, Protocol buffers [WWW Document]. Google Developers, 2019. https://developers.google.com/protocol-buffers (accessed October 25, 2019).

96 Allalouf, M., Gershinsky, G., Lewin-Eytan, L., and Naor, J. (2011). Data-quality-aware volume reduction in smart grid networks. In: *2011 IEEE International Conference on Smart Grid Communications (SmartGridComm)*, 120–125. IEEE https://doi.org/10.1109/SmartGridComm.2011.6102302.

97 Ho, Q.D., Gao, Y., and Le-Ngoc, T. (2013). Challenges and research opportunities in wireless communication networks for smart grid. *IEEE Wireless Communication* 20: 89–95. https://doi.org/10.1109/MWC.2013.6549287.

98 Tyagi, S., Ansari, A.Q., and Khan, M.A. (2010). Dynamic threshold based sliding-window filtering technique for RFID data. In: *2010 IEEE 2nd International Advance Computing Conference (IACC)*, 115–120. Patiala, India: IEEE https://doi.org/10.1109/IADCC.2010.5423025.

99 Tang, L.-A., Han, J., and Jiang, G. (2014). Mining sensor data in cyber-physical systems. *Tsinghua Science and Technology* 19: 225–234. https://doi.org/10.1109/TST.2014.6838193.

100 Smith, S. (2013). *Digital Signal Processing: A Practical Guide for Engineers and Scientists*. Elsevier.

101 Khandeparkar, K., Ramamritham, K., and Gupta, R. (2017). QoS-driven data processing algorithms for smart electric grids. *ACM Transactions on Cyber-Physical Systems* 1, pp. 14:1–14:24 https://doi.org/10.1145/3047410.

102 Sang, Y., Shen, H., Inoguchi, Y. et al. (2006). Secure data aggregation in wireless sensor networks: a survey. In: *2006 Seventh International Conference on Parallel and Distributed Computing, Applications and Technologies (PDCAT'06)*, 315–320. Taipei, Taiwan: IEEE https://doi.org/10.1109/PDCAT.2006.96.

103 Hong, K., Lillethun, D., Ramachandran, U. et al. (2013). Mobile fog: a programming model for large-scale applications on the internet of things. In: *Proceedings of the Second ACM SIGCOMM Workshop on Mobile Cloud Computing, MCC '13*, 15–20. New York, NY, USA: ACM https://doi.org/10.1145/2491266.2491270.

104 OpenNebula, OpenNebula – flexible enterprise cloud made simple, 2019. https://opennebula.org (accessed October 26, 2019).

105 Canonical, Linux containers [WWW document], 2019. https://linuxcontainers.org (accessed October 26, 2019).

106 Docker, Docker [WWW document], 2019. Docker. https://www.docker.com (accessed October 26, 2019).

107 Kozuch, M. and Satyanarayanan, M. (2002). The internet suspend/resume (ISR). In: *4th IEEE Workshop Mobile Computing Systems and Applications*. IEEE CS Press.

108 Clark, C., Fraser, K., Hand, S. et al. (2005). Live migration of virtual machines. In: *Proceedings of the 2Nd Conference on Symposium on Networked Systems Design & Implementation – Volume 2, NSDI'05.*, 273–286. Berkeley, CA, USA: USENIX Association.

109 Kozuch, M., Satyanarayanan, M., Bressoud, T. et al. (2004). Seamless mobile computing on fixed infrastructure. *Computer* 37: 65–72. https://doi.org/10.1109/MC.2004.66.

110 Ha Kiryong and Mahadev Satyanarayanan, OpenStack++ for cloudlet deployment [WWW document], 2015. (accessed October 26, 2019).

111 Cisco, IOx [WWW document], 2019. https://developer.cisco.com/site/iox/docs (accessed October 26, 2019).

112 Willis, D., Dasgupta, A., and Banerjee, S. (2014). ParaDrop: a multi-tenant platform to dynamically install third party services on wireless gateways. In: *Proceedings of the 9th ACM Workshop on Mobility in the Evolving Internet Architecture, MobiArch '14*, 43–48. New York, NY, USA: ACM https://doi.org/10.1145/2645892.2645901.

113 Rodrigo Duro, F., Garcia Blas, J., Higuero, D. et al. (2015). CoSMiC: a hierarchical cloudlet-based storage architecture for mobile clouds. *Simulation Modelling Practice and Theory* 50: 3–19. https://doi.org/10.1016/j.simpat.2014.07.007.

114 Ismail, B.I., Goortani, E.M., Karim, M.B.A. et al. (2015). Evaluation of Docker as edge computing platform. In: *2015 IEEE Conference on Open Systems (ICOS)*, Bandar Melaka, Malaysia, 130–135. IEEE https://doi.org/10.1109/ICOS.2015.7377291.

115 G. Peng, CDN: Content Distribution Network, 2004. arXiv:cs/0411069

116 Ahlgren, B., Dannewitz, C., Imbrenda, C. et al. (2012). A survey of information-centric networking. *IEEE Communications Magazine* 50: 26–36. https://doi.org/10.1109/MCOM.2012.6231276.

117 T.H. Luan, L. Gao, Z. Li et al., Fog computing: focusing on mobile users at the edge, 2015. ArXiv150201815 Cs.

118 Bastug, E., Bennis, M., and Debbah, M. (2014). Living on the edge: the role of proactive caching in 5G wireless networks. *IEEE Communications Magazine* 52: 82–89. https://doi.org/10.1109/MCOM.2014.6871674.

119 Raghupathi, W. and Raghupathi, V. (2014). Big data analytics in healthcare: promise and potential. *Health Information Science and Systems* 2 https://doi.org/10.1186/2047-2501-2-3.

120 Isaai, M.T. and Cassaigne, N.P. (2001). Predictive and reactive approaches to the train-scheduling problem: a knowledge management perspective. *IEEE Transactions on Systems, Man, and Cybernetics Part C: Applications and Reviews* 31: 476–484. https://doi.org/10.1109/5326.983931.

121 Huang, Y., Pardalos, P.M., and Zheng, Q.P. (2017). Deterministic unit commitment models and algorithms. In: *Electrical power unit commitment: Deterministic and two-stage stochastic programming models and algorithms, SpringerBriefs in Energy. Springer USA, Boston, MA* (eds. Y. Huang, P.M. Pardalos and Q.P. Zheng), 11–47. https://doi.org/10.1007/978-1-4939-6768-1_2.

122 Uyar, H. and Türkay, B. (2008). Evolutionary algorithms for the unit commitment problem. *Turkish Journal of Electrical Engineering* 16 (3).

123 Erickson, V.L., Carreira-Perpiñán, M.A., and Cerpa, A.E. (2011). OBSERVE: occupancy-based system for efficient reduction of HVAC energy. In: *Proceedings of the 10th ACM/IEEE International Conference on Information Processing in Sensor Networks*, 258–269. Chicago, IL, USA: IEEE.

124 Molderink, A., Bakker, V., Bosman, M.G.C. et al. (2010). Management and control of domestic smart grid technology. *IEEE Transactions on Smart Grid* 1: 109–119. https://doi.org/10.1109/TSG.2010.2055904.

125 Erickson, V.L., Lin, Y., Kamthe, A. et al. (2009). Energy efficient building environment control strategies using real-time occupancy measurements. In: *Proceedings of the First ACM Workshop on Embedded Sensing Systems for Energy-Efficiency in Buildings, BuildSys '09*, 19, 2009–24. New York, NY, USA: ACM https://doi.org/10.1145/1810279.1810284.

126 IEEE 802.15.4 – IEEE Standard for Information technology – Local and metropolitan area networks – Specific requirements – Part 15.4: Wireless Medium Access Control (MAC) and Physical Layer (PHY) Specifications for Low Rate Wireless Personal Area Networks (WPANs) [WWW document], 2006. https://standards.ieee.org/standard/802_15_4-2006.html (accessed February 14, 2019).

127 Freschi, V., Delpriori, S., Klopfenstein, L.C. et al. (2014). Geospatial data aggregation and reduction in vehicular sensing applications: the case of road surface monitoring. In: *2014 International Conference on Connected Vehicles and Expo (ICCVE)*, 711–716. Vienna, Austria: IEEE https://doi.org/10.1109/ICCVE.2014.7297643.

7

Harnessing the Computing Continuum for Programming Our World

Pete Beckman[1], Jack Dongarra[2,4], Nicola Ferrier[1], Geoffrey Fox[3], Terry Moore[2], Dan Reed[5], and Micah Beck[2]

[1] *Argonne National Laboratory, Lamont, IL, USA*
[2] *University of Tennessee, Knoxville, TN, USA*
[3] *Indiana University, Bloomington, IN, UA*
[4] *Oak Ridge National Laboratory, Oakridge, TN, USA*
[5] *University of Utah, Salt Lake City, UT, USA*

7.1 Introduction and Overview

The number of network-connected devices (sensors, actuators, instruments, computers, and data stores) now substantially exceeds the number of humans on this planet. This is a tipping point, and the societal and intellectual effects of this are not yet fully understood. Billions of things that sense, think, and act are now connected to a planet-spanning network of cloud and high-performance computing (HPC) centers that contain more computers than the entire Internet did just a few years ago. We are now critically dependent on this expanding network for our communications and social discourse; our food, health, and safety; our manufacturing, transportation, and logistics; and our creative and intellectual endeavors, including research and technical innovation. Despite our increasing dependence on this massive, interconnected system of systems in nearly every aspect of our social, political, economic, and cultural lives, we lack ways to analyze its emergent properties, specify its operating constraints, or coordinate its behavior.

Simply put, today we have the tools to instrument and embed intelligence in everything, and we are doing so at a prodigious pace. Although we are the globally distributed designers, builders, and users of this immense, multilayered

Dr. Beck is an associate professor at University of Tennessee, Knoxville. He is currently on detail to the National Science Foundation in the Office of Advanced Cyberinfrastructure. The work discussed herein was completed prior to his government service and does not reflect the views, conclusions, or opinions of the National Science Foundation or of the US government.

Fog Computing: Theory and Practice, First Edition.
Edited by Assad Abbas, Samee U. Khan, and Albert Y. Zomaya.

environment, we are not truly its masters. Each of us manages only some of the networks components, and we can neither predict its aggregate behavior nor easily specify our intensional goals in intuitive language. For all of us collectively, and each of us individually, this must change. Today, we program in the relatively small confines of a single node, defining individual device, instrument, and computing element behaviors, and we are regularly confounded by unanticipated outcomes and unexpected behaviors that result once this individual node/device is exposed to the network collective. As consumers, we want our Internet-capable environmental devices (e.g. thermostats, lighting, and entertainment preferences) to adapt seamlessly to our changing roles and expectations, regardless of location. And yet, rather than specifying the ends we seek, we must specify detailed behaviors for home, office, car, and transient locale. In environmental health, we build and deploy arrays of wireless environmental sensors and edge devices when our goal may really be to "reprioritize edge resources to search for mosquitoes, given a statistically significant change in seasonal temperature and humidity across the nearby river basin." In disaster planning, when satellites show hurricane formation, we manually redirect data streams and simulation software stacks, when our goal is really to "retarget advanced computing resources to predict storm surge levels along the eastern seaboard." In science, when the Laser Interferometer Gravitational-Wave Observatory (LIGO) detects a gravity wave, we scramble to reposition the global network of telescopes to capture multispectral data and launch simulations, when our true goal is to "identify and analyze correlated transient phenomena."

Whatever the desires of consumers, companies, governments, and scientific researchers may be, we continue to build this increasingly digital world with only ad hoc, experiential, and intuitive expectations for the efficacy of alternative design choices. More perniciously, once these choices have been made, modifying or reversing them is often impossible. In large measure, this is because two constraints – resource capabilities and desired outcomes – are convolved, artificially and unnecessarily, on two time scales – design and deployment. The first of these is at the time of design and construction; the second is during outcome. At either time, the resource components may change (e.g. due to availability or failure) or expectations may shift (e.g. due to addition of new instruments or new questions). Moreover, the lifetime of many computations is not minutes or hours, but often days, months, or years. As Figure 7.1 shows, these components vary dramatically in capabilities and numbers but in aggregate define a complex collective that we call the "computing continuum." While significant research and development exists at specific places along this continuum (i.e. focus on HPC, or cloud, or Internet of Things [IoT]), we seek to develop approaches that include the entire computing continuum as a collective whole. Just as early experimentalists who studied molecular behavior in isolation struggled for want of a predictive theory

The Computing Continuum

IoT/Edge — Fog — HPC/Cloud

Size	Nano	Micro	Milli	Server	Fog	Campus	Facility
Example	Adafruit Trinket	Particle.io Boron	Array of Things	Linux Box	Colocated Blades	1000, node cluster	Datacenter
Memory	0.5K	256K	8GB	32GB	256G	32TB	16PB
Network	BLE	WiFi/LTE	WiFi/LTE	1 GigE	10GigE	40GigE	N*100GigE
Cost	$5	$30	$600	$3K	$50K	$2M	$1000M

Count = 10^9
Size = 10^1

Count = 10^1
Size = 10^9

Figure 7.1 The Computing Continuum: Cyberinfrastructure that spans every scale. Components vary from small, inexpensive devices with limited computer resources (IoT) to modestly priced servers with mid-range resources to expensive high performance computers with extensive compute, storage, and network capabilities. This range of capabilities, cost, and numbers forms a continuum.

of gases and materials properties, so we struggle in the absence of a specification methodology and predictive understanding of this new computing continuum. We need a conceptual breakthrough for continuum programming that elevates specifications from components and behaviors to systems and objectives.

This paper outlines a vision for how best to harness the continuum of interconnected sensors, actuators, instruments, and computing systems, from small numbers of very large devices to large numbers of very small devices. Our hypothesis is that only via a continuum perspective can we intentionally specify desired continuum actions and effectively manage outcomes and systemic properties – adaptability and homeostasis, temporal constraints and deadlines – and elevate the discourse from device programming to intellectual goals and outcomes.

7.2 Research Philosophy

As the deployment of intelligent network-connected devices accelerates, so does the urgency with which this research area must be addressed. Industry analysts have begun predicting that "the edge will eat the cloud" [1]. The challenge, however, is not that one form will supplant another, but that we lack a programming and execution model that is inclusive and capable of harnessing the entire computing continuum to program our new intelligent world. Thus, development of a framework for harnessing the computing continuum would catalyze new consumer services, business processes, social services, and scientific discovery.

We believe programming the continuum is not only possible, but realizable, though breakthroughs in the concepts and abstractions are needed for its coordination and management. Our common models of computation assume enumerated resources and predictable computing capabilities. However, in the continuum, the capabilities and numbers of components change dramatically over time, and aggregate functions can be long running – lasting months or years. In such contexts, intelligent, adaptive, closed-loop operations are de rigueur.

To be adopted, any broadly applicable continuum framework must be domain independent and exploit and interconnect extant, lower-level programming models and software systems. It is neither possible nor practical to obviate extant tools and techniques. Thus, the challenge lies in developing new annotations and composable abstractions for continuum computation and data movement, where none currently exist.

Table 7.1 highlights a range of scientific examples where the fabric of computing spans many orders of magnitude, and complex, open, and closed-loop behavior is required [2]. For example, ecologists should be able to link ecosystem monitoring with cloud-based simulations. Edge analysis routines, written to process telescope data, could be connected with HPC work flows that retrain machine learning algorithms automatically; the results of this machine learning could then be used to reposition observational assets based on the detection of new phenomenon. Similarly, materials scientists could write programs linking live instrument analysis at the edge with predictive models running on large-scale computing systems to detect experimental configuration errors.

Table 7.1 Exemplar continuum computing science applications.

Project	Description
Array of Things Urban Science Instrument	Environmental sensors, computer vision, deep learning inference, triggered weather, and traffic computations
Atmospheric Radiation Measurement Climate Research Facility	Software-controlled radar, user-provided sensors, data archive, climate models
Large Synoptic Survey Telescope	Transient phenomena detection and multispectral image correlation
National Ecological Observatory Network	Field-deployed instrumented towers and sensor arrays, sentinel measurements, specimen collection protocols, remote sensing capabilities, natural history archives
Precision Weather Forecasting and Sustainable Agriculture	Inexpensive environmental sensors and citizen science drive customized simulation models for microclimate weather forecasts and aquifer depletion reduction
Intelligent Manufacturing	Sensor measurement, modeling (digital twin), analysis, and control

Quite clearly, however, the merit of advances in continuum programming is not limited to science but will be a catalyst for exploration and advancement in almost every human endeavor.

7.3 A Goal-oriented Approach to Programming the Computing Continuum

As Figure 7.1 shows, along this continuum the product of device count and device size is roughly constant. At either extreme, the scale is large, the resources are geographically distributed, their availability varies over time, and they frequently span multiple control domains. Thus, the computing continuum future should integrate two primary activities with multiple subsidiary goals: (1) developing a goal-oriented annotation and high-level programming model that specifies desired outcomes for the aggregate, rather than a collection, of component behaviors; (2) building mapping tools, a run-time system, and an execution model for managing continuum resources as an abstract machine that also monitors behavior and triggers remappings when necessary.

In one sense, this approach maps loosely to a classic view of computation: a program, a run-time system, and an abstract machine. However, this traditional motif for computation evolved from our experiences directly specifying the actions of hardware under our control. Even the word "program" evokes the notion that we provide commands via a sequence of steps.

However, in the computing continuum, we cannot uniformly command all components to do our bidding – the building blocks are too diverse, the scale is too large, and the component owners and operators are sometimes unknown. Hence, we must begin by describing the goal.

7.3.1 A Motivating Continuum Example

To highlight the breakthrough we envision, consider a realistic scenario from the south Chicago neighborhood of Chatham, which suffers from the highest level of home flooding and associated insurance claims in the city. Chicago's infrastructure is a "combined system," where stormwater and raw sewage share the same underground structures, and when rainwater exceeds capacity, sewage enters homes, endangering health in addition to ruining personal property. Local residents are eager to see sensors, which we actually have deployed in Chatham, linked with weather models to provide warnings for impending danger.

The science/public policy problem can be succinctly stated as: "Eight hours before anticipated rainfall [...] predict underground infrastructure responses and trigger intelligent reactions and warnings if greater than 5% of homes flood or if traffic capacity falls below 70% of normal; then monitor and dynamically adapt urban controls to reduce harm." Decomposing this example, we can sketch the components and control flow, as below.

Eight hours before anticipated rainfall. On an HPC system, periodically run focused weather forecasts. As the risk of rainfall increases, run the forecast more frequently and with a finer resolution.

Predict underground infrastructure responses. Three cloud/HPC models are coupled to predict the response of the storm infrastructure: (1) a regional floodplain model predicts waterway inputs to the neighborhood; (2) a computational hydrology model predicts the absorption of incoming water by soil and infrastructure; and (3) a model of the underground infrastructure predicts when capacity is exceeded. Inputs for the linked models come from other parts of the continuum: (1) live measurements from soil moisture monitors, sump pumps, basement water, and sewer levels provide immediate hyper-local data; and (2) Array of Things (AoT) nodes use intelligent libraries and edge computing resources to analyze camera images of rising floodwater in rivers, on streets, and in rain gardens and detention ponds [3].

Trigger intelligent reactions. Based on a computational sewer model and learned responses to historical actuator settings (e.g. the behavior of pumps and valves under stress), reconfigure local fog/edge components to autonomously respond resiliently, adjusting flow rates and motors within the disrupted network environment.

Trigger public warnings if greater than 5% of homes flood or if traffic capacity falls below 70% of normal. Using cloud/HPC models, calculate the impact of the potential flooding and alert city officials and residents as needed. Furthermore, as new flood data become available from home and citizen science sensors, use image data from consumer cameras and sensors to push computer vision algorithms along optimized intelligent libraries and new learning models to edge resources to identify and then predict when bridges, underpasses, or quickly flowing water will disable vehicles.

Monitor and dynamically adapt urban controls to reduce harm. As the flood danger increases, reprioritize available edge and fog resources to detect, predict, and report on events like rising waters, stalled vehicles, and water erupting from manholes and drains.

In this example, realizing end-to-end data capture, modeling, and analysis, and timely response (closed loop) requires systemic coordination between sensors (edge), behavior detection (fog), models (cloud/HPC), and actuators (edge). Notice that it would be necessary to push new computational elements from the cloud to edge/fog resources during the storm – the continuum is bidirectional, and code is dynamic. Traditional computational work-flow systems like Pegasus

[4] and Kepler [5] provide mechanisms for triggering data movement and computation. However, our goal is much broader, as we intend to program the continuum so that new algorithms and deep learning models can be pushed to appropriate locations (i.e. edge, fog, cloud, and/or HPC computing resources) using a simple lambda (function as a service) [6, 7] abstraction. Ideally, both the annotation and the high-level mapping would be portable to other contexts, spanning cities – Portland to Barcelona – and rural areas, even when the edge hardware and cloud providers differ significantly. For rural and agricultural areas, the challenges shift to dam spillway management, agricultural runoff and water supply quality assessment, and community evacuation.

Realizing and implementing this continuum programming model requires balancing conflicting constraints and translating the high-level specification into a form suitable for execution on a unifying abstract machine model. In turn, the abstract machine must implement the mapping of specification demands to end-to-end resources.

7.3.2 Goal-oriented Annotations for Intensional Specification

This approach to programming the continuum stems from the fluid nature of the underlying resources, especially those at the edge. If every program component and its behavior were static, an imperative behavioral specification could be mapped directly to resources, assuming the resources themselves were static. For static cases, and when developers wish to implement all dynamic management, one could expose the resource-demand graph specifications and control mechanisms for direct use. However, this is rarely possible or practical, as the following examples illustrate:

When weather models predict roadways will degrade, adjust traffic signaling and preferred routing based on local conditions to optimize safety for both pedestrians and vehicles.

When waterway sensors detect increased phosphates, use edge device sensor data and satellite image analysis as inputs to simulations that can predict harmful algae blooms.

After significant seismic activity, reprioritize edge computation to detect smoke, distress calls, and natural gas leaks. Based on air quality and local weather, predict location of the source(s).

Simply put, the programming equivalent to source routing, where the packet originator completely determines the route, is rarely possible in the continuum.

For dynamic cases – the majority – the community must devise an approach that is more declarative and constraint based. It must succinctly describe the aim and enable efficient mapping (and remapping) to disjointed, heterogeneous, shifting resources that behave more like independent agents than a single, cohesive machine. This is analogous to packet routing, which describes where the data should be sent but not how it should be sent.

Attacking this problem requires a two-pronged approach. First, one must devise languages for describing the resources of the continuum, including intelligent libraries, sensors, instruments, and cloud services. Tightly coupled with this, our abstract machine and runtime system will keep historical metrics of performance, interconnection bandwidth, and computational capacity that can be used for building execution graphs. Such work could build on the World Wide Web Consortium's (W3C's) specifications for the Semantic Web [8], including a resource description framework (RDF), web ontology language (OWL), and a query language for RDF (SPARQL). Although the Semantic Web has been slow to evolve, several well-developed technologies for describing resources, data, and ontologies have been deployed.

This first prong is insufficient to fully harness the continuum, for one must also specify goals and desired outcomes. Thus, we must also link the exciting and intense resurgence in autonomous agent research, fueled by advances in machine learning, to build goal-based specifications that can be mapped to resources and computation. The foundations for this field were built decades ago [9–11]. Today, we see successful work in a wide range of fields – from goal-based, human-machine teaming to flocks of autonomous drones.

By fusing the research from these domains with our novel work on edge computing, intelligent libraries, data logistics, and HPC, we believe the community could revolutionize the computing continuum.

7.3.3 A Mapping and Run-time System for the Computing Continuum

Realizing the continuum programming model requires translation of high-level specifications into a form suitable for execution on underlying resources. Cleanly separating the intensional specification from the execution strategies is key to managing temporally varying application demands and shifting resource availability and capability, which is a defining element of the continuum. To accomplish this, we would need to translate goal-oriented application specifications into an annotated resource demand graph and a set of constraints. In turn, continuum resources will be represented by an annotated resource capabilities graph with its own set of constraints (an abstract machine). This abstract machine must instantiate specification demands on continuum resources. An intelligent run-time system is then responsible for mapping and adaptively remapping the resource demands to continuum capabilities. As Figure 7.2 illustrates, these elements define the programming model for the computing continuum.

With this backdrop, consider our motivating example once again: "Eight hours before anticipated rainfall, predict underground infrastructure responses

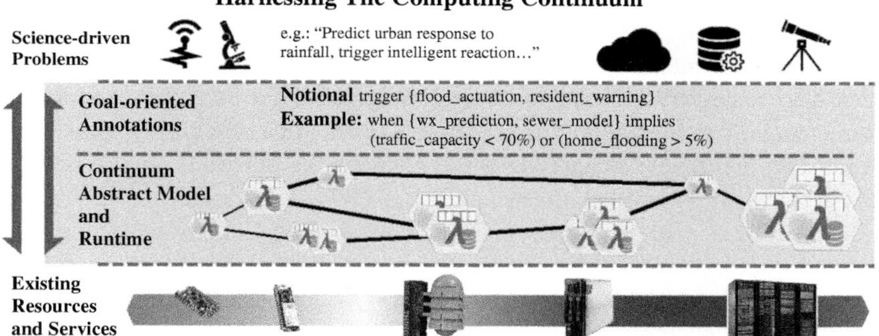

Figure 7.2 Continuum Computing Research Areas: A pictorial depiction of the computer science areas that require research to successfully program the computing continuum. To address problems, such as the sciences examples given in Table 7.1, using existing resources and services, we need an abstract programming model with goal-oriented annotations, along with a run-time system and an execution model. (*See color plate section for the color representation of this figure*)

and trigger intelligent reactions and public health warnings if highway traffic capacity is significantly impacted or if more than 5% of homes flood, then monitor and adapt controls during the storm." This high-level specification results in a resource demand graph of (1) an HPC-based mesoscale weather simulation, (2) water level sensors, (3) roadway flooding detection via image analysis, and (4) public health models and warning and evacuation models. Constraints include computationally intense models, geographic sensor dispersion, and real-time adaptation and actuator control.

Similarly, the resource capabilities graph for the sensor set, edge devices, actuators, and cloud or HPC resources would include annotations that specify: (1) performance characteristics such as node interconnection bandwidth and connectivity, storage capacity, and computation speed; (2ii) programmability (i.e. fixed function device or "over the air" programmable); (3) multiplicity (i.e. an estimate of the number of instances, recognizing these vary over time); (4) control span (i.e. single or shared function and ownership); and (v) domain-specific constraints (e.g. geographic location, power limitations, or maximum usage frequency).

The mapping function would then instantiate environmental monitoring by tapping data streams from a statistical sample of the available water sensors, reprioritize flood image analysis on AoT sensors and fog devices, and then launch a weather model parameterized by a terrain model with a real-time constraint on prediction cycles. Because these resource demands may conflict with or sometimes exceed resource availability, and the resources themselves may shift over time (e.g. due to sensor loss or replacement), any mapping is necessarily imprecise. Limited cloud or HPC availability might force a reduction in forecast accuracy and redeployment of alternative library versions to meet deadline constraints. Thus, the execution system must monitor the efficacy of each mapping and adapt accordingly.

It is also necessary to explore several techniques for mapping annotated resource demand graphs, mapping constraints to resource capability graphs, and learning and adaptively remapping these elements as demands and resources shift. As shown in Figure 7.3, these techniques include, but need not be limited to: (1) the intensional, goal-oriented program specification, translated to an annotated resource demand graph via autonomous agents and machine learning; (2) a resource registry that contains a time-varying list of available resources, attributes, and constraints; (3) optimized, multiversion libraries suitable for deployment on continuum resources, spanning computation-limited sensors, sophisticated edge devices, HPC systems, and cloud resources; (4) temporal "fuzzy logic" [12, 13] for qualitative constraint specification – an approach we used successfully in real-time adaptive control of parallel scientific applications [14, 15]; (5) machine learning [16] and auto-tuning that exploit behavioral data and temporal fuzzy logic constraint specifications to map and remap resource demands to extant resources; and (6) very low overhead, distributed behavioral monitoring tools [17], a behavioral repository, and data sharing via Message Queuing Telemetry Transport (MQTT) [18, 19].

7.3.4 Building Blocks and Enabling Technologies

There are many potential building blocks and enabling technologies for building the continuum. Several of these have been developed by the authors.

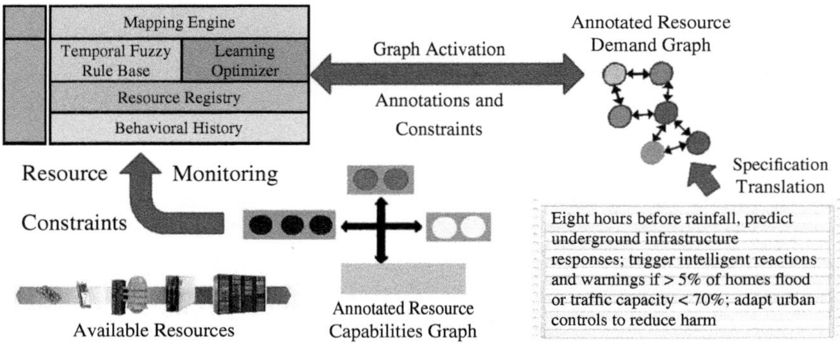

Figure 7.3 Continuum mapping and execution: Research is needed to explore techniques for mapping goal-oriented specifications to the available resources. This graphic shows a possible technique that translates the goal specification into an annotated resource demand graph that is used by the mapping engine, along with resource capability, availability and constraint information, multiversion libraries, and behavioral history.

7.3.4.1 The Array of Things (AoT)

AoT [3] is an NSF-sponsored project at the University of Chicago, Northwestern University, and Argonne National Laboratory focused on supporting urban sciences. AoT is deploying 500 sensor nodes in Chicago, with more than 100 already operational. The core AoT hardware/software platform, Waggle [20, 21], is also being used by a diverse community of scientists for things like exploring the hydrology of a pristine prairie owned by the Nature Conservancy and for studying pollen and asthma in Chattanooga, Tennessee. These sophisticated sensors as well as their lightweight, battery-powered, thumb-sized cousins, support edge computation – the ability to directly analyze the data stream in situ. In addition to air-quality sensors, the nodes support edge computing for in-situ computer vision, machine learning, and audio analysis. For example, the number of pedestrians using a crosswalk can be calculated by GPU-based computer vision algorithms using a combination of Kalman filters and deep learning. The computed results are pushed into the cloud.

7.3.4.2 Iowa Quantified (IQ)

A complementary project, when Reed was at the University of Iowa, has been designing inexpensive battery and solar-powered wireless sensors with multiple wireless protocols (WiFi, LoRaWAN [22]) that are integrated with cloud and data analytics services. Configurable sensors span a wide range of environmental data, including temperature, humidity, gases, and particulate matter. This Iowa Quantified (IQ) project has deployed multidepth soil sensors to enable understanding of moisture dynamics and agricultural productivity [23], and other projects are being planned around microscale weather forecasting and aquifer depletion minimization for precision agriculture.

7.3.4.3 Intelligent, Multiversion Libraries

Across the continuum, the efficient application of scientific computing techniques requires specialized knowledge in numerical analysis, architecture, and programming models that many working researchers do not have the time, energy, or inclination to acquire. With good reason, scientists expect their computing tools to serve them and not the other way around. Unfortunately, the highly interdisciplinary problems of programming the continuum, using more and more realistic simulations on increasingly complex computing platforms, will only exacerbate the problem. To address this, we believe a two-pronged strategy is needed that leverages expertise in developing optimized, automatically tuned libraries that can be deployed dynamically, based on the goal-based annotation and mapping described above. At a low level, the community needs operations that closely mimic the building blocks of the Basic Linear Algebra Subroutines (BLAS) [24] but perform well across the continuum. Studying different approaches to optimization (e.g. mixed-precision algorithms and empirical auto-tuning for different platforms on the continuum) will enable us to provide the intelligent libraries needed to meet the performance constraints of the target application scenarios. At a high level, defining a common name space for different types of operations and services (e.g. "compress," "merge," and "SVD") will enable us to move the locus of computation, in a seamless fashion, between the cloud and the edge. To integrate the bidirectional workflow across the variety of actors, the run-time system should match the resource requirements of the named components of a prototype library with the resources that are actually available when the work flow executes.

7.3.4.4 Data Flow Execution for Big Data

Currently, we are exploring the Twister2 environment [25, 26], which is an early prototype of some of the ideas expressed above. Twister2 has a modular implementation with separate components summarized in Table 7.2, which include the key features of existing "big data" programming environments and have been designed as a toolkit so that one can pick and choose capabilities from different sources. Further, all components have been redesigned as necessary to obtain high performance. Twister2 contains three separate communication packages: (1) one aimed at classic parallel programming, (2) Twister:Net [27] aimed at distributed execution, and (3) publish-subscribe messaging as seen in Apache Storm and Heron to connect edge to cloud.

The first of these communication environments addresses problems in parallel programming that are well known from MPI, and implements bulk synchronous processing; it extends MPI by adding some collectives required for solving big data problems, including topic modeling and graph analytics [28–30]. However, the leading programming environments Spark, Storm, Heron, and Flink that target big data offer as default, a different communication model built around data flow. Twister2 has implemented this as a separate communication system, Twister:Net, which offers a user friendly data application programming interface (API) rather than a messaging API. Twister:Net automatically breaks data bundles into messages and, if necessary, uses disk storage for large volume transfers. A second and more critical feature is that the communication system supports distinct source and target tasks; therefore, Twister:Net can be used in a fully distributed environment, as seen in edge-cloud environments. Twister:Net also implements a rich set of collectives such as keyed reduction [31] where all the famous MPI and MapReduce operations (gather, scatter, shuffle, merge, join, etc.) can be considered as particular versions of "keyed collectives." Here we define a collective operation as a combined communication and computation (as in reduced) operation that involves some variant of all to all linkage of the source and target tasks. The final communication subsystem in Twister2 also supports different source and target

Table 7.2 Twister2 components and status.

Component	Area	Current implementation	Future implementation
Connected Data Flow	Internal fine-grain data flow or external data flow (workflow)	Dynamic internal data flows	Ongoing and add coarse-grain external data flow
High Level API's	Distributed Data Set, SQL, Python, Scala, Graph	Tsets (Twister2 implementation of RDD and Streamlets), Java	Data flow optimizations, SQL, Python, Scala, Graph
Task System	Task migration	Not started	Streaming job task migrations
	Streaming	Streaming execution of task graph	FaaS Function as a Service
	Task execution	Process, threads	More executors
	Task scheduling	Dynamic scheduling, static scheduling; pluggable Scheduling algorithms	More algorithms
	Task graph	Static graph, dynamic graph generation	Cyclic graphs for iteration as in Timely
Communication	Data flow communication DFW	Twister:Net. MPI Based or TCP. Batch and streaming operations	Integrate to other big data systems, Integrate with RDMA
	BSP communication	Conventional MPI, Harp with extra collectives	Native MPI Integration
Job Submission	Job submission (dynamic/static) Resource allocation	Plugins for Slurm, Mesos, Kubernetes, Aurora, Nomad	Yarn, Marathon
Data Access	Static (batch) data	File systems, including HDFS	NoSQL, SQL
	Streaming data	Kafka connector for Pub-Sub Communication, Storm API	RabbitMQ, ActiveMQ

tasks, but uses publish-subscribe messaging. It is more suitable than Twister:Net for unreliable links, such as those found in edge-to-cloud communication.

Twister2 also has a carefully designed data flow execution model that supports linking intelligent nodes; this feature enables support for fault tolerance (i.e. as in the creation of resilient distributed datasets [RDDs] used in Apache Spark) and allows wrapping nodes with rich tools such as learning systems. Further, Twister2 recognizes that some coarse-grain data flow nodes (e.g. those seen in job work flows) are not performance sensitive, but other finer-grain nodes need low overhead implementations with, in particular, use of streaming data without the overhead associated with reading and writing of intermediate files. Note currently if you use Apache Storm or Spark to link distributed subsystems – data center to data center or edge-to-fog to data center – different jobs in each subsystem must be linked. Twister2 offers the possibility of invoking a single data flow across all parts of a distributed system.

7.4 Summary

Modern society is now critically dependent on a global and pervasive network of intelligent devices, both large and small, that are themselves connected to a planet-spanning network of cloud and HPC centers. Despite this dependence, analyzing this burgeoning network's emergent properties and coordinating its integrated behavior is not straightforward. This paper outlines an ambitious plan to elevate programming, coordination, and control of the continuum from ad hoc component assembly to intensional specification and intelligent, homeostatic control.

References

1 Bittman, T. (2017). *Maverick* Research, The edge will eat the cloud.* Gartner Research, Online https://www.gartner.com/doc/3806165/maverick-research-edge-eat-cloud.

2 Stuart Anderson, Ewa Deelman, Manish Parashar et al., Report from the NSF Large Facilities Cyberinfrastructure Workshop, September 2017. http://facilitiesci.org.

3 Array of Things, March 2019. www.arrayofthings.org.

4 Deelman, E., Vahi, K., Rynge, M. et al. (2016). Pegasus in the cloud: science automation through workflow technologies. *IEEE Internet Computing* 20 (1): 70–76.

5 Altintas, I., Berkley, C., Jaeger, E. et al. (2004). Kepler: an extensible system for design and execution of scientific workflows. In: *Proceedings of 16th International Conference on Scientific and Statistical Database Management*, 423–424. IEEE.

6 Apache OpenWhisk, 2018 (accessed April 2018). https://openwhisk.apache.org.

7 Amazon Web Services, Lambda features, 2018 (accessed April 2018). https://aws.amazon.com/lambda/features.

8 Berners-Lee, T., Hendler, J., and Lassila, O. (2001). The semantic web. *Scientific American* 284 (5): 34–43.

9 Bradshaw, J.M., Johnson, M., Uszok, A. et al. (2004). Kaos policy management for semantic web services. *IEEE Intelligent Systems* 19 (7): 32–41.

10 Kagal, L., Finin, T., and Joshi, A. (2003). A policy language for a pervasive computing environment. In: *Proceedings Policy 2003. IEEE 4th International Workshop on Policies for Distributed Systems and Networks*, 63–74.

11 Hexmoor, H., Castelfranchi, C., and Falcone, R. (2003). *Agent Autonomy: Multiagent Systems, Artificial Societies, and Simulated Organizations*. Springer.

12 Zadeh, L.A. (1986). Commonsense reasoning based on fuzzy logic. In: *Proceedings of the 18th Conference on Winter Simulation, WSC '86*, 445–447. New York, NY: ACM.

13 Frigeri, A., Pasquale, L., and Spoletini, P. (2014). Fuzzy time in linear temporal logic. *ACM Transactions on Computational Logic* 15: 4, pp. 30:1–30:22.

14 Droegemeier, K.K., Gannon, D., Reed, D. et al. (2005). Service-oriented environments for dynamically interacting with mesoscale weather. *Computing in Science and Engineering* 7 (6): 12–29.

15 Ramakrishnan, L. and Reed, D.A. (2008). Performability modeling for scheduling and fault tolerance strategies for scientific workflows. In: *Proceedings of the 17th International Symposium on High Performance Distributed Computing, HPDC '08*, 23–34. New York, NY: ACM.

16 Collins, A., Fensch, C., and Leather, H. (2012). Masif: machine learning guided auto-tuning of parallel skeletons. In: *Proceedings of the 21st International Conference on Parallel Architectures and Compilation Techniques, PACT '12*, 437–438. New York, NY: ACM.

17 Gamblin, T., de Supinski, B.R., Schulz, M. et al. (2010). Clustering performance data efficiently at massive scales. In: *Proceedings of the 24th ACM International Conference on Supercomputing, ICS '10*, 243–252. New York, NY: ACM.

18 Andrew Banks, Ed Briggs, Ken Borgendale, and Rahul Gupta (eds.), MQTT Version 5.0, OASIS Committee Specification 01, December 2017. http://docs.oasisopen.org/mqtt/mqtt/v5.0/cs01/mqtt-v5-0-cs01.html.

19 Manzoni, P., Hernandez-Orallo, E., Calafate, C.T., and Cano, J.-C. (2017). A proposal for a publish/subscribe, disruption tolerant content island for fog computing. In: *Proceedings of the 3rd Workshop on Experiences with the Design and Implementation of Smart Objects, SMARTOBJECTS '17*, 47–52. New York, NY: ACM.

20 Waggle Project, March 2019. http://wa8.gl.

21 Beckman, P., Sankaran, R., Catlett, C. et al. (2016). Waggle: an open sensor platform for edge computing. In: *2016 IEEE Sensors*, 1, 3. IEEE.

22 LoRaWAN Wireless Specification, 2018. https://lora-alliance.org.

23 Lee, K.-P., Kuhl, S.J., Bockholt, H.J. et al. (2018). A cloud-based scientific gateway for internet of things data analytics. In: *Proceedings of the Practice and Experience in Advanced Research Computing, July 22–26.*

24 Dongarra, J.J., Du Cruz, J., Hammarling, S., and Duff, I.S. (1990). Algorithm 679: a set of level 3 basic linear algebra subprograms: model implementation and test programs. *ACM Transactions on Mathematical Software* 16 (1): 18–28.

25 Ekanayake, J., Li, H., Zhang, B. et al. (2010). Twister: A runtime for iterative MapReduce. In: *Proceedings of the First International Workshop on MapReduce and Its Applications of ACM HPDC 2010 conference, June 20–25, 2010*. ACM.

26 Supun Kamburugamuve, Kannan Govindarajan, Pulasthi Wickramasinghe et al., Twister2: design of a big data toolkit, Concurrency and Computation, EXAMPI 2017 workshop at SC17 conference, 2019.

27 Kamburugamuve, S., Wickramasinghe, P., Govindarajan, K. et al. Twister:Net – communication library for big data processing. In: *HPC and cloud environments, in Proceedings of Cloud 2018 Conference*. IEEE.

28 Qiu, J. Harp-DAAL for high-performance big data computing. In: *The Parallel Universe*, No. 32, 31–39. https://software.intel.com/sites/default/files/parallel-universe-issue-32.pdf.

29 Harp-DAAL high-performance data analytics framework website. https://github.com/DSC-SPIDAL/harp, https://dsc-spidal.github.io/harp/docs/harpdaal/harpdaal. Accessed September 2018.

30 Chen, L., Peng, B., Zhang, B. et al. (2017). Benchmarking Harp-DAAL: High performance hadoop on KNL clusters. In: *IEEE Cloud 2017 Conference*. IEEE.

31 Fox, G. (2019). Big data overview for Twister2 tutorial. In: *BigDat2019. 5th International Winter School on Big Data*. Cambridge, UK, January 7–11.

8

Fog Computing for Energy Harvesting-enabled Internet of Things

S. A. Tegos[1], P. D. Diamantoulakis[1], D. S. Michalopoulos[2], and G. K. Karagiannidis[1]

[1] *Electrical and Computer Engineering Department, Aristotle University of Thessaloniki, GR-54124, Thessaloniki, Greece*
[2] *Nokia Bell Labs, Munich, Germany*

8.1 Introduction

With the current advance of the Internet of Things, mobile devices have a significant impact on our lives, such as healthcare, entertainment, daily life, etc. Nevertheless, the successful integration of mobile devices in these applications depends on their capability to execute a massive number of intensive tasks quickly, despite the potentially limited computing resources. Moreover, emerging wireless technologies are characterized by versatile, yet stringent energy efficiency requirements of the deployed devices. A novel paradigm for network decentralization is fog computing, which offers the capability to offload computation-heavy applications, whereas a promising direction to achieve the energy sustainability of the end nodes is based on the utilization of energy harvesting (EH).

Regarding fog computing, it offers the capability to release the mobile devices from heavy computation workloads by offloading tasks to a fog server. Compared with the conventional remote cloud, offloading workload to a fog server reduces the data traffic, the energy consumption of users, and the transmission latency, as it is placed closer to end users. It is noted that the computation offloading can be applied in two ways, i.e. binary and partial offloading [1].

1. *Binary offloading*. The computation task should be offloaded as one unit, as it cannot be partitioned.
2. *Partial offloading*. The task is partitionable, but only one of the two parts can be offloaded.

Fog Computing: Theory and Practice, First Edition.
Edited by Assad Abbas, Samee U. Khan, and Albert Y. Zomaya.

The utilization of the fog computing technique is particularly useful to the real-time execution of computation tasks requested by low-power devices. Therefore, it has received considerable attention in both academia and industry.

Low-cost mobile devices with small battery capacity may obtain satisfactory computation performance by integrating EH techniques into fog computing. When EH is used, energy can be harvested by ambient energy sources, such as wind energy, solar power, human kinetic energy, and ambient radio frequency (RF) signals. Moreover, wireless power transfer (WPT), which utilizes RF signals, is considered an efficient way to increase the battery life of the devices. Specifically, a dedicated RF energy transmitter, which can charge the battery of remote devices, is utilized in this technique. A significant advantage of WPT is that it is controllable in comparison with harvesting from ambient sources. Currently, commercial WPT transmitters can deliver at the order of 10 μW RF power to a larger distance of 10 m, which can provide adequate power for the operation of EH devices [2].

In the existing literature, there are several works that investigate the integration of fog computing and EH. Single EH user fog systems with binary offloading are considered [3, 4]. Furthermore, fog systems consisting of two EH users adopting time division multiple access (TDMA) are examined where the near user operates as a relay for the far user [5, 6]. Moreover, multiuser fog systems adopting TDMA are considered where both binary and partial offloading are utilized [7, 8], whereas nonorthogonal multiple access (NOMA) has also been investigated in similar systems [9].

In this chapter, the integration of fog computing and EH is considered, according to which tasks can either be executed locally or offloaded to a fog server. More specifically, both EH from ambient sources and WPT are examined. Furthermore, a review of resource allocation problems is presented, emphasizing their categorization, comparison, e.g. in terms of a common objective, and the identified challenges. Moreover, some future research challenges are investigated.

8.2 System Model

In this section, a fog computing system with EH mobile users (MUs) is examined, which is depicted in Figure 8.1. MUs can execute their tasks locally or/and offload them to the fog server. The computing capability of the fog server can improve significantly the computation experience in each MU [10–12].

It is assumed that time is divided into time slots with length T. More specifically, $\mathcal{N} \triangleq \{1, \ldots, n, \ldots, N\}$ and $\mathcal{T} \triangleq \{1, \ldots, t, \ldots, T\}$ are used to denote the index sets of MUs and time slots, respectively. The wireless channel is considered to be independent and identically distributed (i.i.d.) block fading. The channel

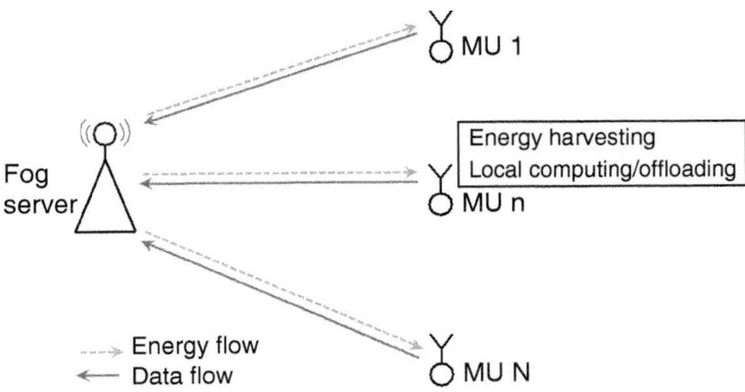

Figure 8.1 Fog system.

power gain in the tth time slot is denoted as h^t, and $h^t \sim f_H(x)$, $t \in \mathcal{T}$, where $f_H(x)$ is the probability density function (PDF) of h^t.

8.2.1 Computation Model

We use $A_n(L_n, \mathcal{T}_n^d)$ to represent a computation task of the nth MU, where L_n denotes the input size of the task, which is measured in bits, and $\mathcal{T}_n^d$ denotes the execution deadline. The computation tasks requested at each MU are modeled as an i.i.d. Bernoulli process, i.e. at the beginning of each time slot, there is a task request with probability ρ, while no task is requested with probability $1 - \rho$. The variable ζ_n^t is used as follows:

$$\zeta_n^t = \begin{cases} 1, & \text{task request in the } t\text{th slot by the } n\text{th MU}, \\ 0, & \text{otherwise}, \end{cases} \tag{8.1}$$

i.e. $\mathbb{P}(\zeta_n^t = 1) = 1 - \mathbb{P}(\zeta_n^t = 0) = \rho$, $t \in \mathcal{T}$, $n \in \mathcal{N}$. It is assumed that all tasks are delay-sensitive, i.e. their execution deadline is lower or equal to the time slot length $\mathcal{T}^d \leq \mathcal{T}$ and also that no buffer is available for queuing the task requests.

Each MU can either execute the computation tasks locally, and/or offload them to the fog server. However, it is possible that neither of these methods is feasible, i.e. when there is not sufficient energy at the MU and, thus, the task will be dropped.

Binary Offloading. The set of the computation mode indicators is denoted as

$$\boldsymbol{K}_{\text{n,binary}}^t \triangleq \{K_{\text{n,j}}^t = \{0,1\}, j = \{m,s,d\}, t \in \mathcal{T}, n \in \mathcal{N}\}, \tag{8.2}$$

where $K_{\text{n,m}}^t = 1$ and $K_{\text{n,s}}^t = 1$ indicate that the task of the nth MU in the nth time slot is executed locally at the tth MU and offloaded to the fog server, respectively, and $K_{\text{n,d}}^t = 1$ indicates that the task is dropped. Therefore, the selection of $\boldsymbol{K}_{\text{n,binary}}^t$ should satisfy the following equation:

$$K_{\text{n,m}}^t + K_{\text{n,s}}^t + K_{\text{n,d}}^t = 1, t \in \mathcal{T}, n \in \mathcal{N}. \tag{8.3}$$

Partial Offloading. The above indicators can be used for the case of partial offloading by extending the above set as

$$\boldsymbol{K}^t_{\text{n,partial}} \triangleq \{K^t_{\text{n,j}} \in [0,1], j = \{m,s,d\}, t \in \mathcal{T}, n \in \mathcal{N}\}. \tag{8.4}$$

$K^t_{\text{n,m}}$ and $K^t_{\text{n,s}}$ denote the part of the computation task of the nth MU in the tth time slot that is executed locally at the nth MU and offloaded to the fog server, respectively, and $K^t_{\text{n,d}}$ indicates the part of the computation task that is dropped. Moreover, (8.3) should be satisfied.

8.2.1.1 Local Execution Model

Let X denote the number of central processing unit (CPU) cycles required for the processing of a one bit input. X differs in each application and can be acquired through off-line measurements [13]. Furthermore, $W_n = L_n X$ CPU cycles are needed in order for the nth MU to successfully execute the task $A_n(L_n, \mathcal{T}_n^d)$. The set of the frequencies scheduled for the W_n CPU cycles in the tth time slot is denoted as $\boldsymbol{f}^t_n \triangleq \{f^t_w, w = 1, \dots, W_n, t \in \mathcal{T}, n \in \mathcal{N}\}$, which can be implemented by adjusting the chip voltage with dynamic voltage and frequency scaling (DVFS) techniques. Consequently, the delay for the local execution of the computation task requested in the tth time slot at the nth MU is given by

$$D^t_{\text{n,local}} = \sum_{w=1}^{W_n} (f^t_w)^{-1}. \tag{8.5}$$

Subsequently, the energy consumption for executing the task in the tth time slot locally at the nth MU can be expressed as

$$E^t_{\text{n,local}} = \mathcal{K} \sum_{w=1}^{W_n} (f^t_w)^2, \tag{8.6}$$

in which $\mathcal{K}$ denotes the effective energy coefficient that depends on the chip architecture. Furthermore, the CPU-cycle frequencies are considered to be constrained by $f^{\max}_{\text{CPU}}$, i.e.

$$f^t_w \le f^{\max}_{\text{CPU}}, w = 1, \dots, W_n, t \in \mathcal{T}, n \in \mathcal{N} \tag{8.7}$$

8.2.1.2 Fog Execution Model

If a MU offloads a task to the fog server, we assume that adequate computational power is available at the fog server and, thus, the corresponding computation latency is ignored [14–16]. The transmit power of the nth MU is denoted as p^t_n and it should not exceed the maximum transmitted power $p^{\max}$.

To this end, considering the existence of multiple users, an access protocol should be adopted in order for the users to transmit their tasks to the fog server. In the existing literature, most of the works consider TDMA due to its simple implementation. Nevertheless, NOMA is also investigated, as it can offer an improved performance compared to TDMA.

TDMA. The achievable rate of the nth MU in the tth time slot, adopting TDMA, can be expressed as

$$r^t_{\text{n,TDMA}} = \omega\tau^t_n \log_2\left(1 + \frac{h^t_n p^t_n}{\sigma^2}\right), \tag{8.8}$$

where ω denotes the available bandwidth, τ^t_n is the portion of the tth time slot allocated to the nth MU and σ^2 is the variance of the equivalent noise at the receiver.

NOMA. Adopting uplink NOMA in the offloading phase allows multiple MU to multiplex on the same time slot for offloading their tasks so that the system performance can be massively enhanced. Assuming perfect successive interference cancellation (SIC) at the fog server, the achievable rate of the nth MU in the tth time slot is given by

$$r^t_{n,\text{NOMA}} = \omega\log_2\left(1 + \frac{h^t_n p^t_n}{\sum_{n'=n+1}^{N} h^t_{n'} p^t_{n'} + \sigma^2}\right), \tag{8.9}$$

where it is assumed that $h^t_1 \geq \dots \geq h^t_n \geq \dots \geq h^t_N$ for fairness reasons.

It is further assumed that the computational complexity is relatively low, thus, if the computation task is executed by the fog server, the execution delay is equal to the time needed for the transmission of the specific task and is given by

$$D^t_{\text{n,server}} = \frac{L}{r^t_n}, \tag{8.10}$$

where r^t_n can be the achievable rate adopting either TDMA or NOMA. More-over, the energy consumed by the nth MU for offloading the tasks to the fog server can be expressed as

$$E^t_{\text{n,server}} = p^t_n D^t_{\text{n,server}} = p^t_n \frac{L}{r^t_n}. \tag{8.11}$$

8.2.2 Energy Harvesting Model

Fog systems with EH MUs call for redesigned offloading policies, which become more complex than the ones of mobile cloud computing systems. Consequently, an optimal computation offloading strategy should compromise between the computation performance of the current and future tasks, as the battery level in each time slot obtained by the battery level and the consumption in the previous time slot.

8.2.2.1 Stochastic Process

Each MU is equipped with an EH circuit, which is capable of harvesting energy from environments such as human kinetic energy, wind energy, and solar power.

Each MU is assumed to be powered only by the harvested renewable energy from the EH circuit. To model the EH process, it is assumed that energy packets successively arrive. To this end, let $E^t_{\mathrm{n,N}}$ denote the units of energy that arrive at the nth MU in the beginning of the tth time slot. It is assumed that energy packet arrivals in different time slots are i.i.d. with the maximum value of $E^{\mathrm{max}}_{\mathrm{n,H}}$. Even though the i.i.d. model is simple, it captures the intermittent and stochastic nature of the renewable energy processes [17, 18]. In a practical scenario, only a part of the arrived energy can be stored into the battery. In the tth time slot energy e^t_n will be harvested and stored in the battery of the nth MU and will be available for either computation offloading or local execution starting from the next time slot. Thus, the following equation should be satisfied:

$$0 \le e^t_n \le E^t_{\mathrm{n,H}}, t \in \mathcal{T}, n \in \mathcal{N}. \tag{8.12}$$

Furthermore, the battery capacity is considered sufficiently large. Moreover, the energy level of the nth MU's battery in the beginning of the tth time slot is denoted as B^t_n. Without loss of generality, we assume that $B^0 = 0$ and $B^t_n < \infty, t \in \mathcal{T}, n \in \mathcal{N}$. Furthermore, for the sake of simplicity, energy consumed for purposes other than task offloading and local computation is ignored.

The energy consumed by the nth MU in the tth time slot, denoted as $\mathcal{E}(\boldsymbol{K}^t, \boldsymbol{f}^t, p^t)$, which depends on the selected computation mode, allocated transmit power and scheduled CPU cycle frequencies, is given by

$$\mathcal{E}_n(\boldsymbol{K}^t_n, \boldsymbol{f}^t_n, p^t_n) = K^t_{\mathrm{n,m}} E^t_{\mathrm{n,local}} + K^t_{\mathrm{n,s}} E^t_{\mathrm{n,server}}, \tag{8.13}$$

which needs to satisfy the following constraint:

$$\mathcal{E}_n(\boldsymbol{K}^t_n, \boldsymbol{f}^t_n, p^t_n) \le B^t_n, t \in \mathcal{T}, n \in \mathcal{N}. \tag{8.14}$$

Therefore, the energy level of the nth MU's battery is modified according to the following equation:

$$B^{t+1}_n = B^t_n - \mathcal{E}_n(\boldsymbol{K}^t_n, \boldsymbol{f}^t_n, p^t_n) + e^t_n, t \in \mathcal{T}, n \in \mathcal{N}. \tag{8.15}$$

8.2.2.2 Wireless Power Transfer

The main disadvantage of conventional EH methods is their dependence on ambient energy sources, which are unpredictable and uncontrollable. Subsequently, harvesting energy from intentionally generated RF signals proves to be an interesting alternative. To this end, WPT has attracted growing interest in wireless communication systems. Nevertheless, WPT creates important challenges in the design of communication systems, as a conflict with the information transmission emerges, since nodes cannot receive information and harvest energy simultaneously [19–21]. In this framework, the users deploy the harvested power to sequentially transmit their information utilizing the harvest-then-transmit protocol [22–24], which is also the core idea investigated in the considered system model.

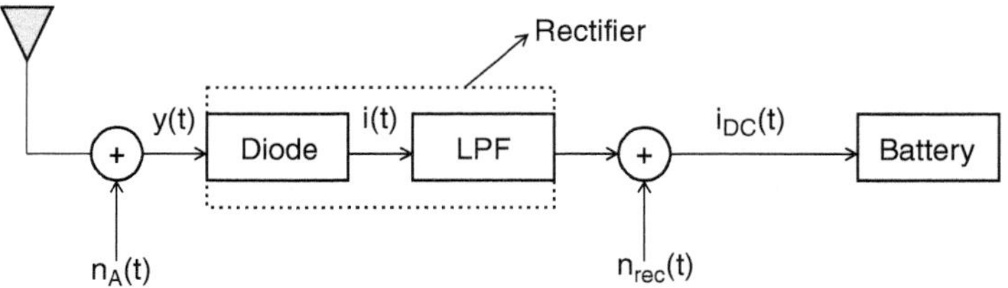

Figure 8.2 Energy receiver.

Figure 8.3 Time slot.

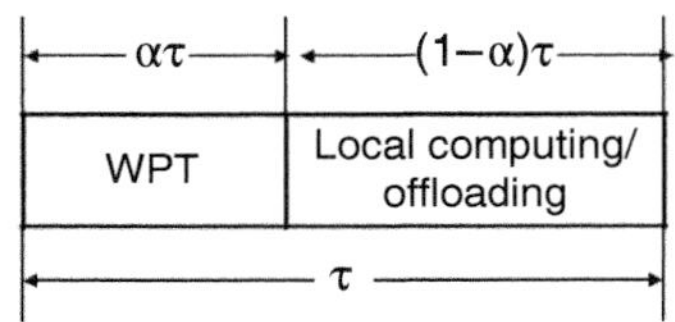

To perform WPT, each MU is equipped with an energy receiver, which is shown in Figure 8.2. In this receiver, the received RF signal is converted to a direct current (DC) signal by a rectifier consisting of a Schottky diode and a passive low pass filter, which is then stored to the battery.

In the considered system model, we assume that in the beginning of each time slot and for time $a_n\tau$, where $a_n \in [0, 1]$ is the portion of the time slot that is used for EH by the nth MU, the fog server, which is also utilized as base station, transmits power to the nth MU which is utilized by them for EH. The transmitted power by the fog server to the nth MU in the tth time slot is denoted by P^t_n. In the remaining portion of the time slot with duration $1 - \mathrm{a}_n$, the nth MU executes locally its computation tasks or/and offloads them to the fog server. This time allocation is depicted in Figure 8.3. In case that it offloads its tasks to the fog server, the harvest-then-transmit(offload) protocol is utilized.

Depending on the relationship between the harvested energy and the DC signal, two EH models are used in the literature, i.e. the linear EH model and the nonlinear EH model.

Linear EH Model. When the linear EH model is adopted, the converted energy in the energy receiver of the nth MU is assumed to be linearly proportional to the expected value of the DC signal, with a conversion efficiency $0 < \eta_n \leq 1$. Moreover, the harvested energy due to the equivalent processing noise is not considered. Therefore, the energy harvested by the nth MU in the tth time slot can be expressed as

$$e^t_{\mathrm{n,L}} = a_n \tau \eta_n h^t_n P^t_n. \tag{8.16}$$

It is observed in the above equation that the harvested energy in this EH model is linear with respect to the portion a_n of the time slot, which is dedicated to harvesting and also to the transmitted power.

Nonlinear EH Model. Generally, the harvested energy is nonlinear function of the received RF signal, due to the nonlinearity of the rectifier and more specifically the Schottky diode. To this end, in the literature [25–30], there have been several works considering practical nonlinear EH models. Considering a specific widely used nonlinear EH model, the harvested energy is given by [28]

$$\begin{aligned} e^t_{\text{n,NL}} &= a_n\tau\frac{\frac{P^t_{\text{n,s}}}{1+e^{-A(h^t_nP^t_n-B)}}-\frac{P^t_{\text{n,s}}}{1+e^{AB}}}{1-\frac{1}{1+e^{AB}}} \\ &= a_n\tau\left(\frac{P^t_{\text{n,s}}(1+e^{AB})}{e^{AB}(1+e^{-A(h^t_nP^t_n-B)})}-\frac{P^t_{\text{n,s}}}{e^{AB}}\right), \end{aligned} \tag{8.17}$$

where $P^t_{\text{n,s}}$ denotes the maximum harvested power of the nth user in the t- th time slot considering the EH circuit is saturated. Moreover, A is a positive constant, which reflects the nonlinear charging rate with respect to threshold. Given the EH circuit, the parameters $P^t_{\text{n,s}}$, A, and B can be determined by the curve fitting.

8.3 Tradeoffs in EH Fog Systems

In the existing literature, some tradeoffs are investigated mainly in the form of optimization problems. Two characteristic examples are presented in this section, i.e. the tradeoff between the energy consumption and the latency and the one between the execution delay and the task dropping cost.

8.3.1 Energy Consumption vs. Latency

The tradeoff between the energy consumption and the latency is discussed in [8]. More specifically, a multiuser fog system with WPT is investigated. Also, TDMA and partial offloading are deployed. The aim of the formulated optimization problem in this work is the minimization of the energy consumption at the fog server considering both the energy consumption for computation and for WPT, assuring that the N users' computation tasks are successfully executed. The energy consumption is minimized with the optimization of the number of offloaded bits, the local CPU frequencies and the time allocation among all users. The local computing latency and the EH constraints are also taken into account.

The problem is optimally solved by standard convex optimization techniques and some interesting conclusions are extracted. First, if the EH constraint is not tight for a user, i.e. this user harvests adequate wireless energy and should compute all the tasks locally. Moreover, it is always beneficial to process some bits locally, which means that offloading all of them to the fog server is not optimal. Furthermore, the more stringent the EH constraint is, the more bits should be offloaded to the fog with a smaller rate, as more time should be allocated for EH. Also, adopting NOMA instead of TDMA in the considered system may be able to improve the system performance [9]. Finally, using the more practical nonlinear EH model, described in (8.17), instead of the ideal linear one, may result in worse performance and higher complexity optimization problem and, thus, it would be an interesting challenge.

8.3.2 Execution Delay vs. Task Dropping Cost

In fog computing systems without EH, a key metric to evaluate the performance is execution delay, which is also utilized for the optimization of the offloading policy of the considered system [14, 16, 31–33]. However, taking into account the stochastic and intermittent nature of the harvested energy, some of the requested tasks cannot be executed and, thus, they are dropped. This phenomenon occurs when the MU has harvested a small amount of energy, because of the fading in the wireless channel between the fog server and the mobile device. In [3], where the considered system consists of the fog server and one MU, a new useful metric, namely execution cost, has been introduced. It is expressed as the weighted sum of the task dropping cost and the execution delay. Taking into account the fact that some tasks may be dropped, each dropped task is penalized by a unit of cost. Considering the earlier, the execution cost in the tth time slot is given by

$$C^t = \mathcal{D}(\boldsymbol{K}^t, \boldsymbol{f}^t, p^t) + \phi \boldsymbol{l}(\zeta^t = 1, K_d^t = 1), \tag{8.18}$$

where φ (in seconds) denotes the weight of the task dropping cost, $\boldsymbol{l}(\cdot)$ is the indicator function, and $\mathcal{D}(\boldsymbol{K}^t, \boldsymbol{f}^t, p^t)$ is given by

$$\mathcal{D}(\boldsymbol{K}^t, \boldsymbol{f}^t, p^t) = \boldsymbol{l}(\zeta^t = 1)(K_m^t D_{\text{local}}^t + K_s^t D_{\text{server}}^t). \tag{8.19}$$

Moreover, it is assumed that the successful execution of a task is preferred to dropping a task, i.e. $\tau^d \leq \phi$. Binary offloading is also deployed. Furthermore, a task, which is decided to be executed, should be completed before the deadline $\mathcal{T}^d$, which means that the following equation should be satisfied:

$$\mathcal{D}(\boldsymbol{K}^t, \boldsymbol{f}^t, p^t) \leq \mathcal{T}^d, t \in \mathcal{T}. \tag{8.20}$$

Therefore, considering the earlier definitions and assumptions, the aim of the optimization problem is the minimization of the sum of the execution costs in all time slots. Regarding the constraints of the problem, neither fog server execution nor local execution should be performed, unless there is a task request i.e.,

$$K_m^t + K_s^t \leq \zeta^t, t \in \mathcal{T}. \tag{8.21}$$

Moreover, in each time slot, the amount of consumed battery energy cannot exceed either $E_{\max}$, which is necessary for limiting the overdischarge of the battery [34], or the battery energy level in the specific time slot B^t, $t \in \mathcal{T}$. Furthermore, the whole amount of the arrived energy cannot be harvested and only a part of it is stored, which means that

$$0 \leq e^t \leq E_H^t, t \in \mathcal{T}. \tag{8.22}$$

Also, the transmit power and the CPU-cycle frequency are positive values and cannot exceed $p^{\max}$ and $f_{\mathrm{CPU}}^{\max}$, respectively. Finally, it should be considered that the computation mode indicators are binary variables, because binary offloading is assumed and (8.3) should be satisfied.

A Lyapunov optimization framework is used to solve the above optimization problem, according to which an algorithm is deployed that dynamically computes the optimal offloading policy. The proposed algorithm asymptotically achieves the optimal performance of the optimization problem. However, the optimal solution is achieved at the price of slow convergence and a higher battery capacity requirement. Hence, we can balance the system performance and the required battery capacity/convergence time, by adjusting the control variables.

8.4 Future Research Challenges

The integration of EH and WPT in fog computing is a topic that has attracted a lot of interest recently. Although some useful contributions have been made in this direction, there are still several research challenges that need to be addressed.

In practical systems, a large number of offloading tasks may lead the fog server to overloading, i.e. to a point that its computing power has to be allocated among the received offloading tasks. Subsequently, not only the task offloading time, but also the computation delay at the fog server should be taken into consideration.

Moreover, it is useful to extend EH fog systems to fading channels. It should be noted that it is not necessary to locally compute or offload its tasks in each time slot. In this case, it can be preferable for a MU to remain idle in some time slots and solely save the received energy in its energy storage device. This approach would result in a more complicated analysis, as the time allocation problem would be high dimensional.

Another challenge can be cooperative fog EH systems consisting of more than two users. To this end, users can be allowed to share both the radio and the computing resource. In other words, the users with higher computation power can assist weaker users to execute their tasks. In such systems, the optimization is more complicated, as the energy consumption for executing tasks for other MUs and transmitting the results back to them needs to be estimated and compared with that for offloading the tasks to the fog server.

Furthermore, another research challenge is to combine the concepts of WPT and EH utilizing renewable sources by utilizing a power beacon that is co-located with the fog server. The advantage of this deployment is that the limited availability of energy due to the utilization of renewable sources could be mitigated by the controllable transmission of RF signals.

Finally, it is highlighted that the existing works [5–9] that investigate WPT in fog systems consider various idealizing assumptions ignoring the effect of practical constraints, e.g., nonlinear EH characteristics and hardware impairments, which have not been investigated. To this end, nonlinear EH models as the one presented in the previous section can be utilized.

Acknowledgment

This work was supported by Nokia Bell Labs through the global donation program for Wireless Powered Remote Patient Monitoring (SPRING).

References

1 Mao, Y., You, C., Zhang, J. et al. (2017). A survey on mobile edge computing: the communication perspective. *IEEE Communication Surveys and Tutorials* 19 (4): 2322–2358. ISSN 1553-877X, doi: https://doi.org/10.1109/COMST.2017.2745201.

2 Bi, S. and Zhang, R. (2016). Placement optimization of energy and information access points in wireless powered communication networks. *IEEE Transactions on Wireless Communications* 15 (3): 2351–2364. ISSN 1536-1276, doi: https://doi.org/10.1109/TWC.2015.2503334.

3 Mao, Y., Zhang, J., and Letaief, K.B. (2016). Dynamic computation offloading for mobile-edge computing with energy harvesting devices. *IEEE Journal on*

Selected Areas in Communications 34 (12): 3590–3605. ISSN 0733-8716, doi: https://doi.org/10.1109/JSAC.2016.2611964.

4 Zhang, G., Zhang, W., Cao, Y. et al. (2018). Energy-delay trade-off for dynamic offloading in mobile-edge computing system with energy harvesting devices. *IEEE Transactions on Industrial Informatics* 14 (10): 4642–4655. ISSN 1551-3203, doi: https://doi.org/10.1109/TII.2018. 2843365.

5 Hu, X., Wong, K., and Yang, K. (2018). Wireless powered cooperation-assisted mobile edge computing. *IEEE Transactions on Wireless Communications* 17 (4): 2375–2388. ISSN 1536-1276, doi: https://doi.org/10.1109/TWC.2018. 2794345.

6 Ji, L. and Guo, S. (2019). Energy-efficient cooperative resource allocation in wireless powered mobile edge computing. *IEEE Internet of Things Journal* 1 ISSN 2327-4662, doi: https://doi.org/10.1109/JIOT.2018.2880812.

7 Bi, S. and Zhang, Y.J. (2018). Computation rate maximization for wireless powered mobile-edge computing with binary computation offloading. *IEEE Transactions on Wireless Communications* 17 (6): 4177–4190. ISSN 1536-1276, doi: https://doi.org/10.1109/TWC.2018.2821664.

8 Wang, F., Xu, J., Wang, X., and Cui, S. (2018). Joint offloading and computing optimization in wireless powered mobile-edge computing systems. *IEEE Transactions on Wireless Communications* 17 (3): 1784–1797. ISSN 1536-1276, doi: https://doi.org/10.1109/TWC.2017.2785305.

9 Wang, F. and Zhang, X. (2018). Dynamic computation offloading and resource allocation over mobile edge computing networks with energy harvesting capability. In: *Proceedings of 2018 IEEE International Conference on Communications (ICC), Kansas City, MO*, 1–6. https://doi.org/10.1109/ICC.2018.8422096.

10 Barbarossa, S., Sardellitti, S., and Di Lorenzo, P. (2014). Communicating while computing: distributed mobile cloud computing over 5G heterogeneous networks. *IEEE Signal Processing Magazine* 31 (6): 45–55. ISSN 1053-5888. doi: https://doi.org/10.1109/MSP.2014.2334709.

11 Kumar, K. and Lu, Y. (2010). Cloud computing for mobile users: can offloading computation save energy? *Computer* 43 (4): 51–56. ISSN 0018-9162, doi: https://doi.org/10.1109/MC.2010.98.

12 Satyanarayanan, M., Bahl, P., Caceres, R., and Davies, N. (2009). The case for VM-based cloudlets in mobile computing, 8. *IEEE Pervasive Computing* (4): 14–23. ISSN 1536-1268, doi: https://doi.org/10.1109/MPRV.2009.82.

13 Miettinen, A.P. and Nurminen, J.K. (2010). Energy efficiency of mobile clients in cloud computing. *HotCloud* 10: 4–4.

14 Kwak, J., Kim, Y., Lee, J., and Chong, S. (2015). Dream: dynamic resource and task allocation for energy minimization in mobile cloud systems. *IEEE Journal on Selected Areas in Communications* 33 (12): 2510–2523. ISSN 0733-8716, doi: https://doi.org/10.1109/JSAC.2015.2478718.

15 You, C., Huang, K., and Chae, H. (2016). Energy efficient mobile cloud computing powered by wireless energy transfer. *IEEE Journal on Selected Areas in Communications* 34 (5): 1757–1771. ISSN 0733-8716, doi: https://doi.org/10.1109/JSAC.2016.2545382.

16 Zhang, W., Wen, Y., Guan, K. et al. (2013). Energy-optimal mobile cloud computing under stochastic wireless channel. *IEEE Transactions on Wireless Communications* 12 (9): 4569–4581. ISSN 1536-1276, doi: https://doi.org/10.1109/TWC.2013.072513.121842.

17 Lakshminarayana, S., Quek, T.Q.S., and Poor, H.V. (2014). Cooperation and storage tradeoffs in power grids with renewable energy resources. *IEEE Journal on Selected Areas in Communications* 32 (7): 1386–1397. ISSN 0733-8716, doi: https://doi.org/10.1109/JSAC.2014.2332093.

18 Mao, Y., Zhang, J., and Letaief, K.B. (2015). A Lyapunov optimization approach for green cellular networks with hybrid energy supplies. *IEEE Journal on Selected Areas in Communications* 33 (12): 2463–2477. ISSN 0733-8716, doi: https://doi.org/10.1109/JSAC.2015.2481209.

19 Krikidis, I., Timotheou, S., Nikolaou, S. et al. (2014). Simultaneous wireless information and power transfer in modern communication systems. *IEEE Communications Magazine* 52 (11): 104–110. ISSN 0163-6804, doi: https://doi.org/10.1109/MCOM.2014.6957150.

20 Liu, Y., Wang, L., Elkashlan, M. et al. (2016). Two-way relay networks with wireless power transfer: design and performance analysis. *IET Communications* 10 (14): 1810–1819. ISSN 1751-8628, doi: https://doi.org/10.1049/iet-com.2015.0728.

21 Zhong, C., Suraweera, H.A., Zheng, G. et al. (2014). Wireless information and power transfer with full duplex relaying. *IEEE Transactions on Communications* 62 (10): 3447–3461. ISSN 0090-6778, doi: https://doi.org/10.1109/TCOMM.2014.2357423.

22 Diamantoulakis, P.D., Pappi, K.N., Karagiannidis, G.K. et al. (2017). Joint downlink/uplink design for wireless powered networks with interference. *IEEE Access* 5: 1534–1547. ISSN 2169-3536, doi: https://doi.org/10.1109/ACCESS.2017.2657801.

23 Ju, H. and Zhang, R. (2014). Throughput maximization in wireless powered communication networks. *IEEE Transactions on Wireless Communications* 13 (1): 418–428. ISSN 1536-1276, doi: https://doi.org/10.1109/TWC.2013.112513.130760.

24 Wu, Q., Tao, M., Kwan Ng, D.W. et al. (2016). Energy-efficient resource allocation for wireless powered communication networks. *IEEE Transactions on Wireless Communications* 15 (3): 2312–2327. ISSN 1536-1276, doi: https://doi.org/10.1109/TWC.2015.2502590.

25 Clerckx, B. (2018). Wireless information and power transfer: nonlinearity, wave-form design, and rate-energy tradeoff. *IEEE Transactions on Signal Processing* 66 (4): 847–862. ISSN 1053-587X, doi: https://doi.org/10.1109/TSP.2017.2775593.

26 Xiong, K., Wang, B., and Liu, K.J.R. (2017). Rate-energy region of swipt for MIMO broadcasting under nonlinear energy harvesting model. *IEEE Transactions on Wireless Communications* 16 (8): 5147–5161. ISSN 1536-1276, doi: https://doi.org/10.1109/TWC.2017.2706277.

27 Wang, S., Xia, M., Huang, K., and Wu, Y. (2017). Wirelessly powered two-way communication with nonlinear energy harvesting model: rate regions under fixed and mobile relay. *IEEE Transactions on Wireless Communications* 16 (12): 8190–8204. ISSN 1536-1276, doi: https://doi.org/10.1109/TWC.2017.2758767.

28 Kang, J., Kim, I., and Kim, D.I. (2018). Wireless information and power transfer: rate-energy tradeoff for nonlinear energy harvesting. *IEEE Transactions on Wireless Communications* 17 (3): 1966–1981. ISSN 1536-1276, doi: https://doi.org/10.1109/TWC.2017.2787569.

29 Clerckx, B., Zhang, R., Schober, R. et al. (2019). Fundamentals of wireless information and power transfer: from RF energy harvester models to signal and system designs. *IEEE Journal on Selected Areas in Communications* 37 (1): 4–33. ISSN 0733-8716, doi: https://doi.org/10.1109/JSAC.2018.2872615.

30 Tegos, S.A., Diamantoulakis, P.D., Pappi, K.N. et al. (2019). Toward efficient integration of information and energy reception. *IEEE Transactions on Communications* 67 (9): 6572–6585. https://doi.org/10.1109/TCOMM.2019.2916831.

31 Chen, X. (2015). Decentralized computation offloading game for mobile cloud computing. *IEEE Transactions on Parallel and Distributed Systems* 26 (4): 974–983. ISSN 1045-9219, doi: https://doi.org/10.1109/TPDS.2014. 2316834.

32 Huang, D., Wang, P., and Niyato, D. (2012). A dynamic offloading algorithm for mobile computing. *IEEE Transactions on Wireless Communications* 11 (6): 1991–1995. ISSN 1536-1276. doi: https://doi.org/10.1109/TWC.2012.041912.110912.

33 Muñoz, O., Pascual-Iserte, A., and Vidal, J. (2015). Optimization of radio and computational resources for energy efficiency in latency-constrained application offloading. *IEEE Transactions on Vehicular Technology* 64 (10): 4738–4755. ISSN 0018-9545, doi: https://doi.org/10.1109/TVT.2014.2372852.

34 Sun, S., Dong, M., and Liang, B. (2016). Distributed real-time power balancing in renewable-integrated power grids with storage and flexible loads. *IEEE Transactions on Smart Grid* 7 (5): 2337–2349. ISSN 1949-3053, doi: https://doi.org/10.1109/TSG.2015.2445794.

9

Optimizing Energy Efficiency of Wearable Sensors Using Fog-assisted Control

Delaram Amiri[1], *Arman Anzanpour*[2], *Iman Azimi*[2], *Amir M. Rahmani*[4], *Pasi Liljeberg*[2], *Nikil Dutt*[3], *and Marco Levorato*[3]

[1] *Department of Electrical Engineering and Computer Science, University of California Irvine, Irvine, CA, USA*
[2] *Department of Future Technologies, University of Turku, Turku, Finland*
[3] *Department of Computer Science, University of California Irvine, Irvine, CA, USA*
[4] *School of Nursing, University of California Irvine, Irvine, CA, USA*

9.1 Introduction

Continuous clinical-level monitoring of patients conditions is imperative in an ample range of medical applications. For instance, monitoring post-operative patients to detect early signs of health deterioration can significantly improve care outcomes. Current technologies can only provide clinical-level monitoring in hospital settings, where the patient is in a controlled environment and traditional sensors can be used. However, once discharged, patients are left in a vulnerable position. Achieving clinical-level monitoring in everyday settings would have a tremendous impact on patients health, but is a technological challenge that has not been solved yet.

Internet of Things (IoT) technologies have been recently widely used to build systems capable of continuously monitoring subjects, acquiring a variety of biosignals [1–3]. The healthcare IoT paradigm proves a way to ubiquitous and personalized monitoring of individual's conditions in everyday settings. However, these technologies have several limitations that make continuous high-quality sensing di cult. In fact, these sensors have limitations in terms of storage, computation load and energy supply. Furthermore, different from hospital environments, in everyday settings the activities the monitored subject engages may cause a degradation of the quality of the signals due to the movement between the skin and the sensor. Some wearable sensors, such as photoplethysmogram (PPG), electrocardiogram

Fog Computing: Theory and Practice, First Edition.
Edited by Assad Abbas, Samee U. Khan, and Albert Y. Zomaya.

(ECG), and electromyography (EMG) are particularly influenced by this effect and may suffer a significant loss of accuracy.

Interestingly, different activities may cause a different degree of degradation. For instance, motion artifacts affecting the measured signal when the subject is "Running" are much larger compared to those generated when the person is "Sitting" or "Sleeping." Achieving the same signal-to-noise ratio (SNR) in all the activities and, thus, the same detection quality, requires the tuning of the sensing power to that of the noise. In other words, higher sensing energy could be used when the patient is Running, and a lower energy could be used when the patient is Sitting. Nonetheless, wearable sensors are manufactured in industry for worst case scenario in terms of noise level, corresponding to a large sensing power, which results in a high energy consumption and a short battery lifetime. The layered and pervasive IoT infrastructure can be used to support solutions enabling the adaptation of sensing parameters to the joint person-technology system state. The ability of the system to track its own state and determine optimal parameters can be connected with the general concept of "self-awareness." Herein, we extend this notion across the layers of the IoT infrastructure, and focus patient's activity as a main driver of adaptation due to the specific application domain.

In the computing literature, the general notion of context captures the state of the system, including any descriptor of the user's state. Users' activity is considered as a subclass of context information, whose estimation and tracking requires the acquisition of signals from sensors and the implementation of inference algorithms. Intuitively, imposing a further burden on the sensor to have them connect with each other and implement algorithms is not a suitable option. Therefore, the needed "semantic" support necessarily comes from the higher layers of the IoT infrastructure, which has sufficient resources to perform signal fusion and processing. In particular, fog and edge computing resources [4, 5], which are connected to the mobile devices through one-hop low-latency wireless links, are particularly indicated to host compute-intense tasks informing system-level control. Implementing optimization algorithms controlling the sensor layer on the fog layer can avoid imposing overheads to the sensor layer while being able to rapidly respond to changes thanks to the local control. In this chapter, we propose approaches using the real-time connection between the sensors and the fog layer to build a context-aware and self-aware control loop determining the sensing power used by wearable sensors.

The rest of the chapter is organized as follows: Section 9.2 provides a background in remote patient monitoring. Section 9.3 discusses researches done in energy-efficient sensor control. Section 9.4 discusses the design challenges in healthcare IoT. Proposed energy-efficient fog computing is defined in Section 9.6. Section 9.7 concludes the chapter.

9.2 Background

Remote health monitoring is a promising approach to extend reactive and proactive healthcare solutions for populations at risk beyond traditional clinical settings. Such a service allows continuous monitoring of patients in their daily routines, enabling early intervention services in case of health deterioration [2, 6]. Moreover, it has the potential to alleviate medical costs and hospital visits for patients, improving their quality of life as well as independent living. IoT as an advance network of objects can be advantageously applied in such applications.

IoT-based systems leverage a variety of sensors, communication infrastructures, and computing resources to deliver monitoring solutions [1, 7, 8]. In the context of remote health monitoring in everyday settings, these systems demand continuous data acquisition with high-level quality attributes, where various vital signs should be collected seamlessly while end-users are involved in daily routines.

In this context, photoplethysmography (PPG) is a promising noninvasive mechanism that captures various vital signs for users in everyday settings. PPG is a nonobtrusive optical method employed to measure blood volume variations in the microvascular bed of tissues [9]. The blood variations are associated with cardiac and respiratory activities. They also reflect an estimate of arterial oxygen saturation. Accordingly, the PPG method is widely used in several wearable and IoT-based systems, through which different vital signs such as heart rate, respiration rate, and oxygen saturation (i.e. SpO_2) are extracted. The PPG method often contains two light sources along with one light sensor placed on a body organ such as a fingertip (see Figure 9.1). The light sources emit red and infrared lights to the tissue, while the light sensors absorb the light reflection from the tissue, capturing the PPG waveforms, which correspond to the amount of oxygenated hemoglobin molecules in the veins.

Different techniques have been proposed to obtain heart rate and respiration rate from the PPG waveform (i.e. red or infrared waveform) [10, 11]. Feature-based techniques are designed to obtain the vital signs by extracting certain features (e.g. maximum intensity of the pulse and baseline variations) from the signal [12]. These techniques, nevertheless, are susceptible to the presence of motion artifacts and surrounding noises, which distort the features.

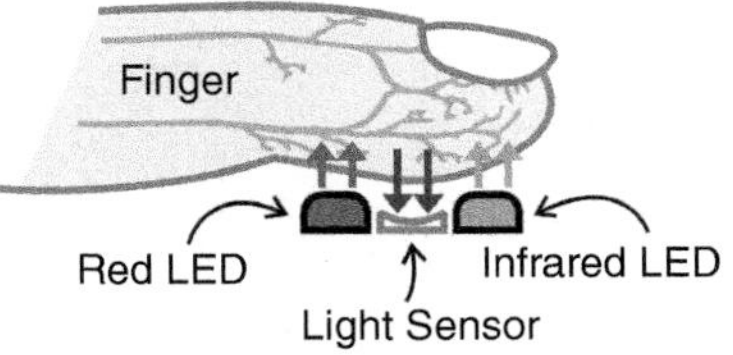

Figure 9.1 PPG sensors consisting of two light sources and one light sensor.

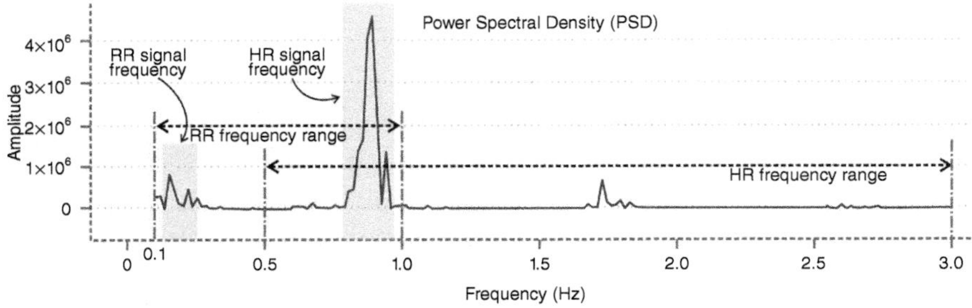

Figure 9.2 Power spectral density (PSD) of one-minute PPG signal.

Alternatively, filter-based techniques extract the vital signs, leveraging band-pass filters [13]. In this context, two band-pass filters are designed according to the respiration and heartbeat frequency ranges. Initially, the cutoff frequencies are set to 0.1–1 Hz (6–60 breath rate/minute) and 0.5–180 Hz (30–180 beats/minute) for the respiration rate and heart rate extraction, respectively. The boundaries are, then, narrowed down to the vital signs' frequencies to mitigate the noise. In this regard, the cutoff frequencies are dynamically selected, exploiting the peak values in the power spectral density (PSD) of the signal [14] (see Figure 9.2). Leveraging the band-pass filters, the respiratory and heartbeat signals are extracted. Then, a peak detection algorithm is performed to obtain local maximum points in the derivative of the biosignals. The time interval between two consecutive peaks indicate the heart rate and respiration rate values.

Despite the heart rate and respiration rate, the SpO_2 is derived from both infrared and red waveforms. As shown in Figure 9.3, four features (i.e. AC_{IR}, AC_{RED}, DC_{IR}, and DC_{RED}) are first extracted. Then, the SpO_2 is determined using the following

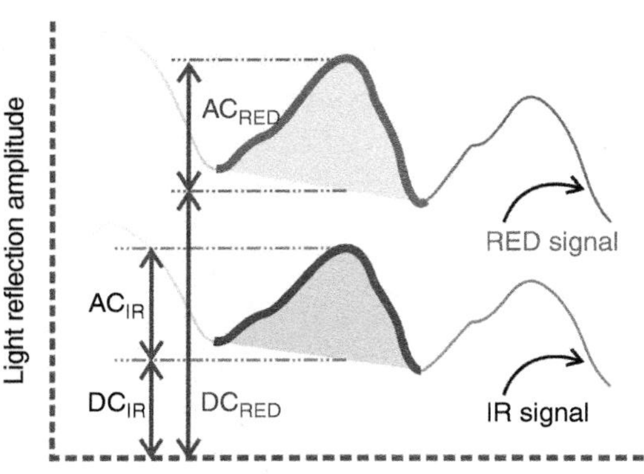

Figure 9.3 PPG waveforms and the four features extracted for SpO_2 calculation.

equations:

$$R = \frac{AC_{RED} \cdot DC_{IR}}{AC_{IR} \cdot DC_{RED}} \tag{9.1}$$

$$SpO_2 = \alpha R^2 + \beta R + \gamma \tag{9.2}$$

where α, β, and γ are constants obtained from the sensor's specification [15].

9.3 Related Topics

In healthcare IoT applications, sensors are frequently energy bottlenecks of the system. Algorithms to reduce energy consumption of sensors have been widely studied. Different approaches have been proposed including sensor sleep scheduling [16] and power aware cognitive communication protocols [17]. Specifically, in healthcare applications, one of the most important considerations is designing energy efficient protocols.

For instance, interconnecting biosensors mounted or embedded on a body via wireless body area network (WBAN) can be implemented with energy efficient multiple access control (MAC) protocol. MAC layer is responsible to manage data packet transmissions from the sensors across the network. Characteristics like scheduling duration of sensor sleep time, path routing, and scheduling duty cycles are MAC protocol adjustment methods that can extend the battery life of wearables [18].

Chang et al. proposed a routing protocol for WBAN considering expected transmission count and residual energy metrics for optimal path selection [19]. Pradhan et al. compared energy consumption in four protocols, 802.15.4, IEEE 802.15.6, SMAC, and TMAC as hybrid MAC protocols in healthcare [20].

Scheduling sleep duration of the sensors are proposed to enhance the energy efficiency by reducing the unnecessary idle listening. Scheduling sleeping periods to match the needs of different applications in sensor networks are proposed [21–24]. In 2017, Kaur et al. [25] proposed a solution to determine sleep intervals of sensors based on their remaining battery level, usage history, and quality of the measured signal. Zhang et al. [26] reduced the computational complexity by formulating the sleep scheduling while also paying attention to the network reliability in WBANs [26].

Typically, these methods are utilized to improve energy efficiency in a system level, turning attention away from the context. For applications in the healthcare domain, sensory data type, biosignals, sent to the gateway magnifies the importance of data accuracy impacted by both the context of environment and the dynamics of the system itself. Received data in the gateway can be managed using fog-assisted energy efficient optimization algorithms to find optimal solutions for the sensor's configurations to maximize the sensor's battery life while maintaining reliability.

Studies conducted by Zois et al. [27–29] focused on detecting activity of an individual by gathering partial observations from the sensors considering a gateway as the energy bottleneck. However, the objective of these frameworks, based on Markov decision process (MDP) theory, is that of detecting the activity itself, and the focus is on the optimization of data transfer to a sink. In contrast, we propose using the activity as a context to control sensing accuracy in an edge-based architecture to extend the lifetime of sensors.

9.4 Design Challenges

As we discussed earlier, a remote patient monitoring system is expected to make the patient's vital signs available and collectible for healthcare professionals wirelessly. Furthermore, it was mentioned that the PPG signal is a good source for at least three vital signs.

Even though the use of PPG signal is a good source of data, it poses several challenges to the system. The first challenge is the behavior of the PPG signal in different contexts. The typical PPG signal displayed in Figure 9.3 is recordable in the hospital setting when the patient is sitting or lying on the bed without movement. This typical PPG signal consists of two parts: AC part, which oscillates with each heartbeat, and DC part, which forms the baseline of the signal. The AC part is the result of the changes in the amount of oxygenated hemoglobin in the blood and the changes in the DC part are due to the pressure applied from other body tissues to the blood vessels. For example, during the respiration process, the change in the size of lungs applies pressure to blood vessels and causes a low-frequency oscillation in the DC part. These changes in the DC part enable us to measure the respiration rate from the DC part oscillation frequency by filtering the AC part. Such changes in the DC part creates a challenge in signal processing when the patient has some activities. Each body movement causes a change in the DC part and more intense activities make larger changes in the signal baseline. When the amplitude of changes is larger than the amplitude of AC part, the detection of heartbeat peaks would be more difficult. In addition, when the body movements are rhythmic (e.g. walking, jogging, running) with a frequency close to respiration, the calculation of respiration rate is also rather difficult.

The other challenge in the PPG signal acquisition is the amount of noise in the signal. Ambient light diffuses to the exposed body tissues close to the sensor spot and causes a level of noise to the recorded signal. Although increasing the brightness of light-emitting diodes (LEDs) in the PPG sensor reduces the effect of ambient light noise, it increases also the power consumption in the sensor node.

The last PPG-related challenge is that most parts of signal processing are not possible to carry out with low power microcontrollers of wearable sensor nodes. The sensor node should send the raw signal to a gateway or cloud server for further processing. This, in turn, requires more power for radio transmission. In the following sections, we describe the solutions to cover the mentioned challenges.

9.5 IoT System Architecture

IoT is a term for describing methods that enable us to sense and control a variety of parameters and objects wirelessly through the Internet. Developing a system based on IoT has several benefits in comparison to traditional wireless sensors and remote control approaches. The most significant characteristic of IoT is its well-organized distribution of energy sources, data storage, and processing power. As shown in Figure 9.4, a common architecture of IoT-based remote health monitoring application consists of three layers. The first layer belongs to the sensor nodes that are recording and reporting the patient state including medical parameters, activity level, posture, location, and environment properties. Sensor nodes send the collected data to one or more gateway devices, which are placed close to the patients so that energy cost of radio communication between sensors and gateways remains at a low level. In the gateway layer, the devices have their own storage and processing units powerful enough to perform preprocessing and fog-computing actions before transferring data to the cloud server. The preprocessing actions may include data filtering, data fusion, data analysis, compression, and encryption. Fog-computing methods may be used to offload a portion of cloud-based tasks to the gateway device at the edge of the network to reduce the bandwidth need, data size, and server load. In the cloud layer, the server receives collected data from several gateways and stores the data recorded from all patients. Such a huge amount of data enables the cloud server to compare patients with their earlier conditions, with other patients in the same condition, and the

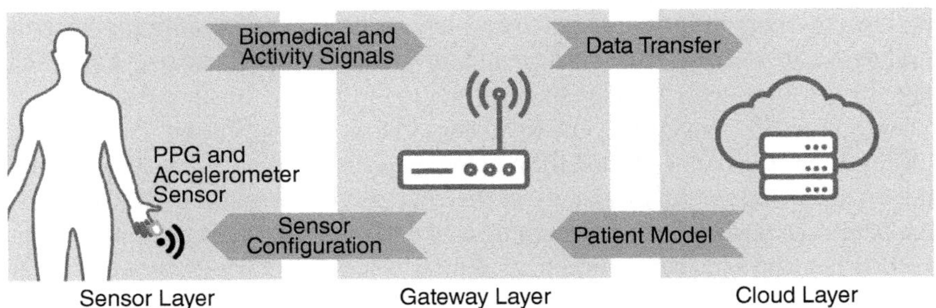

Figure 9.4 IoT system architecture.

consequences of the current condition in other patients. The server then would be able to learn from the patients' history and predict the future of patients' health.

In our setup, we use a battery-powered microcontroller connected to a digital I2C PPG sensor, a 3D accelerometer sensor, a temperature sensor, and a Wi-Fi transmission module as our sensor devices. The amount of power required to drive LEDs in the PPG sensor is configurable, the sensors node is able to receive configurations from the gateway device remotely. The other configurable parameters are the recording and hibernation durations. The gateway is a Wi-Fi-enabled Linux machine, which receives the recorded data, performs fog-assisted optimization algorithms, decides about the configuration of the next recording period, and sends the new settings back to the sensor device. The cloud server gets updated with all collected data and updates the root mean squared error (RMSE) values for the gateway over longer periods of time.

9.5.1 Fog Computing and its Benefits

Fog computing leverages the concept of geo-distribution of networks at the edge, enabling local/hierarchical data analysis, decision making, and storage [4, 5, 30]. Prevalence of connected devices and IoT-based systems demands an intermediary layer of computation, in which the local solutions provide low-latency responses, load balancing, and adaptivity for system behavior.

With the growth of IoT-based systems, a rapid increase in the number of connected devices has led to a massive volume of data that needs to be processed [31, 32]. Cloud computing has, thus far, provided scalable and on-demand storage and processing resources to fulfill the requirement of IoT. However, the most recent IoT-based applications require mobility, low-latency response, and location-awareness [33]. Moreover, the latency of data transmission between the edge and the cloud is unsatisfactory especially in latency-sensitive systems such as health monitoring [32]. In this regard, Cisco states that "Today's cloud models are not designed for the volume, variety, and velocity of data that the IoT generates" [34]. Therefore, fog computing can be considered as a complementary solution for the cloud computing paradigm to enable such latency demanding applications [35] as it can relocate location dependent, time-dependent, massive scale, and latency-sensitive tasks from the cloud server to the edge of the network [36].

Fog computing provides several lightweight services at the edge of the network, locally analyzing data collected from heterogeneous connected devices. Depending on the computational capacity of the edge servers or gateway devices, such fog-based services can include not only conventional tasks such as protocol conversion but also local data processing applications, some of which are outlined as follows. There is a variety of applications such as data filtering and data fusion to ensure high-level data quality at the edge, improving the data accuracy and performing data abstraction [4, 37]. Such applications can decrease the amount of data that should be sent to the cloud server and subsequently save external bandwidth. Moreover, local decision making is a solution at the edge by which the system's availability and reliability are increased particularly when the Internet connection is poor [38]. Adaptive sensing and actuation is another application that intelligently tunes the system's configuration at the edge according to the context information [39]. Such a dynamic reconfiguration can considerably improve the system-driven quality attributes such as energy efficiency. Certain security-related services can be also performed at the edge, protecting data from unauthorized access (e.g. authentication, data encryption/decryption, anomaly detection) [4, 40].

In summary, fog computing provides several benefits to the system. The benefits include (but are not limited to) (1) local processing, notification, and actuation with short latency, (2) interoperability and reconfigurability for the system, (3) energy efficiency for sensor nodes, and, (4) mobility support for the users, and (5) reliability and availability of the service. Storing data close to the sensor network and processing it locally leads to quick notification and rapid response. Acting as a power manager, fog computing can reduce and optimize the energy consumption of the sensor nodes while not imposing the management overheads to the sensors.

9.6 Fog-assisted Runtime Energy Management in Wearable Sensors

Based on the design challenges discussed in Section 9.4, it is desired to create a platform that leverages portable sensors coupled with a layered communication and processing architecture for local control purposes. Accommodating runtime optimization algorithms at the edge enables the deployment of smart solutions to control the settings of the sensors. Configurations/knobs such as sampling frequency and transmission rate can also be used to control the communication delay and bandwidth between sensor and fog layer. However, in this chapter, we only focus on controlling the sensor's sensing power to minimize energy, which does not influence the communication characteristics of the system (i.e. the volume of transmitted data is unchanged). Energy efficient algorithms need to consider the environmental situation to find a robust and optimal solution. Daily activities, in particular strenuous ones, impact the quality of sensor's measurements. For instance, the noise caused by body movements during running is strong enough to significantly distort the PPG signal. In contrast, physical activities such as sleeping or sitting impose less motion artifacts. Therefore, smaller energy budgets can be assigned to the sensor to satisfy a similar threshold of accuracy. This highlights the necessity of context awareness in managing quality and energy of characteristics of sensors at the edge.

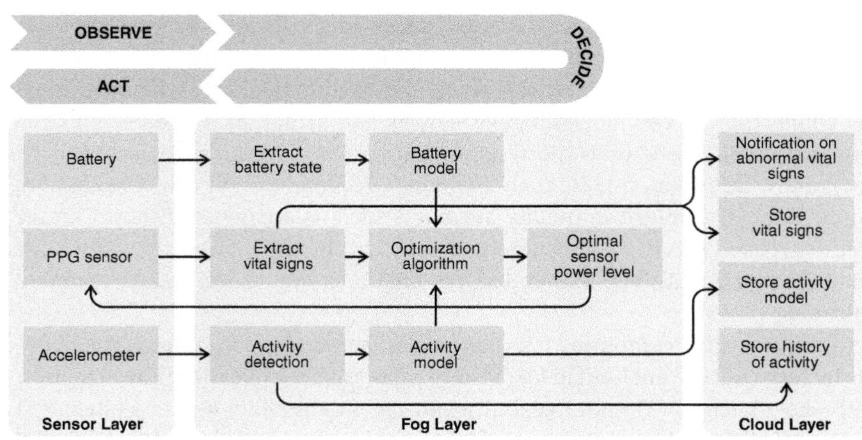

Figure 9.5 The high level system architecture.

The high-level view of our proposed fog-assisted architecture is shown in Figure 9.5. The architecture consists of three different phases known as Observe, Decide, and Act.

- *Observe.* Collecting data from a PPG sensor and an accelerometer along with the battery state of charge of the sensor node and transmitting it wirelessly to the fog layer.
- *Decide.* (1) Preprocessing and filtering the data received from the sensor layer, (2) extracting vital signs from the PPG signal, (3) detecting activity of a user from collected accelerometer data, (4) modeling the statistics of activity dynamics, (5) calculating the transition probabilities between activity states, (6) modeling transition probabilities between battery states during charging and discharging modes, and (7) implementing optimization algorithms to utilize the user's activity model, battery transitioning model, and vital signs to find an optimal solution for the sensor's power level.
- *Act.* Regulating the sensor node's power level based on the optimal solution derived in the fog layer.

In addition, the activity model, vital signs, and the user's history of activity are stored in the cloud for further analysis. Notifications on detecting abnormalities in vital signs are transmitted to the cloud to warn the clinicians. Section 9.6.1 discusses the concept of computational self-awareness in the proposed IoT architecture, using the three processes of Observe, Decide, and Act, while Section 9.6.2 presents the proposed energy optimization algorithms to control the sensor layer.

9.6.1 Computational Self-Awareness

Computational self-awareness enables a computing system to act reliably, optimally, and adaptively, despite the radical changes in its own state and environment [41–44]. Computational self-awareness therefore requires the computing system to be empowered by knowledge about both itself as well as the surrounding environment. By exploiting computational self-awareness principles, the system's dynamic behavior can be managed to provide a high level of quality of service, allowing accurate optimization schemes. Computational self-awareness has thus far been investigated in various applications including cyberphysical systems, remote health monitoring, and mobile applications [41, 45, 46].

Within healthcare applications, the system leverages semantic information including the synergy between the system, the individual, and the surrounding environment. In this regard, self-awareness is performed in computing systems via a closed-loop framework where three different phases – observe, decide, and act (ODA) – are deployed together with reflective models of the system [47, 48]. (Figure 9.5) shows the application of the ODA loop, together with reflective models of the battery and activity, in the IoT-based health-monitoring system architecture. Using the self-aware enhanced ODA-loop, the cognitive fog layer adaptively configures the sensor nodes to provide a high-level of accuracy as well as enhancing the energy and bandwidth efficiency of the system.

In each iteration, first, data collection is performed via the sensor network, obtaining internal (e.g. system status) and external (e.g. user's health status) data. Second, the collected data are fed to data analytic approaches and models are created. Then, the best behavior (i.e. the system's configuration) is determined according to the recent observation. The selected configuration should satisfy both the system's and application's requirements. Third, if adaptation in system behavior is needed, the changes are applied and the selected configuration is implemented.

9.6.2 Energy Optimization Algorithms

In our proposed IoT system architecture, the fog layer is equipped with a cognitive optimization algorithm that tracks the dynamics of the sensor and user's activity to offer an optimal solution for the sensor's energy level. Our optimization formulation centers on the minimization of energy expenditure of a sensor node while satisfying the application requirements in terms of measurement accuracy. Measurement accuracy is directly influenced by the power level chosen in the sensor. Figure 9.6 demonstrates the process of modeling accuracy measurement as a function of power level in a PPG sensor. We first start by introducing the error in the accuracy of measurements. We use RMSE between vital signs extracted

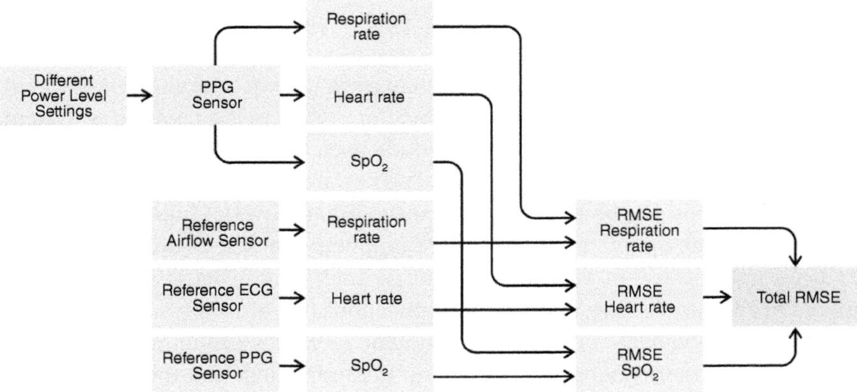

Figure 9.6 Modeling accuracy of measurements (e.g. $\varepsilon(X, U)$) in PPG sensor using *RMSE*.

from PPG signal (e.g. SpO_2, heart rate, and respiration rate) and the true values of the extracted features. The ground-truth features are extracted from the following three sensors:

1. A chest band using an ECG sensor to collect the reference heart rate
2. An airflow sensor placed on the user's upper lip to measure the true value of respiratory rate
3. A precise PPG sensor with higher power consumption to extract SpO_2

We set the PPG sensor's power level (e.g. U) to all possible values. Higher power levels lead to more energy consumption as well as higher accuracy of measurements. In order to incorporate context-awareness in our problem formulation, we assume users are engaged in different activities. Based on the intensity of the activity, a different amount of motion artifact will be observed in the signal. We consider the following list of activities: "sleeping," "sitting," "walking," "jogging," and "running." $RMSE$ is calculated for different combinations of the sensor's power level for different user's activities (e.g. X). Comparing the true values of the vital signs as references to the measured values is a proper metric to model the accuracy. Therefore, the proposed approach can model the error in the accuracy of extracted vital signs (e.g. $\varepsilon(X, U)$) as a function of sensing power level in the sensor and the activity of an individual.

Consider that the probability density function (PDF) of error denoted by $\rho(\varepsilon(X, U))$ to follow a Gaussian distribution. Therefore, we can define the probability of error as the tail probability of normal distribution when the user's activity $X = x$ and the power level of the sensor is $U = u$ where

$$\mathcal{P}_{\mathcal{T}}(X, U) = \int_{-\infty}^{\mathcal{T}} \rho(\varepsilon(X, U) \mid X = x, U = u) d\varepsilon. \tag{9.3}$$

The threshold $\mathcal{T}$ determines the maximum threshold in error tolerance. For instance, larger values of $\mathcal{T}$ show that larger values of *RM SE* are acceptable in calculating the error probability $\mathcal{P}_{\mathcal{T}}(X, U)$. We now define the regions of abnormal vital signs by marking the vital signs y as abnormal and assign a Gaussian PDF with calculated mean m_a and variance σ_a (e.g. $f_a(y \mid X) \sim \mathcal{N}(m_a, \sigma_a)$). Herein, the probability of abnormal vital signs ubiquity can be distinguished from the normal vital signs with threshold θ. Considering the user's activity $X = x$, the probability of existence of abnormal vital signs can be written as

$$\mathcal{P}_\theta(X) = \int_\theta^\infty f_a(y \mid X = x)dy. \tag{9.4}$$

Equation (9.3), along with the marked abnormal vital signs in Equation (9.4), can be used to determine the probability of abnormality misdetection. The probability of abnormality misdetection is a metric to determine the possibility that abnormal vital signs are not detected due to error in sensor's measurements. In fact, probability of misdetection can be defined as a joint event of ubiquity of abnormal vital signs and sensor's error tolerance. We proved in [39] that the upper bound for probability of misdetection in abnormality (e.g. $\mathcal{P}_{\mathrm{D}}$) can be written as

$$\mathcal{P}_{\mathrm{D}}(X, U) = \mathcal{P}_\theta(X)\mathcal{P}_{\mathcal{T}}(X, U). \tag{9.5}$$

We use the formulation of $\mathcal{P}_{\mathrm{D}}(X, U)$ as the main presentation of deterioration risk factor. After modeling the probability of misdetection, we can define the important factors to be optimized. On the one hand, lower power levels in the sensor consume less energy leading to the ability to monitor a patient for a longer time.

However, the signal captured by the sensor is distorted by noise due to low SNR. On the other hand, using higher energy levels in the sensor increases the energy consumption leading to shorter battery life, but enhances the accuracy of the extracted vital signs. Therefore, it is important to find an optimal solution as a trade-off between energy consumption (e.g. $C_{\mathrm{TX}}(U)$) as a result of choosing power level U and desired level of satisfactory in probability of misdetection. The optimization problem can be defined as:

(a) Minimizing cost function of energy consumption over constraints of probability of misdetection;

$$\begin{aligned} \underset{U}{\text{minimize}} \quad & C_{\mathrm{TX}}(U) \\ \text{subject to} \quad & \mathcal{P}_{\mathrm{D}}(X, U) \le \eta \\ & \text{or, equivalently,} \\ & \mathcal{P}_{\mathcal{T}}(X, U) \le \frac{\eta}{\mathcal{P}_\theta(X)} = \zeta \end{aligned} \tag{9.6}$$

or

(b) Defining total cost function of energy consumption and probability of misdetection with parameter $0 \leq \omega \leq 1$ to control the importance weight of the two factors;

$$C_{\text{total}}(X, U) = \omega \mathcal{P}_{\text{D}}(X, U) + (1 - \omega)C_{\text{TX}}(U). \tag{9.7}$$

Therefore, the optimization problem can be written as:

$$\underset{U}{\text{minimize}} \quad C_{\text{total}}(X, U) \tag{9.8}$$

Defining an optimization problem based on the cost function requires finding an optimal solution for the power level of the sensor. Solutions to the optimization problems can be proposed to minimize the accumulated cost function over a finite time horizon or to minimize the instantaneous cost. Therefore, both methods find an optimal power level for the sensor. Figure 9.7 demonstrates the process of implementing the optimization algorithm.

Two approaches are proposed in this chapter to solve the optimization problem, the (1) myopic strategy and (2) MDPs strategy. In addition, the performance of aforementioned strategies is compared.

9.6.3 Myopic Strategy

Optimizing the instantaneous cost function results in implementing real-time algorithms with linear time complexity known as the myopic strategy. Intuitively, myopic strategy finds an optimal solution for the sensor's power level based on instantaneous activity states. Myopic strategy can find the lowest power level in Equation (9.6) that satisfies the maximum probability of misdetection. This strategy can be implemented on the edge devices with limited memory and computations to calculate real-time solution for the sensor's power level. In order to find a solution based on a longer perspective to avoid possible outage and performance reduction, a strategy that evaluates the outcome of all possible actions can be proposed. The next section discusses MDP strategy that allows optimization of power level over a temporal horizon.

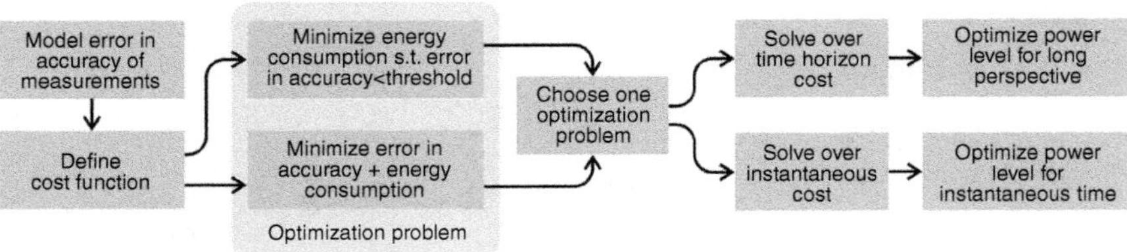

Figure 9.7 Optimization algorithm implementation.

9.6.4 MDP Strategy

Optimizing the accumulated cost over time horizon can be achieved by tracking the history of dynamics of the system as well as exploring different possible actions to achieve longer perspective of solutions. MDP is a strategy that calculates the optimal power level in the sensor by using a sliding window over a finite time horizon by tracking the stochastic state of the system. Stochastic model in MDP consists of user's activity as the contextual state and battery's state of charge as the self-state. We consider that the battery is quantized as distinct battery states. We model the transitioning probabilities between battery states depending on a possible power level chosen as the action. Finite battery states can be defined as $Q = \{Q_1, Q_2, \ldots, Q_K\}$ with Q_1 as the battery state completely discharged and Q_K as the fully charged state. We can define transitioning probabilities from battery state Q_k to $Q_{k'}$ with the possible discrete power levels $U \in \{U_0, U_1, U_2, \ldots, U_m\}$,

$$q(k' \mid k, U) \triangleq \mathcal{P}\ (Q'_k \mid Q_k, U) \tag{9.9}$$

Considering the transition probabilities defined in Equation (9.9), the dynamics of the battery can be modeled as a Markovian process with the states transitioning shown in Figure 9.8.

In addition, the Markov model tracks the changes in activity of a user as state space X in – sleeping, sitting, walking, jogging, running. Note that activities during the day change over time and building different models is a necessity. Therefore, we uniformly break down the daily activity into smaller periods. We consider Markov model corresponding to ith period in $\{1, 2, \ldots, n\}$ each with duration of $\frac{24}{n}$ hours.

Figure 9.9 shows the example of Markov chain for activities during one period. This type of model is updated to the change of subject's context over time and personalized based on the daily activity. We build the Markov chain transitioning from activity X_j to $X_{j'}$ during period i with the following transition probabilities,

$$p_i(j' \mid j) = \mathcal{P}(X_{ij'} \mid X_{ij}) \tag{9.10}$$

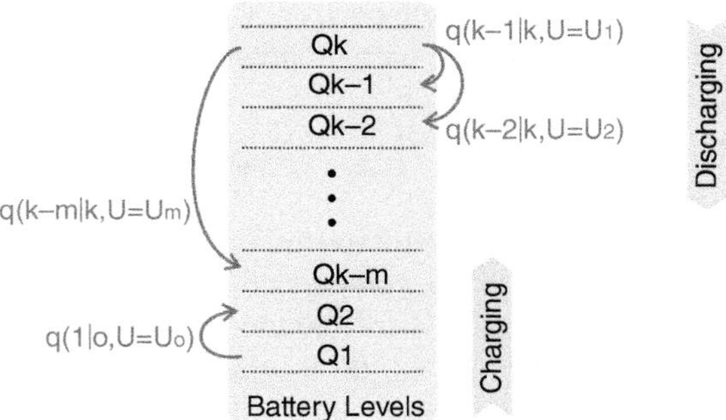

Figure 9.8 Markov chain of battery states during charging and discharging.

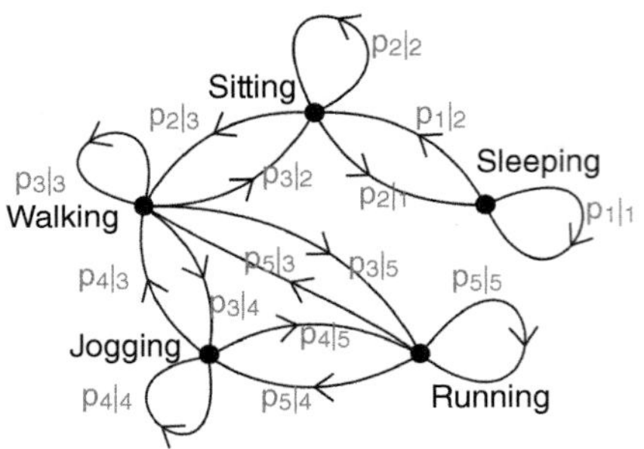

Figure 9.9 Markov chain of activities of an individual during one period.

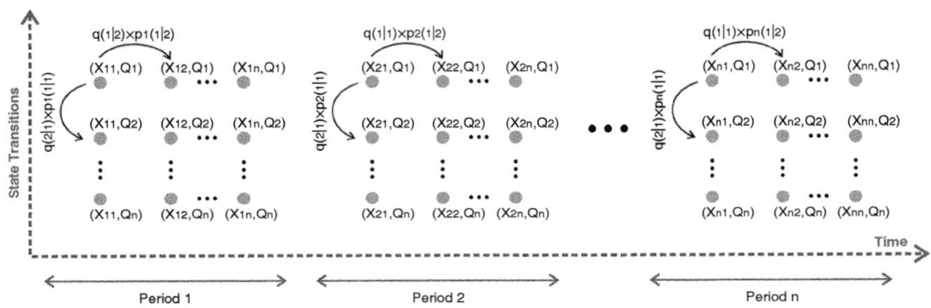

Figure 9.10 Markov chain of joint battery and activity states during period 1 to *n*.

The total state of the system can be written as the joint event state space of battery state Q_k and activity state X_{ij}. We prove that the transition probabilities in joint states is the multiplication of battery and activity transition probabilities [49]. The resulting Markov chain with proposed state transitions considering the changes of activity as a function of the time of the day is shown in Figure 9.10.

We take advantage of the nonhomogeneous Markovian model, where transition probabilities are calculated depending on the period of the day. The optimization algorithm then chooses an optimal action U^* such that the accumulated cost function over the finite time horizon N can be calculated based on the instantaneous cost function. Herein, with problem formulation in Equation (9.7) we have:

$$U^* = \underset{U}{\operatorname{argmin}} \mathbb{E} \left(\sum_{t=0}^{N} \gamma^t [C_{\text{total}}] \right) \tag{9.11}$$

where γ is the discount factor taking values commonly between 0.9 and 1. Figure 9.11 shows the comparison between two strategies myopic with problem formulation presented in Equation (9.6) and MDP with problem formulation shown in Equation (9.8) for a subject during 24 hours of monitoring. The daily activity of the subject is divided into four periods of six hours (e.g. $i = \{1, 2, 3, 4\}$).

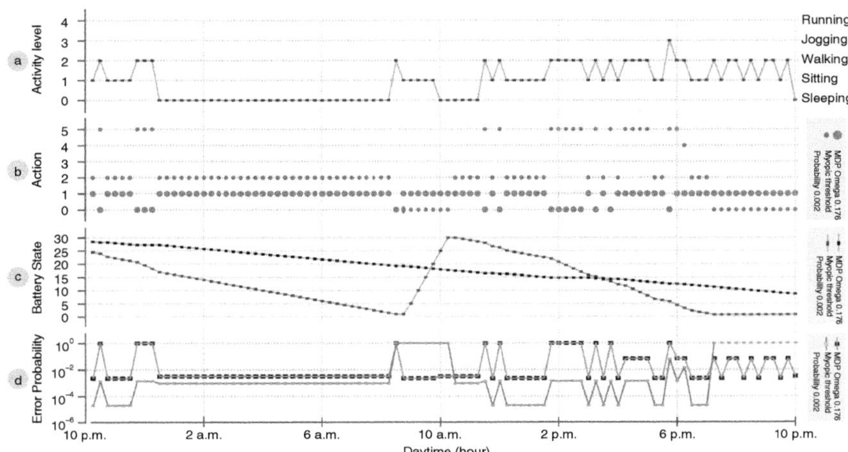

Figure 9.11 24-hour health monitoring of a healthy person. (a) User's activity level. (b) Optimal sensor's power level. (c) battery state tracking based on sensor's power level. (d) Probability of error expected regarding the user's activity. The red line and gray line indicate the myopic and MDP methods, respectively.

For each period, the transition probability between activity is trained based on the history gathered from one week of activity monitoring. Markov chain for each period is modeled based on the transition probabilities shown in Figure 9.9. Since the activity model of a user changes over time, the transition probabilities need to be updated regularly. We calculated the average transition probabilities for each period over one week of activity data. The trained model is used for the following week to find the optimal solution. In addition, we evaluated our results for activities of 14 subjects. For each subject, the model is updated weekly to adapt the system to its change throughout the time. The results are evaluated for four weeks of data to calculate average energy consumption in the PPG sensor layer using myopic and MDP strategies. The joint battery and activity states are modeled and the corresponding Markovian model based on the period of day is used to find an optimal solution in MDP strategy.

We compared MDP and myopic strategies by setting the parameter ω in Equation (9.7) to 0.17 and the threshold for probability of error to 0.002 (e.g. ζ in Equation (9.6)). For the sake of fair comparison, the parameters ω and ζ are chosen in the way that average probability of misdetection over four weeks for both myopic and MDP methods become the same. Parameter ω determines the importance of optimizing total cost function in MDP. For values close to 0, the energy consumption will have higher impact on the cost function. Therefore, MDP chooses policies to minimize energy consumption in the sensor. Whereas, ω close to 1 stresses the importance of probability of misdetection in Eq. (9.7). In this case, MDP chooses the optimal actions to fulfill minimization of probability of misdetection. During one month of monitoring, the fixed average probability of error will achieve 0.26 by setting $\omega = 0.17$ in MDP and $\zeta = 0.002$ in myopic strategies. Figure 9.11a shows the activity of this subject for 24 hours starting at 10 p.m. to 10 p.m. of the following day. Figure 9.11b shows a comparison of optimal actions taken using the two proposed methods. This sensor's sensing power level could be chosen in five different settings, $U \in \{U_0, U_1, U_2, U_3, U_4, U_5\}$. Power levels in aforementioned PPG sensor can be used with corresponding current levels (e.g. $U_1 = 0.8\,\text{mA}$, $U_2 = 3.5\,\text{mA}$, $U_3 = 6.3\,\text{mA}$, $U_4 = 9.2\,\text{mA}$, and $U_5 = 12\,\text{mA}$). For each current level, the power consumption is specified. For instance, current level U_0 indicates the sensor is in the sleeping mode. The power consumption is {69.3, 73.26, 79.86, 84.15, 89.43 mW}, respectively, for the rest of the current levels. We set the battery states so that the PPG sensor reduces 1 level of state during one hour of sensing when the current level U_1 is chosen. The highest current level U_5 consumes 5 levels of battery during one hour. Therefore, the battery is discharged after 30 hours of using the lowest current level U_1. Figure 9.11c shows the battery state tracking using the optimal actions in both strategies with $Q = \{Q_1,, Q_2, \ldots, Q_{30}\}$ as the battery states. The optimal action changes the battery state over time. It is shown that MDP strategy can extend the battery lasting twice more compared to the *myopic* method. This is useful since the frequency of charging the battery is less, resulting in planning wisely on using the energy resources. Figure 9.11d demonstrates the probability of error with respect to the action taken in both methods. Note that it is assumed that the sensor will be forced to go to sleeping mode U_0 when the battery is discharged, which results in probability of error to be 1. The *myopic* method finds an optimal solution just based on the current activity of a user without considering that the chosen action will discharge the battery faster and this will increase the likelihood the battery drainage during vigorous activities including, walking, jogging, and running, which requires accurate monitoring. *Myopic* has a linear time complexity leading to solving the optimization problem in real time. This method specifically is practical when the model of activity of a subject is not available for training Markovian model. During one month of monitoring, we observed the average of 12% reduction in energy consumption comparing MDP with *myopic* fulfilling the same probability of error. The increase in battery lasting was observed with 2× in MDP.

Figure 9.12 compares two methods of MDP, *myopic* and selection of constant power level. In this experiment, average error probability is calculated as a function of average energy consumption during one month for 14 subjects. Results indicate that in these three methods, choosing parameters in MDP strategy can save energy compared to *myopic* method. For instance, for a fixed average error probability of 0.32, MDP consumes 3.8 KJ, while *myopic* consumes 4.46 KJ and

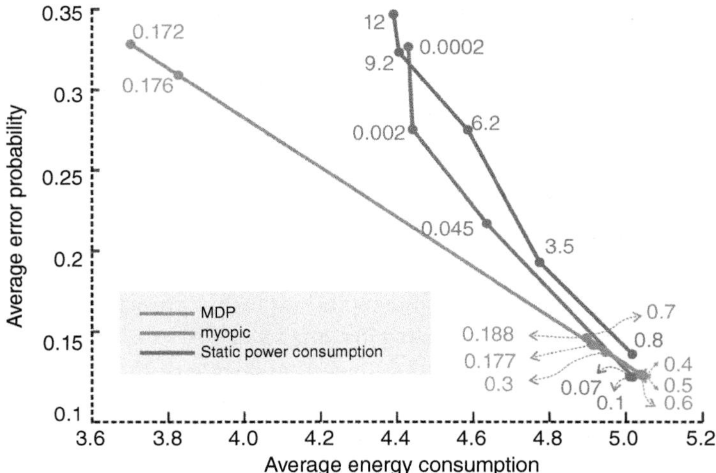

Figure 9.12 Average probability of error as a function of energy consumption (KJ) in four weeks averaged over 14 subjects. Comparison between three methods of MDP, myopic, and static power consumption. The myopic method is evaluated based on $\zeta \in \{0.0002, 0.002, 0.045, 0.07, 0.1\}$ in Equation (9.6). MDP is evaluated based on $\omega \in \{0.172, 0.176, 0.177, 0.188, 0.3, 0.4, 0.5, 0.6, 0.7\}$ in Equation (9.8). Static power consumption is calculated based on $U \in \{0.8, 3.5, 6.2, 9.2, 12$ mA$\}$.

static power level with $U_4 = 9:2$ mA. Therefore, MDP can observe an average of 12% reduction in energy consumption while fulfilling the same error probability. In addition, highest power level $U_5 = 12$ mA consumes slightly less power while the average error probability of 0.25 is achieved. Our model takes into consideration that the user charges the battery with probability of one only when they are sleeping or sitting but not walking, jogging, or running. Choosing high power levels leads to a faster drainage of battery meaning that the user needs to recharge the battery more frequently. During the charging process, the probability of error is 1. Therefore, the average probability of error in $U_5 = 12$ mA is higher than using $U_4 = 9:6$ mA during the one month of monitoring the 14 subjects.

9.7 Conclusions

The three-layer IoT paradigm has opened new avenues of monitoring patients out of hospitals using wearable sensors. Such a system includes wearables with tight energy constraints making it critical to empower these sensors with run-time energy management approaches. User context can be exploited to minimize the measurement energy with a minimal loss of measurement accuracy, however, such algorithms require contextual information to control the energy budget in the sensor to monitor subjects accurately imposing energy overhead on the sensor layer. Energy-constrained wearable sensors cannot often afford such an overhead, however, if a proper architecture is used, the overhead can be migrated to the next layer (i.e. smart gateways at the fog layer) enabling local fog-assisted control of sensors. The fog layer provides an opportunity to perform real-time control of the sensor's configurations. In this chapter, we demonstrated optimization methods to address a twofold goal, minimizing the energy consumption while fulfilling a satisfactory threshold of probability of misdetection in abnormality. We used the key idea of context-awareness to bring accurate solutions to the optimization algorithms. We proposed two methods: optimizing instantaneous cost function known as myopic strategy, and MDP as the solution to accumulated cost function over finite time horizon. We compared results of MDP and myopic during monitoring a subject for 24 hours and we observed average of $2x$ battery life extension in MDP strategy compared with myopic strategy.

Acknowledgment

This material is based upon work supported partially by the US National Science Foundation (NSF) WiFiUS grant CNS-1702950 and Academy of Finland grants 311764 and 311765.

References

1 (2018). Internet-of-Things and big data for smarter healthcare: from device to architecture, applications and analytics. *Future Generation Computer Systems* 78: 583–586.

2 (2017). The Internet of Things for basic nursing care:a scoping review. *International Journal of Nursing Studies* 69: 78–90.

3 Islam, S.M.R., Kwak, D., Kabir, M.D. et al. (2015). The Internet of Things for health care: a comprehensive survey. *IEEE Access* 3: 678–708.

4 Rahmani, A.M., NguyenGia, T., Negash, B. et al. (2018). Exploiting smart e-health gateways at the edge of healthcare internet-of-Things: a fog computing approach. *Future Generation Computer Systems* 78: 641–658.

5 Rahmani, A.M., Liljeberg, P., Preden, J.-S., and Jantsch, A. (2017). *Fog computing in the Internet of Things: intelligence at the edge*. Springer.

6 Baig, M.M. and Gholamhosseini, H. (2013). Smart health monitoring systems: an overview of design and modeling. *Journal of Medical Systems* 37 (2): 9898.

7 Gubbi, J., Buyya, R., Marusic, S., and Palaniswami, M. (2013). Internet of Things (IoT): a vision, architectural elements, and future directions. *Future Generation Computer Systems* 29 (7): 1645–1660.

8 Catarinucci, L., De Donno, D., Mainetti, L. et al. (2015). An IoT-aware architecture for smart healthcare systems. *IEEE Internet of Things Journal* 2 (6): 515–526.

9 Allen, J. (2007). Photoplethysmography and its application in clinical physiological measurement. *Physiological Measurement* 28 (3): R1.

10 Charlton, P.H., Bonnici, T., Tarassenk, L. et al. (2016). An assessment of algorithms to estimate res-piratory rate from the electrocardiogram and photoplethysmogram. *Physiological Measurement* 37 (4): 610.

11 Pimentel, M.A.F., Charlton, P.H., and Clifton, D.A. (2015). Probabilistic estimation of respiratory rate from wearable sensors. In: *Wearable Electronics Sensors* (ed. S.C. Mukhopadhyay), 241–262. Springer.

12 Karlen, W., Raman, S., Ansermino, J.M., and Dumont, G.A. (2013). Multiparameter respiratory rate estimation from the photoplethysmogram. *IEEE Transactions on Biomedical Engineering* 60 (7): 1946–1953.

13 Garde, A., Karlen, W., Ansermino, J.M., and Dumont, G.A. (2014). Estimating respiratory and heart rates from the correntropy spectral density of the photoplethysmogram. *PLoS One* 9 (1): e86427.

14 Lindberg, L.-G., Ugnell, H., and Oberg, P.A. (1992). Monitoring of respiratory and heart rates using a bre-optic sensor. *Medical and Biological Engineering and Computing* 30 (5): 533–537.

15 Maxim Integrated, High-sensitivity pulse oximeter and heart-rate sensor for wearable health, https://www.maximintegrated.com/en/products/sensors/MAX30102.html (accessed November 1, 2018).

16 Wang, L. and Yang, X. (2006). A survey of energy-efficient scheduling mechanisms in sensor networks. *Mobile Networks and Applications.*

17 Aijaz, A. and Aghvami, A.H. (2015). Cognitive machine-to-machine communications for internet-of-things: a protocol stack perspective. *IEEE Internet of Things Journal* 2 (2): 103–112.

18 Bhatt, C., Dey, N., and Ashour, A.S. (2017). *Internet of Things and Big Data technologies for next generation healthcare.* Springer.

19 Chang, L.-H., Lee, T.-H., Chen, S.-J., and Liao, C.-Y. (2013). Energy-efficient oriented routing algorithm in wireless sensor networks. In: *2013 IEEE International Conference on Systems, Man, and Cybernetics (SMC)*, 3813–3818. IEEE.

20 Pradhan, G., Gupta, R., and Biswasz, S. (2018). Study and simulation of WBAN MAC protocols for emergency data traffic in healthcare. In: *2018 Fifth International Conference on Emerging Applications of Information Technology (EAIT)*, 1–4. IEEE.

21 Chen, Y., Zhao, Q., Krishnamurthy, V., and Djonin, D. (2007). Transmission scheduling for optimizing sensor network lifetime: a stochastic shortest path approach. *IEEE Transactions on Signal Processing* 55 (5).

22 Williams, J.L., Fisher, J.W., and Willsky, A.S. (2007). Approximate dynamic programming for communication-constrained sensor network management. *IEEE Transactions on Signal Processing.*

23 Wang, K., Wang, Y., Sun, Y. et al. (2016). Green industrial Internet of Things architecture: an energy-effcient perspective. *IEEE Communications Magazine* 54 (12): 48–54.

24 Zhu, C., Leung, V.C.M., Yang, L.T., and Shu, L. (2015). Collaborative location-based sleep scheduling for wireless sensor networks integrated with mobile cloud computing. *IEEE Transactions on Computers* 64 (7): 1844–1856.

25 Kaur, N. and Sood, S.K. (2017). An energy-efficient architecture for the internet of things (IoT). *IEEE Systems Journal* 11 (2): 796–805.

26 Zhang, R. et al. (2018). Energy-efficient sleep scheduling in WBANs: from the perspective of minimum dominating set. *IEEE Internet of Things Journal.*

27 Zois, D.-S., Levorato, M., and Mitra, U. (2012). A POMDP framework for heterogeneous sensor selection in wireless body area networks. In: *2012 Proceedings IEEE INFOCOM*, 2611–2615. IEEE.

28 Zois, D.-S., Levorato, M., and Mitra, U. (2014). Active classification for POMDPS: a Kalman-like state estimator. *IEEE Transactions on Signal Processing* 62 (23): 6209–6224.

29 Zois, D.-S., Levorato, M., and Mitra, U. (2013). Energy-efficient, heterogeneous sensor selection for physical activity detection in wireless body area networks. *IEEE Transactions on Signal Processing* 61 (7): 1581–1594.

30 Flavio Bonomi, Rodolfo Milito, Jiang Zhu, and Sateesh Addepalli, Fog computing and its role in the internet of things, Proceedings of the first edition of the MCC workshop on mobile cloud computing, 2012.

31 Cortes, R., Bonnaire, X., Marin, O., and Sens, P. (2015). Stream processing of healthcare sensor data: studying user traces to identify challenges from a Big Data perspective. *Procedia Computer Science* 52: 1004–1009.

32 Dastjerdi, A.V. and Buyya, R. (2016). Fog computing: helping the Internet of Things realize its potential. *Computer* 49 (8): 112–116.

33 Sarkar, S. and Misra, S. (2016). Theoretical modelling of fog computing: a green computing paradigm to support IoT applications. *IET Networks* 5 (2): 23–29.

34 Cisco Systems, Fog computing and the Internet of Things: extend the cloud to where the things are, Available online: www.cisco.com (2015) (accessed March 21, 2019).

35 Liu, Y., Yu, F.R., Li, X. et al. (2018). Distributed resource allocation and computation offloading in fog and cloud networks with non-orthogonal multiple access. *IEEE Transactions on Vehicular Technology* 67 (12): 12137–12151.

36 Hong, H.u.-J. (2017). From cloud computing to fog computing: unleash the power of edge and end devices. In: *2017 IEEE International Conference on Cloud Computing Technology and Science (CloudCom)*, 331–334.

37 Bellavista, P., Berrocal, J., Corradi, A. et al. (2018). A survey on fog computing for the internet of things. *Pervasive and Mobile Computing* 52: 71–99.

38 Azimi, I., Anzanpour, A., Rahmani, A.M. et al. (2017). HiCH: hierarchical fog-assisted computing architecture for healthcare IoT. *ACM Transactions on Embedded Computing Systems (TECS)* 16 (5s): 174.

39 Delaram Amiri, Arman Anzanpour, Iman Azimi et al., Edge-assisted sensor control in healthcare IoT, *2018 IEEE Global Communications Conference* (GLOBECOM), pp. 1–6. IEEE, Abu Dhabi, United Arab Emirates, 2018.

40 Shirazi, S.N., Gouglidis, A., Farshad, A. et al. (2017). The extended cloud: review and analysis of mobile edge computing and fog from a security and resilience perspective. *IEEE Journal on Selected Areas in Communications* 35 (11): 2586–2595.

41 Lewis, P.R., Platzner, M., Rinner, B. et al. (2016). *Self-Aware Computing Systems*. Springer.

42 Lewis, P.R., Chandra, A., Parsons, S. et al. (2011). A survey of self-awareness and its application in computing systems. In: *2011 Fifth IEEE Conference on Self-Adaptive and Self-Organizing Systems Workshops*, 102–107. IEEE.

43 Preden, J.S., Tammemäe, K., Jantsch, A. et al. (2015). The benefits of self-awareness and attention in fog and mist computing. *Computer* 48 (7): 37–45.

44 Dey, A.K., Abowd, G.D., and Salber, D. (2001). A conceptual frame-work and a toolkit for supporting the rapid prototyping of context-aware applications. *Human Computer Interaction* 16 (2–4): 97–166.

45 Anzanpour, A., Azimi, I., Götzinger, M. et al. (2017). Self-awareness in remote health monitoring sys-tems using wearable electronics. In: *Proceedings of the Conference on Design, Automation & Test in Europe*, 1056–1061. European Design and Automation Association.

46 Azimi, I., Anzanpour, A., Rahmani, A.M. et al. (2017). Self-aware early warning score system for IoT-based personalized healthcare. In: *eHealth 360*, 49–55. Springer.

47 Henry Hoffmann, Martina Maggio, Marco D. Santambrogio et al., SEEC: a framework for SElf-awarE Computing, Computer Science and Artificial Intelligence Laboratory Technical Report MIT-CSAIL-TR-2010-049, 2010.

48 Dutt, N., Jantsch, A., and Sarma, S. (2016). Toward smart embedded systems: a self-aware system-on-chip (soc) perspective. *ACM Transactions on Embedded Computing Systems (TECS)* 15 (2): 22.

49 Amiri, D., Anzanpour, A., Azimi, I. et al. (2019). Context-aware sensing via dynamic programming for edge-assisted wearable systems. *ACM Transactions on Computing for Healthcare.*

10

Latency Minimization Through Optimal Data Placement in Fog Networks

Ning Wang and Jie Wu

[1]*Department of Computer Science, Rowan University, Glassboro, NJ, USA*
[2]*Center for Networked Computing, Temple University, Philadelphia, PA, USA*

10.1 Introduction

With the wide availability of smart devices, people are able to access social websites, videos, and software from the Internet anywhere at any time. According to [1], in 2016, Netflix had more than 65 million users, with a total of 100 million hours spent streaming movies and TV shows daily. This accounted for 32.25% of the total downstream traffic during peak periods in North America. Facebook had 2.19 billion monthly active users up to the first quarter of 2018 [2]. The traditional mechanism is to store content into the central server, which leads to a relatively long data retrieving latency. To alleviate this issue, Fog Networks (FNs) are proposed and applied in commercial companies. According to Cisco's white paper [3], global streaming traffic is expected to account for 32% of all consumer traffic on the Internet by 2021, and more than 70% of this traffic will cross FNs up, an increase from 56% in 2016.

To solve the problem of the long content retrieving latency when storing data in the center, FNs can geographically store partial or entire data at fog servers deployed close to clients. This will then reduce network and center server loading times and provide a better quality of service, i.e. low latency to end-clients [4, 5]. Due to the relatively large storage space and bandwidth needs of retrieving data, an FN typically employs a limited number of fog servers, each of which has medium or low storage capacity. Such servers can collectively store a larger number of objects and serve clients from all networks with larger aggregate bandwidth needs.

An illustration of the FN network model is shown in Figure 10.1. There are five fog server nodes where data can be stored to support two different data requests,

Fog Computing: Theory and Practice, First Edition.
Edited by Assad Abbas, Samee U. Khan, and Albert Y. Zomaya.

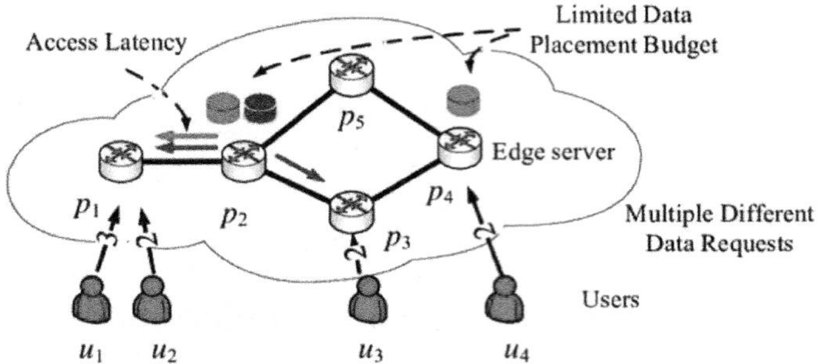

Figure 10.1 An illustration of a typical fog network.

represented by red and blue colors. Note that there is a capacity limit for each server node. In a given time period, we can assume that there are different sets of data requests that arrive based on the prediction. If the data request cannot be satisfied in the local server node, it will search the nearby servers in FNs to find the requested data and there is a corresponding access latency. For example, user u_2 needs to retrieve data from the server p_2, which incurs a transmission from p_2 to p_1. The access latency between a pair of nodes is assumed to be constant over a relatively long period. In addition, there may be multiple identical data requests, for instance, trending news. In Figure 10.1, users u_1 and u_4 request the same data and users u_2 and u_3 request another data. The bandwidth and server processing speed are assumed to be sufficient, and thus, multiple identical data requests can retrieve data from the same server node.

In this chapter, we address the latency minimization problem in FNs through data placement optimization with limited total placement budget, called the multiple data placement with budget problem (MDBP). It is motivated by the fact that the data deployment cost is the major cost in FNs [6]. However, MDBP is nontrivial. For example, in Figure 10.1, only servers p_2, p_3, and p_4 have available storage and the available slots are 2, 1, and 1. The total budget is 3 and all link latencies are 1. The first challenge is due to the network topology. For example, it is easy to see that p_5 is a bad placement location since it is far from all users. However, p_2 is a good location for user u_1, but bad for user u_4. Therefore, for each data placement, we need to jointly consider its influence to all corresponding requests. The competition between two data requests further complicates this challenge. The second challenge is due to limited budget and request demands. In this example, users u_1 and u_4 have relatively larger demands; these high demand users should have low access latency but it does not mean that we need to always assign them more data copies. For example, it may be the case all the high demand requests are close to each other. The third challenge is due to copy distribution. If we have multiple data for a request, its decision should be further revised. For example, if we use two data copies for data request 1, we can place data at locations p_2 and p_4 to satisfy users u_1 and u_4 so that they both have low access latency. In addition, the limited storage size of each server location further complicates the proposed problem. It is because we cannot simply apply the individual solutions of each single data request since they may violate the server storage constraint.

We discuss the solution for the MDBP in two different cases depending on whether there is data replication. In the first case, we do not consider data replication so that we can focus on optimizing data placement. We find that we can transfer the MDBP into a min-cost flow formulation and thus find the optimal solution. We also propose a local information collection method to reduce the time complexity of the min-cost flow transformation in the tree topology, the common topology of FNs. In a general case with data replication, the MDBP turns out to be NP-complete even when there is only one type of data request. We propose a dynamic programming solution to find the optimal solution in the single request scenario with line topology. In the general case, we first propose a heuristic algorithm. However, its performance is related to the system environments. In addition, we further propose a novel assignment approach based on the rounding technique. It first finds the optimal fractional solution and then gradually transfers into the integer solution without avoiding the server capacity constraint. The proposed approach is proved to have an approximation ratio of 10.

The contributions of this chapter are summarized as follows:

- To our best knowledge, we are the first to consider the optimal data placement with budget in FNs for multiple data requests under the server capacity constraint.
- We find the optimal solution of MDBP in the scenario where there is no data replication through min-cost flow formulation, and propose a local information collection method to reduce the time complexity.
- In the general case, we prove that the MDBP is NP-complete. A novel rounding approach is proposed, and it has a constant approximation ratio of 10.
- We verify the effectiveness of the proposed approaches using the PlanetLab trace, which is a worldwide Internet trace.

The remainder of the chapter is organized as follows: the related works are in Section 10.2. The problem statement is in Section 10.3. The min-cost flow formulation is provided in Section 10.4 for the data placement without replication. In the general scenario, we first prove that the proposed problem is NP-complete and then propose a novel rounding solution with approximation analysis in Section 10.5. The experimental results from the PlanetLab trace are shown in Section 10.6, and we conclude the chapter with Section 10.7.

10.2 Related Work

The data placement problem [7, 8] in FNs is a fundamental problem. The existing work can be briefly categorized into two types based on whether the data placement changes with time.

10.2.1 Long-Term and Short-Term Placement

Early work in [9] studied the k-cache placement problem in a given network with special topologies, i.e. line and ring. The difference between [9] and this chapter is that in [9], each data can fractionally support clients, but this assumption is not true in this chapter. The work in [10] studied the trade-off between selecting a better traffic-delivery path and increasing the number of FN servers. In recent work in [6], the authors considered the dynamic network demand and found that deploying data close to the end-users might not always be the optimal solution. In [1], to find the optimal on-demand content delivery in the short term (e.g. one-week), the authors formulated and solved the content placement problem with the constraints of storage space and edge bandwidth. In [11], the authors discussed the optimal data update frequency decision in data placement. In [12], the authors provided a hierarchical data management architecture to maximize the traffic volume served from data and minimize the bandwidth cost (Table 10.1).

10.2.2 Data Replication

The work in [13] addressed the facility location problem with different client types, which is equivalent to the case of data placement without replication since there

Table 10.1 Summary of symbols.

Symbol	Interpretation
m	Total type of data request
n	Number of fog servers
p_i	Server i
s_i	Server p_i's storage size
l_{ij}	Access latency to between servers p_i to p_j
d_{ij}	Demand of the data request j at server i
x_{ij}	Server node i has data j
θ	Data placement budget
$\mathbb{X}$	Overall data placement planning
$c(d_{ij}, \mathbb{X})$	Latency of d_{ij} with a placement planning $\mathbb{X}$

is only one facility for a type of client. Authors in [14] discuss the optimal single data request placement with replication in the tree topology. The optimal solution is obtained through the dynamic programming technique. In [15], data can be transferred between a set of proxy servers and thus the authors optimized the data transformation cost. In [16], authors jointly considered optimization of data replication costs and access latency. In [17], they jointly consider the data receiving for multiple different data. Therefore, the co-locations of data are important.

This chapter addresses long-term data placement with data replication. Our work differs from existing works by considering multiple data requests with different demands and limited total placement budgets.

10.3 Problem Statement

In this section, we discuss the proposed network models, followed by the problem formulation and challenges.

10.3.1 Network Model

An FN is an overlay network over the Internet; it is composed of a set of fog servers. The server nodes are potential places for data deployment. Without loss of generality, we model the topology of the overlay network with a connected undirected graph $G(V, E)$, where V is the set of vertices denoting the servers, and E is the set of edges denoting the data transmission between servers in the network and the edge weight is the corresponding latency. We assume that the topology of the target network is known in advance and there is a total of n server nodes in the FN, i.e. $|V| = n$. Note that there is a storage limit in each server, denoted by s_i.

This chapter considers an off-line scenario. We assume that there are m different data in total. For each data, there can be one or multiple data requests (up to n). A data request is denoted by d_{ij}, which means there is a request from location i for the data j and the value of d_{ij} is the total demand at that location. Let $\mathbb{X} = \{x_{11}, x_{12}, \cdots, x_{nm}\}$ be a feasible placement plan. For each data request, it will be satisfied by the nearest server that stores the data in the FN. The nearest data can be found through the shortest-path algorithms. Therefore, for a data request d_{ij}, its latency for a given placement planing $\mathbb{X}$ is $c(d_{ij}, \mathbb{X})$, which is

$$c(d_{ij}, \mathbb{X}) = d_{ij} \times \min_{i' \in [1,n] \& x_{i'j}=1} l_{ii'}. \tag{10.1}$$

Currently, we assume that all data requests are the same size. In the future, we might extend it to a more general case where different data requests might have different sizes.

10.3.2 Multiple Data Placement with Budget Problem

To meet all data requests and minimize data access latency, the system provider can place data in the server nodes of FNs. However, the data placement can be costly. Therefore, we propose the MDBP to minimize the average data access latency for the FN. Specifically, the MDBP is as follows – we would like to plan the data placement locations to minimize the user's data access latency while avoiding the server's capacity. Based on whether there is replication, we consider two versions: (1) there is only one data for each data request in the network. (2) There may be multiple data for a data request, but there is an overall data placement budget. The problem can be mathematically formulated as follows.

$$\min \sum_{i=1}^{n} \sum_{j=1}^{m} c(d_{ij}, \mathbb{X}) \tag{10.2}$$

$$\text{s.t.} \sum_{j=1}^{m} x_{ij} \leq s_i \quad \forall i, \tag{10.3}$$

$$\sum_{i=1}^{n} \sum_{j=1}^{m} x_{ij} \leq \theta \quad \forall i, j, \tag{10.4}$$

$$x_{ij} \in \{0, 1\}, \quad \forall i, j, \tag{10.5}$$

where $x_{ij} \in \mathbb{X}$ is a placement decision. The first constraint, i.e. Equation (10.3), means that each server cannot exceed its capacity. The second constraint, i.e. Equation (10.4), is the total budget constraint where θ is the given placement budget. The last constraint, i.e. Equation (10.5), means that data cannot be partitioned.

Note that in the aforementioned formulation, the MDBP does not consider the users' mobility [17], that is, a data request represents a user. However, it can be easily extended to a case that considers users' mobility. The idea of the conversion is that we can transfer a mobile user into multiple static users in the MDBP, where the demand of each static user is the percentage of staying at that mobile node [18]. An example is shown in Figure 10.1, where there are two users. User 1 spends 60% of its time in p_1, and 40% of its time in p_4. User 2 spends 50% of its time in p_1 and p_3. After the conversion, we can apply the solution of MDBP to the extended model.

10.3.3 Challenges

The proposed MDBP is nontrivial even when the user request pattern is given/predicted, because we need to jointly consider the data placement distribution for multiple data requests. We cannot simply consider each data request individually and then combine the solutions. The reason is that it might be the case that there are optimal locations for multiple data requests, but the server storage size is not large enough to hold all of them. Given the data replication,

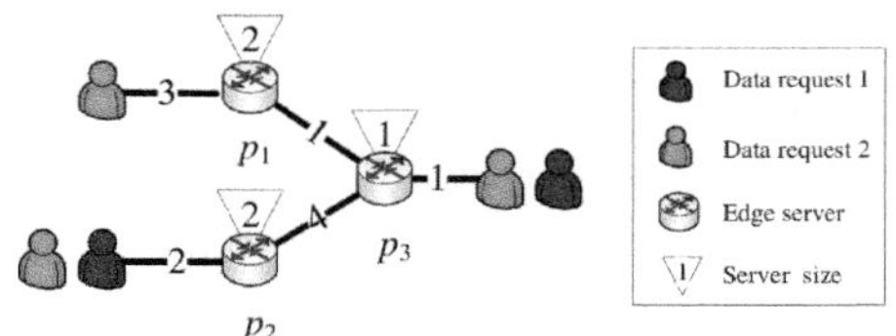

Strategy	Distribution	Placement	Latency Sum
1	$(1d_1, 3d_2)$	p_1:$\{d_1,d_2\}$, p_2:$\{d_2\}$, p_3:$\{d_2\}$	15
2	$(1d_1, 3d_2)$	p_1:$\{d_2\}$, p_2:$\{d_1,d_2\}$, p_3:$\{d_2\}$	13
3	$(2d_1, 2d_2)$	p_1:$\{d_1\}$, p_2:$\{d_1,d_2\}$, p_3:$\{d_2\}$	11
4	$(2d_1, 2d_2)$	p_1:$\{d_2\}$, p_2:$\{d_1,d_2\}$, p_3:$\{d_1\}$	10

Figure 10.2 An illustration of the problem's challenges.

how to jointly distribute the data in each type of request another challenge. An illustration of the challenges of the MDBP is shown in Figure 10.2, where there are two different types of data requests and three available servers. We ignore the access servers in this example. The weights on the edges represent the corresponding communication latency. In this toy example, the storage size of each of these three servers is 2, 2, and 1, respectively, and the total data budget in the network is 4. In this example, all data requests have the unit demand to simplify the illustration. We propose four different strategies to minimize the overall latency and show the placement challenge. First, if we place one data copy for data request 1 and three data copies for data request 2, we find that improper placement leads to large latency. In strategy 1, the latency for data request 1 is $7 + 2$, and the latency for data request 2 is $3 + 2 + 1$. Therefore, the overall latency is 15. In this toy example, strategies 1 and 2 and strategies 3 and 4 have the same data distribution but different latencies. In addition, this example shows that different data distributions influence on the latency, i.e. strategies 3 and 4 have lower latency compared to strategies 1 and 2.

10.4 Delay Minimization Without Replication

In this section, we focus on the storage placement optimization without replication. The problem formulation is introduced first, followed by the min-cost flow solution.

10.4.1 Problem Formulation

The MDBP can be simplified since different data have no influence on each other in terms of the latency and thus, Equation (10.1) can be reformulated as follows.

$$\min \sum_{i=1}^{n} \sum_{j=1}^{m} \sum_{k=1}^{n} l_{ik} d_{kj} x_{ij} \tag{10.6}$$

$$\text{s.t.} \sum_{i=1}^{n} x_{ij} = 1, \quad \forall j, \tag{10.7}$$

$$\sum_{j=1}^{m} x_{ij} \leq s_i \quad \forall i, \tag{10.8}$$

$$x_{ij} \in \{0, 1\}, \quad \forall i, j, \tag{10.9}$$

where x_{ij} is a decision variable to show that the server p_i has a data j. The first constraint is that each data can be placed at one location and only one location in the network and therefore, there is no replication. The second constraint is that the total amount of data placed at a server node cannot exceed its capacity constraint.

10.4.2 Min-Cost Flow Formulation

In this subsection, we would like to explain that the MDBP can be solved by the min-cost flow formulation.

Theorem 10.1 *The MDBP problem is equivalent to the min-cost flow problem.*

Proof: We prove it by showing the step-by-step transformation. The solution of the MDBP can be considered as a matching process. For each type of data, we generate a demand node, and thus, there are m demand nodes in total, i.e. $\{r_1, r_2, \cdots, r_m\}$. Similarly, we generate a node for each server node and thus, there are n server nodes $\{p_1, p_2, \cdots, p_n\}$. For each demand node r_i, there are n edges for all server nodes. The corresponding weight of a node represents a possible placement decision to place data into that server in the MDBP. The unit edge cost, C_{ij} is the corresponding latency summation to store the data in that location for all corresponding types of data request(s), $C_{ij} = \sum_{i=1}^{n} \sum_{k=1}^{n} l_{ik} d_{kj}$. The edge capacity is 1. In addition, there is a virtual source and a virtual destination. Each demand node has an edge with the virtual source, where the corresponding unit edge cost is 0 and the edge capability is 1. Each server node has an edge with the virtual destination, where the corresponding unit edge cost is 0 and the edge capability is s_i. The demand of the min-cost flow is set as the total number of data request, i.e. m, in the network and the demand is generated at the virtual source. All demands are consumed at the virtual destination. According to the max-flow theory, if all edges' capacities are integral, the final solution is always an integer solution based on the augmenting path method [19]. Then, if there is a flow in an edge that connects the data request and server location, we can use the same data placement between the data node and the server node in the MDBP. Since this assignment leads to the same latency cost as that in the corresponding min-cost flow problem, the total demand of the virtual sink ensures that each data type has to be matched

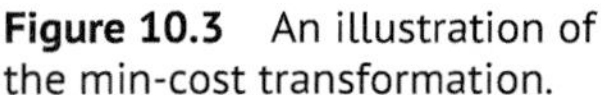

Figure 10.3 An illustration of the min-cost transformation.

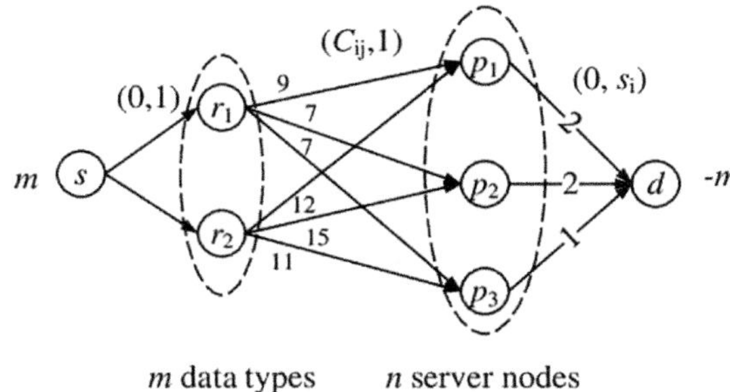

in the MDBP. The capacity of links between the server node and the destination node ensures that each server cannot exceed its corresponding storage capacity in the MDBP.

Figure 10.3 shows the min-cost flow formulation for the example in Figure 10.2. Each data request can be assigned to any server. The capacity constraint of each server location is controlled by the edge flow constraint between the server node and virtual destination in the min-cost flow formulation. For example, the cumulated latency for data request 1 in three servers is 9, 7, and 7, respectively. The min-cost flow can be solved by the successive shortest path [20].

10.4.3 Complexity Reduction

In this subsection, we design an efficient local information collecting method that can reduce the overall time complexity of cumulated latency calculation in min-cost flow calculation. The time complexity of calculating the cumulated latency of each type of data request in all server locations is $O(n^2)$ in the simple approach since we need to traverse the network for each data request. Here, we show that the overall time complexity can be reduced to $O(n)$ in the tree topology, which is the common topology in FNs [21].

Theorem 10.2 *The cumulated latency calculation of a data request can be finished in $O(n)$ in the tree topology.*

Proof: We prove this theorem by introducing the $O(n)$ calculating method.

The procedure is as follows: We transfer G into a tree by considering an arbitrary node as the root node. Then, we aggregate the latency cost gradually from the leaves to the root of the networks. For each node i, it keeps a vector of C_i, $C_i = \{c_1, c_2, \cdots, c_k\}$, where k is the total number of data requests. It records its distance to all the data request access locations. In Figure 10.4, $k = 2$. Initially, all elements in

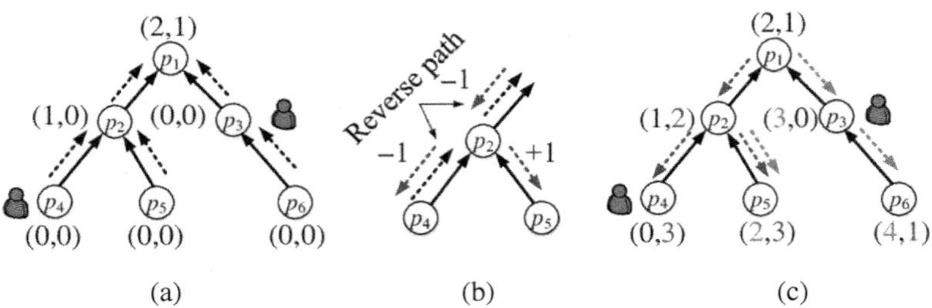

Figure 10.4 An illustration of latency updating procedure: (a) demand collection, (b) updating rule, (c) demand updating.

the vector are 0. Then, we calculate the total latency of the data placement in any node by using the following two steps:

1. *Upward information collection*. We gradually update the latency value of each element c_j in C_i from leaf nodes to the root node as follows. If we have seen the corresponding data request from a successor node i', then $c_j = c_j + l_{ii'}$. Otherwise, c_j keeps its original value.
2. *Downward information updating*. We gradually update the latency value in the revised visiting order to summarize the information from different branches. If the branch is used in the information collection and the corresponding data is seen again, then $c_j = c_j - l_{ii'}$. Otherwise, $c_j = c_j + l_{ii'}$.

After the information updating is done, the placement cost of a location is $\sum_{i=1}^{k} c_i$. Since the each node will be only visited twice, the overall time complexity is $O(n)$.

An illustration is shown in Figure 10.4, where all edge weights are units. Since the calculation is independent for different data requests, we use a data request to illustrate the procedure. In Figure 10.4, the data requests generate at server nodes p_3 and p_4. There are two data requests in this example. Each server node keeps a bracket, where each element records the distance from the current location to the corresponding data request location. Note that in the upward information collection procedure, each node only collects the distance information of the data requests in its subtree. Therefore, at the end, only the root node has the information of all data requests. In downward information updating, the information collected by the root is distributed in all branches. The data updating of the node p_2 illustrates one case, where the corresponding distance decreases by 1. The data updating of the node p_4 illustrates two cases: (i) 3 increases by 1 for data request 1, and (ii) 0 increases by 1 for data request 2, since the distance to data requests 1 and 2 increases. Note that here, we use a binary tree as an example, but this method works in any tree topology.

10.5 Delay Minimization with Replication

In this section, the problem hardness in the general case is discussed first, followed by the optimal solution in a line topology, and the solution in the general scenario.

10.5.1 Hardness Proof

Theorem 10.3 *The proposed MDBP is NP-complete.*

Proof: We prove the proposed MDBP is NP-complete by first proving that it belongs to the NP-class. For a given placement, we can verify whether all constraints are satisfied simultaneously in polynomial time. Now, we show its NP-completeness by a reduction of the k-median problem [22].

The k-median problem is as follows: in a metric space, $G(E, V)$, there is a set of clients $C \in V$, and a set of facilities $F \in X$. We would like to open k facilities in F, and then assign each client node j to the selected center that is closest to it. If location j is assigned to a center i, we incur a cost c_{ij}. The goal is to select k centers so as to minimize the sum of the assignment cost.

The reduction from a special case of the MDBP to the k-median problem is as follows: in a special case of the MDBP, we assume that there is only one type of data request. In this setting, there is no capacity constraint since each available server node should have at least one storage space. Assume that the total budget is k. We need to determine the data placement location, which is equivalent to finding the centers (facilities) in the k-median problem. Its weight is the total latency cost for the data request at the corresponding server location. Clearly, if we set the latency as the corresponding c_{ij} in the k-median problem, we can apply the solution of the MDBP to the k-median problem.

10.5.2 Single Request in Line Topology

We have proven that the MDBP problem is NP-complete in the general graph even in the simple request case. However, when the FN has some particular topology, i.e. the line topology, we can determine the optimal placement for that user by using dynamic programming in the single data request scenario. Specifically, we first sort all data requests directionally. Then, we can define a state called $opt[i,j]$, which represents the minimum cost for that data request up to the first i requests with j copies. Then, we can update $opt[i,j]$ as follows.

$$opt[i,j] = \min \begin{cases} opt[i,j-1] \\ opt[i',j-1] + \min\limits_{p_k \in [p_{i'+1}, p_i]} c_k[i'+1, i], \forall i' < i, \end{cases} \tag{10.10}$$

Algorithm 10.1 Dynamic Programming for Single Request

Input: Job information and rental cost function
Output: The placement result $\mathbb{X}$.

Initialize $opt[i, j] = \infty$ except $opt[0, 0] = 0$.
for Each request i from 0 to m **do**
  for Each request i' from 0 to i **do**
    for Each server location between i' and i **do**
      Update $opt[i, j]$ based on Eq. 10.
return Return the assignment of $opt[n, \theta]$

where i' is a request prior to request i. The $c_k[i' + 1, i]$ is the latency for assigning a new data copy at server p_k, which is between $p_{i'}$ and p_i to cover data requests in this range. The idea behind Equation (10.10) is that the optimal solution always falls into one of two cases: (1) we have moved to the location i without adding one more data copy to get the optimal result; (2) we have moved to the location i and we can use one more data copy to reduce the overall latency.

A toy example can be used to illustrate the proposed dynamic programming in Figure 10.5. In this example, let us assume that $\theta = 2$, which means that we can use two data copies at most. Initially, we can use only one data to cover three users. After calculating the four available server locations, $opt[1, 1] = 0$, $opt[1, 2] = 3$, and $opt[1, 3] = 4$, which achieve the optimal value when the data is put at p_1, p_2, and p_3. Then, we add one more data copy into the network. It is easy to calculate $opt[2, 1] = 0$, $opt[2, 2] = 0$, $opt[2, 3] = 1$. Here is an example of a calculation of $opt[2, 3]$.

$$opt[i, j] = \min\{opt[0, 0] + c_3[1, 3], opt[1, 1] + c_3[2, 3], \\ opt[2, 1] + c_4[3, 3]\}, \tag{10.11}$$

In Equation (10.11), we use the $\min_{p \in [i'+1, i]} c_p[i' + 1, i]$, and ignore the calculation procedure. It is clear that $opt[1, 1] + c_3[2, 3] = 1$ is the minimum in this example.

10.5.3 Greedy Solution in Multiple Requests

If there are multiple different requests, we cannot simply apply the optimal solution for each request because it might not make a feasible solution if we add them together.

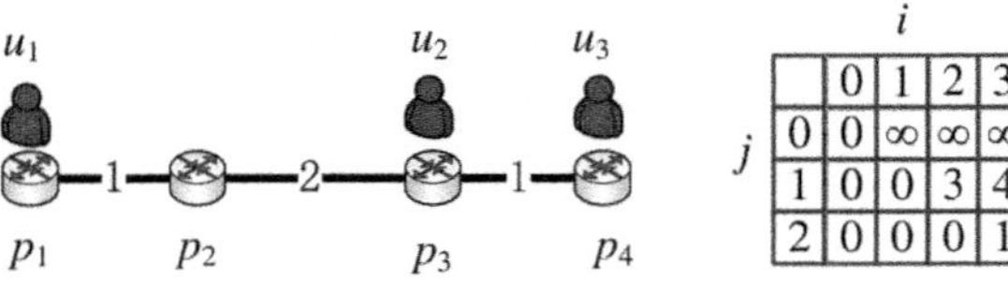

j \ i	0	1	2	3
0	0	∞	∞	∞
1	0	0	3	4
2	0	0	0	1

Figure 10.5 The dynamic programming in the line topology.

To address this problem, we first propose the following heuristic algorithm as shown in Algorithm 10.2. Initially, when the data request is not covered, we go through all types of data requests and calculate their optimal data placements so far. The data whose placement leads to the minimum latency in each round will be selected to put into the network. It is shown in lines 1–4. After that, there is one data for any data request in the network to ensure the coverage constraint. Then, for each round, we pick the data, which can reduce the latency maximum if that data can change its location once, shown in lines 5–8. However, the proposed heuristic may be far from optimal. An example is shown in Figure 10.6, where there is one slot fog server location 1 and 3 to store data. The greedy solution is shown in Figure 10.5. It will select the blue data request in the first round due to the smallest latency increase. However, this option leads to the large latency at the second round. The optimal solution is shown in Figure 10.5, where the placement decision jointly considers two data requests.

Theorem 10.4 *The proposed heuristic algorithm does not have an approximation ratio smaller than 3ρ, where ρ is the maximum data request ratio in the network over the time.*

Proof: We can prove Theorem 10.4 through an extreme example. We propose a contradiction example in the line topology, where each server location has the unit storage capacity. Assume that there is a set of data requests at locations $\{p_1, p_2, p_3, \cdots, p_k\}$, and for each data request, there is only one location. Therefore, we use $\{p_1, p_2, p_3, \cdots, p_k\}$ to denote the data request locations in this proof. In addition, we assume that there exist server locations in $\{p_0, (p_1+p_2)/2, (p_2+p_3)/2, \cdots, (p_k+p_{k+1})/2\}$ where $p_0/2$, and $p_{p+1}/2$ are locations to the left of p_1, and right of location p_k, respectively. In addition, $p_1 - p_0 < (p_2 - p_1)/2$, and $p_{k+1} - p_k > (p_k - p_{k-1})/2$. Therefore, according to the heuristic algorithm, we will put all the data into the first server at the right except the last one. The overall latency is

$$\sum_{i=1}^{k} (p_{i+1} - p_i)/2 + p_k - p_0 = (p_{k+1} - p_1 - p_0)/2 + p_k \tag{10.12}$$

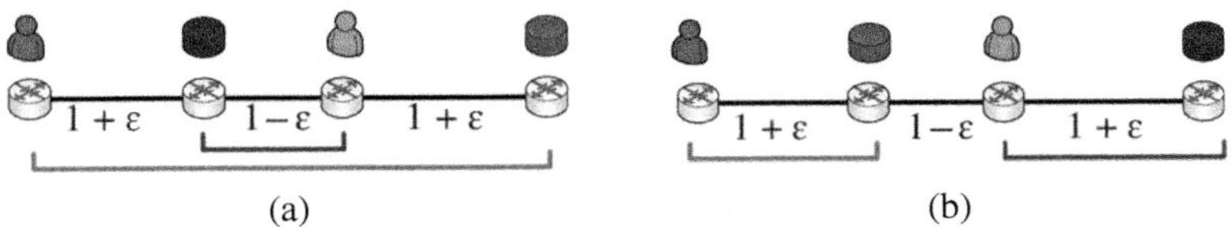

Figure 10.6 An illustration of the greedy algorithm: (a) greedy, (b) optimal.

Instead, if we assign each data request to the first server location at the left, the overall latency is

$$\sum_{i=1}^{k+1} (p_i - p_{i-1})/2 = (p_{k+1} - p_0)/2 \tag{10.13}$$

Then, $\frac{p_{k+1}-p_0}{2} < \frac{p_{k+1}-p_1-p_0}{2} + p_k \approx 3\frac{p_{k+1}-p_0}{2}$ when k is a large number. The number of data requests at a time also has an influence on the ratio and the ρ, which is the maximum number of requests in the network over time. Therefore, the overall approximation ratio is 3ρ.

According to Theorem 10.4, we know that the proposed heuristic algorithm cannot achieve good performance even in the line topology. Therefore, it is necessary to propose an approach that can guarantee adequate performance in different network environments.

10.5.4 Rounding Approach in Multiple Requests

To improve the performance of the greedy solution, we propose a two-step rounding solution. In the first step, we relax the proposed problem into the linear programming (LP) and obtain the lower bound of the MDBP. Then we propose a novel rounding technique, which first rounds a half-integral solution through the min-cost flow. Then, we can further convert the half-integral solution to an integer solution.

10.5.4.1 Generating Linear Programming Solution

The linear programming formulation of the MDBP problem is,

$$\min \sum_{i=1}^{n} \sum_{j=1}^{m} \sum_{k=1}^{h} l_{ij} d_k y_{ijk} \tag{10.14}$$

$$\text{s.t.} \sum_{i=1}^{n} y_{ijk} \geq 1, \forall j, \tag{10.15}$$

$$y_{ijk} \leq z_{ij}, \quad \forall i, j \tag{10.16}$$

$$\sum_{j=1}^{m} z_{ij} \leq s_i, \quad \forall i, \tag{10.17}$$

$$\sum_{i=1}^{n} \sum_{j=1}^{m} z_{ij} \leq \theta, \quad \forall i, \tag{10.18}$$

$$y_{ijk}, z_{ij} \in [0, 1], \quad \forall i, j \tag{10.19}$$

Algorithm 10.2 Min-Volume Algorithm

Input: Data request distribution and network configuration
Output: The placement result $\mathbb{X}$.

while uncovered data request **do**
  for data request i from 1 to m **do**
    for server location j from 1 to n **do**
      check the placement cost of x_{ij}.
  $x_{ij} = \arg\min_{x} \; c(d_{ij}, \mathbb{X}) - c(d_{ij}, \mathbb{X} \cup x_{ij}) - \mathbb{X})$
while The movement budget θ is not reached do
  for data requst i from 1 to m **do**
    for server location j from 1 to n **do**
      check the placement cost of x_{ij}.
  $x_{ij} = \arg\min_{x} \; c(d_{ij}, \mathbb{X}) - c(d_{ij}, \mathbb{X} \cup x_{ij}) - \mathbb{X})$
Return the placement result.

Algorithm 10.3 Rounding Algorithm

Input: Data request distribution and network configuration
Output: The placement result $\mathbb{X}$.

Calculate the linear programming solution of MDBP.
Aggregate demand into centers based on the geodistance.
Convert the fractional solution to the half-integer solution.
Convert the half-integer solution to the integer solution.
Return the placement result.

which is formulated from the angle of each data request. Therefore d_k is the demand for a data request k. Let us assume that the total number of data requests is h. Note that $h \geq m$ due to the repeated data requests. In this formulation, y_{ijk} means the server node i has data j and covers data request k fractionally, z_{ij} means that server i has z_{ij} amount of data j. Equations (10.15, 10.16) ensure that each data request has to be satisfied and the assignment is feasible. Equation (10.17) is the capacity constraint for each server, and Equation (10.18) is the total placement budget constraint.

10.5.4.2 Creating Centers

Since it is hard to directly convert the fractional solution to integral solution, we would like to aggregate the assignment into several "center" servers, so that each center has at least a half-data. To simplify the following explanation, we create the notation $L_k, L_k = \sum_{i=1}^{n} l_{ij}y_{ijk}$, which is the unit demand cost of the LP solution for data request k. Let us consider all the data requests that need data j. We sort them in increasing order of the C_k. Then, for each data request k, if there is an another data request k' and $l_{kk'} < 4max(L_{k'}, L_k) = 4L_k$, we would like to consolidate the demand on data request d_k to d'_k, that is, $d'_k = d'_k + d_k$. We apply this procedure to all data requests; the remaining servers with nonzero demands are called center servers. Since each data moves at most $4L_k$, it is clear that any solution can incur an additive factor of at most $4OPT$ (Figure 10.7).

10.5.4.3 Converting to Integral Solution

After we get the data center servers, an important issue is how we can assign data requests so that the result is an integral solution that doesn't violate the server's capacity constraint. We refer to [22] to propose a two-phase solution where the first step is to build a half-integral solution. This ensures that the distance between a data request and the server serving it is fractionally bounded by the access cost. Therefore, the fractional solution is equivalent to a feasible flow to a min-cost flow problem with integral capacities. Note that to apply the solution in [23] to the MDBP, we need to add a virtual node before going to the destination with the link capacity as the total budget. With the property of the min-cost flow, we can always find an integer solution of no greater cost. By applying the min-cost flow transformation in [23], it has an approximation ratio of 6.

The proposed algorithm has a constant approximation ratio of 10. The center creation incurs at most $4OPT$. The half-integral solution transformation introduces at most $3OPT$. The integer solution conversion further leads to $2OPT$. Therefore, the overall cost is $4 + 2 \times 3 = 10OPT$.

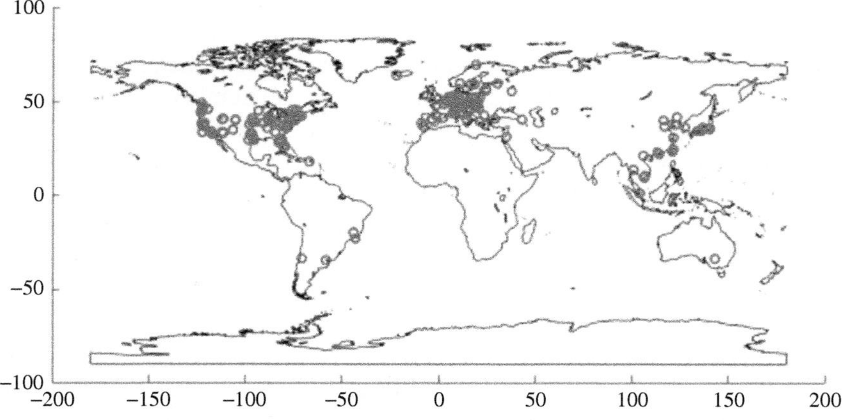

Figure 10.7 Server distribution.

10.6 Performance Evaluation

We evaluate the performance of the proposed solution in this section. The compared algorithms, the trace, the simulation settings, and the evaluation results are presented as follows.

10.6.1 Trace Information

In this chapter, we use the PlanetLab trace [24] generated from the PlanetLab testbed. PlanetLab is a global research network that supports the development of new network services. It contains a set of geo-distributed hosts running PlanetLab software worldwide. In this trace, the medians of all latencies, i.e. round trip time (RTT), between nodes are measured through the King method. 325 PlanetLab nodes and 400 other popular websites are measured (Figure 10.8).

In the PlanetLab trace, the domain of each node is provided. Each node's geometric location is retrieved through the domain-to-IP database and the IP-to-Geo database, provided by [25, 26], respectively. Some domains are no longer in service. In the end, there are 689 nodes. It is reported that the mapping error is within 5 mi and can be ignored in our experiments. Figure 10.9 shows the trace visualization results.

10.6.2 Experimental Setting

We conduct experiments on two scales, i.e. the world and the United States. At the United States scale, we use the nodes on the west coast to simulate the line-topology. The number of data requests, m, changes from 2 to 5. The number

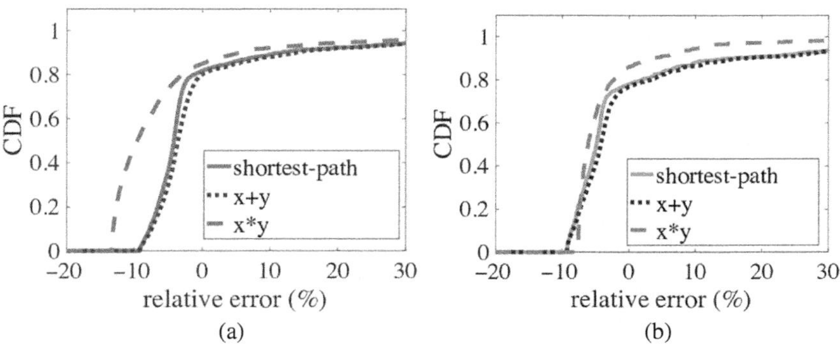

Figure 10.8 Latency-distance mapping: (a) greedy, (b) optimal.

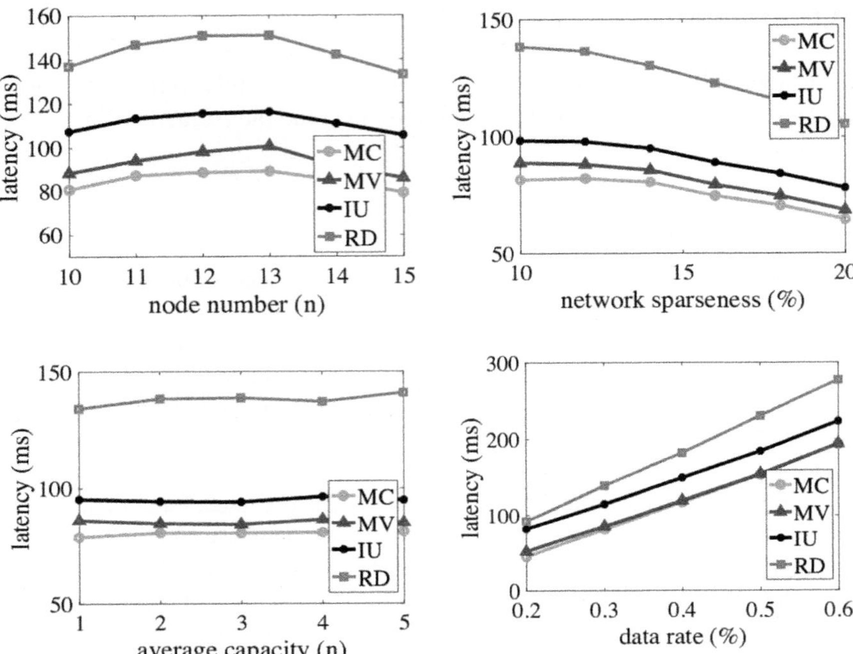

Figure 10.9 Performance comparison without data replication.

of users changes from 10 to 20, which are randomly selected from the first 325 nodes in the PlanetLab trace. The data request location is randomly generated in each round. The pair latency is known and therefore, the topology is not important in the experiments. The number of data placement budgets also changes in the experiments from m to $2m$. We change the following four settings in the experiments: (1) the number of data budgets, (2) the number of data, (3) the number of the data requests, and (4) different server capacities.

10.6.3 Algorithm Comparison

We compare five algorithms in the experiments:

1. *Random (RD) algorithm.* It is a benchmark algorithm. During the first m rounds, a type i data is randomly selected and placed. After that, a data is randomly selected and put into the network.
2. *Min-volume (MV) algorithm.* The data whose placement leads to the minimal cost increase is selected. Specifically, in the first m rounds, if a data has been selected, it cannot be selected again.

3. *Iteration updating (IU) algorithm.* The IU algorithm places different data requests in an order so that the location that leads to the minimal cost increase is selected in each round.
4. *Min-cost (MC) algorithm.* The MC algorithm is proposed in this chapter and it is explained in Section 10.4 when there is no data replication.
5. *Rounding (RO) algorithm.* The RO algorithm is proposed in this chapter and it is explained in Section 10.5.4.

In a special scenario without data replication, the IU algorithm doesn't work since each data always has one data copy, and is thus removed in this case. Besides, the RD algorithm is not necessary since the MC algorithm is optimal.

10.6.4 Experimental Results

10.6.4.1 Trace Analysis

We verify the access latency between servers and their corresponding distances. The mapping result is shown in Figure 10.10. We use three different distance measurement methods, that is, the shortest path, which is the geodistance of the corresponding GPS coordinates, the sum of latitude and longitude distances, and the

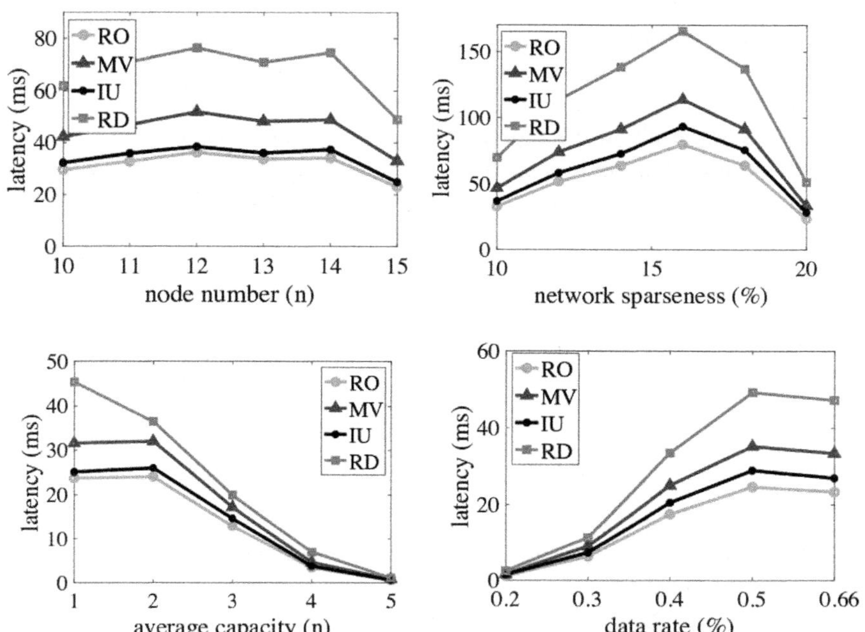

Figure 10.10 Performance comparison with data replication.

area between two nodes in terms of latency estimation. Figure 10.9 shows the cumulative distribution functions of three distance measurements, i.e. Figure 10.9 shows that using the shortest path has a relatively low estimation error, i.e. 10% for 80% of nodes when using the shortest path to estimate the latency between a pair of servers.

10.6.4.2 Results Without Data Replication

Figure 10.9 shows the results of the performance of proposed algorithms in the case, where there is no data replication. The results clearly show that the proposed MC algorithm achieves the lowest latency, followed by the MV algorithm. The RD algorithm's performance is the worst, which demonstrates the necessity of data placement optimization. In Figure 10.9, we change the number of servers in the experiments, i.e. the size of the FNs. The result shows that when the FNs have a larger size, improper data placement leads to bad performance. The RD algorithm has more than 100% of the MC algorithm's latency in Figure 10.9. In Figure 10.9, we gradually increase the average number of edges in the network with a certain amount of servers. The results show that when the network is sparse, there is a large performance difference between algorithms. In the experiments, the average latency decreases around 20% with a lower network spareness level. The IU algorithm has similar performance to MV since they are both greedy algorithms that cannot generate the optimal data placement order.

In Figure 10.9, we change the average storage size of each server. The results show that the average latency is relatively stable. A possible reason is that the data is generated uniformly in the experiments and thus, the placement of data into different locations has minimum influence on the final result. Figure 10.9 shows the results of different data request rates. When the data request rate increases, all algorithms have a larger latencies.

10.6.4.3 Results with Data Replication

Figure 10.10 shows the performance of proposed algorithms in a case where there is data replication. The budget number is two times the number of data request in the experiments. The results clearly show that the proposed RO algorithm achieves the lowest latency, followed by the IU, MV, and RD algorithms. The RD algorithm's performance is the worst, which demonstrates the necessity of data placement optimization.

In Figure 10.10, we change the number of servers in the experiments, i.e. the size of the FNs. The results show that when the FNs have a larger size, improper data placement leads to bad performance. The RD algorithm has more than 150% the latency of the IU algorithm in Figure 10.10. Compared with the results in Figure 10.9, all algorithms have better performance due to a greater amount of data placed in the network. In Figure 10.10, we gradually increase the average number of edges in the network with a certain amount of servers. The results show that when the network is sparse, the performance difference between different algorithms is large. An interesting result is that the average latency first increases, then later decreases with an increase in the network sparseness level.

In Figure 10.10, we change the average storage size of each server. The results show that along with an increase in average storage size, the latency decreases very quickly due to increased placement flexibility. In Figure 10.9, the average latency is reduced by more than 20% with one additional storage capacity in the server node. Figure 10.10 shows the results of different data request rates. When the data request rate increases, all algorithms have a larger latency. However, the RD algorithm has the fastest speed in terms of latency increasing, which demonstrates the effectiveness of the data placement optimization.

10.6.4.4 Summary

Based on the experiments, we find that data placement optimization is very necessary. The average performance can be improved by more than 50% in most cases by comparing the proposed algorithms with the random placement. In a general case with data replication, the proposed RO algorithm has significant improvement compared to other algorithms, which indicates that both the data budget distributions in each data request and their corresponding placements are very important.

10.7 Conclusion

In this chapter, we consider the data placement issue in the FNs so that users have low access delay. Specifically, we discuss the MDBP, whose objective is to minimize the overall access latency. We begin with a simple case, where there is no data replication. In this case, we propose a min-cost flow transformation and thus find the optimal solution. We further propose efficient updating strategy to reduce the time complexity in the tree topology. In a general case with data replication, we prove that the proposed problem is NP-complete. Then, we propose a novel rounding algorithm, which incurs a constant-factor increase in the solution cost. We validate the proposed algorithm by using the PlanetLab trace and the results show that the proposed algorithms improve the performance significantly.

Acknowledgement

This research was supported in part by NSF grants CNS 1824440, CNS 1828363, CNS 1757533, CNS 1629746, CNS-1651947, CNS 1564128.

References

1 Gomez-Uribe, C.A. and Hunt, N. (2016). The Netflix recommender system: algorithms, business value, and innovation. *ACM Transactions on Management Information Systems* 6 (4): 13.

2 Statistica, Facebook statistics, 2018 [Online]. Available: https://www.statista.com/statistics/264810/number-of-monthly-active-facebook-users-worldwide

3 Cisco Visual Networking Index: Forecast and methodology, 2016–2021 [Online]. Available: https://www.cisco.com/c/en/us/solutions/collateral/service-provider/global-cloud-index-gci/white-paper-c11-738085.html

4 Ghose, D. and Kim, H.J. (2000). Scheduling video streams in video- on-demand systems: a survey. *Multimedia Tools and Applications* 11 (2): 167–195.

5 Claeys, M., Bouten, N., De Vleeschauwer, D. et al. (2015). An announcement-based caching approach for video-on-demand streaming. In: *Proceedings of the 2015 11th International Conference on Network and Service Management (CNSM)*. IEEE, November 9–13.

6 Tang, G., Wu, K., and Brunner, R. (2017). Rethinking CDN design with distributee time-varying traffic demands. In: *Proceedings of the IEEE INFOCOM*. IEEE, May 1–4.

7 Sahoo, J., Salahuddin, M.A., Glitho, R. et al. (2016). A survey on replica server placement algorithms for content delivery networks. *IEEE Communications Surveys & Tutorials* 19 (2): 1002–1026.

8 N. Wang and J. Wu, Minimizing the subscription aggregation cost in the content-based pub/sub system, in Proceedings of the 25th International Conference on Computer Communications and Networking (ICCCN), 2016.

9 Krishnan, P., Raz, D., and Shavitt, Y. (2000). The cache location problem. *IEEE/ACM Transactions on Networking* 8 (5): 568–582.

10 M. Yu, W. Jiang, H. Li, and I. Stoica, Tradeoffs in CDN designs for throughput oriented traffic, in Proceedings of the 8th International Conference on Emerging Networking Experiments and Technologies (CoNEXT '12), ACM, December 10–13, 2012.

11 Applegate, D., Archer, A., Gopalakrishnan, V. et al. (2010). Optimal content placement for a large-scale vod system. *IEEE/ACM Transactions on Networking* 24 (4, ACM): 2114–2127.

12 S. Borst, V. Gupta, and A. Walid, Distributed caching algorithms for content distribution networks,in Proceedings of the 29th Conference on Information Communications (INFOCOM '10), IEEE, San Diego, CA, March 14–19, 2010.

13 Wang, L., Li, R., and Huang, J. (2011). Facility location problem with different type of clients. *Intelligent Information Management* 3 (03): 71.

14 B. Li, M. J. Golin, G. F. Italiano, X. Deng, and K. Sohraby, On the optimal placement of web proxies in the Internet, in Proceedings of the IEEE INFOCOM '99, IEEE, March 21–25, 1999.

15 Xu, J., Li, B., and Lee, D.L. (2002). Placement problems for transparent data replication proxy services. *IEEE Journal on Selected Areas in Communications* 20 (7): 1383–1398.

16 Hu, M., Luo, J., Wang, Y., and Veeravalli, B. (2014). Practical resource provisioning and caching with dynamic resilience for cloud-based content distribution networks. *IEEE Transactions on Parallel and Distributed Systems* 25 (8): 2169–2179.

17 K. A. Kumar, A. Deshpande, and S. Khuller, Data placement and replica selection for improving co-location in distributed environments, arXiv preprint arXiv:1302.4168, 2013.

18 Tärneberg, W., Mehta, A., Wadbro, E. et al. (2017). Dynamic application placement in the mobile cloud network. *Future Generation Computer Systems* 70: 163–177.

19 L. R. Ford Jr, and D. R. Fulkerson, A simple algorithm for finding maximal network flows and an application to the Hitchcock problem, Technical Report, 1955.

20 A. Goldberg and R. Tarjan, Solving minimum-cost flow problems by successive approximation, in Proceedings of the ACM STOC, 1987.

21 L. Gao, H. Ling, X. Fan, J. Chen, Q. Yin, and L. Wang, A popularity-driven video discovery scheme for the centralized P2P-VoD system, in Proceedings of the WCSP 2016, Yangzhou, China, IEEE, 2016.

22 M. Charikar, S. Guha, É. Tardos, and D. B. Shmoys, A constant-factor approximation algorithm for the k-median problem, in Proceedings of the Thirty-first Annual ACM Symposium on Theory of Computing (SOTC '99), Atlanta, GA, May 1–4, 1999.

23 Baev, I., Rajaraman, R., and Swamy, C. (2008). Approximation algorithms for data placement problems. *SIAM Journal on Computing* 38 (4): 1411–1429.

24 C. Lumezanu and N. Spring, Measurement manipulation and space selection in network coordinates,in Proceedings of the 28th International Conference on Distributed Computing Systems (ICDCS 2008), IEEE, Beijing, China, June 17–20, 2008.

25 InfoByIP, Domain and IP bulk lookup tool [Online]. Available: https://www.infobyip.com/ipbulklookup.php.

26 Maxmind [Online]. Available: www.maxmind.com.

11

Modeling and Simulation of Distributed Fog Environment Using FogNetSim++

Tariq Qayyum[1], Asad Waqar Malik[1,2], Muazzam A. Khan[1], and Samee U. Khan[3]

[1]Department of Computing, School of Electrical Engineering and Computer Science, National University of Sciences and Technology (NUST), Islamabad, Pakistan
[2]Department of Information System, Faculty of Computer Science & Information Technology, University of Malaya, Kuala Lumpur, Malaysia
[3]Electrical and Computer Engineer, North Dakota State University, Fargo, ND, USA

11.1 Introduction

Extending the cloud services near to the end-user is the basic concept of fog computing [1]. Fog computing is the emerging technology where intelligence is pulled toward the edge of the network, closer to the devices where data is being generated. According to Gartner[1] (2017), there will be 8.4 billion devices connected to the Internet at the end of 2017. Further, the network is not designed to support communication of such a large number of devices. The fog computing concepts and network parameters can play a vital role in supporting the delay sensitive applications. Thus, simulators can be used to understand the behavior of these devices and fog environments. In fog environment with respect to Internet of Thing (IoT), there are three main layers. These layers are the cloud layer, fog layer, and the device layer.

Cloud Layer – This layer is composed of a conventional cloud. A larger number of cloud data centers configured in a geographically distributed manner to work together to provide storage, computation, and networking services. A caching and processing model is also combined with this layer to provide different services that can be accessed globally. The cloud can be used to gather useful information from different fog devices to facilitate other services. It also incorporates with other sources of information for advancing the perception of business intelligence [2]; for instance, network resource optimization [3], smart

1 https://www.gartner.com/newsroom/id/3598917

Fog Computing: Theory and Practice, First Edition.
Edited by Assad Abbas, Samee U. Khan, and Albert Y. Zomaya.

power distribution [4], and the healthcare system [1]. Cloud layer also provides a guideline to improve the lower layer, e.g. fog layer for delay sensitive applications.

Fog Layer – Fog layer is composed of the network tools, like routers, switches, gateways, and base stations (BS). It can also include other devices, for instance, mobile phones, servers, industrial controllers, and video cameras, and roadside unit (RSU). Fog devices can also be configured and deployed in two ways: (1) on IoT devices or (2) in a stratified fashion between the IoT devices and the cloud layer. The fog layer stretches the cloud services toward edge/devices. This stretching helps in improving the computational overhead and reduces network delay.

Apart from traditional packet forwarding and routing, the computation support for real-time and delay-sensitive applications can be installed at the fog layer. The fog devices can provide services like content dissemination [5], vehicular navigation [6], smart traffic lights management [7], and temporary caching services.

Device Layer – Mainly, there are two types of devices: static and mobile. Static devices are installed at a particular location like radio-frequency identification (RFID) tags, forest fire detectors, and air quality sensors. The devices are configured to send a particular type of data to the fog layer in a near real time. On the other hand, mobile devices can change their location. These also include mobile phones, and wearable devices, such as smart-watches, smart-clothes, cameras, vehicles, and fitness shoes [8]. However, the devices with less distance between them can communicate with each other. These devices are battery operated and provide limited storage, caching, and computation capability [9]. The core feature of such devices is to gather information and forward it to the fog or cloud layer.

The next section covers the basic concept and features of existing fog simulators. The rest of this chapter will help you understand the fog computing using a framework, known as FogNetSim++. Using this framework has a number of advantages over other frameworks. In this framework, you can simulate realistic network characteristics like packet loss, bandwidth, error rate, and many more. It supports a number of mobility models; you can study these models thoroughly from INET reference.[2] A very fine-grained handover mechanism is implemented in the framework to avoid data loss in case of mobile nodes.

11.2 Modeling and Simulation

Modeling and simulation have been widely adopted in academic and industry to benchmark new protocols and system designs. Almost every area of science

2 https://inet.omnetpp.org/docs/users-guide/ch-mobility.html

is taking advantage of modeling and simulation. The overall process can be explained as models and entities of the real world. The simulation is performed to test new models and entities to observe its behavior with varying system dynamics [10].

Modeling and simulation have evolved as very emerging platforms to test and analyze the existing or new systems and protocols Moreover, the existing modules are used to simulate the more complex system. Thus, reusability is the key parameter of open-source simulators. The domain of existing simulator models ranges from architectural modeling to the flight trajectory, and weapon explosion models, where the behavior of weapons can be simulated prior to the real testing.

Mostly the existing simulator is a discrete event (DES), where the time is advanced in discrete intervals. The DES controls events that are time-dependent, is moreover categorized into event- (ESS) and time-stepped simulation (TSS) models. The TSS is applied where monitoring of the system is required after a fixed time interval. On the other hand, in ESS, simulation time jumped to the time of the next unprocessed event.

Fog Simulators – Fog computing has appeared as an emerging technology. Researchers are exploring different aspects of fog technology such as fog communication protocols, architecture, data exchange format, and simulation frameworks. To support their proposal, various simulators have been developed by academia, which is based on event-based time advancement. Some of the simulators are open-sourced where researchers and scientists can easily incorporate their work to benchmark the performance. A comparison of different simulators is given in Figure 11.1. The table summarizes the main features offered by existing simulators. In the next section, we discuss the FogNetSim++ architecture design and its working.

Simulators	**Prog. Language**	**Platform**	**Network Configuration**	**Open Source**	**Mobile Nodes**	**Customize Mobility models**	**Scheduling Algorithms**	**Device Handover**	**Energy Module**
MobIoTSim (8)	Java	Linux	No	Yes	Yes	No	No	No	No
SimpleIoTSimulator (10)	Java	Unix	No	No	Yes	No	No	No	No
IBM BlueMax (11)	Java/Python	Cloud	No	No	Yes	No	No	No	No
Google IoT Sim (13)	NA	Cloud	No	No	Yes	No	No	No	No
iFogSim/MyiFogSim (14)	Java	All	No	Yes	Yes	No	Yes	No	No
Cooja (17)	C	Linux	No	Yes	Limited	No	No	No	No
FogTorch (18)	Java	All	No	Yes	No	No	No	No	No
RECAP simulator (19)	N/A	–	Limited	N/A	No	No	No	N/A	N/A
EmuFog (20)	Python	All	Yes	Yes	No	No	No	No	No
EdgeCloudSim (6)	Java	All	No	Yes	Yes	No	Yes	No	No
Edge-Fog cloud (21)	Python	All	No	Yes	Limited	No	No	No	No
Mobile Fog (22)	N/A	–	No	No	Yes	No	No	No	No
FogNetSim++	C++	All	Yes	Yes	Yes	Yes	Yes	Yes	Yes

Figure 11.1 A comparison of different simulators for fog computing environments.

11.3 FogNetSim++: Architecture

FogNetSim++ is built at the top of OMNET++. The high level of architecture is shown in Figure 11.2. FogNetSim++ is composed of various modules. Figure 11.3 shows the graphical user interface of FogNetSim++ where a single broker (task scheduler) is connected with a number of fog nodes. There is a number of mobile users to seek computation resources. A working model is presented in Figure 11.4 where a number of static and mobile users requesting resources from a broker. If a broker doesn't have the required amount of resources, it will forward the request

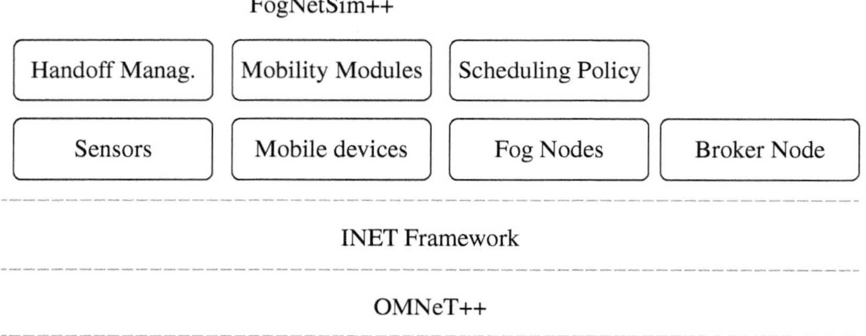

Figure 11.2 FogNetSim++ high-level architecture.

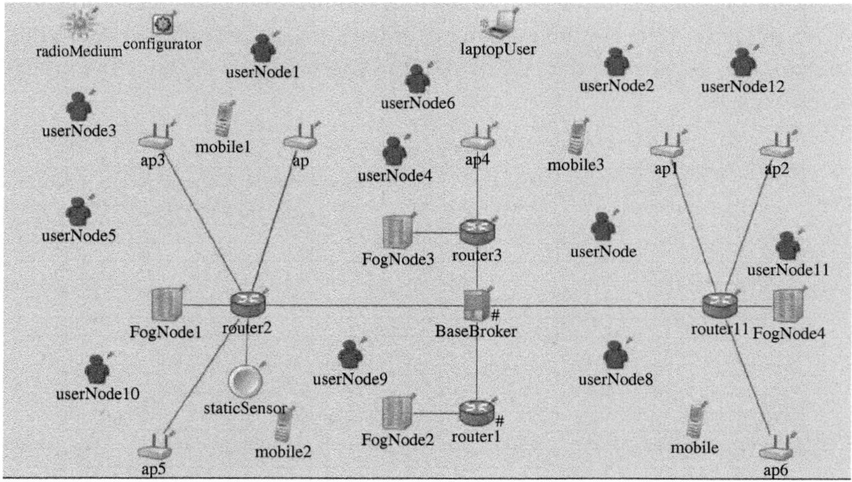

Figure 11.3 FogNetSim++: Graphical user interface, showing static, mobile, and fog computing nodes. (*See color plate section for the color representation of this figure*)

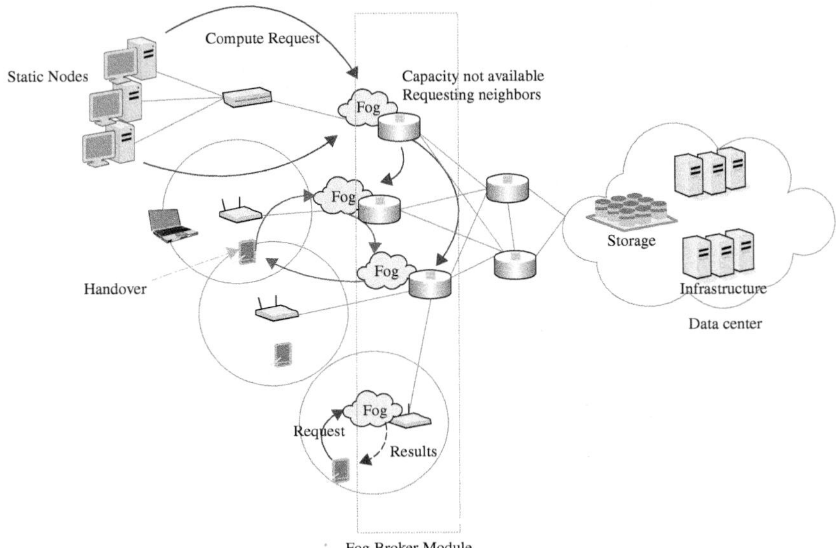

Figure 11.4 FogNetSim++: showing the handover features managed through single broker node. (*See color plate section for the color representation of this figure*)

to the neighboring fog nodes. The request can forward to backend cloud if no fog resource is available nearby. During the task execution, if the user leaves the premises of an access point, the broker node is responsible for tracking the user and deliver the response of execution through a different base station. The main modules of FogNetSim++ are explained below:

Broker Node – Broker node is the core module of FogNetSim++. It is responsible for managing connections, computation requests, handover, and fog nodes. User nodes and fog nodes establish a connection with the broker. Initially, the fog nodes update the broker about the available resources. The broker node maintains the list of all the online fog nodes. When the request arrives at the broker, the broker assigns the request to the best available fog resource according to the scheduling algorithm at the broker. The broker also starts tracking the user to ensure safe handover. The FogNetSim++ is open-source, so researchers can easily replace the existing resource sharing algorithm with other algorithms. Figure 11.5 shows the internal structure of a broker node, where existing modules of OMNeT++ are used in addition to Message Queuing Telemetry Transport (MQTT) to meet the desired functionality. Thus, the broker can establish a connection with other devices using UDP, TCP, SCTP, and MQTT.

User Nodes – FogNetSim++ supports both static and mobile nodes. Static nodes are the nodes connected through an Ethernet cable with the broker, and mobile

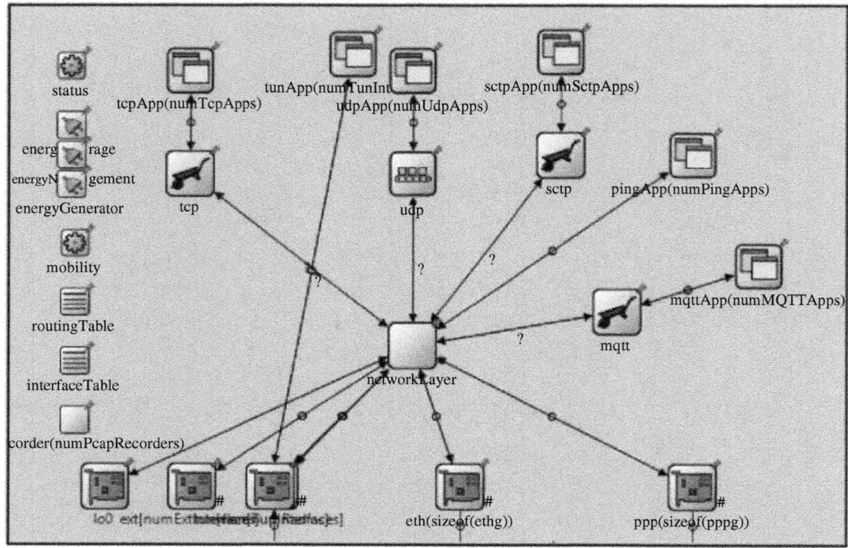

Figure 11.5 FogNetSim++: Internal structure of broker node.

nodes are the wireless nodes and sensors. FogNetSim++ supports a number of mobility models at wireless devices. The detailed explanation of the mobility model is given in the INET manual.[3]

11.4 FogNetSim++: Installation and Environment Setup

This section describes the installation procedure and environment setup on Ubuntu 16.04 LTS or Ubuntu 18.04 LTS. As FogNetSim++ is built over OMNeT++; there-fore, installation of OMNeT++ is the first step to perform. OMNeT++ requires a number of packages as listed below.[4]

11.4.1 OMNeT++ Installation

The OMNeT++ can installed by following the steps mentioned here. First update the system by typing.

```
$ sudo apt-get update
```

3 https://inet.omnetpp.org/docs/users-guide/ch-mobility.html
4 https://doc.omnetpp.org/omnetpp/InstallGuide.pdf

Once the system is updated install the packages by typing the following command in terminal.

```
$ sudo apt-get install build-essential gcc g++ bison flex perl
python python3 qt5-default libqt5opengl5-dev tcl-dev tk-dev
libxml2-dev zlib1g-dev default-jre doxygen graphviz
libwebkitgtk-1.0
```

OMNeT++ provides a very efficient visualization of network objects and flow. To use this feature in Ubuntu 16.04, install the following package.

```
$ sudo add-apt-repository ppa:ubuntugis/ppa
$ sudo apt-get update
$ sudo apt-get install openscenegraph-plugin-osgearth libosgearth-dev
```

Optional parallel simulation support (optional) can also be activated by running the following command.

```
$ sudo apt-get install openmpi-bin libopenmpi-dev
```

Now download omnet++4.6 (generic archive) from http://omnetpp.org. Place the archive at any location, usually /Home/[omnetpp], change directory to "/Home" and run the following command:

```
$ tar xvfz omnetpp-5.4.1-src.tgz After that
```

After that run following commands:

```
$ cd omnetpp
$ gedit    /.bashrc
Add this line in the file, then save it:
export PATH=HOME=omnet pp      4:6=bin :PATH
and run these commands to install downloaded
omnet++:
$ ./configure
$ make
Once its done, run the omnet++ by typing
$ omnetpp
```

Now the OMNeT++ installation is complete. The next step is to download and add INET package from OMNeT++ website. There are two ways to link INET with OMNeT++, (1) download INET 3.3.0 from http://inet.omnetpp.org and import in OM-NeT++ as a project, (2) inside OMNeT++ GUI goto "Help ! simulation models," you will see INET in the list, install it from here. Now, you can run the existing samples of INET. In the next section, we discuss the FogNetSim++ environment setup.

11.4.2 FogNetSim++ Installation

To install FogNetSim++, goto GitHub project[5] github.com/rtqayyum/fognetsimpp, and download two archives with names ("fognetsimpp.zip" and "mqttapp.zip"). Once the download is completed, follow these steps:

Step 1	Extract "mqttapp.zip" and place it in following directory of Inet inet/applications/ (directory might not be same, just expand INET project, expand src, and then expand applications. This is the directory where you need to paste the "mqttapp" folder), you can also see other application folders.
Step 2	Import "fognetsimpp.zip" as OMNeT++ project. Right click at this project, and add reference of INET.

11.4.3 Sample Fog Simulation

This section explains the working of FogNetSim++ using a sample simulation where a mobile node sends a task for computation to the broker. The broker assigns that task to an available fog node.

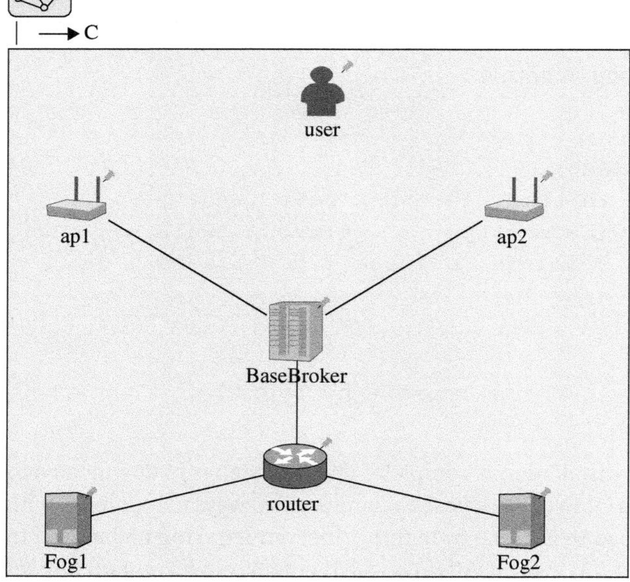

Figure 11.6 GUI – Sample fog simulation.

5 https://github.com/rtqayyum/fognetsimpp

Step 1 Open OMNeT++ and create a project from File ! New ! OMNeT++ Project, name it test.

Step 2 Right click at this “test” project ! Properties ! Project References ! select “INET” from list, and press “ok.”

Step 3 Right click at this “test” project ! New ! Network Description File, and name it “sample.ned.” Add the following code in this file as shown below.

```
import inet.node.inet.WirelessHost;
import inet.node.inet.StandardHost;
import inet.node.inet.Router;
import inet.node.wireless.AccessPoint;
import ned.DatarateChannel;
import inet.networklayer.configurator.ipv4.
IPv4NetworkConfigurator;
import inet.physicallayer.ieee80211.
packetlevel.Ieee80211ScalarRadioMedium;
import inet.common.lifecycle. LifecycleController;
network WirelessNetwork
f
types:
channel C extends DatarateChannel
f
datarate = 100Mbps;
delay = 0.1us;
g
submodules:
Fog1: StandardHost f
@display("p=412,616;i=device/router");
g
Fog2: StandardHost f
@display("p=582,616;i=device/router");
g
BaseBroker: StandardHost f
@display("p=582,424;i=device/server;");
g
user: WirelessHost;
router1: Router f
@display("p=582,552;q=Tasks");
g
configurator: IPv4NetworkConfigurator f
@display("p=222,50");
```

```
33. g
34. radioMedium: Ieee80211ScalarRadioMedium f
35. @display("p=60,50");
36. g
37. ap1: AccessPoint f
38. @display("p=997,172");
39. g
40. ap2: AccessPoint f
41. @display("p=133,172");
42. g
43. lifecycleController:LifecycleControllerf
44. @display("p=50,50");
45. g
46. connections:
47. Fog1.ethg++ <--> C <--> router.ethg++;
48. Fog2.ethg++ <--> C <--> router.ethg++;
49. router.ethg++ <--> C <-->
50. BaseBroker.ethg++ <--> C <--> ap1.ethg++;
51. BaseBroker.ethg++ <--> C <--> ap2.ethg++;
52. Fog1.ethg++ <--> C <--> router.ethg++;
53. Fog1.ethg++ <--> C <--> router.ethg++;
54 g
```

The lines from 1 to 8 are the references for the module we are going to use in the simulation. The modules are already defined in INET. Lines from 9 to 45 describes the network, with one BaseBroker, two fog nodes, two wireless access points, one router, one radio medium, and communication protocol is IPv4 configurator.

Lines from 46 to 53 show the connections among these modules using data channel "C," and the data channel is defined in line 12. The GUI of the sample can be seen in Figure 11.6.

Step 4 Right click at project "test" and create another file of type ".ini" and name it "omnet.ini." Add the following code in this file.

```
network =WirelessNetwork
tkenv-plugin-path = ../../../etc/plugins
**.constraintAreaMinX = 0m
**.constraintAreaMinY = 0m
**.constraintAreaMinZ = 0m
**.constraintAreaMaxX = 1000m
```

```
**.constraintAreaMaxY = 1000m
**.constraintAreaMaxZ = 0m
**.user.mobilityType = "CircleMobility"
**.user.mobility.cx = 300m
**.user.mobility.cy = 300m
**.user.mobility.r = 250m
**.user.mobility.speed = 40mps
**.user.mobility.startAngle = 180deg
# mobility
**.user.mobilityType = "LinearMobility"
**.user.mobility.speed = 20mps
**.user.mobility.angle = 0
**.user.mobility.acceleration = 0
**.user.mobility.updateInterval = 100ms
**.radio.transmitter.power = 3.5mW
# nic settings
**.wlan.mac.EDCA = false
**.wlan.mgmt.frameCapacity = 10
**.wlan.mac.maxQueueSize = 14
**.wlan.mac.rtsThresholdBytes = 3000B
**.wlan.mac.bitrate = 54Mbps
**.wlan.mac.basicBitrate = 6Mbps  24Mbps
**.wlan.mac.retryLimit = 7
**.wlan.mac.cwMinData = 31
**.wlan.mac.cwMinBroadcast = 31
# Queues
**.eth[*].queueType = "DropTailQueue"
**.eth[*].queue.dataQueue.frameCapacity = 40
 # mqtt apps
**.user.numUdpApps = 1
**.user.udpApp[*].typename = "mqttApp2"
**.user.udpApp[*].messageLength = 1024B
**.user.udpApp[*].localPort = 1000
**.user.udpApp[*].destAddresses = "BaseBroker"
**.user.udpApp[*].destPort = 1001
**.user.udpApp[*].sendInterval = 1.5s
```

```
**.user.udpApp[*].startTime = 0.0s
**.user.udpApp[*].stopTime = 300s
**.user.udpApp[*].publishToTopics = "test topic 1"
**.user.udpApp[*].publish = true
**.user.udpApp[*].taskSize = 1500
# basebroker apps
**.BaseBroker.numUdpApps = 1
**.BaseBroker.udpApp[*].typename = "BrokerBaseApp3"
**.BaseBroker.udpApp[*].localPort = 1001
**.BaseBroker.udpApp[*].MIPS = 0
# computeBroker apps
**.Fog*.numUdpApps = 1
**.Fog*.udpApp[*].typename = "ComputeBrokerApp3"
**.Fog*.udpApp[*].localPort = 1001
**.Fog*.udpApp[*].MIPS = 1000
**.Fog*.udpApp[*].destAddresses = "BaseBroker"
**.Fog*.udpApp[*].destPort = 1001
**.Fog*.udpApp[*].messageLength = 100 Byte
**.Fog*.udpApp[*].sendInterval = 1s
**.Fog*.udpApp[*].startTime = 0.0s
# visualizer
*.visualizer.energyStorageVisualizer. displayEnergy Storages = true
# status
*.user.hasStatus = true
# power
*.user.energyStorage.typename = "SimpleEpEnergyStorage"
*.user.wlan[*].radio.energyConsumer. typename = "StateBasedEpEnergyConsumer"
*.user.energyManagement.typename = "SimpleEpEnergy Management"
*.user.energyStorage.nominalCapacity = 0.05J
*.user.energyManagement. nodeShutdown Capacity = 0.1 * energyStorage.nominalCapacity
*.user.energyManagement.nodeStartCapacity = 0.5 * energyStorage.nominalCapacity
*.user.energyStorage.initialCapacity = uniform(0.1 * energyStorage.nominalCapacity, energyStorage.nominalCapacity)
```

```
75. *.user.energyGenerator.typename = "AlternatingEpEnergy
Generator"
76. *.user.energyGenerator.powerGeneration = 4mW
1. *.user.energyGenerator.sleepInterval = exponential(25s)
77. *.user.energyGenerator.generationInterval =
exponential(25s)
```

In line 1, the name of the network is defined. The area is specified in lines 3–8. Two kinds of mobility models are applied at the user in lines 9–14 and 15–20, circular mobility and linear mobility, respectively. You need to choose one at a time and comment the other mobility.

The wireless properties are given in lines 21–31. The queue is defined in lines 33–34.

FogNetSim++ uses different applications at a different kind of host. For instance, at user, we have used "mqttApp2," at baseBroker, we used "Base-BrokerApp3," and at fog nodes, we used "ComputeBrokerApp3" as shown in lines 35–62.

The configuration about energy is applied at lines 62–77.

Step 5 Once both the files are ready. Right click at this newly created file "om-net.ini" ! Run As OMNeT++ Simulation. A new window will open, here click at the run button.

In order to run the existing FogNetSim++ sample simulation to understand the framework and its basic functionalities. The sample simulation is located under FogNetSim++ ! simulations ! testing.

11.5 Conclusion

Fog computing is an emerging area of research. The significance of fog computing can be seen with applications it supports. Many core technologies are using the concept of edge/fog computing. It is important to know the working of existing simulators so that researchers can extend this framework by adding new modules and test cases. The fog computing can be used in healthcare, to support big data applications where rapid response is very crucial. In this chapter, we have covered the most comprehensive fog simulator, its core modules, architecture, and also demonstrate the simple test application.

References

1 Bonomi, F., Milito, R., Zhu, J., and Addepalli, S. (2012). Fog computing and its role in the Internet of Things. In: *Proceedings of the First Edition of the MCC Workshop on Mobile Cloud Computing*, 13–16. ACM.

2 Cisco, The Internet of Things: extend the cloud to where the things are, white paper, 2015. Available: https://www.cisco.com/c/dam/enus/solutions/trends/iot/docs/computingoverview.pdf.

3 Peng, M., Yan, S., Zhang, K., and Wang, C. (2016). Fog-computing-based radio access networks: issues and challenges. *IEEE Network* 30 (4): 46–53.

4 Jalali, F., Hinton, K., Ayre, R. et al. (2016). Fog computing may help to save energy in cloud computing. *IEEE Journal on Selected Areas in Communications* 34 (5): 1728–1739.

5 Ahmed, E. and Rehmani, M.H. (2017). Mobile edge computing: opportunities solutions and challenges. *Future Generation Computing Systems* 70: 59–63.

6 Ni, J., Lin, X., Zhang, K., and Shen, X. (2016). Privacy-preserving real-time navigation system using vehicular crowdsourcing. In: *IEEE 84th Vehicular Technology Conference (VTC-Fall)*, 1–5.

7 Shropshire, J. (2014). Extending the cloud with fog: security challenges & opportunities. In: *Proceedings of the 20th Americas Conference of Information Systems (AMCIS 2014)*, 1–10.

8 Sehgal, V.K., Patrick, A., Soni, A., and Rajput, L. (2015). Smart human security framework using Internet of Things cloud and fog computing. In: *Intelligent Distributed Computing*, 251–263. Cham, Switzerland: Springer.

9 Gazis, V. (2016). A survey of standards for machine-to-machine and the Internet of Things. *IEEE Communications Surveys and Tutorials* 19 (1): 482–511, 1st Quart.,.

10 Buss, A. and Jackson, L. (1998). Distributed simulation modeling: a comparison of HLA, CORBA, and RMI. In: *Proceedings of the 30th Conference on Winter Simulation (WSC '98)*, Washington, DC, USA, 819–825, December 13–16.

11 Pflanzner, T., Kertész, A., Spinnewyn, B., and Latré, S. (2016). Mobiotsim: towards a mobile IOT device simulator. In: *2016 IEEE 4th International Conference on the Future of the Internet of Things and Cloud Workshops (FICloudW)*, 21–27. IEEE.

12 Simplesoft simpleiotsimulator, http://www.smplsft.com/SimpleIoTSimulator.html (accessed August 10, 2018).

13 IBM BlueMix platform, https://console.ng.bluemix.net (accessed August 10, 2018).

14 Google Cloud Platform. [Online]. Available: https://cloud.google.com/solutions/iot/ (accessed August 10, 2018).

15 Gupta, H., Dastjerdi, A.V., Ghosh, S.K., and Buyya, R. (2017). iFogSim: A toolkit for modeling and simulation of resource management techniques in the Internet of Things, edge and fog computing environments. *Software Practice and Experience* 47 (9): 1275–1296.

16 Contiki Cooja [Online]. Available: http://www.contiki-os.org/start.html (accessed August 10, 2018).

17 Brogi, A., Forti, S., and Ibrahim, A. (2017). How to best deploy your fog applications, probably. In: *Proceedings of the 1st IEEE International Conference on Fog and Edge Computing, ICFEC 2017*, 105–114.

18 Byrne, J., Svorobej, S., Gourinovitch, A. et al. (2017). RECAP simulator: simulation of cloud/edge/fog computing scenarios. In: *Proceedings of the Winter Simulation Conference (WSC)*, 4568–4569.

19 Mayer, R., Graser, L., Gupta, H. et al. (2017). EmuFog: Extensible and scalable emulation of large-scale fog computing infrastructures. In: *Proc. IEEE Fog World Congr. (FWC)*, 1–6.

20 Sonmez, C., Ozgovde, A., and Ersoy, C. (2017). EdgeCloudSim: an environment for performance evaluation of edge computing systems. In: *Proceedings of the Second International Conference on Fog and Mobile Edge Computing (FMEC)*, Valencia, Spain, 39–44.

21 Mohan, N. and Kangasharju, J. (2016). Edge-fog cloud: a distributed cloud for Internet of Things computations. In: *Proceedings of the Cloudification of the Internet of Things*, 1–6. CIoT.

22 Hong, K., Lillethun, D.J., Ramachandran, U. et al. (2013). Mobile fog: A programming model for large-scale applications on the Internet of Things. In: *Proceedings of MCC@SIGCOMM*, 15–20.

23 Qayyum, T., Malik, A., Khan, M.A. et al. (2018). FogNetSim++: A toolkit for modeling and simulation of distributed fog environment. *IEEE Access* 6: 63570–63583.

Part II

Fog Computing Techniques and Applications

12

Distributed Machine Learning for IoT Applications in the Fog

Aluizio F. Rocha Neto[1], *Flavia C. Delicato*[2], *Thais V. Batista*[1], *and Paulo F. Pires*[2]

[1] *DIMAp, Federal University of Rio Grande do Norte, Natal, Brazil*
[2] *PGC/IC/UFF, Fluminense, Federal University, Niterói, Brazil*

12.1 Introduction

The term Internet of Things (IoT) was coined by the British Kevin Aston in 1999 [1] and since then we have seen such a paradigm come about thanks to the evolution of several enabling technologies. IoT advocates a reality in which the physical and virtual worlds mingle through rich interactions. Its boundaries become almost invisible, so as to be possible on the one hand, the augment of the physical world with virtual information, and, on the other hand, the extension of the virtual world to encompass concrete objects. To do so, the first step is to instrument physical entities, through sensors, capable of acquiring various types of environmental variables, and actuators, capable of changing the state of physical objects. By instrumenting objects, they become endowed with the ability of perceiving the surrounding world and are generally denoted as intelligent or smart objects. However, the perception capability is only a small portion of what we call intelligence. Data on perceived phenomena can be partially processed locally (in the smart objects themselves) and then transmitted (usually through wireless interfaces) until eventually they are available on the Internet via virtual representations of the perceived physical phenomenon. This local processing characterizes another facet of intelligence in IoT objects, and its degree of complexity depends entirely on the amount of resources available in the device. Physical devices can communicate with their connected neighbors in order to exchange data on the monitored (common) phenomena, enriching the information before transmitting

Fog Computing: Theory and Practice, First Edition.
Edited by Assad Abbas, Samee U. Khan, and Albert Y. Zomaya.

it. Once entering the virtual world, the data captured by the objects can serve as input to the most sophisticated processing and analysis systems in order to extract useful knowledge for a variety of purposes. This complex processing characterizes the ultimate degree of intelligence related to IoT systems.

Given the number of physical objects on Earth, if only a tiny fraction of them are instrumented by sensors/actuators, IoT will certainly have the potential to be the largest source of Big Data on the planet and as such, brings numerous research challenges to be solved to enjoy all the possible benefits of this paradigm. In this context, it is important to understand where these benefits really come from, and what challenges and avenues of research need to be explored to concretize the view of a fully connected, virtualized and Smart world.

First of all, although the IoT paradigm is focused on the connection of objects, its real potential lies not in the objects themselves, but in the ability to generate valuable knowledge from the data extracted from these objects. It can be said that IoT actually is not about things but about data. Therefore, the process of transforming the raw data collected through the perception of the physical environment in value-added information and, ultimately, in high-level knowledge capable of optimizing decision making, is crucial to obtaining the real benefit of IoT. At the heart of this transformation process are computational intelligence (CI) techniques. In particular, machine learning (ML) techniques are promising to process the data generated in order to transform them into information, knowledge, to predict trends, produce valuable insights, and guide automated decision-making processes. However, the use of machine learning (ML) techniques in IoT brings up several challenges, especially regarding the computational requirements demanded by them.

As discussed earlier, the IoT devices themselves may be part of this data transformation process. However, with their limited resources, they generally cannot perform very complex processing. With their vast computing capabilities, cloud platforms are the natural candidates as backend of IoT systems and perform computationally intensive analysis and long-term data storage. However, with the increasing growth in data generated by sensors and IoT devices, their discriminated transmission to the cloud began to generate a set of problems. Among them, the excessive use of network bandwidth has brought congestions and poor performance in the communication infrastructure. In addition, several IoT applications have critical response time requirements, and the high and unpredictable latency of the Internet for accessing remote data centers in the cloud is not acceptable in this context.

The challenge of minimizing network latency and improving computational performance in the context of IoT environments has been recognized among the main concerns of the IoT research community. In this context, the concept of *fog computing* was suggested by the industry and further extended by academic

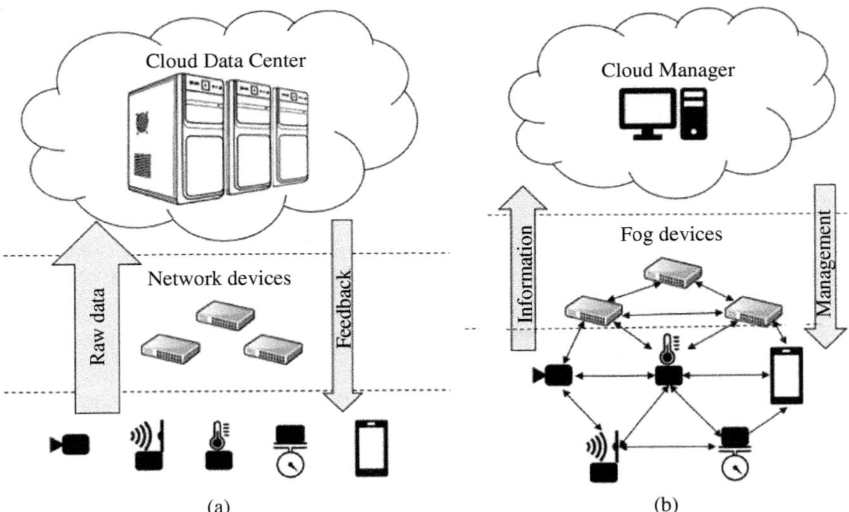

Figure 12.1 Three-tier architecture for IoT (a) and two-tier cloud assisted IoT (b).

researchers. Cisco – one of the main pioneers on fog computing – states that the fog infrastructure (composed by smart switches, routers, and dedicated computing nodes) brings the cloud closer to the things that produce IoT data. The main idea in this approach is to put the computation resources closer to the data source – the IoT devices – so that they require less network traffic and are more cloud independent, reducing the application overall delays.

In a similar way, the concept of *edge computing* is also used to refer to pushing the computation down to the end devices, i.e. toward the "edge" of the network. However, according to [2], edge is the first hop from the IoT devices. It enables that computations be executed on small datacenters close to the edge device itself. In this way, some advanced data analytics and processing can be performed near the devices, thus reducing system response time and network traffic. Figure 12.1 presents the differences between the traditional organization of IoT system architecture in three tiers and the novel vision, introduced by fog/edge computing paradigms. In the three-tier architecture (Figure 12.1a), IoT devices have to send a large amount of raw data to be processed on servers hosted typically in cloud data centers and wait for some feedback. With the two-tier fog/edge computing paradigm (Figure 12.1b), the edge and network devices collaboratively process the raw data and send only high-level information to a manager in the cloud.

In this two-tier architecture, the data produced by the IoT devices can then be processed: (1) in the data source itself (also called the Things tier/layer), (2) in cloud platforms, and (3) in the intermediate layer of fog/edge. Depending on the application requirements (such as response time) and the complexity of the required processing technique, each tier will be more suitable for performing the computation. The overhead that current ML techniques imposes on devices of the fog and things tiers, in which resources are scarce, hinders their widespread adoption in this context. In addition to the challenges of the processing complexity required by the ML techniques, the specific features of the data produced by IoT devices raise further requirements. Typically, data produced by IoT are time-series values, which are gathered during a period and transmitted to a gateway for further processing. Compared with Big Data produced by other sources, such as social data, sensor data streams have additional features that call for a shifting from the Big Data paradigm to the Big Data stream paradigm, which we discuss throughout this chapter.

Motivated by all the aforementioned issues, this chapter aims at presenting the reader with a broad view on the challenges of using ML techniques at the edge of the network, taking advantage of the capabilities of this tier while dealing with the computational constraints of fog devices. Due to the constant evolution of IoT devices and ML techniques, recent published researches in these fields are prone to become out of date quickly. Besides, there is a great rush in the technological scientific community to apply modern artificial intelligence techniques in their systems in order to increase their efficiency and provide better solutions for their problems. The ultimate goal of CI applied in context of IoT systems is to build valuable knowledge for supporting better decision-making processes in a wide range of application domains.

The next section presents the main challenges in processing IoT data, considering its peculiar features and the intersection of IoT and the Big Data paradigm. Section 12.3 gives a bird's-eye view on the employment of CI techniques at the edge of the networks, as a strategy to extract useful information from raw data and reduce the network overhead, as proposed by the fog computing paradigm. Section 12.4 describes the main challenges related to the execution of ML techniques within the resource constrained fog devices, analyzing different categories of devices according to their resources. Finally, we present some approaches to distribute intelligence on fog devices, with focus on distributed processing and information sharing.

12.2 Challenges in Data Processing for IoT

In recent years, the number of IoT devices/sensors has increased exponentially and there is an estimation that by 2020 it will reach 26 billion.[1] This network

---

1 http://www.gartner.com/newsroom/id/2636073

of devices/sensors encompasses an infrastructure of hardware, software, and services integrating the Internet with the physical world. This change in the physical world is giving a new shape in the way that individuals observe and interact with the environment. A plethora of applications can emerge, exploiting the diversity of real-time monitored data, and combining them in different ways to provide value-added information to users. Examples of advanced and intelligent applications are smart home, smart building, and smart city. The *SmartSensor infrastructure* [3] is an example of recent research initiatives aimed to support these applications and it was developed to manage a set of wireless sensor networks (WSNs), where sensors work collaboratively, extracting environmental data to be further processed and augmented with semantics information to provide value-added services to users.

At the heart of IoT ecosystems are the tiny sensors spread around the physical environment, possibly deployed in a large number, of the order of hundreds to thousands, depending on the size of the monitored region. Such sensors will create a data deluge encompassing different kinds of information with different values. How to handle these data and how to dig out the useful information from them are issues being tackled in recent researches. In [4], the authors suggest a simple taxonomy to classify the types of data in the IoT. According to the authors, "data about things" refers to data that describe the objects/things themselves (e.g. state, location, identity), while "data generated by things" refer to data generated or captured by things (e.g. room temperature, people counting, traffic condition). Data *about* things can be used to optimize the performance of the systems and infrastructures, whereas data *generated by* things contains useful information about interaction between humans and systems and among systems themselves; such data can be used to enhance the services provided by IoT [5].

In this section, we discuss the challenges introduced by IoT systems in three different dimensions: the generation, transmission, and processing of data. Such dimensions are typically related to and addressed by three well-known research fields, namely Big Data, Big Data stream, and data stream processing. Figure 12.2 shows a taxonomy for these challenges imposed by IoT system and the respective aspects addressed by each of the three dimensions.

12.2.1 Big Data in IoT

Big Data refers to voluminous and complex data sets that require advanced data processing application software. It involves several challenges that encompass capturing, storing, sharing, searching, transferring, analyzing, visualizing, updating, and protecting data. The IoT environment is one of the key areas responsible for this trend of data explosion. High data rate sensors, such as surveillance cameras, are one of the most relevant in this context.

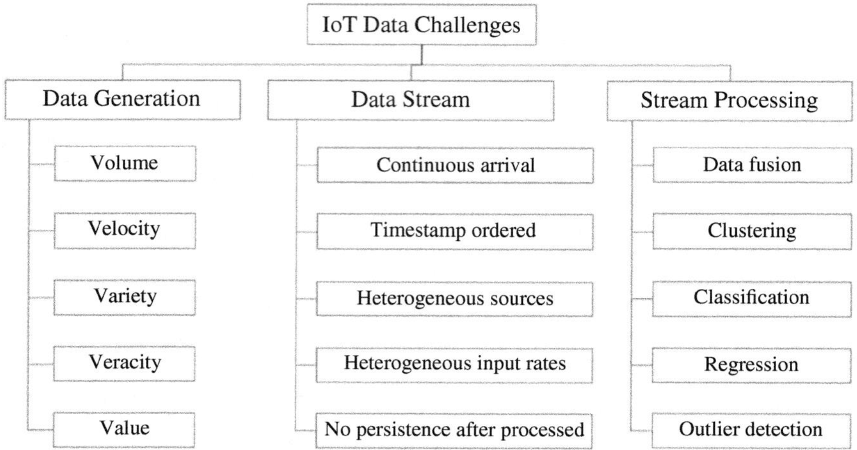

Figure 12.2 IoT Data Challenges in three dimensions: generation, transmission, and processing of data.

Data scientists have defined six features to address all the issues related to Big Data:

1. *Volume.* This is the key aspect to consider a data set to be "big," as the size of the data is measured as volume.
2. *Velocity.* This is related to the rate of data generation and transmission. The huge amount and speed of IoT Big Data coming from several sources demands the support for big data in real-time.
3. *Variety.* With increasing volume and velocity comes increasing variety, which describes the huge diversity of data types any IoT system can produce.
4. *Veracity.* Refers to how accurate and how truthful is the data set. It is more than reliability of the data source, it also involves the degree of confidence in the analysis of the data. This feature is quite important in IoT system that uses crowd-sensing data.
5. *Variability.* This feature refers to the different rates of data flow, which may vary from time to time or place to place. For instance, in an IoT application for traffic management using sensors, the speed at which data is loaded in a database can have expressive variations depending on the heavy traffic hours.
6. *Value.* Data must have value, i.e. all the infrastructure required to collect and interpret data on an IoT system must be justified by the insights you can gather to improve the user experience and services.

Another important aspect for the value of data is its *temporariness.* In many smart IoT systems that are time-sensitive, such as traffic accident monitoring or river flooding prediction, and where the system has to take some automatic action in a real-time manner, the potential value of data depends on data freshness that needs to be processed. Otherwise, the processing results and actions become less valuable or even worthless.

Qin et al. [6] present an IoT data taxonomy classifying the IoT data into three categories, regarding different characteristics of data: data generation, data quality, and data interoperability. Data generation is related to the Big Data features aforementioned such as velocity, scalability, dynamics, and heterogeneity. Data quality is related to the following characteristics:

- *Uncertainty*. In IoT data, uncertainty can refer to variance from the expected states of the data, due to several reasons such as transmission errors or sensing precision.
- *Redundancy*. Errors in sensor reading logic can produce multiple data that can interfere in the semantic of sensor data. For example, in radio-frequency identification (RFID) data, the same tag can be read multiple times at the same location when there are multiple RFID readers at the same spot to improve the sensing.
- *Ambiguity*. Refers to different ways of interpreting data from diverse things to distinct data consumers. This misunderstanding may interfere with the overall inference about the sensed environment.
- *Inconsistency*. Such as ambiguity, inconsistency may occur in the interpretation of data that may produce misunderstandings. It is common to detect inconsistency in sensing data in case of several sensors monitoring an environment.

Data interoperability is addressed by the following characteristics:

- *Incompleteness*. In many IoT applications that detect and react to events in real time, the combination of data from different sources is important to build a complete picture of the environment. However, in this cooperation of mobile and distributed things who are generating the data, some problem related to the data quality from any source may negatively interfere in the whole system.
- *Semantics*. Semantic technologies have been used to endow machines with the ability of processing and interpreting IoT data.

Despite presenting all the aforementioned characteristics, the data produced by IoT devices have additional features, not always observed in Big Data. Typically, data produced by IoT sensors consist of time-series values, which are sampled over a defined period and then transmitted to a gateway for further processing. Compared with Big Data produced by other sources, such as social data, sensor data streams have additional features that call for a shifting from the Big Data paradigm to the Big Data Stream paradigm, which we discuss in the next section.

12.2.2 Big Data Stream

Time-series data are not processed in the same way as the typical Big Data nor as easily interpretable such as, for instance, a document, video, or other data available on the Internet. Data items arrive continuously and sequentially as a stream and are usually ordered by a timestamp value besides including other additional attribute values about the data item. Differently from Big Data that is typically produced in controlled and owned data warehouses, data streams are usually generated by heterogeneous and scattered sources. In general, IoT systems do not have direct access or control over such data sources. Moreover, the input characteristics of a data stream are usually not controllable and typically unpredictable. Data streams are provided in different rates, for instance, from small number of bytes per second to several gigabits. The inherent nature of the input does not allow one to easily make multiple passes over a data stream while processing (and still retaining the usefulness of the data).

Finally, numerous important IoT scenarios, such as logistics, transportation, health monitoring, emergence response, require predictable latency, real-time support, and could dynamically and unexpectedly change their requirements (e.g. in terms of data sources). Big Data stream–oriented systems need to react effectively to changes and provide smart, semi-autonomous behavior to allocate resources for data processing, thus implementing scalable and cost-effective services. The inherent inertia of Big Data approaches, that commonly rely on batch-based processing, are not proper for data processing in IoT contexts with dynamism and real-time requirements.

The Big Data stream paradigm enables ad-hoc and real-time processing to connect streams of data and consumers, benefiting scalability, dynamic configuration, and management of heterogeneous data formats. In summary, although Big Data and Big Data stream cope with massive numerous heterogeneous data, Big Data centers on the batch analysis of data, Big Data stream deals with the management of data flows and real-time data analysis. This feature has an impact also on the data that are considered relevant to consumer applications. For instance, while for Big Data applications it is important to keep all sensed data in order to be able to perform any required computation, Big Data stream applications might decide to perform real-time data processing on the raw data produced by IoT devices to reduce the latency in transmitting the results to consumers, with no need to persist such raw data. IoT platforms such as Xively1[2] (former Cosm), EcodiF [7], or Nimbits2[3] are focused on handling Big Data stream produced by IoT devices. However, they basically allow only publishing and visualization of streaming data from

2 https://xively.com
3 https://www.crunchbase.com/organization/nimbits

sensor devices; they lack processing and analytic features. The data remain in the same raw state thus hindering the detection of valuable, insightful information, which could effectively produce knowledge for decision-making processes.

In this context, the issue of translating the huge amount of sensor generated data streams from their raw state into higher-level representations and making them accessible and understandable for humans or interpretable by machines and decision-making systems remains open. One promising approach to tackle this issue is by applying techniques of CI. CI refers to the ability of a computer system to learn a specific task from data or experimental observation. The methods used in CI are close to the human's way of reasoning, i.e. it combines inexact and incomplete knowledge to produce control actions in an adaptive way. The vast research on CI techniques can be of great help to generate useful knowledge from such massive amounts of data generated by a world of interconnected IoT resources.

12.2.3 Data Stream Processing

Data stream processing is a paradigm that deals with a sequence of data (a stream) and a series of operations being applied to each element in the stream [8]. The typical real-time stream processing should include solutions for five items:

1. Integration with several data sources
2. Live data discovery and monitoring
3. Anomalies detection
4. Events detection (collecting, filtering, prediction, etc.)
5. Performance, scalability and real-time responsiveness

Item (1) is particularly relevant for time-sensitive IoT applications that deal with data provided by multiple data sources, and a timely fusion of data is needed to bring all pieces of data together. In this context, information fusion and sharing play a critical role for fast analysis and consequently providing reliable and accurate actionable insights [9]. According to [10], *Information Fusion* deals with three levels of data abstraction: measurement, feature, and decision, and it can be classified into these categories:

- *Low-level fusion.* Also referred to as *signal (measurement) level fusion.* Inputs are raw data that are joined with a new piece of more accurate data (with reduced noise) than the individual inputs.
- *Medium-level fusion.* Attributes or features of an entity (e.g. shape, texture, position) are fused to obtain a feature map that may be used for other tasks (e.g. segmentation or detection of an object). This type of fusion is also known as *feature/attribute level fusion.*

- *High-level fusion.* Also known as *symbol* or *decision level fusion.* Inputs are decisions or symbolic representations that are joined to produce a more confident and/or a global decision.
- *Multilevel fusion.* This fusion involves data of different abstraction levels, i.e. multilevel fusion occurs when both input and output of fusion can be of any level (e.g. a measurement is fused with a feature to provide a decision).

In [11], the authors propose Olympus, a framework for supporting the paradigm known as Cloud of Sensors (CoS). According to the authors, CoS is a type of ecosystem within the broader domain of Cloud of Things (CoT). CoS systems exploit the potential advantages of WSNs along with smart sensors' specific features to provide *Sensing as a Service* to client applications. A CoS comprises virtual nodes built on top of physical WSN nodes and provides support to several applications following a service-based and on-demand provision model. In this context, Olympus is a virtualization model for CoS that uses information fusion for reducing data transmission and increasing the abstraction levels of the provided data for aiding application decision processes. In Olympus decentralized virtualization model, physical sensors and actuators are linked to virtual nodes (at the measurement level according to the earlier classification), which perform the low-level fusion of the sensing data. Theses nodes send the reduced data to the feature level virtual nodes, which in turn produce the feature data to the high level fusion virtual nodes in order to perform the decision-making processes.

Data stream mining is a technique that can extract useful information and knowledge from a sequence of data objects of which the number is potentially *unbounded* [12]. This mining (analysis) is important to address the aforementioned items of live data discovery (2) and anomalies detection (3) in stream processing. Some typical tasks for data stream mining are:

- *Clustering.* This task involves grouping a set of objects putting similar objects in a same group (called a cluster). This is a technique commonly used for statistical data analysis, in several fields, including pattern recognition, image analysis, information retrieval, data compression, and computer graphics.
- *Classification.* In statistics, classification is the task of identifying which, from a set of categories (subpopulations), a new observation belongs to, on the basis of a training set of data containing observations (or instances) whose category membership is known. In other words, classification uses prior knowledge to guide the partitioning process to construct a set of classifiers to represent the possible distribution of patterns. This means that each example fed into the algorithm is classified into a recognizable data class or type.
- *Regression.* In statistics, regression is a technique that allows exploring and inferring the relationship of a dependent variable (response variable) with

specific independent variables (explanatory variables or "predictors"). Regression analysis can be used as a descriptive data analysis method (e.g. curve fitting) without requiring any assumptions about the processes that allowed data to be generated. Regression also designates a mathematical equation that describes the relationship between two or more variables.

- *Outlier and anomaly detection.* This is the task of finding data points that are the most different from the remaining points in a given data set. In general, outlier detection algorithms calculate the distance between every pair of points in a two-dimensional space. Outliers are the data points most distant from all other points.

Typically, in IoT applications, data streams from different sensors are real-time events that frequently form complex patterns. Each pattern represents a unique event that must be interpreted with minimal latency, to enable decision making in the context of current situation. The requirements of analyzing heterogeneous data streams and detecting complex patterns in near real time have formed the basis of a research area called complex event processing (CEP). CEP encompasses several tasks such as processing, analyzing, and correlating event streams from different data sources to infer more complex events almost in real time [13]. CEP lacks the predictive power provided by many ML data analysis methods. This is because most CEP applications relies on predefined rules to offer reactive solutions that correlate data streams but do not deal with historical data due to its limited memory. On the other hand, for some IoT applications, such as traffic management, prediction of a forthcoming event, for example, a congestion, is clearly more useful than detecting it after it has already occurred. This case is more notorious in applications for predicting natural disasters and epidemic diseases. ML techniques can complement the CEP data analysis methods to improve the inference and prediction process for such applications.

Finally, another important issue in the data stream processing is the real-time responsiveness (item 5 earlier) to perform data analysis with CEP and ML techniques. With its huge computing capacity and massive storage capacity, the cloud has traditionally been used to run the complex ML and event processing algorithms. However, it is well known that data transfer to the cloud and return of results are subject to unpredictable and potentially high latency of the Internet. To reduce the delay in processing all data streams and taking timely decisions, a promising approach is to perform the processing and decision tasks within end devices (IoT sensors and mobile devices), or at edge devices, thus leveraging their computational capabilities as proposed by the fog computing paradigm. The next section discusses the challenges to run CI techniques for processing IoT data in the fog context.

12.3 Computational Intelligence and Fog Computing

As mentioned before, in order to minimize the communication latency and improve the computational performance of IoT applications, the emerging paradigms of fog and edge computing have been advocated as potential and promising solutions. In this context, the compute nodes of the edge or fog tier will perform all or part of the various data stream processing steps. To tackle the issues related to performance, scalability, and real-time responsiveness, and considering the massive amount of data produced by large IoT systems, researchers have investigated the deployment of ML techniques in the fog computing paradigm.

12.3.1 Machine Learning

Recently, ML, a research field of CI, have been widely implemented in a number of domains that depend on complex and massive data processing such as medicine, biology, and engineering, providing solutions to gather the information hidden in the data [6]. The central goal of CI is to provide a set of algorithms and techniques that can be used to solve problems that humans perform intuitively and near automatically but are otherwise very challenging for computers. Combining ideas from many different fields, such as optimization theory, statistics, and cognitive science, ML algorithms are constantly improving. In addition, ML has benefited from the evolution in the capacity of the computational resources currently available even in IoT scenarios.

ML can be defined as the creation and use of *models* that are *learned from data*. In this context, a model is a specification of a mathematical (or probabilistic) relationship that exists between variables. Typically, in IoT scenarios the aim of adopting ML techniques is using historical and stream data to develop models able to *predict* various outcomes for new data or to provide *insights*. Examples of usage are below, just to name a few:

- Predicting the increase rate of energy consumption based on the current temperature and power status of an IoT system
- Predicting whether a set of events represents a dangerous situation in some environment (for instance a smart building)
- Predicting an intersection congestion when the number of vehicles increases in a smart road application

ML techniques can be divided into three subdomains: supervised learning, unsupervised learning, and reinforcement learning. The difference among them lies in the learning process of the intelligent algorithm. When the input data can be labeled with the desired outputs of the algorithm, supervised learning is applied because it is possible to supervise the accuracy of the training process.

Table 12.1 Comparison of machine learning technologies [14].

Learning types	Data processing tasks	Methodology	Learning algorithms	Example of IoT applications
Supervised learning	Classification/ Regression/ Estimation	Statistical classifiers	K-nearest neighbors Naïve Bayes Hidden Markov model Bayesian networks	Fruit identification through classification of data patterns
		Computational classifiers	Decision trees Support vector machine	
		Connectionist classifiers	Neural networks	
Unsupervised learning	Clustering/ Prediction	Parametric	K-means Gaussian mixture model	Anomaly detection in IoT healthcare system
		Nonparametric	Dirichlet pr. mix model X-means	
Reinforcement learning	Decision-making	Model-free	Q-learning R-learning	Urban traffic prediction
		Model-based	TD learning Sarsa learning	

Unsupervised learning is applied when you can interpret data based only on input data trying to find some hidden pattern or intrinsic structure. Unsupervised learning is useful whenever one wants to explore the available data but does not have yet a specific goal or is not sure what information the data contains. Reinforcement learning differs from normal ML because it does not use training data set, but interactions with the external environment to constantly adapt and learn on given points as a kind of feedback. Supervised and unsupervised learning is suitable for data analysis while reinforcement learning is adequate for decision-making problems [14]. Table 12.1 presents these three classes of ML methods/techniques from different perspectives. Any of these ML techniques can be applied to a wide variety of IoT applications as we see in the next paragraph.

In [15], the authors analyze some of existing supervised and unsupervised learning techniques from the perspective of the challenges posed by processing IoT data. Their work aimed at helping to choose a suitable algorithm for some data analytic tasks. For example, to find unusual data points and anomalies in smart data, the support vector machine (SVM) algorithm is suggested. To predict the categories (labels) of data, neural networks use approximation functions to obtain the desired output. To find hidden patterns or intrinsic structure in data, unsupervised learning applies the clustering technique. K-means is the most widely used clustering algorithm for exploratory data analysis.

The K-nearest neighbors (KNN) is one of the most naive ML algorithms being used for classification and regression problems. For each new data point, KNN searches through all the training data for the K closest entries (nearest neighbors). Once the K closest data points are identified, a function will assess the distance (or degree of separation) between the new data point and its neighbors to determine the class of this new data point. Unlike KNN, Naïve Bayes (NB) does not need to store the training data. The NB algorithm learns the statistics for each predictor (attribute or feature) of the training data and applies the Bayes' Theorem (*all predictors are independent from each other*) to build the *predictor-class probability*. A mathematical function uses these probabilities to determine the class of the new data point according to its predictors. In [16], the author uses a Raspberry Pi to run the KNN and NB algorithms to predict the condition of a vehicle considering the information about the engine, coolant, etc.

Decision tree algorithms are used to develop classification models in the form of a tree structure. They are quite popular among data analysts because they are easy to read and understand. A decision tree algorithm chooses the outputs according to the possible *decision paths* represented in a hierarchical tree structure. To construct the tree structure, a data set is decomposed into smaller subsets based on the available predictors. The goal is to obtain subsets that contain only one class of outcomes. Unlike the other models, decision trees can easily handle a mix of numeric (e.g. room temperature) and categorical (e.g. user's perception of temperature – hot/cold) predictors and can even classify data for which attributes are missing. As an example of application, in [17], the authors used a wearable device equipped with motion and orientation sensors (accelerometer and gyroscope) to detect several activities of a person when in the living room. Some activities were: watching TV, reading newspaper, chatting, taking a nap, listening to music, and walking around. They used the decision tree and Hidden Markov model (HMM) algorithms to build the model for activity detection.

An artificial neural network (ANN), or neural network for short, is a technique inspired by the way the brain operates – a collection of neurons wired together. The neurons that represent processing units are organized in layers – *input layer*, one or more *hidden layers* and *output layer*. Each neuron in a layer receives the outputs of the previous layer neurons, does a calculation with an activation function, and then forwards the results to the next layer neurons. Figure 12.3 shows an example of feed-forward neural network with two hidden layers. The algorithms used in

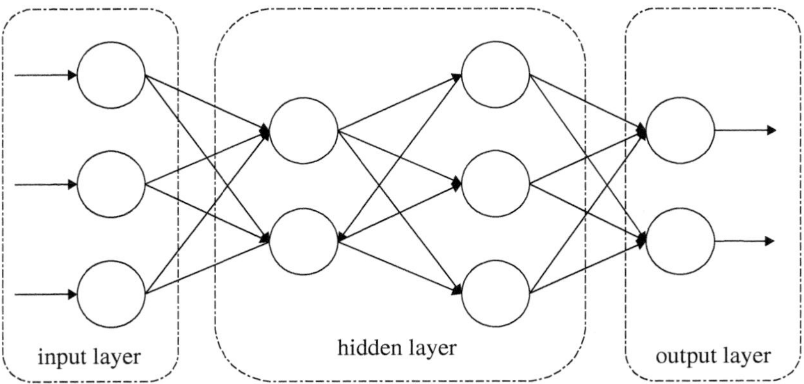

Figure 12.3 A multilayer feed-forward neural network.

neural networks form a framework for many ML techniques capable of dealing with complex data inputs. An example of application of this technique is described in [18]. The authors developed an intelligent IoT-based hydroponics system for tomato plants. The system gathers several data about the monitored plant such as pH, temperature, humidity, and light intensity. Then, employing ANN to analyze such data, the system can predict the appropriate actions to automatically control the hydroponic system.

Reinforcement learning is an ML method based on trial and error technique. A processing unit called *agent* develops his learning experience according to the actions he takes in an interactive *environment*, which returns a value indicating whether an action was right or wrong. Unlike supervised learning, where the desired output is the correct answer for performing a data analysis, reinforcement learning uses rewards and punishment as signals for positive and negative behavior. In the IoT context, for example, there are various researches on applying reinforcement learning to actuators embedded in urban traffic signals with the purpose of alleviating traffic congestion in the Smart City domain application. The authors in [19] present a thorough literature survey of adaptive traffic signal control (ATSC) approaches using three reinforcement learning algorithms, namely: TD-learning, Q-learning, and Sarsa learning.

An important issue for applying ML in the IoT context regards the capacity of the algorithms to learn from data streams that evolve over time and are generated by nonstationary distributions. Many ML algorithms are designed for data processing tasks in which data is completely available and can be loaded any time into memory. Such a condition does not hold most cases of IoT systems that produce large data streams. Such a problem requires *incremental learning algorithms*, which can modify the current model whenever new data is obtained [20]. In this case, the

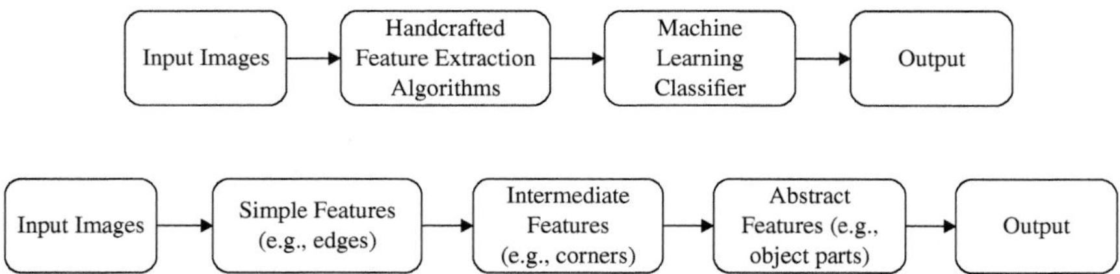

Figure 12.4 Traditional machine learning (a) and deep learning (b) approaches to image classification.

objective is to permanently maintain a precise decision model consistent with the current state of the process that generates data. In [21], the authors identify the following four computational model properties for such learning systems:

1. Incremental and real-time learning;
2. Capacity to process data in constant time and using limited memory;
3. Limited access to data already processed;
4. Ability to detect and adapt the decision model to concept changes.

In [22], the authors evaluate a lightweight method based on incremental learning algorithms for fast classification in real-time data stream. They use the proposed method to do outlier detection via several popular learning algorithms, like decision tree, Naïve Bayes, and others.

12.3.2 Deep Learning

In recent years, with the increasing computational capacities, new generation IoT devices have been able to apply more advanced neural networks to capture and understand their environments. For example, in a smart home application, the security system can unlock the door when it recognizes its user's face. Using a cascade of nonlinear processing units (layers) for feature extraction and transformation, a *deep learning* (DL) method can extract from an image a hierarchy of concepts that correspond to different levels of abstraction. Figure 12.4 depicts the main aspects of this technique that distinguish it from traditional ML feature extraction and classification methods [23].

DL architectures such as Deep Neural Network (DNN) and Convolutional Neural Network (CNN) have been applied to many fields including computer vision, audio recognition, speech transcription, and natural language processing, where they have produced results comparable to human experts. A DNN is basically a feed-forward neural network with several hidden layers of interconnected neurons (or processing units). Collectively, the neurons of the hidden layers are responsible for transforming the data of the input layer into the desired output

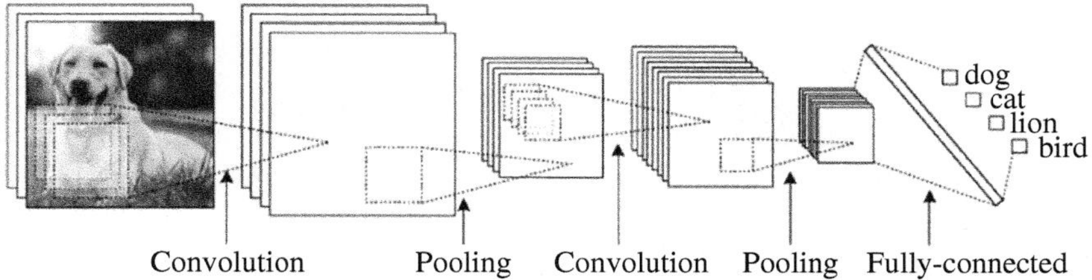

Figure 12.5 Typical CNN architecture [25].

data (inference classes) of the output layer. Each hidden layer neuron has a vector of weights with a size equal to the number of inputs as well as a bias. These weights are optimized during the training process. The hidden layer neuron receives the inputs from the previous layer, calculates a weighted summation of these inputs, and calls an activation function to produce the output, passing the resulting sum as a parameter. These weights are what make each neuron unique. They are fixed during the test of the network, but during the supervised training these weights are changed in order to *teach* the network how to produce the expected output for the respective data inputs.

A CNN is a class of ANN and an alternative to DNN when applied to analyzing visual imagery due to scalability issues (e.g. high resolution images imply thousands of neurons) and the translation-invariance property of such models. DNN might not learn the features that undergo minor changes in some images, like a rotation of the face. CNN has solved these problems by supporting *sparse interactions*, *parameter sharing*, and *translation-equivariance computations* [24]. As any ANN, a CNN also consists of input layer, multiple hidden layers, and an output layer. The hidden layers of a CNN typically consist of one or more convolutional and pooling layers, and finally a fully connected layer equivalent to those used in DNNs. Convolutional layers apply convolution operations to the input, called *filters*, extracting the features from data, reducing the number of parameters and passing the result to the next layer. In the training process, a CNN automatically learns the values for these filters, which represent hierarchical feature maps (e.g. edges, corners and object parts) of the input images. The pooling layers operate on the feature maps produced by the filters combining the outputs of neuron clusters at one layer into a single neuron in the next layer. *Max* and *average pooling* uses the maximum and average value, respectively, from output of a cluster of neurons at the prior layer causing dimensionality reduction and the representations to be invariant to translations. The last layers (i.e. a DNN fully connected) are applied to complete classification. Figure 12.5 presents a typical CNN architecture for image classification.

Many modern devices, such as smart phones, drones, and smart cars, are equipped with cameras. The deployment of embedded CNN architecture on such devices is enabling a variety of smart applications. For example, using the smart phone's camera to detect plant disease, drone aerial images to detect traffic congestion, and collision prediction using vehicle's camera, just to name a few. In [9], the authors present a complete survey on applying DL for IoT data and streaming analytics, analyzing different DL architectures for different IoT scenarios. They also analyze how DL was embedded in solutions using intelligent IoT devices, as well as compare deployment approaches in the fog and cloud.

Besides the complexity of ML algorithms, other issues to be addressed are where and how to deploy the CI techniques considering the high volume of data and computational resources required to run the algorithms, especially on resource constrained fog devices. The next section presents some approaches recently researched on this regard.

12.4 Challenges for Running Machine Learning on Fog Devices

Traditionally, ML algorithms run on powerful computers hosted in cloud data centers to accommodate their high demand for computation, memory, and power resources. In the context of IoT and fog computing, data streams are generated (and eventually processed) by devices located at the edge of the network that have limited power and hardware resources. Smart applications that use large sets of sensing data, such as audio and images, aggravate the memory demand and bandwidth consumption. Therefore, applying modern data analytics processing using ML techniques without imposing communication overheads is a challenge, and researchers have recently investigated how to deploy CI on fog devices in an efficient way.

In this context, the challenges for running any resource demanding program on fog devices are related to the following aspects: processing time, memory demand, bandwidth, and power consumption. Generally, processing time and bandwidth consumption are directly related to the performance and the energy required to run the application, so any reduction in these two factors is important to save energy and increase the overall efficiency of the system. Besides, there are many other aspects to be analyzed when designing a hardware solution for an IoT system. Energy efficiency is one of the key characteristics for IoT products (nobody wants to recharge a smartwatch three times a day). System boards that consume a lot of power hamper battery autonomy and impose an increase in product size. On the other hand, low power system boards that are generally low performance may not be able to run applications that are sensitive to delays. Thus, the hardware industry has continually researched how to balance these various aspects to better deliver system boards for future technological solutions.

In 2015, the authors in [26] published an initial study aimed to analyze the development of embedded and mobile devices that could run some DL techniques. They experimented some device kits running two DNNs and two CNNs models to process audio and image sensor data, respectively. The DNNs assessed in the experiments were: Deep Keyword Spotting (KWS) [27], which was used to perceive voice specific phrases (e.g. "Hey Siri"), and DeepEar (Lane et al., [28]), which was composed of three coupled DNNs offering separate recognition tasks – emotion recognition, speaker identification, and ambient sound classification. The two CNNs were: AlexNet [29], an object recognition model supporting more than 1,000 object classes (e.g. dog, car), which in 2012 obtained state-of-the-art levels of accuracy for the well-known data set ImageNet[4]; and SVHN [30], a specialized model for recognizing numbers in complex and noisy scenes like Google Streetview images. They tested these models in three device kits:

1. *Intel Edison*. The smallest and the weakest computing power of all experimented kit (500 MHz dual-core Atom "Silvermont" central processing unit [CPU] assisted by a 100 MHz Quark processor, 1 GB of RAM).
2. *Qualcomm Snapdragon 800*. Widely used in smartphones and tablets (three processors: a Krait 4-core 2.3 GHz CPU, an Adreno 330 GPU and a 680 MHz Hexagon DSP, 1 GB of RAM).
3. *Nvidia Tegra K1*. Developed for extreme GPU performance in IoT context (Kepler 192-core GPU, 2.3 GHz 4-core Cortex CPU, 2 GB of RAM).

The authors measured the execution time and energy consumption of each hardware platform for every model performing 1,000 separate inference tasks. Table 12.2 shows the average execution time for each model using the processing units of the hardware platforms. The first insight is about the feasibility for running such models on these devices, specially the weakest of the processors – Intel Edison. AlexNet has 60.9 million parameters, five convolutional layers, three pooling layers, and two hidden layers summing 4,096 neurons. Running AlexNet on Edison and Snapdragon toke 238 and 159 seconds, respectively, and consumed 110 and 256 J, respectively, from a 2,000 mAH battery. On the other hand, running AlexNet on Tegra's GPU took only 49 ms and 0.232 J from the battery. Other good results were the Deep KWS, the smaller model, consuming only 12.1 mJ on Edison and 10.2 Mj on Snapdragon's DSP with execution times below 64 ms.

Those experiments showed that running a heavy DL model like AlexNet on low-capacity computer is still possible, but the response time and energy consumption will probably make the solution infeasible for most Smart IoT applications.

4 http://www.image-net.org

Table 12.2 Execution time (ms) for running DL models in three hardware platforms.

				Tegra		Snapdragon		Edison
Model	**Type**	**Qty. parameters**	**Architecture[a)]**	**CPU**	**GPU**	**CPU**	**DSP**	**CPU**
DeepKWS	DNN	241 K	h:3; n:128	0.8	1.1	7.1	7.0	63.1
DeepEar	DNN	2.3 M	h:3; n:512 or 256	6.7	3.2	71.2	379.2	109.0
SVHN	CNN	313 K	c:2; p:2; h:2; n:1728	15.1	2.8	1615.5	—	3352.3
AlexNet	CNN	60.9 M	c:5; p:3; h:2; n:4096	600.2	49.1	159 383.1	—	283 038.6

a) c: convolution layers; p: pooling layers; h: hidden layers; n: neurons in hidden layers.

One way to overcome this problem is to use GPUs. CNN algorithms require a huge number of mathematical operations using large matrices of floating-point numbers that are intensive time-consuming if performed in pipeline mode by the CPU. However, the hundreds or thousands of processing cores of GPUs can easily speed up the execution of a huge set of mathematical operations such as the ones required by CNN algorithms, as we can see in experiments using AlexNet running on Tegra's GPU compared to its CPU.

Considering the continuous evolution of the IoT and Edge devices and how they can be deployed for executing various applications that demand intelligent processing, our research group developed a study about the hardware resources found in most of these devices, analyzing their capacities for running ML algorithms [14]. The result of this study was a classification that matches hardware resources with learning algorithms. Table 12.3 presents the proposed classes of devices with the respective suitable algorithms. Devices class 1 are the smallest and least computational powerful of them, and its role is only to produce data. From this perspective, class 1 devices are known as the IoT sensors by the research community, with some basic computational resources. At the other end, class 5 devices are the highest powerful computational equipment and can be considered as the unlimited resources available in cloud data centers. The experiments done with the devices of class 2, 3, and 4 showed that they have capacity to perform the three levels of information fusion: measurement, feature, and decision, respectively. Unlike class 1 devices that have microcontrollers, they have true CPU, RAM memory and a multitask operating system. Devices class 3 and 4 are the ones really used for running ML algorithms on the fog. The main difference between them is that class 4 devices have a powerful GPU for parallel processing, which is very important for the execution of algorithms for training DNNs, for instance. Class

Table 12.3 Classification of smart IoT devices according to their capacities [23].

Class	Hardware capacity	Power consumption	Suitable algorithms	Main application
1	No storage Low CPU and memory	≤1 W	Basic computation	Data generation
2	Storage ≤4 GB Memory ≤512 MB CPU single-core	≤2 W	Basic statistic	Measurement level fusion
3	Storage ≤8 GB Memory ≤2 GB CPU quad-core	≤4 W	Classification/ Regression / Estimation	Feature level fusion
4	Storage ≥16 GB Memory ≥4 GB CPU and GPU	≤8 W	Prediction/ Decision-making	Decision level fusion
5	Very large	Very high	Any	Autonomous system

3 device can use a trained neural network to make the mid-level data fusion and extract the relevant features of the data, like a detection of an object. Class 4 device can go further and taking decisions using a set of higher-level data fusion.

Another important aspect is the power consumption of the devices. On the one hand, IoT devices are to be deployed in nonrefrigerated environments and with limited power sources, less computation requires less power and yields less heat. On the other hand, devices with high computational power are more expensive, require a lot of energy to work and yield more heat, so that they should be used with care in IoT projects. The next section presents some solutions brought by industry to address these issues in IoT system design.

12.4.1 Solutions Available on the Market to Deploy ML on Fog Devices

Thinking about the potential that ML can bring to IoT applications, the industry has created new hardware specifically tailored to develop ML-based solutions. For example, OpenMV[5] is a tiny open hardware kit for IoT developers with an embedded camera that can detect face and find eyes using built-in feature detection algorithms and expending less than 200 mA of energy. This kit facilitates, for

5 https://openmv.io

Figure 12.6 OpenMV, a machine vision kit for IoT developers.

instance, the development of smart applications that require authenticating users by their faces.

The OpenMV kit runs an operating system based on MicroPython,[6] a version of Python 3 to run on microcontrollers and constrained devices that includes a small subset of the Python standard library. It also comes with an IDE (integrated development environment) that organizes in tiled windows the code, the image from camera, some graphs and a terminal to interact with its operating system. Figure 12.6 shows one of the OpenMV model kits.

In [31], the authors presented a method to improve the line-finding algorithm to control a self-tracking smart car using OpenMV as the main development platform. The improved method uses Hough line transformation to plan the best path solution based on detecting two black lines under white background as the driving road. The OpenMV used was equipped with an OV7725 camera that is capable of taking 640 × 480 8-bit images at 60 FPS (frames per second) or 320 × 240 at 120 FPS, which is very suitable for smart car algorithms based on image processing. The results showed that the improved algorithm performs better than the traditional one, especially when cruising a continuous bend road.

Recently, the Raspberry Pi,[7] a palm-sized single-board computer (SBC), has been used as the edge device to process IoT complex data and streams. In [32], the authors developed a facial recognition system for use by police officers in smart cities. This system uses a wireless camera mounted on the police officer's uniform to capture images of the environment, which is passed to a Raspberry Pi 3 in the officer's car for face detection and recognition. To accomplish the facial recognition, they initially use the Viola-Jones object detection framework[8] to find all the faces present in the live video stream sent by the camera. Then, the ORB[9] (Oriented FAST and Rotated BRIEF) method is executed to extract from the identified faces

6 http://micropython.org
7 https://www.raspberrypi.org
8 https://en.wikipedia.org/wiki/Viola-Jones_object_detection_framework
9 https://en.wikipedia.org/wiki/Oriented_FAST_and_rotated_BRIEF

Figure 12.7 NCS based on Intel Movidius Myriad 2 Vision Processing Unit (VPU).

all their features, which are transmitted to a trained SVM classifier in the cloud. For the purpose of saving transmission power and bandwidth, only the features extracted from the face are sent to the cloud.

In order to run DL models on devices like Raspberry Pi, Movidius, a company acquired by Intel in 2016 that designs specialized low-power processor chips for computer vision, has launched the Neural Compute Stick (NCS) [33],[10] a USB device (Figure 12.7) containing Intel's Myriad 2 VPU (vision processing unit). The Neural Compute Engine in conjunction with the 16 SHAVE[11] cores and an ultra-high throughput memory fabric makes Myriad VPU, a consistent option for on-device DNNs and computer vision applications. Most drones in the market have used this technology to interpret the images they capture and trigger actions like following an object in movement or take a picture when the user makes some gestures. Intel has distributed the OpenVINO toolkit[12] for developers to profile, tune, and deploy CNNs on DL applications that require real-time inferencing.

To probe the power of this NCS as a replacement of a GPU card, the authors in [34] designed a CNN model to detect humans in images for a low-cost video surveillance system. To run the pretrained model on a desktop PC, they tested two SBC: Raspberry Pi 3 (1GB RAM) and Rock64[13] that has the same CPU (ARM Cortex-A53) but 4GB RAM. The model used to extract the features and classify the images was the MobileNet [35] that uses the technique known as Single Shot MultiBox Detector (SSD) [36]. MobileNet was proposed by Google as a lightweight CNN suitable for Android mobile devices and embedded systems. It uses an architecture called *depthwise separable convolutions*, which significantly reduces the number of parameters when compared to the network with normal convolutions with the same depth in the neural networks. They used the software development kit (SDK) of NCS to load the pretrained CNN onto the USB device in order to speed up the human detection processing of the images captured from a camera connected to the SBC. The detected image of a human is then sent to another CNN model to distinguish specific persons from strangers. The image data set used in this second model has 10,000 photos of five persons in different angles. This second

10 https://developer.movidius.com
11 https://en.wikichip.org/wiki/movidius/microarchitectures/shave_v2.0
12 https://software.intel.com/openvino-toolkit
13 https://www.pine64.org

model was created and pretrained in Tensorflow [37] on the desktop PC as well. In the end, they present the execution time of human detection (first CNN used) on Rock64 and Raspberry Pi 3, first using only the CPU and later with the NCS running the model. On Raspberry Pi 3, the execution time is five times faster with the NCS, reducing from 0.9 to 0.18 seconds. On Rock64 this time drops from 0.5 to 0.15 seconds with the NCS, six times faster.

12.5 Approaches to Distribute Intelligence on Fog Devices

In a large-scale system, such as Smart Cities, thousands of IoT devices will continuously generate huge amounts of data. In order to extract valuable information for applications, all this data has to be collected and processed in a timely and efficient way. As we pointed out in the Section 12.1, in the traditional three-tier architecture for IoT systems, data must be sent and processed in cloud data centers, what can be infeasible especially for time-sensitive multimedia applications due to the overhead of communications and delays. Therefore, applying ML techniques close to the data sources for extracting useful information and reducing the bandwidth consumption and delay is a promising approach to mitigate these issues. Other important advantages related to the distributed data learning process are presented by [38]:

- In a large-scale system, a possible approach to improve the accuracy in inferences is adopting different learning processes to train multiple classifiers in distributed data sets. Each classifier will be using different learning biases and joining them in the same inference process can soften the inefficient characteristics of these biases.
- For large-scale smart applications where the data volume is huge as well as processing is challenging, the divide-and-conquer approach may be the best strategy. Due to memory, storage, and processing time limitations, each processing unit can work collaboratively with different data partitions, thus reducing the overall computational cost. In addition, this strategy improves the availability of classifiers since smaller processing units can be more easily replaced in case of failures.

To investigate the advantages of distributed learning for extracting knowledge from a data set collected from sensors in multiple locations, the authors in [39] proposed and developed a novel framework. Specifically, they exploit the technique called hypothesis transfer learning (HTL) [40] where each data producer performs an initial learning process yielding a partial model (knowledge) from its own data set. In order to obtain a unique model (global inference), the partial models have to be shared among the data producers to refine their partial models "embedding knowledge" from the other partial models. On the one hand, the benefit of this approach is that the amount of data to be transmitted over the network is minimized since the partial models are much smaller than the raw data. On the other hand, the accuracy of the final model has to be as close as possible to a model trained directly over all data sets. In order to experiment the approach, they used a data set of human activities (standing, sitting, lying, walking, stand-to-sit, etc.) collected through smart-phones' sensors (accelerometer and gyroscope) wore on the waist of 30 volunteers. The results showed that they reached both goals of minimizing the network traffic while keeping a high accuracy. With the HTL solution, the accuracy of the final model using only smart-phones was comparable to a cloud-trained model with all data sets, but with the network overhead having a drastic reduction up to 77%. In a later article [47], the same authors explore the three-tier architecture of fog computing introducing data collectors (DCs) to perform the partial learnings on the fog. The aim of the research was to evaluate the trade-off between accuracy and network traffic when varying the number of data collectors (DCs) located on the fog. The obtained results showed that increasing the number of data collectors decreases the network traffic and the accuracy of the HTL solution does not vary significantly with the number of DCs. Therefore, they conclude that a distributed HTL solution using the fog computing approach is the most promising technique.

In [41], the authors introduced EdgeSGD, a variation of the stochastic gradient descent (SGD) algorithm to provide analytics and decentralized ML on nodes at the edge of the network. With this algorithm, they could estimate on each node the feature vector for linear regression using multiple training subset of the entire data set. They used the proposed algorithm to learn and predict seismic anomalies through seismic imaging from sensors placed in different locations. They evaluated the performance of the algorithm on an edge computing testbed – a cluster composed of 16 SBC BeagleBone.[14] They also compared the performance of the proposed solution with other distributed computation frameworks such as MapReduce, decentralized gradient descent (DGD) [42], and EXTRA [43], examining the aspects of fault tolerance and communication overhead. Such a comparative analysis showed that edge processing with EdgeSGD outperformed the other frameworks in terms of execution time while having the same robustness of them.

There is a consensus that with the rapid evolution of hardware, which is being smaller and more energy-efficient, edge/fog devices will be increasingly powerful. In [44], the authors try to explore this potential by creating an architecture for distributed stream processing. The proposed architecture creates a cluster of edge devices to process stream images (video) in parallel for a faster response time when

14 https://beagleboard.org/black

compared to off-loading data to cloud platforms. They implement the proposed solution using the open source framework Apache NiFi,[15] where a *data flow* represents a chain of processing tasks with data streamed in a "flow-based programming" (FSB). The processing units are called *processors*, and each processor can perform any data operation and has an input and an output port, which serve to interconnect the processors creating a data flow topology. There are more than 100 built-in processors in NiFi and it is possible to develop and plug in custom-made processors. To probe the proposed approach, the authors applied a CCTV-based intelligent surveillance system (ISS) to recognize persons by their faces from the camera images. For this application, they developed three processors:

1. *CaptureVideo* (CV). The data flow starts with this processor. It uses a camera to capture the video stream continuously. At a given frequency, this processor splits the video stream into separate frames and sends them to cluster nodes. Thus, this processor orchestrate the tasks off-loaded to the cluster as a coordinator;
2. *DetectFaces* (DF). For each frame received from the CV processor, this processor execute a facial detection algorithm, cropping the detected faces and placing them on other objects. Then, these objects are sent to the next processor;
3. *RecogniseFaces* (RF). Upon arrival of a new object containing a face, this processor performs a facial recognition algorithm to identify the detected person. This processor is trained with a data set with a small group of human faces and outputs a prediction value (10,000 indicates a match) for each face in this group.

They deployed the application on a laptop (CPU Intel i7-4510U quadcore, 16GB RAM) running the CV processor and a local virtual cluster consisting of five VirtualBox instances of the Raspberry Pi acting as edge nodes. Each node is an virtual machine running the Raspbian OS with 1GB RAM and up to 20% of the host CPU. The edge nodes execute the DF and RF processors. In the first tested topology, all edge nodes have the same training data set and the coordinator sends each frame to a unique cluster node for face detection and further recognition. Another topology tested was splitting the training images in separate subsets given to each cluster node and the coordinator sends the frames to all nodes, expecting at least one of them to recognize the face in each image. This second topology was designed to bring parallelism and some performance benefits compared to the first one, since the larger the training set, the greater the time to reach an inference. They also run the experiment with a third setup running a software to detect and recognize faces on a cloud-based computer containing the whole data set of the trained images. For each experiment they measure the *time delay* – the total time for the coordinator to receive an inference (result of the face recognition task) starting when a

15 https://nifi.apache.org

frame is captured by CV processor. According to the authors, this period of time encompasses all possible delays related to network latency and data serialization. The solution designed with the proposed architecture of an edge cluster had a far superior performance compared to processing in the cloud. The results shown that, when using the complete training set, it performed up to five times faster, and nine times faster with the split training set. Thus, they conclude that splitting the training set is a good strategy for gaining performance since with a training set five times smaller they have achieved a performance 40% faster (compared to cloud-only solutions).

Analyzing this last discussed work [44] and aiming at improving some aspects of their proposed architecture, our research group developed a novel approach to define the topology of Fog devices for executing ML techniques. As we discussed in Section 12.4, the hardware resources required by different ML algorithms can vary from a single CPU and few megabytes of RAM to a powerful GPU and gigabytes of RAM. Our approach hierarchically organizes the fog devices in three clusters considering the classification of the devices presented in Table 12.3. Combining the devices of classes 2, 3, and 4 with the three levels of information fusion (presented in Section 12.2.3) – measurement, feature, and decision – each cluster of devices will extract valuable information from IoT data streams through these respective data abstraction layers (Figure 12.8). On the bottom of the hierarchy there are the IoT devices with their sensors generating data. This raw data is collected by the fog device responsible for measurement level fusion (class 2) that, by running some statistical algorithms, remove the noise and outputs to the upper layer new pieces of data that are more accurate/valuable than its data input. In the next layer, the class 3 device extracts the feature maps of the data using some

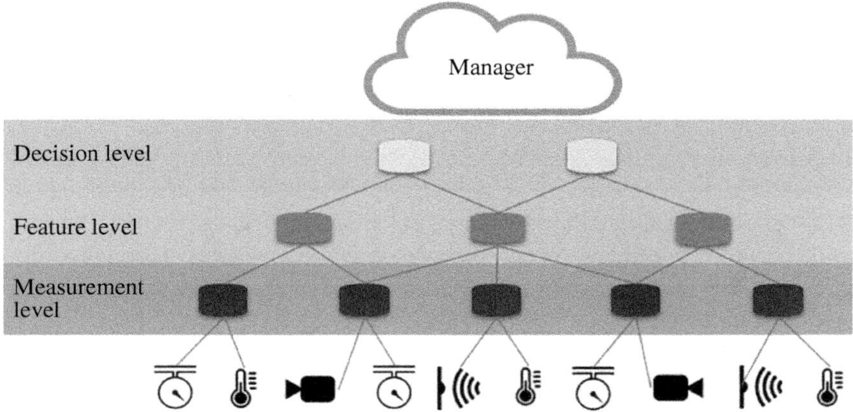

Figure 12.8 Fog topology with devices grouped by levels.

classification, regression, or estimation algorithm passing them to the upper layer. In the decision level, a prediction algorithm yields knowledge from its input data allowing the device to execute decision-making processes in order to take some actions.

To illustrate our approach, let us consider, for example, its application to an ISS in a Smart City scenario. The CCTV-cameras are the class 1 devices generating the videos as the raw data. Devices of classes 2, 3, and 4 can be single-board computers (e.g. Raspberry Pi, Nvidia Jetson,[16]) placed in the network close to the cameras (i.e. in the fog layer) in order to reduce bandwidth and latency. A class 2 device in this scenario is responsible for detecting objects in the video through their movements in the images (i.e. pattern variation in the pixels of the background image) taking snapshots of them, which represents the more valuable data. In turn, a class 3 device receives the pictures with the objects detected and runs a DNN or CNN to classify them based on their feature maps. Finally, a class 4 device can run a deeper classifier to identify the object within a set of objects of interest. To speed up the inference process running the DL algorithm a Raspberry Pi could be equipped with the NCS presented in Section 12.4.1. Figure 12.9 presents a possible example of this multilevel data fusion for video processing from CCTV-cameras in a Smart City ISS application.

This distributed processing approach considerably reduces the amount of data to be sent over the network, and, in each level, the information yielded can feed different applications in the same context. For instance, using the example of ISS presented in Figure 12.8, the class 3 device could feed an application that counts the number of cars, buses, motorcycles passed by the camera region, whereas the class 4 device could feed another application by checking whether the detected car is a stolen vehicle. This way, all data exchanged among devices must be tagged with the sensor identification along other metadata of application control (e.g. direction of the object movement).

In terms of processing load, the devices of classes 3 and 4 are the ones that will run the DL models, and so will be the most powerful and consume more energy but will be in smaller numbers. This does not seem to be a problem, since class 3 and 4 devices can serve more than one device of the lower class in sporadic classification tasks. If the volume of classifications increases greatly, new instances of the devices should be added to address the new demand. A micro-service execution platform that can be instantiated on-demand would be a good way to deliver this distributed processing approach.

16 https://www.nvidia.com/en-us/autonomous-machines/embedded-systems-dev-kits-modules

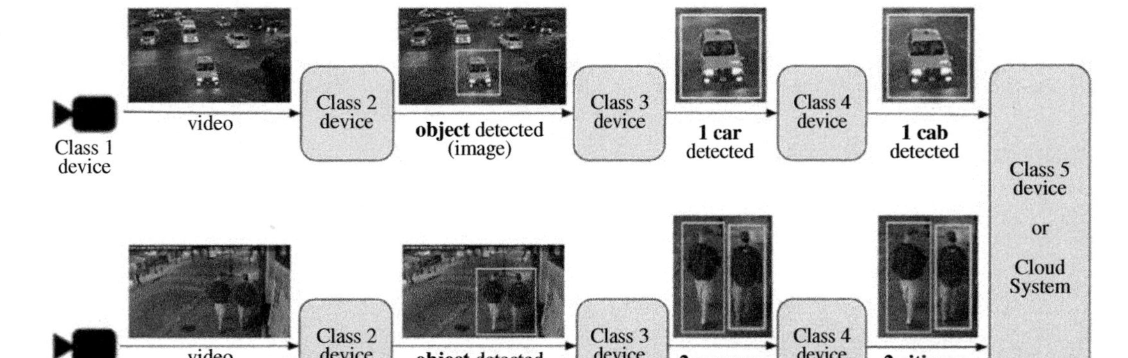

Figure 12.9 Multilevel data fusion for video processing.

12.6 Final Remarks

In the research community, there is a consensus that IoT is the leading technology to enable advanced, ubiquitous and intelligent systems. Using all kinds of data, provided by a growing number of heterogeneous devices (from wearables to cameras), a huge diversity of information can be extracted and interpreted to enhance the user experience. The gathering and analysis of this Big Data is one of the main challenges faced by the community nowadays. As we have seen throughout this chapter, ML techniques have been applied to perform advanced data analysis in a very efficient way in order to allow for valuable inferences even in resource-constrained devices. Looking at the opportunities of a new market, the industry has developed new hardware to overcome the limitations of running heavy ML algorithms on edge devices.

However, the limited hardware of edge/fog device is only part of the problem, and there are further issues that have been addressed by researchers developing IoT intelligent systems:

- *Accuracy* vs. *energy consumption.* When designing applications to run on devices with limited power, energy consumption is always a concern. In [45], the authors analyzed several recent and advanced DNN models to investigate how their accuracy, memory demands, and the time for the inference may interfere with energy consumption. They demonstrated that there is a hyperbolic relationship between the inference time (processing time of the entire neural network to reach a classification/prediction) and the inference accuracy. Some minor increments in accuracy may lead to long processing times and consequently a substantial increase in energy consumption. Thus, this indicated that energy constraint will limit the best achievable accuracy. They also pointed out that there is a linear relationship between the number of operations a neuron does and the inference time. Therefore, there is an obvious trade-off that must be carefully handled.
- *Lack of complete training data set for each domain.* It is well known that a neural network model is only really good when it has a large training data set as close as possible to all data to be interpreted. Having a complete training set is not an easy task especially in applications that use audio and image data sets from different domains. There are initiatives to create public data sets, like the ImageNet with 14 million images of 21,000 categories, to facilitate the deployment and tests of the ML algorithms developed by community. But in the context of a Smart City application, for example, car images in ImageNet are often useless because they do not reflect the images captured by the cameras normally positioned on poles. In voice recognition applications, a single human voice can vary greatly or be very similar to the voice of others, which makes their neural network training difficult.

- *Security and privacy*. The processing of personal data captured by the user's device sensors always opens up a discussion about users' privacy and security. Mechanisms to let the user know that their sensitive or personal data is being used to improve their experience are not always well understood. The main concern is the capture of this data by hackers and attackers who may exploit vulnerabilities in communications or processing and then blackmail users. Recently, Chinese startups have developed computer vision systems to capture and identify the face of any citizen in public spaces, demanded by the Chinese government.[17] This has sparked controversy in other countries regarding the privacy of people by governments and companies.

Finally, another important issue in developing smart IoT systems is the topology and communication protocols suite of the devices and networks. In [46], the authors present Edge Mesh as a novel paradigm to enable distributed intelligence in IoT through edge devices. This new paradigm integrates the best features of cloud, fog, and cooperative computing into a mesh network of edge devices and routers. Despite the advances toward a collaborative approach at the edge, the authors claim that there are many open research challenges for implementing Edge Mesh. Among them, we can mention the communication between different types of devices, to promote data sharing and to allow executing the intelligence of the application at the edge of the network.

Therefore, we can conclude with a final thought that the ML algorithms will evolve along with IoT devices, pushing forward the development of new hardware, software, and network protocols to address all the challenges presented in this chapter.

Acknowledgments

This work is partially supported by São Paulo Research Foundation – FAPESP, through grant number 2015/24144-7. Flávia C. Delicato, Paulo F. Pires, and Thais Batista are CNPq Fellows.

References

1 K. Ashton, That "Internet of Things" thing: in the real workd things matter more than ideas, *RFID Journal*, https://www.rfidjournal.com/articles/view?4986, 2009.

---

17 https://www.businessinsider.com/china-facial-recognition-tech-company-megvii-faceplusplus-2018-5

2 Yousefpour, A., Fung, C., Nguyen, T. et al. (2019). All one needs to know about fog computing and related edge computing paradigms: a complete survey. *Journal of Systems Architecture*, 98: 289–330.
3 Delicato, F.C., Pires, F.P., and Batista, V.T. (2013). *Middleware Solutions for the Internet of Things*. Springer.
4 Ali, N. and Abu-Elkheir, M. (2012). Data management for the Internet of Things: green directions. In: *IEEE Globecom Workshops*, 386–390.
5 Tsai, C., Lai, C., Chiang, M., and Yang, L.T. (2014). Data Mining for Internet of things: a survey. *IEEE Communication Surveys and Tutorials* 16 (1): 77–97.
6 Qiu, J., Wu, Q., Ding, G. et al. (2016). A survey of machine learning for big data processing. *EURASIP Journal on Advances in Signal Processing* 1: 67.
7 P. F. Pires, E. Cavalcante, T. Barros et al., A platform for integrating physical devices in the Internet of Things, 12th IEEE International Conference on Embedded and Ubiquitous Computing, 1. Washington, DC, 2014.
8 Garofalakis, M.N., Gehrke, J., and Rastogi, R. (2016). *Data Stream Management - Processing High-Speed Data Streams*. Springer.
9 Mohammadi, M., Al-Fuqaha, A., Sorour, S., and Guizani, M. (2018). Deep learning for IoT big data and streaming analytics: a survey. *IEEE Communication Surveys and Tutorials* 20 (4): 2923–2960.
10 Nakamura, E.F., Loureiro, A.A., and Frery, A.C. (2007). Information fusion for wireless sensor networks: methods, models, and classifications. *ACM Computing Surveys* 39 (3): 9.
11 Santos, I.L., Pirmez, L., Delicato, F.C. et al. (2015). Olympus: the cloud of sensors. *IEEE Cloud Computing* 2 (2): 48–56.
12 Qin, Y., Sheng, Q.Z., Falkner, N.J. et al. (2016). When things matter: a survey on data-centric internet of things. *Journal of Network and Computer Applications* 64: 137–153.
13 Akbar, A., Khan, A., Carrez, F., and Moessner, K. (2016). Predictive analytics for complex IoT data streams. *IEEE Internet of Things Journal* 4 (5): 1571–1582.
14 Neto, A.R., Soares, B., Barbalho, F. et al. (2018). Classifying smart IoT devices for running machine learning algorithms. In: *45° Seminário Integrado de Software e Hardware 2018 (SEMISH)*, 18–29. Natal – Brazil.
15 Mahdavinejad, M.S., Rezvan, M., Barekatain, M. et al. (2018). Machine learning for Internet of Things data analysis: a survey. *Digital Communications and Networks* 4 (3): 161–175.
16 Srinivasan, A. (2018). IoT cloud based real time automobile monitoring system. In: *2018 3rd IEEE International Conference on Intelligent Transportation Engineering (ICITE)*, 231–235. Singapore.

17 Lee, S.-Y. and Lin, F.J. (2016). Situation awareness in a smart home environment. In: *2016 IEEE 3rd World Forum on Internet of Things (WF-IoT)*, 678–683.

18 Mehra, M., Saxena, S., Sankaranarayanan, S. et al. (2018). IoT based hydroponics system using deep neural networks. *Computers and Electronics in Agriculture* 155: 473–486.

19 El-Tantawy, S., Abdulhai, B., and Abdelgawad, H. (2014). Design of Reinforcement Learning Parameters for seamless application of adaptive traffic signal control. *Journal of Intelligent Transportation Systems*: 227–245.

20 Faceli, K., Lorena, A.C., Gama, J., and Carvalho, A.C. (2015). *Inteligência Artificial: Uma Abordagem de Aprendizado de Máquina*. Rio de Janeiro: LTC.

21 Domingos, P. and Hulten, G. (2001). A general method for scaling up machine learning algorithms and its application to clustering. In: *Proceedings of 18th International Conference on Machine Learning*, 106–113. San Francisco, CA: Morgan Kaufmann.

22 Fong, S., Luo, Z., and Yap, B.W. (2013). Incremental learning algorithms for fast classification in data stream. In: *2013 International Symposium on Computational and Business Intelligence*, 186–190.

23 A. Rosebrock, Deep learning for computer vision with Python, PyImageSearch, http://PyImageSearch.com, 2018.

24 Goodfellow, I., Bengio, Y., and Courville, A. (2016). *Deep Learning*, www.deeplearningbook.org:. MIT Press.

25 G. Carlsson, Using topological data analysis to understand the behavior of convolutional neural networks. Retrieved from AYASDI: https://www.ayasdi.com/blog/artificial-intelligence/using-topological-data-analysis-understand-behavior-convolutional-neural-networks, June 21, 2018.

26 N. D. Lane, S. Bhattacharya, P. Georgiev et al., An early resource characterization of deep learning on Wearables, smartphones and internet-of-things devices, pp. 7–12, Seoul, South Korea: ACM, 2015.

27 Chen, G., Parada, C., and Heigold, G. (4091, 2014). Small-footprint keyword spotting using deep neural networks. In: *2014 IEEE International Conference on Acoustics, Speech and Signal Processing (ICASSP)*, 4087. IEEE.

28 Lane, N.D., Georgiev, P., and Qen, L. (2015). DeepEar: robust smartphone audio sensing in unconstrained acoustic environments using deep learning. In: *International Joint Conference on Pervasive and Ubiquitous Computing*, 283–294. Osaka, Japan: ACM.

29 Krizhevsky, A., Sutskever, I., and Hinton, G.E. (2012). ImageNet classification with deep convolutional neural networks. In: *Advances in Neural Information Processing Systems 25: 26th Annual Conference on Neural Information Processing Systems*, 1106–1114. Lake Tahoe, Nevada.

30 Y. Netzer, T. Wang, A. Coates, R. Bissacco, B. Wu, and A. Y. Ng, Reading digits in natural images with unsupervised feature learning, NIPS Workshop on Deep Learning and Unsupervised Feature Learning, 2011.

31 Mu, Z. and Li, Z. (2018). Intelligent tracking car path planning based on Hough transform and improved PID algorithm. In: *2018 5th International Conference on Systems and Informatics (ICSAI)*, 24–28.

32 Sajjad, M., Nasir, M., Min Ullah, F. et al. (2019). Raspberry Pi assisted facial expression recognition framework for smart security in law-enforcement services. *Inf. Sci.* 479: 416–431.

33 Intel Developer Zone, Neural compute stick, Retrieved from Intel Movidius: https://developer.movidius.com, 2018.

34 Yang, L.W. and Su, C.Y. (2018). Low-cost CNN Design for Intelligent Surveillance System. In: *2018 International Conference on System Science and Engineering (ICSSE)*, 1–4.

35 A. G. Howard, M. Zhu, B. Chen et al., MobileNets: Efficient Convolutional Neural Networks for Mobile Vision Applications, arXiv:1704.04861, 2017.

36 Liu, W., Anguelov, D., Erhan, D. et al. (2016). SSD: single shot multiBox detector. In: *Computer Vision – ECCV 2016*, 21–37.

37 M. Abadi, A. Agarwal, P. Barham et al., TensorFlow: large-scale machine learning on heterogeneous distributed systems, *CoRR*, 2016.

38 Peteiro-Barral, D. and Guijarro-Berdiñas, B. (2013). A survey of methods for distributed machine learning. *Progress in Artificial Intelligence* 2 (1): 1–11.

39 Valerio, L., Passarella, A., and Conti, M. (2016). Hypothesis transfer learning for efficient data computing in smart cities environments. In: *2016 IEEE International Conference on Smart Computing (SMARTCOMP)*, 1–8. IEEE Computer Society.

40 Kuzborskij, I., Orabona, F., and Caputo, B. (2015). Transfer learning through greedy subset selection. In: *International Conference on Image Analysis and Processing*, 3–14. Springer.

41 Kamath, G., Agnihotri, P., Valero, M. et al. (2016). Pushing analytics to the edge. In: *GLOBECOM*, 1–6. IEEE.

42 Yuan, K., Ling, Q., and Yin, W. (2016). On the convergence of decentralized gradient descent. *SIAM Journal on Optimization*: 1835–1854.

43 Ling, Q., Wu, G., and Yin, W. (2015). EXTRA: an exact first-order algorithm for decentralized consensus optimization. *SIAM Journal on Optimization* 25 (2): 944–966.

44 Dautov, R., Distefano, S., Bruneo, D. et al. (2017). Pushing intelligence to the edge with a stream processing architecture. In: *2017 IEEE International Conference on Internet of Things (iThings) and IEEE Green Computing and*

Communications (GreenCom) and IEEE Cyber, Physical and Social Computing (CPSCom) and IEEE Smart Data (SmartData), 792–799. Exeter, United Kingdom: IEEE Computer Society.

45 Canziani, A., Paszke, A., Culurciello, E. An analysis of deep neural network models for practical applications, *CoRR*, 2016.

46 Sahni, Y., Cao, J., Zhang, S., and Yang, L. (2017). Edge mesh: a new paradigm to enable distributed intelligence in internet of things. *IEEE Access* 5: 16441–16458.

47 L. Valerio, A. Passarella and M. Conti, Accuracy vs. traffic trade-off of learning IoT data patterns at the edge with hypothesis transfer learning, 2016 IEEE 2nd International Forum on Research and Technologies for Society and Industry Leveraging a better tomorrow (RTSI), Bologna, 2016, 1–6. doi: 10.1109/RTSI.2016.7740634.

13

Fog Computing-Based Communication Systems for Modern Smart Grids

Miodrag Forcan and Mirjana Maksimović

Faculty of Electrical Engineering, University of East Sarajevo, 71123 East Sarajevo, Bosnia and Herzegovina

13.1 Introduction

Traditional and aging power grids are not able to successfully deal with the present increasing demands for the delivery of high-quality power in a secure, safe, reliable, and affordable manner. Growing energy demands, depletion of primary energy resources and increased utilization of renewable energy sources, sustainability, secure supply, and competitive energy prices are major driving forces for the conventional electric power grid evolution [1]. The recent technology development has influenced the traditional power grid, at both the transmission levels and distribution levels, making it smarter than ever before and capable of overcoming the existing challenges. This results in the appearance of the energy network integrated with the information and communication technologies (ICTs), known as Smart Grid (SG), that revolutionize generation, delivery, and use of electricity [2, 3].

The conventional electrical power grid is established on the communication infrastructure that transmits data gathered from a variety of sensing devices to a central control place for its processing and storage. This refers to the supervisory control and data acquisition (SCADA) system. The main drawback of existing SCADA systems is the lack of real-time control for the power distribution network. In the majority of distribution networks around the world, the concept of distribution system state estimation (DSSE) is not even deployed. Moving from the traditional power grid, where the demand predictions and supply monitoring mainly rely on historical data, toward the SG, which relies on the utilization of real-time data, demands a dynamic architecture, intelligent algorithms, and efficient mechanisms [4]. Hence, the SG is established on a huge number of a variety of smart

Fog Computing: Theory and Practice, First Edition.
Edited by Assad Abbas, Samee U. Khan, and Albert Y. Zomaya.

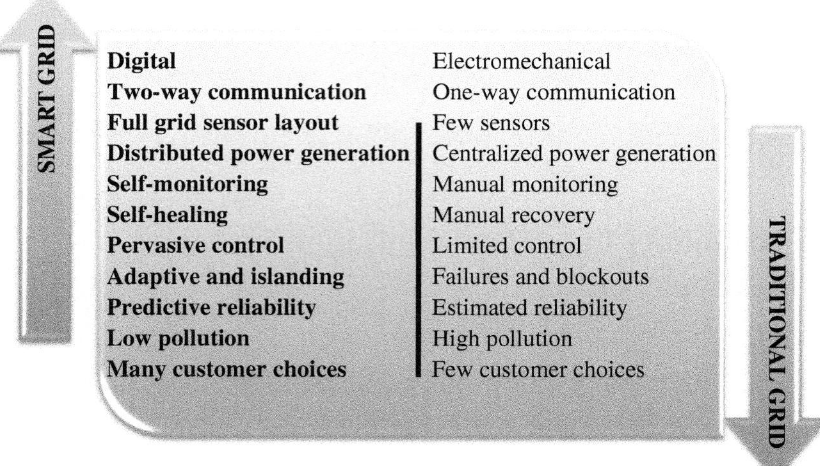

SMART GRID
Digital
Two-way communication
Full grid sensor layout
Distributed power generation
Self-monitoring
Self-healing
Pervasive control
Adaptive and islanding
Predictive reliability
Low pollution
Many customer choices
Electromechanical
One-way communication
Few sensors
Centralized power generation
Manual monitoring
Manual recovery
Limited control
Failures and blockouts
Estimated reliability
High pollution
Few customer choices
TRADITIONAL GRID

Figure 13.1 Comparison between the Smart Grid and the traditional grid.

devices, intelligent sensing, control methods, and integrated communications [5]. Combination of two-way of electricity flow with the fast and fully integrated bidirectional or two-way communication makes SG more secure, resistant, reliable, manageable, efficient, and sustainable than existing outdated and overburdened power grids (Figure 13.1).

The SG consists of the smart energy system (power grid, transmission grid, distribution grid, micro grid), smart information system (smart sensors, smart meters, phasor measurement units [PMUs], information metering, and measurement), and smart communication system. Compared to the traditional grid, the SG has an immense number of sensing and controlling devices deployed. Numerous sensing and monitoring devices deployed at the generation, transmission, distribution, and consumers' sides, produce an immense quantity of a heterogeneous and dynamic data in near real time, which must be processed and analyzed in an appropriate manner to obtain valuable insights and knowledge on which basis accurate and timely actions and predictions related to the power generation, supply, and demand can be performed. The SG can be looked at like a network of micro grids connected to each other via a cloud that, based on the collected and processed data, can run, disconnect, or self-heal [6, 7]. Hence, the successful operation of SGs that consists in decreased energy use, optimal work and coordination between all SGs' elements [8] depends on real-time, bidirectional, reliable, secure, and adaptive communication.

In other words, the advanced ICTs play a key role in the SG deployment and applications and are essential for the better understanding of the numerous aspects of SG and the overall energy sector. To achieve SG communication demands, there is a need to successfully overcome certain challenges such as bandwidth issues, storage capacity, information flow rate, advanced data analytics, security, etc.

This chapter presents an attempt to highlight existing communication technologies deployed in power grids and to introduce a new idea of the feeder-based communication system using fog/cloud computing for distribution management system (DMS). In the first section, an overview of communication technologies in SG is given and Internet of Things (IoT) application perspective is considered. The second section introduces the proposed feeder-based DMS using fog/cloud computing. The proposed communication scheme is tested for near real-time applications in the third section. Finally, useful remarks are given in the conclusion.

13.2 An Overview of Communication Technologies in Smart Grid

The goal of the SG is to enable nonstop, safe, and efficient delivery of high-quality power. This goal is achievable by using numerous smart devices (sensors and meters) and smart and secure communication networks. The communication infrastructure of the SG includes:

- *Advanced Metering Infrastructure (AMI)* – provides near real-time monitoring of power usage through an automated communication system between smart meters and a utility company. The data collected in AMI of the SG is huge, and its transmission to the utility center for processing and storage is quite challengeable. Some functionalities of AMI are near real-time monitoring and control of power quality, energy efficiency improvement, power pricing adaption, the energy consumption management, self-healing, improving the reliability, outages management/alert, etc. AMI doesn't have a standardized architecture, but it usually consists of smart meters, data concentrators, a utility center and bidirectional communication infrastructure [9], including web application with updated online data of consumers and mobile applications. In other words, AMI comprises Wide Area Measurement Systems (WAMSs), Sensor and Actuator Networks (SANETs) that can be grouped under a hierarchical structure on the basis of Home Area Networks (HANs), Neighborhood Area Networks (NANs), and Wide Area Networks (WANs) [4].
- *Wide Area Measurement Systems (WAMS)* – connect PMUs with larger and dynamic coverage, hence providing real-time, continuous, and synchronous monitoring and control of the electrical power grid [4].

- *Sensor and Actuator Networks (SANETs)* – is a network composed of the various devices able to sense and react to their SG environment. Sensors and actuators, controllers, and communication networks are the key components of SANET [4].
- *Home Area Networks (HANs)/Industrial Area Networks (IANs)/Business Area Networks (BANs)* – include home automation and building automation. These networks are deployed and operated within a small area (1–100 m), enable the connections of electrical appliances and other integrated devices/systems to the smart meters (mainly using ZigBee, Power Line Communications [PLC], Bluetooth, or Wi-Fi), supporting data rates up to 100 kbps [9–11].
- *Neighborhood Area Networks (NANs)* – cover transmission and distribution domains (100 m–10 km), enabling the two-way communication between a number of individual smart meters and a data concentrator/substation at data rates from 100 kbps to 10 Mbps [10]. Many technologies and networks, such as Worldwide Interoperability for Microwave Access (WiMAX), cellular communications, PLC, digital subscriber line (DSL), and optical communications, can be used in NAN [9].
- *Wide Area Networks (WANs)* – enable the communication of a number of data concentrators to a utility center. Utility centers consist of meter data management systems (MDMS), geographic information systems (GIS), configuration systems, etc. and their role is to collect, process, and store data from smart meters and interface with the suppliers [9]. WAN covers several hundred kilometers with the typical data rate from 10 to 100 Mbps [10], thus requiring long-range and high-bandwidth communication technologies, such as cellular communications, optical communications, and satellite communications.

The realization of communication tasks in SG is quite complex since the SG can be looked at as the large-scale network of heterogeneous systems. As an essential element of an efficient SG, ICTs must enable interconnection and real-time data exchange between the power utility and consumers, between the power utility and suppliers as well within the power utility. Hence, communication technologies must be chosen in a manner that enables reliable and secure bidirectional flow of near real-time information by satisfying required bandwidth, data rate, transmission range, security, latency, and reliability issues, adaptivity, etc.

The basic categorization of communication technologies is into wired technologies and wireless technologies. Comparison of communication technologies for the SG is shown in Table 13.1.

The wired communication technologies mainly used in the SG are:

- *Power Line Communications (PLC)* – enables two-way communication flow using existing power lines. PLC provides low-cost communication, and ubiquitous solutions, it is scalable and stable without impact on the environment.

Table 13.1 Comparison of communication technologies for the SG [10–13].

Technology	Max. theoretical data rate	Coverage	Applications	Advantages	Disadvantages
PLC	500 kbps (NB-PLC) 200 Mbps (BB-PLC)	150 km and more (NB-PLC) 1.5 km (BB-PLC)	NAN, WAN (NB-PLC) HAN, NAN (BB-PLC)	• Preexisting wide communication infrastructure • Physically separate from other telecommunication networks • Low costs	• High interference • Difficult to achieve high data rates • Complex routing • Slow developments
Fiber optic	40 Gbps	Up to 80 km	WAN	• High data rates • Long distance communications • Immunity to interference • Low latency	• High costs • Difficult to upgrade
DSL	100 Mbps	Up to 5 km	NAN	• Preexisting wide communication infrastructure	• Charges for using the network operator services • Not feasible for network backhaul
ZigBee	250 kbps	Up to 100 m	HAN, NAN	• Low costs • Low power consumption • High number of nodes	• Low data rates • Not scalable to larger networks
Wi-Fi	600 Mbps	Up to 100 m	HAN, NAN	• Low costs • High flexibility	• High interference • Comparatively high power usage

(Continued)

Table 13.1 (Continued)

Technology	Max. theoretical data rate	Coverage	Applications	Advantages	Disadvantages
Bluetooth	1 Mbps	Up to 100 m	HAN	• Low power consumption • Fast data exchange • Ease of access • No configuration requirement	• High interference
WiMAX	75 Mbps	Up to 50 km	NAN, WAN	• Low costs • Low latency • Suitable for thousands of simultaneous users	• Complex network management • The relatively costly terminal equipment • Licensed band requirements
Cellular	>1 Gbps (5G)	Up to 50 km	NAN, WAN	• Preexisting cellular communication infrastructure • Sufficient bandwidth • High data rates • Extensive coverage, • Lower maintenance costs • Strong security • Highly flexible • Reduced interference	• Monthly charges for using the cellular services • Difficult to guarantee the desired delay
Satellite	1 Mbps	Up to 6000 km	WAN	• Long distance • Highly reliable	• The very expensive terminal equipment • High latency

PLC is being used by several SG applications, such as substation automation, automatic meter reading (AMR) and AMI, remote monitoring, a vehicle to grid communications. The main challenges of PLC are related to security issues, near real-time communication, coexistence, and standardization [14]. PLCs are classified into [10, 12, 15]:

- *Narrowband PLC* – uses multicarrier schemes with frequencies in 3–500 kHz band, has a communication range of 150 km and more, and has low data rate (up to 10 kbps) or high data rate (up to 500 kbps), and
- *Broadband PLC* – operates in 2–250 MHz frequency band, has communication range up to 1.5 km, and data rate up to 10 Mbps for long-range communication and up to 200 Mbps for short-range communication.

• *Optical Communications* – are used between substation and utility companies control centers due to their high data rate (tens of Gbps) over longer distances, low latency, and immunity to interference. The main disadvantage of optical communications is the high cost of equipment, installation, and maintenance [10]. Optical networks can be:
  - *Active Optical Networks (AONs)* – use active splitting points, and supports maximal distance of 80 km.
  - *Passive Optical Networks (PONs)* – use passive splitting points and support maximal distance of 10–20 km. PONs are further classified into broadband passive optical network (BPON), Ethernet passive optical network (EPON), and gigabit passive optical network (GPON).

• *Digital Subscriber Line (DSL)* – is a low-cost solution based on data transmission over the traditional telephone lines. It provides maximum speed up to 100 Mbps and has been used for communication from homes to the utilities [16]. Achievement of higher data rates of longer distances is a major challenge DSL faces. There are two categories of DSL [10]:
  - Asymmetric DSL with the higher download speed than upload speed.
  - Symmetric DSL with equal upload and download speed.

On the other side are wireless communication technologies that have lower installation costs, and higher flexibility and scalability compared to the wired communication technologies. In the SG environment, the following wireless technologies can be used:

• *ZigBee* – is a short-range wireless communication technology based on IEEE 802.15.4 standard. It is a low-cost, self-organizing, secure, and fast communication technology with low power consumption and less interference. ZigBee operates at 868 MHz, 915 MHz, and 2.4 GHz, provides data rates 20–250 kbps, has coverage range up to 100 m, and is applicable for home automation, controlling and monitoring of a variety of loads, in AMI, etc. [17]. The main challenges the ZigBee faces are a limited lifetime, storage, data rates, processing capabilities, and interference issues [8].

- *Wi-Fi* – is a wireless communication technology based on IEEE802.11 standards that operates in license-free industrial scientific medical (ISM) frequency bands with low-cost radio interfaces. The highest data rate of Wi-Fi achieves 600 Mbps. Wi-Fi is generally utilized in HAN/local area network (LAN), and in the SG vision plays a key role in connecting smart devices with the Internet and for their energy usage management [10].
- *Bluetooth* – is a low-cost and short-range (typically 10 m but goes up to 100 m) wireless communication technology based on IEEE 802.15.1 standard. Bluetooth operates in ISM frequency band, it consumes low power and achieves data rates of 1 Mbps. In the SG applications, Bluetooth is used for power system monitoring and management, user-vehicle-charging system, etc. [15].
- *Worldwide Interoperability for Microwave Access (WiMAX)* – is a low-cost wireless broadband communication technology based on IEEE 802.16 standard. WiMAX operates at two frequency bands: 2–11 GHz (for non-line-of-sight applications), and 11–66 GHz (for line-of-sight applications), has large coverage area (up to 50 km), and high data rate (up to 75 Mbps). In the SG environment, WiMAX is suitable for NAN and WAN technology [8, 10, 15].
- *Cellular communications* – are widely employed in the SG environment due to well-established infrastructure and low maintenance, high data rates, greater bandwidth, and wider coverage area. The existing cellular communications technologies, such as global system for mobile communications (GSMs), general packet radio service (GPRS), 2G, 3G, 4G (and soon 5G) have been used for data transmission between different components in the SG, in AMI, automated demand response (ADR), outage management, etc. [8, 10].
- *Satellite communication* – with the help of satellites placed in low Earth orbits (LEOs), medium Earth orbits (MEOs), and geostationary Earth orbits (GEOs) enable communication in remote areas where other communication means are not applicable. Traditionally, satellite communication has been used for SCADA systems. High costs and very high latency are the main disadvantages of satellite communications. The satellite communications can be used as a backup in the SG environment [10].

It is also important to highlight some novel and promising technologies that have the potential to be used in the SG. Some of them are [10]:

- *Cognitive Radio (CR)* – will enhance the spectral efficiency, and network security, and improve long-range and large-scale data transmission,
- *Smart Utilities Networks (SUNs)* – hold a potential to be implemented in several SG applications, such as AMR, remote control, outage detection, monitoring reliability, and quality.

- *TV White Spaces (TVWS)* – with the potential to be used for smart metering applications, and transport telematics.

Other emerging technologies suitable for the realization of the IoT are also suitable for the low power wide area networks (LPWANs) that demand the transmissions of a few kb over long distances at low speeds in the most power-efficient way. As such these networks are suitable for smart metering applications in the SG environment [16]. These technologies are:

- *SigFox* – is a narrowband (or ultra-narrowband) technology that uses the 868 MHz band, and has very low noise effect.
- *LoRaWAN* – operates in the 433, 868, and 915 MHz bands, and supports data rates up to 50 kbps.
- *NB-IoT* – a narrowband IoT radio technology deployable in the GSM spectrum, capable of supporting numerous low throughput devices.

The IoT plays a key role in the realization of the SG vision as it has the potential to enable robust communication that is always available, scalable, secure, and interoperable. As an immense number of devices and components, such as sensors, smart meters, customer devices, mobile data terminals, substations, are building blocks of the SG, the data they produce are voluminous, heterogeneous, and dynamic. The data generated in the SG can be classified as [18]: meter usage data, event message data, operational data, non-operational data, and metadata. The effectiveness of the SG based on dynamic and optimized management and control of the energy demanded, consumed, and delivered [6] depends on the two-way flow of near real-time data and timely and adequate data processing. Adequate near real-time monitoring, control, and maintenance of a huge number of smart devices applied in the SG depend on the fast and accurate data collection, transmission, and analysis. This is the point where the concept of Big Data comes in. Artificial intelligence and a variety of data mining techniques enable the extraction of useful information and valuable knowledge from a massive amount of fast-generated and heterogeneous data. The processing of these data can be realized as batch processing (for static and non-real-time applications), stream processing (for real-time applications), and hybrid processing (suitable for real-time and non-real-time applications) [18, 19].

Generally, cloud computing solves many problems related to managing Big Data. In the large-scale systems, such as SG, sending and storing huge amounts of data on the cloud is not practical due to latency and bandwidth issues. When there is a need for a fast response fog computing is a far better approach as it enables decentralized and intelligent processing closer to the place where data are produced and sends only the resulting information to the cloud for further processing and storage. The fog-based architecture of the SG consists of smart meters, fog servers, and cloud data centers. Therefore, fog computing is not a replacement for cloud computing, it only deals with fewer data than the cloud that simplifies analysis and extraction of valuable information. This directly increases response time, reduces the congestion and latency, and successfully deals with the loss of connectivity or poor connection with the cloud [20]. Fog computing in the SG environment shows its power in successfully dealing with the massive amount of data related to energy management, information management, and security [21]. It is important to underline that load balancing and resource utilization between cloud data centers and fog layers is crucial for efficient data analysis and decision making. The inclusion of osmotic computing at the fog layers contributes to faster responses as it enables better distribution, management, and execution of computation tasks at the edge of the network or within cloud data centers.

Communication in the IoT-supported SG environment includes information flow from smart meters and IoT devices, and information flow between smart meters and utility control centers [19]. In the fog-based SG, near real-time bidirectional communication between utilities and consumers is established, enabling customers to monitor and control their electricity through a device connected to the edge of the network [21]. In the fog-based architecture in the SG environment, four types of communication can be defined: smart device to smart device (i.e. via ZigBee or Bluetooth), smart device to fog server (i.e. through Ethernet), fog server to fog server and fog server to cloud server (i.e. through Wi-Fi or cellular communications) [22].

13.3 Distribution Management System (DMS) Based on Fog/Cloud Computing

A DMS is a collection of applications designed to monitor and control the entire distribution network efficiently and reliably [23]. DMS concept emerged as an extension of energy management system (EMS) proposed for transmission networks in the 1970s. The term EMS is closely related to the SCADA computer system widely used in transmission networks for collecting information about power system status via measurements provided by remote terminal units (RTUs). In the early days of DMS, in the 1990s, the metering infrastructure supporting its functions was mainly limited to distribution equipment located in combined transmission and distribution substations [24]. There were no periodic measurements in distribution substations and along the distribution feeders meaning that the observability of distribution networks was very low at the time and very expensive to improve. Improving observability requires the installation of the new measurement units along the entire distribution network and establishing the new communication networks. Today, the fast development of communication systems is taking place, which could provide efficient solutions for wide DMS applications in the near future with lower costs. Integration of renewable energy sources in existing distribution networks increases significance and pushes forward new solutions for DMS [25].

The major functions used in DMS are distribution system state estimation (DSSE), load flow (LF), and Volt-Var Optimization (VVO). Time triggering of the major functions is presented in Figure 13.2 according to the concept from Atanackovic and Dabic [26]. DSSE is used to improve system observability, check for and detect errors in both system measurements and network parameters, and to mitigate against measurement and communication system noise [27]. LF is the procedure used for obtaining the steady state voltages of electric power systems at the fundamental frequency [28]. In advanced DMS, LF will be used on a more frequent basis [29]. VVO is a well-known application for minimization of reactive power losses and for maintaining an optimal voltage profile. According to the concept from Figure 13.2, DSSE calculations should be performed every time new input data is available, while LF calculations should be triggered only when a network topology change is detected. VVO should be also initiated periodically according to the established timetable, but less often than DSSE calculations. Network topology processing (NTP) and observability analysis (OA) are initial prefunctions necessary for the application of previous major functions.

Observabilities of the majority of existing distribution networks in the world are very low. Most of the available input data at the distribution level are measurements (or pseudo-measurements) of power or current injections [27]. Telemetered values from measurement devices located at distribution feeders outside of substations are very rare. Some advanced distribution network utilities, such as BC Hydro (Province of British Columbia), already started implementation of SCADA at the distribution system level [26]. Similar projects can be found around the world; all are based on wired communication technologies (i.e. optical communications), in spite of installation and maintenance high costs. In such systems, DMS

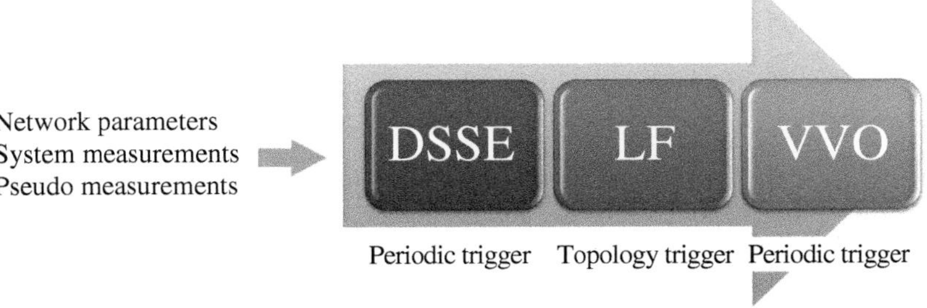

Figure 13.2 Time triggering of the major functions used in DMS.

is controlled by distribution system operator (DSO), which gathers all available measurements and performs the calculation of DSSE and LF. For large distribution networks with hundreds of substations and even more feeders, it could be very challenging to install communication wires. Collecting and analyzing such Big Data at DSO locations is not a simple task. Since at distribution system level communication latency is not a priority, application of modern wireless communication technologies could be a good alternative. The deployment of smart meters and intelligent electronic devices (IEDs) equipped with two-way communications is increasing [30]. These modern devices are capable to wirelessly send and receive data, therefore increasing observability of the distribution network. An innovating feeder-based communication scheme for DMS using fog/cloud computing is presented in Figure 13.3.

Proposed communication schemes use fog servers as data concentrators for distribution feeders. The precondition necessary for deployment of this scheme is the availability of telemetered values along every feeder, which is expected in future SG. Measurements from a low voltage level could be provided by smart meters, while measurements from high voltage levels could be provided by PMUs and IEDs. It is expected that these near real-time measurements enable high observability of the distribution network, which later will improve DMS. Calculation of basic DMS functions, such as NTP, OA, DSSE, and LF, are all performed at

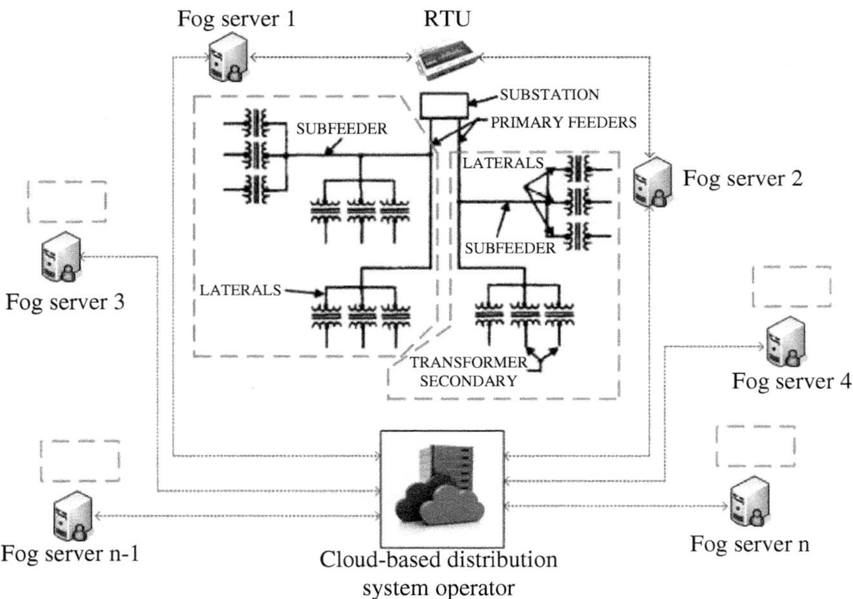

Figure 13.3 Feeder-based communication scheme for DMS using fog/cloud computing. (*See color plate section for the color representation of this figure*)

the fog server, thus significantly lowering data transmission rates necessary for cloud-based DSO, as can be seen in Figure 13.3. Smart meters, IEDs, and PMUs send data to feeder-based fog servers wirelessly. Measurements from smart meters are available every few minutes, thus helping to determine load profiles and lowering uncertainty of pseudo-measurements. Measurements from IEDs and PMUs could be available at every few seconds, thus enabling near real-time calculations of DMS functions at fog servers.

An important advantage of the proposed communication scheme is that periodical Big Data calculations required for DMS are avoided at cloud servers representing DSO. Cloud-based DSO receives calculation results from feeder-based fog servers instead of Big Data from all feeders. Calculations of DMS functions could be performed directly at the cloud server by sending a request to fog servers if necessary in some occasions. Communication between fog servers of distribution feeders at the same substation and existing RTU is also provided. In this way, the transmission system operator (TSO) could request data directly from feeders physically grouped at one substation. The proposed communication scheme is bidirectional, which means that control actions could be transmitted to equipment from fog servers or DSO directly. By enabling fog to fog bidirectional communication micro grid concept could be also deployed.

13.4 Real-time Simulation of the Proposed Feeder-based Communication Scheme Using MATLAB and ThingSpeak

To illustrate the benefits of the proposed feeder-based communication scheme a real-time DMS is simulated using MATLAB [31] and the open source IoT platform ThingSpeak [32]. ThingSpeak platform is used to model fog server communicating with the well-known MATLAB-based model of IEEE 13 Node Test Feeder [33]. This approach is chosen for the real-time simulation of the SG environment as it allows a two-way information flow between the user and the simulated system, and hence makes possible real-time sharing data and remote control.

ThingSpeak IoT platform is secured by usernames and passwords and enables graphical representations of several channels and their fields. Each channel has its own ID, can be public (seen by other users) or private (seen by specific users), and can contain up to eight fields of data (data can be both numeric and alphanumeric), and therefore enables presentation of several parameters at the same time. It is important to highlight that the channels receive sensor data every 15 seconds. Alongside eight fields of data, there are four other fields dedicated for location details: Description, Latitude, Longitude, and Elevation. In addition, every private channel has two keys: API (application programming interface) read key and API write key. These keys are randomly created with a unique alphanumeric string used for authentication [34], enabling the channel to store or retrieve data from the things via the Internet or LAN [35]. ThingSpeak IoT platform enables the data importing, exporting, and saving in the csv or xlsx file format. The access to collected, processed, visualized, saved, and historical data can be realized by using any device that has Internet connectivity (i.e. computers or smartphones), through the user authentication using the username and password. In this way, secure and quick access to accurate, complete, and up-to-date information is enabled [36].

Distribution feeder test cases were designed to evaluate and benchmark algorithms in solving unbalanced three-phase radial systems. Each of these represents reduced-order models of an actual distribution circuit [37]. In this analysis, 13-bus feeder, operating at 4.16 kV, has been used. A detailed three-phase model of the previous feeder with network and consumption data is available in [38]. Communication between the feeder model and ThingSpeak IoT platform is established using MATLAB Simulink Desktop Real-time Toolbox. A simplified communication scheme is presented in Figure 13.4.

The first step in the simulation of DMS is making the feeder model fully observable. Three-phase measurement blocks are placed at every feeder node, while pseudo-measurements are neglected due to simplicity. It is assumed that

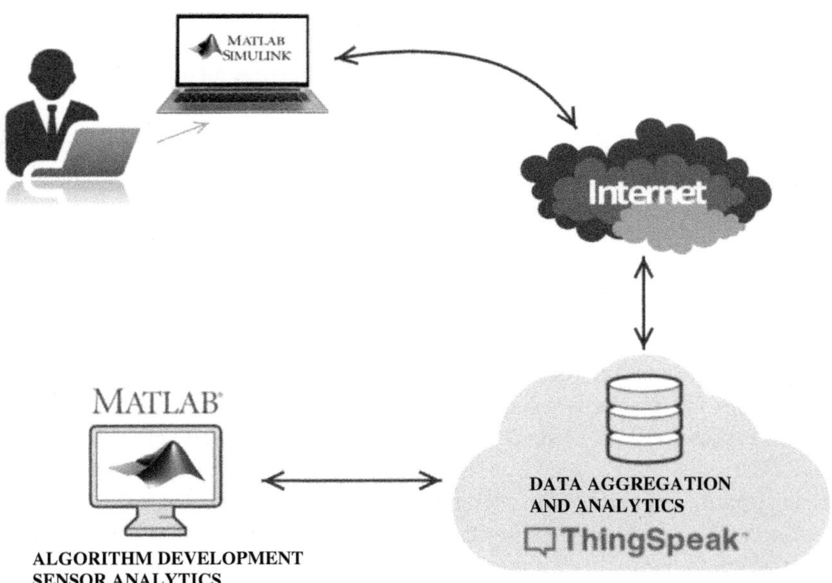

Figure 13.4 Simplified communication scheme connecting MATLAB/Simulink and ThingSpeak IoT platform used for DMS simulation. (*See color plate section for the color representation of this figure*)

DMS functions NTP and OA are already performed at the fog server (ThingSpeak IoT platform). To successfully and periodically perform DSSE function, measurements of node voltage magnitudes and phase angles are being sent to ThingSpeak at every 20 seconds in real time. For every node voltage measurement, one ThingSpeak channel with six fields is assigned (two fields for every phase), which makes 12 channels and 72 fields in total. Distribution feeder topology with channel assignments for node voltage measurements is presented in Figure 13.5.

It can be seen that distribution feeder topology is phase-unbalanced. The majority of network branches are in three-phase, but there are also two- and one-phase branches. Every measurement block sends voltage data to its corresponding ThingSpeak channel, which is determined by the channel's ID number. Data is stored and visualized in real time in ThingSpeak IoT platform, thus enabling real-time monitoring of the voltage profile along the feeder. Voltage profile monitoring is an essential part of DSSE function. DSSE is determined by knowing all node voltage magnitudes or RMS values and phase angles. LF function calculations are performed directly in Simulink and not in ThingSpeak IoT platform in this analysis due to simplicity.

In the first simulation test, the voltage profile of the distribution feeder model is monitored in real time during 200 seconds time interval. Three-phase tripping of network branch 671–692 is simulated after 100 seconds, as can be seen in Figure 13.6.

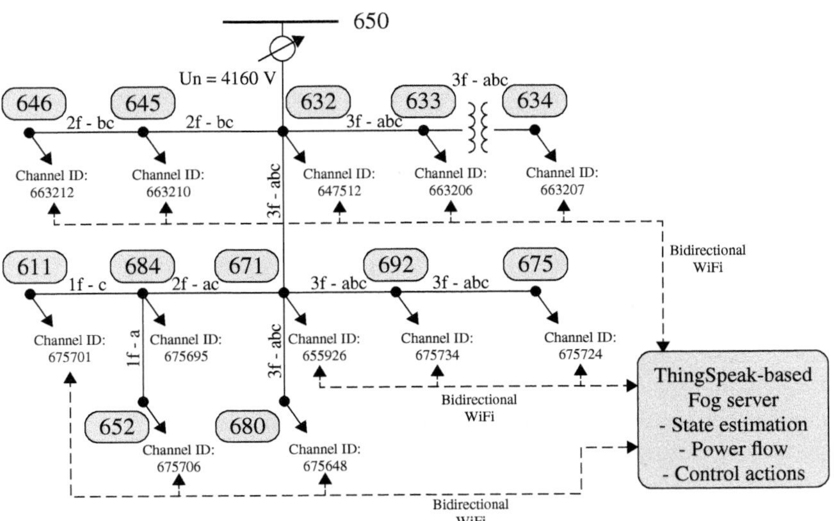

Figure 13.5 Distribution feeder topology with ThingSpeak channel assignments for node voltage measurements.

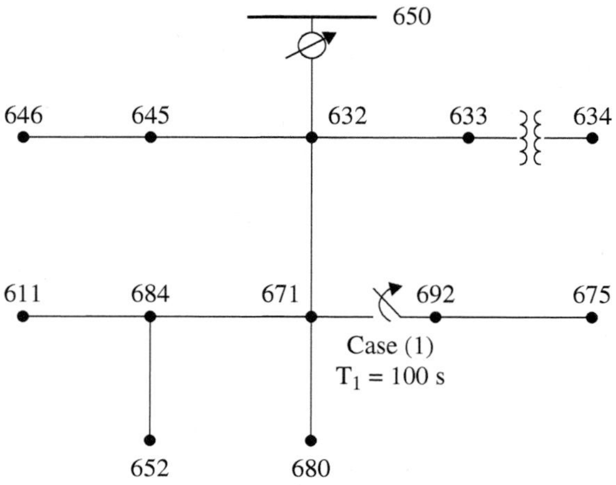

Figure 13.6 Simulation test 1 – Three-phase tripping of network branch 571–692 and real-time monitoring of feeder voltage profile.

Feeder voltage profile is visualized using online MATLAB Visualization Tool integrated into ThingSpeak IoT platform and corresponding results are presented in Figure 13.7. Voltage RMS values are presented only for the most important nodes in this analysis: 671, 684, 611, 652, 692, and 675. According to default feeder model settings, the voltage of node 632 is predefined and kept constant during simulation, meaning that events downstream from the node 632 will not affect voltages of nodes 633, 634, 645, and 646.

According to plots (a), (b), (c), and (d) in Figure 13.7, it can be noticed that RMS values of voltages of "heavily" loaded phases a and c increased after tripping of branch 571–692, while RMS value of the voltage of "low" loaded phase b decreased. From plots (e) and (f) in the same figure, it can be seen that RMS values of node voltages 692 and 675 dropped to zero in all phases, which is a direct consequence of simulated tripping event. According to previous information, the fog server becomes aware of the new voltage profile. Real-time monitoring function successfully detected a change in voltage profile during simulation time. Tolerated phase voltage limits are $\pm 10\%$ and presented in magenta color. Obtained voltage profile should not be sent periodically to cloud-based DSO at every 20 seconds, but only in the cases of change or request.

LF calculations have been performed before and after of the tripping event to monitor per-phase active and reactive powers from node 650 to node 632 (feeder power consumption). Before the tripping event calculated powers were: $P_a = 1.23$ MW, $Q_a = 0.61$ Mvar, $P_b = 0.98$ MW, $Q_b = 0.34$ Mvar, $P_c = 1.31$ MW, $Q_c = 0.59$ Mvar. After the tripping event corresponding calculated powers were: $P_a = 0.7$ MW, $Q_a = 0.45$ Mvar, $P_b = 0.91$ MW, $Q_b = 0.5$ Mvar, $P_c = 0.89$ MW, $Q_c = 0.52$ Mvar.

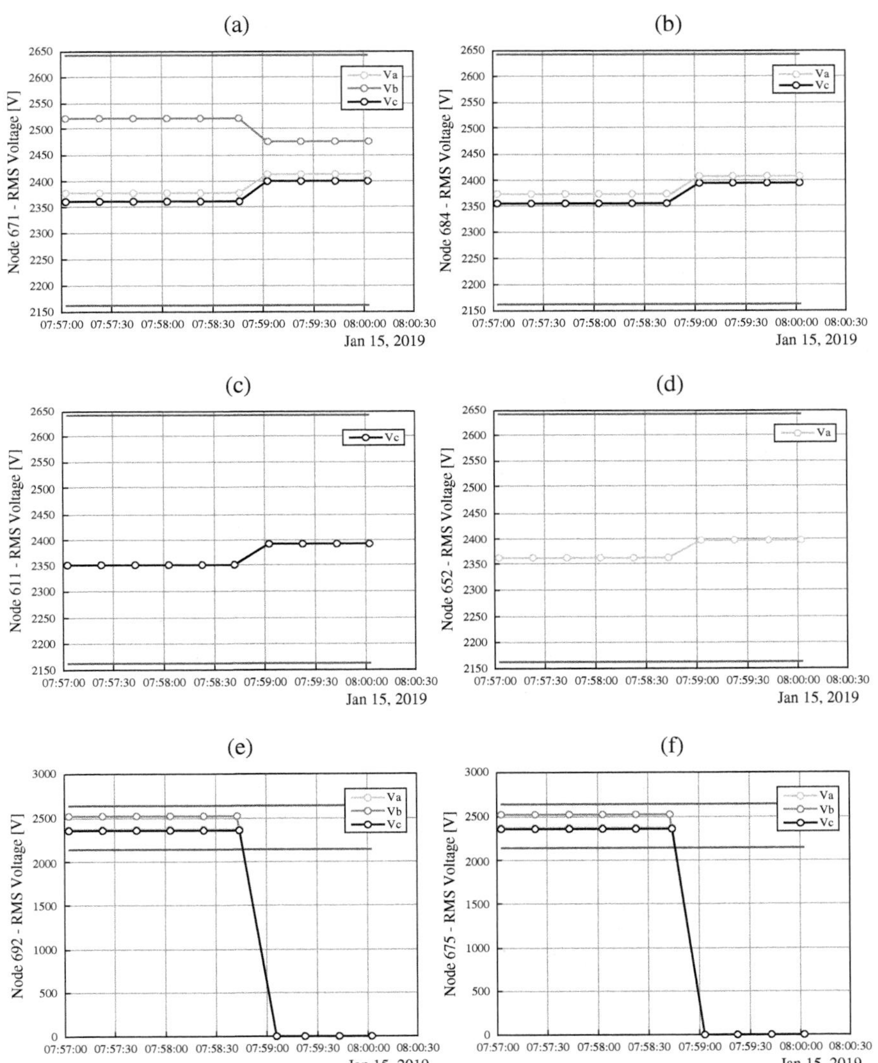

Figure 13.7 Three-phase voltage profile of distribution feeder downstream from the node 632 in the case of simulation test 1: (a) node 671; (b) node 684; (c) node 611; (d) node 652; (e) node 692; (f) node 675.

In the second simulation test, the voltage profile of the distribution feeder model is monitored in real-time during 200 seconds time interval with online Volt-Var control implemented. Phase voltages are compared with corresponding limits in real time. In the case of lower voltage limit violated Volt-Var control is programmed to switch on the battery for reactive power compensation. At first, "heavy" inductive load is switched on at node 675, after 100 seconds from the start of the simulation, as can be seen in Figure 13.8. Later, a control signal for switching the battery on at the same node is anticipated in the case of voltage lower limit violation.

As can be seen in Figure 13.8, "heavy" inductive load is modeled as nonbalanced with per-phase inductive powers: $Q_{LA} = 2.85$ Mvar, $Q_{LB} = 0.9$ Mvar and $Q_{LC} = 3.18$ Mvar, while battery model capacitive power is $Q_C = 3$ Mvar in phases A and C (phase B is not compensated). Similar to the first simulation case, the feeder voltage profile is visualized by using the online MATLAB visualization tool; corresponding results are presented in Figure 13.9. Voltage RMS values of nodes 671, 684, 611, 652, 692, and 675 are monitored in real time.

According to the results from plots in Figure 13.9, it can be noticed that the insertion of a "heavy" inductive load causes a voltage drop in phases A and C. Violation of lower voltage limit is detected by a Volt-Var control tool and the signal for switching the battery on is sent from ThingSpeak to the feeder model in near real time. The program is executed and the battery is switched on, 40 seconds from the detection of voltage drop. The voltage level is then successfully restored within acceptable limits.

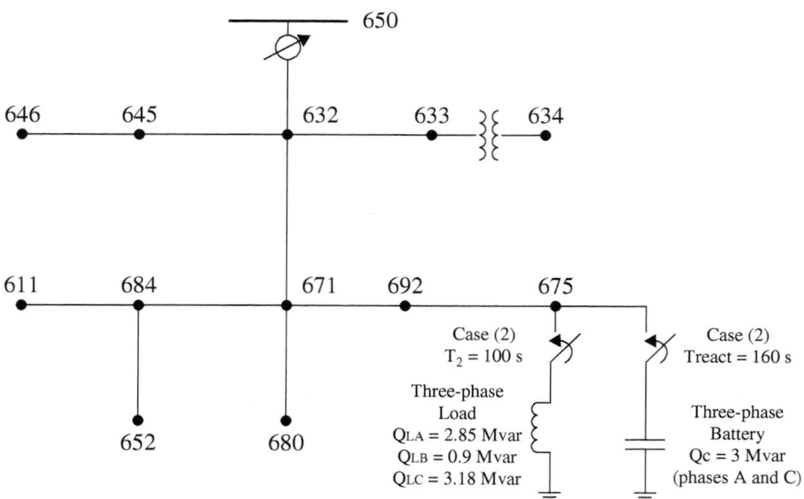

Figure 13.8 Simulation test 2 – "Heavy" loading of the distribution feeder, real-time monitoring of feeder voltage profile and executing real-time Volt-Var control.

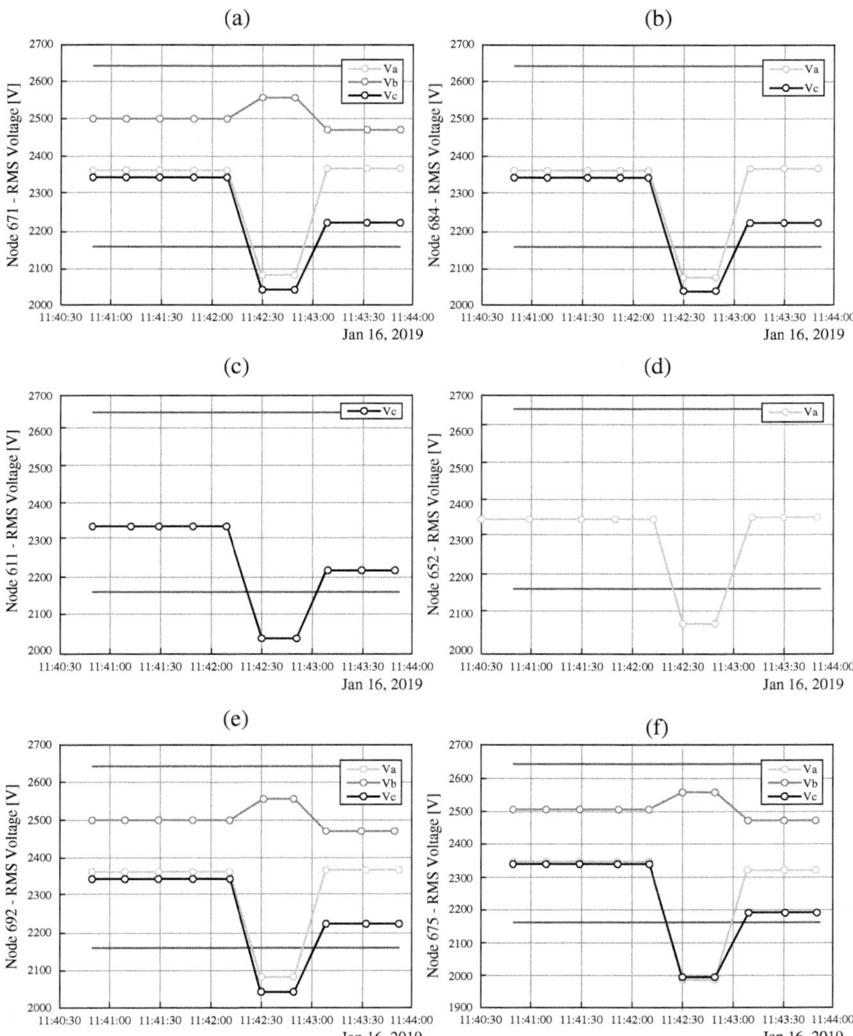

Figure 13.9 Three-phase voltage profile of distribution feeder downstream from the node 632 in the case of simulation test 2: (a) node 671; (b) node 684; (c) node 611; (d) node 652; (e) node 692; (f) node 675.

Simulation test 2 represents a valuable example illustrating real-time monitoring and control of voltage profile using the proposed communication scheme. LF calculations have been also performed for cases with "heavy" inductive load before and after reactive power compensation. Total reactive power consumed by the feeder before the compensation was $Q_{total} = 7.74$ Mvar. While, after the compensation, total measured reactive power is reduced to value $Q_{total} = 2.8$ Mvar, thus lowering feeder voltage drops.

Calculations of LF in a distribution network should be performed at fog servers. Telemetered measurements from every feeder required for calculations of LF are collected by individual fog servers (one fog server per one feeder concept). With this approach, potential Big Data issues are lower when compared with centralized communication schemes. In simulation tests, LF calculations are performed by the local processor, which also performs calculations related to the feeder model. ThingSpeak supports complex calculations using MATLAB tools and it is planned in future research to implement LF algorithms directly on the cloud.

13.5 Conclusion

Existing communication technologies deployed in the power system need to be reconsidered and newly analyzed to determine possibilities of their usage in modern SG. Near real-time data transmission and analysis are required by the new concept. Another challenge to overcome, related to distribution networks, is the requirement of higher observability, which can be achieved by the deployment of SMs, IEDs, and PMUs.

Wired communication technologies are always related to high costs and it is not an easy task to install them in large distribution networks with hundreds of feeders and tens of substations. Wireless communication technologies could be an efficient and cheaper alternative if they properly meet transmission rates and communication latency requirements. One of the key questions is where to perform calculations of Big data related to DMS. Sending large amounts of data directly to cloud-based DSO could impose difficulties in communication systems, especially when it comes to near real-time data transmission.

In this chapter, a new communication scheme is proposed using fog servers as feeder-based calculation units necessary for DMS. The main advantage of the proposed scheme is a significant reduction of data sent to cloud-based DSO. DSO receives data from fog servers only if measurement changes are detected or by sending the request, while fog servers receive data periodically from smart meters, IEDs, and PMUs along the feeder. The majority of the required calculations for DMS functions is performed at fog servers and only results representing feeders near real-time snapshots are sent to the cloud-based DSO. The proposed scheme could be easily connected with existing RTUs at substations, thus enabling, on request, communication with TSO. By expanding the proposed scheme and enabling mutual communication between fog servers micro grids could be formed.

Capability testing of the proposed communication scheme is performed using MATLAB and open source IoT platform ThingSpeak. Two near real-time simulation tests are defined: (1) three-phase tripping of network branch and real-time monitoring of feeder voltage profile, and (2) heavy loading of the distribution feeder, real-time monitoring of feeder voltage profile and executing real-time Volt-Var control. According to obtained results, communication scheme proved to be capable and efficient for supporting major DMS functions, such as DSSE and LF. A second simulation test revealed another important capability of the scheme to support near real-time control actions related to reactive power compensation (Volt-Var control).

References

1 Heirman, D. (2012). What makes Smart Grid – smart – and who is in the game? *IEEE Electromagnetic Compatibility Magazine* 1: 95–99.

2 Department of Energy of USA, Communication requirements of Smart Grid, 2010. [Online]: https://www.energy.gov/sites/prod/files/gcprod/documents/Smart_Grid_Communications_Requirements_Report_10-05-2010.pdf

3 Z. Jiang, F. Li, W. Qiao et al., (2009) A vision of smart transmission grids, IEEE Power & Energy Society General Meeting - PES '09, Canada, 2009.

4 Kayastha, N., Niyato, D., Hossain, E., and Han, Y. (2014). Smart Grid sensor data collection, communication, and networking: a tutorial. *Wireless Communications* 14 (11): 1055–1087.

5 Xi, F., Satyajayant, M., Guoliang, X., and Dejun, Y. (2012). Smart Grid – the new and improved power grid: a survey. *IEEE Communications Surveys & Tutorials* 14 (4): 6–9.

6 Maksimović, M. and Forcan, M. (2019). Internet of Things and Big Data Recommender Systems to support Smart Grid. In: *Big Data Recommender Systems: Recent trends and Advances* (eds. O. Khalid, S.U. Khan and A.Y. Zomaya), 145–172. London: The Institution of Engineering and Technology.

7 Vijayapriya, T. and Kothari, D.P. (2014). Smart grid: an overview. *Smart Grid and Renewable Energy* 2 (4).

8 Baimel, D., Tapuchi, S., and Baimel, N. (2016). Smart Grid communication technologies: overview, research challenges and opportunities. In: *IEEE International Symposium on Power Electronics, Electrical Drives, Automation and Motion (SPEEDAM)*, 116–120. Italy.

9 A. Chasempour, Optimizing the Advanced Metering Infrastructure Architecture in Smart Grid. PhD Thesis. Utah State University, 2016. [Online]: https://digitalcommons.usu.edu/cgi/viewcontent.cgi?article=6061&context=etd

10 Alam, S., Sohail, M.F., Ghauri, S.A. et al. (2017). Cognitive radio based Smart Grid communication network. *Renewable and Sustainable Energy Reviews* 72: 535–548.

11 Kuzlu, M., Pipattanasomporn, M., and Rahman, S. (2014). Communication network requirements for major Smart Grid applications in HAN, NAN and WAN. *Computer Networks* 67: 74–88.

12 Kabalci, Y. (2016). A survey on smart metering and Smart Grid communication. *Renewable and Sustainable Energy Reviews* 57: 302–318.

13 Mahmood, A., Javaid, N., and Razzaq, S. (2015). A review of wireless communications for Smart Grid. *Renewable and Sustainable Energy Reviews* 41: 248–260.

14 Yigit, M., Gungor, V.C., Tuna, G. et al. (2014). Power line communication technologies for Smart Grid applications: a review of advances and challenges. *Computer Networks* https://doi.org/10.1016/j.comnet.2014.06.005.

15 Zafar, R., Mahmood, A., Razzaq, S. et al. (2018). Prosumer based energy management and sharing in Smart Grid. *Renewable and Sustainable Energy Reviews* 82: 1675–1684. https://doi.org/10.1016/j.rser.2017.07.018.

16 Andreadou, N., Olariaga Guardiola, M., and Fulli, G. (2016, 2016). Telecommunication technologies for Smart Grid projects with focus on smart metering applications. *Energies* 9: 375. https://doi.org/10.3390/en9050375.

17 R. Zafar, U. Naeem, W. Ali, and A. Mahmood, Applications of ZigBee in Smart Grid Environment: a review,in Proceedings of the 2nd International Conference on Engineering & Emerging Technologies (ICEET). Lahore, PK: Superior University, 2015.

18 Daki, H., El Hannani, A., Aqqal, A. et al. (2017). Big Data management in Smart Grid: concepts, requirements and implementation. *Journal of Big Data* 4 (13).

19 Y. Saleem, N. Crespi, M. H. Rehmani, and R. Copeland, Internet of Things-aided Smart Grid: technologies, architectures, applications, prototypes, and future research directions, 2017 [online]: https://arxiv.org/ftp/arxiv/papers/1704/1704.08977.pdf

20 F. Y. Okay and S. Ozdemir, A fog computing based Smart Grid model, International Symposium on Networks, Computers and Communications (ISNCC), Tunisia, 2016.

21 Vinueza Naranjo, P.G., Shojafar, M., Vaca-Cardenas, L. et al. (2016). Big Data over Smart Grid – a fog computing perspective. In: *Proceedings of the 24th International Conference on Software, Telecommunications and Computer Networks (SoftCOM)*, 1–6. Split, Croatia.

22 Vatanparvar, K. and Al Faruque, M.A. (2015). Energy management as a service over fog computing platform. In: *Proceedings of the ACM/IEEE Sixth International Conference on Cyber-Physical Systems (ICCPS '15)*, 248–249. New York, NY, USA: ACM.

23 Huang, Y.F., Werner, S., Huang, J.J. et al. (2012). State estimation in electric power grids: meeting new challenges presented by the requirements of the future grid. *IEEE Signal Processing Magazine* 29 (5): 33–43.

24 Cassel, W.R. (1993). Distribution management systems: functions and payback. *IEEE Transactions on Power Systems* 8 (3): 796–801.

25 Schwerdfeger, R. and Westermann, D. (2013). Design of an energy distribution management system for the vertical management of volatile infeed. In: *2013 IEEE Power & Energy Society General Meeting*, 1–5. Vancouver, BC.

26 Atanackovic, D. and Dabic, V. (2013). Deployment of real-time state estimator and load flow in BC Hydro DMS - challenges and opportunities. In: *2013 IEEE Power & Energy Society General Meeting*, 1–5. Vancouver, BC.

27 Hayes, B. and Prodanovic, M. (2014). State estimation techniques for electric power distribution systems. In: *2014 European Modelling Symposium*, 303–308. Pisa.

28 Herraiz, S., Sainz, L., and Clua, J. (2003). Review of harmonic load flow formulations. *IEEE Transactions on Power Delivery* 18 (3): 1079–1087.

29 Borlase, J.S. (2009). The evolution of distribution. *IEEE Power and Energy Magazine* 7 (2): 63–68.

30 Primadianto, A. and Lu, C. (2017). A review on distribution system state estimation. *IEEE Transactions on Power Systems* 32 (5): 3875–3883.

31 MATLAB, The MathWorks, Inc., Natick, MA, US, 2017.

32 ThingSpeak, 2019 [Online]: https://thingspeak.com

33 Schneider, K.P. et al. (2018). Analytic considerations and design basis for the IEEE distribution test feeders. *IEEE Transactions on Power Systems* 33 (3): 3181–3188.

34 M. A. Gómez Maureira, D. Oldenhof, and L. Teernstra, ThingSpeak – an API and web service for the Internet of Things, 2014. [Online]: https://staas.home.xs4all.nl/t/swtr/documents/wt2014_thingspeak.pdf

35 Mohamad, A.A.H., Mezaal, Y.S., and Abdulkareem, S.F. (2018). Computerized power transformer monitoring based on Internet of Things. *International Journal of Engineering & Technology* 7 (4): 2773–2778.

36 Shekhar, J., Abebaw, D., Abebe Haile, M. et al. (2018). Temperature and heart attack detection using IOT (Arduino and ThingSpeak). *International Journal of Advances in Computer Science and Technology* 7 (11): 75–82.

37 IEEE PES, 2019, [Online]: http://sites.ieee.org/pes-testfeeders/resources

38 Mathworks, 2019, [Online]: https://www.mathworks.com/help/physmod/sps/examples/ieee-13-node-test-feeder.html

14

An Estimation of Distribution Algorithm to Optimize the Utility of Task Scheduling Under Fog Computing Systems

Chu-ge Wu and Ling Wang

Department of Automation, Tsinghua University, Beijing, China

14.1 Introduction

Internet of Things (IoT) has gained remarkable and rapid development today and has been widely adopted into industry and daily life [1]. Human-facing and machine-to-machine applications are coming into being, where end-to-end latency are both well worth mentioning. The human-facing applications, such as gaming, virtual reality [2], and multimedia services [3], are needed to guarantee the quality of service (QoS) and quality of experience (QoE) to optimize the user experience and satisfaction. Meanwhile, for the machine-to-machine applications, such as motion control in cars, the requests need to be timely responsive [4]. IoT systems are expected to meet the latency requirements with finite resources to support these real-time applications and to optimize the latency has become a major problem for IoT responsive applications.

Fog computing [5] has been developed to decrease the latency of IoT applications since it is realized near the IoT devices and the data sources. It mainly consists of the basic computation model and the data transmission model to collect, process, and send data accordingly and reduce the data transmission volume over the Internet. In that case, the latency of applications as well as the battery power consumed to send and receive data are decreased, respectively. Meanwhile, the user experience is improved. It is said that responsive IoT applications are considered to be one of the primary reasons for the adoption of fog computing [2]. However, to fully exploit the computing system structure and solve the resource scheduling problem is not a trivial task as the fog computing infrastructure is complex. In addition, different requirements developed by users brings difficulties to model the problem. It's known that the task scheduling on multiprocessors problem is NP-hard when two processors are used [6]. This problem focuses on

Fog Computing: Theory and Practice, First Edition.
Edited by Assad Abbas, Samee U. Khan, and Albert Y. Zomaya.

the latency optimization for responsive IoT applications under the fog computing environment. This problem, which considers more factors, can also be proved to be NP-hard, so there is no known polynomial algorithm for this problem until now.

The remainder of the chapter is organized as follows: Section 14.2 outlines estimation of distribution algorithm (EDA) and its application in the area of scheduling. Section 14.3 reviews the related work about fog computing system and methods to model and reduce the latency of applications. Section 14.4 provides the problem statement including system model, application model, time-dependent utility functions and objectives in mathematic formulas.

And in Section 14.5, we elaborate our proposed algorithm, including decoding and encoding method, EDA scheme and the local search procedure. The simulation environment parameters, comparison algorithms are introduced in Section 14.6 and in Section 14.6.1 the comparison results are given. In the end, we conclude our paper in Section 14.7 with some conclusions and future work.

14.2 Estimation of Distribution Algorithm

The evolutionary optimization algorithms have been proven to be very efficient to solve NP-hard optimization problems in the past two decades. The evolutionary optimization algorithms are good at solving the nonconvex, discrete optimization problems to achieve acceptable solutions in limited time. EDA was a kind of efficient population-based evolutionary algorithm. Statistic theory and methods are the most important part of EDA. A certain probability model is built according to the problem information in EDA and it's adopted to be sampled and produced the population. Meanwhile, the probability model is updated by the elite individuals in the population. Thus, the evolution of EDA is driven by updating the probability model. The basic procedure of EDA is:

1. Initialize the population and the probability model;
2. Sample the probability model randomly to achieve new population and replace the formal population with the new one;
3. Choose the elite solutions in accordance to the objective function;
4. Update the probability value of the model with the elite individuals by certain statistic learning method;
5. If the stopping criterion is met, output the best solution until now; or, turn to step 2.

According to the different types of selection operator for selecting elite solutions, the class of probability models, and the updating procedure of the probability model, EDAs could be divided into different types. Some classical probability models, such as the univariate model, chain model, bayesian network and tree model, have been applied to solve different kinds of problems, such as machine learning [7], bioinformatics [8], and production scheduling problems [9].

For the complex scheduling problems, the EDA scheme embedded with a knowledge-driven local search method performs well as the proposed algorithm and is good at both exploration and exploitation. In our previous work, our proposed EDA performed well in a direct acyclic graph (DAG) scheduling problem for both cloud computing [10] and fog computing infrastructure [11, 12]. The application completion time, the total tardiness of application, energy consumption of cloud or fog computation nodes and the lifetime of IoT system have been considered as the optimization objectives and well optimized in our previous work. Inspired by the successful application of EDA for this problem, the uEDA in this work will be adopted to maximize the task utility to optimize the task scheduling and computation resource allocation.

14.3 Related Work

As IoT has been realized initially and fog computing is put forward, lots of new research focused on the related topics is proposed. The measurement, modeling, and optimization of end-to-end latency of IoT applications are discussed. A detailed introduction to the fog computing and related algorithms is offered in [13]. Resource management, cooperative offloading, and load balancing are all needed to consider the fog computing environment. QoS is considered to be one of the important objectives for the algorithms enabling real-time applications. In [2], the authors examine the properties of task latencies in fog computing by considering both traditional and wireless fog computing execution points. Similarly, the authors consider the fog computing service assignment problem in bus networks [14]. A model minimizing the communication costs and maximizing utility with tolerance time constraints is given to formulate the problem.

To model the latency and optimize the task scheduling, many metrics are developed. Tardiness is a well-known metric to measure how much a task is delayed. Total tardiness, weighted total tardiness, the largest tardiness value and the number of delayed tasks are all designed to measure whether a batch of tasks is processed on time and how much they're delayed. Surveys such as [15, 16] offer total tardiness problems. They mainly introduce the single machine problem as well as the parallel machine and flowshop problem. In [17], the authors survey the weighted and unweighted tardiness problems. On the other hand, the task utility is an efficient performance measure for the real-time systems [18]. Different kinds of time-dependent functions are tailored for tasks with different response time needs. To distinguish the hard and soft deadline constrained tasks, we prefer time-dependent functions to measure the performance of scheduling algorithm rather than weighted total tardiness. In this way, there is no need to ensure the weighted parameter value.

14.4 Problem Statement

A generalized three-tier IoT system [19] is shown in Figure 14.1. This system consists of things, fog, and cloud tiers.

IoT devices, such as sensors and cameras, are deployed to collect data and process it accordingly. To fully cover the corresponding region, each cluster consists of heterogeneous IoT devices and is located into different places. As the different clusters are far away from each other, there is no directed transmission between the clusters and within the same cluster, IoT devices are assumed to be fully connected.

The fog computing tier is located near the IoT devices to reduce the data transmission. Different devices, such as a laptop, Raspberry Pi, and local servers are defined as fog nodes. Each fog node is in charge of a cluster of IoT devices and preprocess and send the computation tasks for the further computation. Within the fog tier, the devices are fully connected.

The cloud tier is the third tier of the system, which is usually far away from the IoT data sources. The containers with computation and storage capabilities are considered as cloud nodes. The cloud nodes are considered more reliable and efficient than things and fog nodes. In our model, the fog tier is connected with the cloud tier and within the cloud tier, the processors are fully connected with each other.

For the IoT applications, it is represented as a DAG because most applications can be modeled as workflow. We use the nodes in DAG to model the computation tasks while the edges to model the precedence constraints. The weight value on the edge represents the data size transferred between the tasks. The

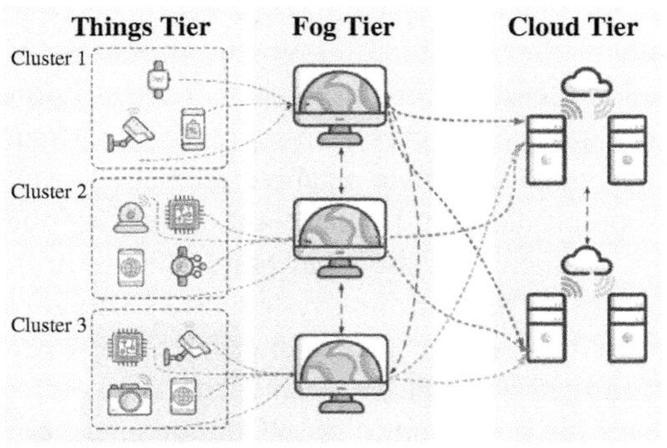

Figure 14.1 Our proposed three-tier IoT system architecture.

communication-to-computation ratio (*CCR*) is calculated as average communication time divided by computation time and is used to describe the application.

To measure the QoE for each task, the utility-related information of each task to reduce the latency of each task and accelerate the completion of the application, to maximize the sum of task utility is adopted as the objective of this work. The utility task j is calculated as follows in this work:

$$U_j := f_j(C_j) \tag{14.1}$$

where time-dependent utility function [18] $f_j(t)$ represents the utility of task j on certain completion time point C_j. According to different needs, different functions are used to describe the requirements on the tasks. In this work, two kinds of tasks are considered. Some tasks must be finished before a certain deadline, such as for the motion control, otherwise it will lead to serious results. For these tasks, they are considered as hard deadline constrained and the step function, shown as (14.2), is used as their utility function. In addition, some tasks are soft deadline constrained and the wait-readily-first function is used, shown as (14.3).

$$fs(t) = \begin{cases} 1, & t \leq T_v \\ 0, & otherwise \end{cases} \tag{14.2}$$

where T_v denotes the hard deadline of the certain task. This function implies that if the task is finished before T_v, its utility is settled as 1, otherwise, the utility turns to be 0.

$$fw(t) = \begin{cases} 1, & t \leq T_e \\ 1 & -\frac{t-T_e}{T_s-T_e}, T_e < t \leq T_s \\ 0, & T_s < t \end{cases} \tag{14.3}$$

where T_e, T_s denotes soft deadline interval of a certain task. If the task is finished before T_e, its utility equals 1, and after T_e, the utility decreases as the finish time increases. If it's finished after T_s, the utility turns to be 0. In this way, the objective of this problem is designed as:

$$U = \max \sum_j f_j(C_j) \tag{14.4}$$

In this work, both hard deadline and soft deadline constrained tasks are given two deadline values (d1j, d2j). For the hard deadline constrained tasks, d1j = d2j = Tvj is settled. And for the soft deadline constrained tasks, we have d1j = Tej, d2j = Tsj. The deadline value information is given in advance.

Because of the precedence constraints, all the tasks can only start being processed after all its input data from its parent tasks are received. And the tasks are nonpreemptive during processing. The communication time between any two connected tasks (i, j) is dependent on the transmission rate $B(m_i, m_j)$ associated with the assigned computing nodes(m_i, m_j), $(m_i \neq m_j)$. Thus, the earliest start time of task j (EST_j) could be indicated as follows:

$$EST_j = \begin{cases} 0, & j = 0 \\ \max_i \left\{ EST_i + \delta(m_i, m_j) \cdot \dfrac{D(i,j)}{B(m_i, m_j)} \right\}, & \forall i \to j \end{cases} \tag{14.5}$$

where task 0 denotes the root task of the DAG, which is a virtual task. Its processing time is set 0 and it's the parent node of the entry nodes. $\delta(m_i, m_j)$ is an indicative function where $\delta(m_i, m_j) = 1$ denotes $m_i \neq m_j$ and $\delta(m_i, m_j) = 0$ denotes $m_i = m_j$. Because the IoT applications begin sensing data by the IoT devices, the entry tasks in DAG are processed in the things tier. It's assumed that the entry tasks of a certain application are allocated in the same cluster.

14.5 Details of Proposed Algorithm

The proposed algorithm that aims to maximize the sum of utility is introduced in this section. First, the encoding and decoding procedure is introduced. After that the uEDA including initialization and repair procedure is presented. Finally, a local search aiming at decreasing total tardiness and increasing successful rate is introduced.

14.5.1 Encoding and Decoding Method

The individuals are encoded as a processing sequence, which is a permutation from 1 to n. The permutation is produced by our proposed uEDA.

When the individuals are decoded, the tasks are considered and assigned according to the processing permutation produced by uEDA. The computation nodes in the three tiers are treated as unrelated processors, where the processor assignment is determined by EFTF rule (earliest finish time first). According to EFTF, a task is assigned to the processor to achieve the minimum tardiness if the deadline cannot be met on any processors. This ensures the greedy optimization of task utility. In addition, if its deadline can be met on more than one processor, EFTF rule ensures that the task is assigned to the processor with minimum completion time.

As the step function for hard deadline constrained tasks turns 0, the hard deadline is missed. For each task violating its hard deadline during the decoding procedure, it's considered to be moved forward. The task will be swapped with its forward tasks in processing permutation until the forward task is hard deadline constrained or there is precedence constraint between the task pair. The swap operator is repeated until the deadline is met or the condition is not satisfied.

14.5.2 uEDA Scheme

To achieve optimal processing permutation, EDA is adopted to produce a range of individuals and learns from the elite individuals. We use a probability model to describe the relationship between the task pairs. Based on this model, the problem information is used to initialize the probability, and the new population is sampled. Population-based incremental learning (PBIL) is used to update the probability values. The details of the proposed algorithm is given as follows.

14.5.2.1 Probability Model and Initialization

A task permutation probability model P, considering the task relative position relationship, is built to describe and determine the task processing order. Each probability is defined as (14.6), where $p_{i,j}(g)$ represents the probability that task i is ahead of task j in the permutation at the gth generation of the evolution. It's known that the probability matrix is symmetrical and the diagonal of the matrix is set as 1 for the convenience of calculation.

$$P(g) = \{p_{i,j}(g)\}_{n\times n} \tag{14.6}$$

For the initialization, if there is a precedence constraint between a pair of tasks, the probability value ($p_{i,j}$) is initialized as (14.7). In this way, the precedence constraint is settled down during the evaluation.

$$p_{i,j}(0) = \begin{cases} 1, & i \rightarrow j \\ 0, & j \rightarrow i \\ 0.5, & otherwise \end{cases} \tag{14.7}$$

14.5.2.2 Updating and Sampling Method

To evaluate the population and produce new individuals, the probability model learns from the elite population and is updated. The elite population consists of the $N \times \eta\%$ best individuals, where $\eta\%$ is the elite population proportion. The updating scheme is presented as (14.8) and (14.9) where a certain task pair (i,j) is recorded as $I_{i,j}^{k}(g)$ as (14.9). The sum of indicators is divided by the amount of elite individuals to achieve the frequency of a certain task pair and it's used to update the former probability as the PBIL method [18]:

$$p_{i,j}(g+1) = (1-\alpha) \times p_{i,j}(g) + \alpha \times \sum_k I_{i,j}^{k}(g)/(N \times \eta\%) \tag{14.8}$$

$$I_{i,j}^{k}(g) = \begin{cases} 1, & \text{task } i \text{ is before task } j \text{ in the } k\text{th individual} \\ 0, & \text{otherwise} \end{cases} \tag{14.9}$$

where $\alpha \in (0, 1)$ represents learning rate and $I_{i,j}^{k}(g)$ is the indicator function corresponding to the k-th individual of the population. To achieve new individuals, a sampling method is designed based on the relative position probability model. And the detailed procedure could be referred in [11].

14.5.3 Local Search Method

To optimize the solutions obtained by uEDA, a problem-specific local search method is designed. Before the stopping criterion is met, the tasks with low utility value is considered first and moved forward until the condition is not satisfied. If the solution is improved, the formal processing permutation is replaced by the new one. A detailed pseudo code of the local search method is presented at first and then it is explained.

Algorithm 14.1 Local Search Method

```
s = 0;
While s < ls
    Calculate task utility value for tasks within processing permutation π;
 For each task t according to increasing order of task utility value
     If t has been moved
          continue;
     Else
         Record that t has been moved;
    Find out t is the i-th task in permutation π;
            If i>1 and π_{i-1} is not precedent of π_i and du(π_{i-1}) ≤ du(t)
             swap( π_{i-1}, π_i) to achieve π';
            Evaluate π' and calculate U'.
            s++;
            If U'>U
                π := π';
                s = s + 5;
                break;
            Else
                i- -;
            End If
          End If
     End If
 End For
  End While
```

14.6 Simulation

14.6.1 Comparison Algorithm

To evaluate the performance of our proposed algorithm, a modified heuristic method solving total tardiness problem are selected and modified as benchmark algorithm. For fair comparison, the algorithms are all coded in C++ language and run under the same running environment.

Some heuristic methods to solve total tardiness problems are efficient at solving the latency optimization scheduling problems. Thus, three heuristic methods, earliest deadline first (EDF), shortest processing time first (SPF), and shortest t-level task first (STF), are adopted to solve this problem where t-level in STF is calculated as the longest path from the entry task to the certain task [2].

EDF sorts the tasks in the ascending sequence of the tasks' due dates. Similarly, SPF and STF sort the tasks in ascending sequence of processing time and t-level values of the tasks, respectively. To make them more efficient for this problem, the hard-deadline tasks are considered before the soft-deadline tasks when scheduling, as the pseudo code of Algorithm 14.2 shows. And the three heuristic algorithms are used altogether and the best solution is adopted as their solution.

Algorithm 14.2 Modified Heuristic Algorithm

```
A := available task set, initialized with the root task;
While A is not empty
     If there are hard ddl constrained tasks in A
          t := EDF/SPF/STF on hard deadline constrained tasks;
     Else
          t := EDF/SPF/STF on soft deadline constrained tasks;
     End If
End While
Find out available tasks in children tasks of t and push into A;
Schedule t onto the system with EFTF.
```

14.6.2 Simulation Environment and Experiment Settings

The testing environment setting and its variables are set as Table 14.1. For each instance, the application is produced by the IoT devices in one cluster and it is processed on the certain one IoT device cluster. Based on the environment settings, the transmission rate between the IoT devices in the same cluster is $1/1 = 1$. Because different clusters of IoT devices are located far away from each other, they send information to each other through its fog nodes and thus the transmission rate between IoT devices in different clusters is $1/1 + 2/1 = 3$. In addition, the transmission rate between IoT devices and its fog node is $1/1 = 1$ and between IoT devices and other fog nodes is $1/1 + 1/5 = 1.2$. And the transmission rate between things and cloud tier, fog nodes, fog and cloud tier and cloud nodes are $1/1 + 1/5 = 1.2$, $1/5 = 0.2$, $1/5 = 0.2$, and $1/10 = 0.1$. The data transmission time could be calculated as data size multiplies the transmission rate.

Table 14.1 Simulation environmental parameters.

Parameter	Implication
Number of clusters in things tier	6, and different number of IoT devices in each cluster ({3, 3, 3, 4, 4, 4});
Number of fog nodes in fog tier	6;
Number of cloud services in cloud tier	2;
Processing speed of IoT, fog, and cloud nodes;	{1, 5, 10};
Bandwidth within IoT, fog, and cloud tier;	{1, 5, 10};
Bandwidth between IoT and fog tier;	1
Bandwidth between fog and cloud tier;	5
Pseudorandom task graph generator	TGFF [19]
Task number n of a DAG	200 ± 100
CCR	{0.1, 0.5, 1.0, 5.0}
Data size	Normal distribution with CCR
Maximum number of successors and processors of a certain task	10
High bound of deadline assignment (Hb)	2.0
Low bound of deadline assignment (Lb)	1.0
Hard deadline constrained tasks proportion	10%
Urgency ratio of tasks	{1.0, 1.5, 2.0}
Stopping criterion generations	30
Population size (N)	10
Learning rate (α)	0.05
Proportion of elite population (η%)	20%
Steps of local search (ls)	$n/3$

The parameters of the utility functions are settled according to DAG information where T_{v} is settled as (14.10). For the soft deadline constrained tasks, T_{e}, T_{s} are settled as (14.11) and (14.12).

$$T_{vi} = ((Hb - Lb) \times rand(0,1) + Lb) \times tlvl_i \tag{14.10}$$

$$T_{ei} = Lb \times tlvl_i \tag{14.11}$$

$$T_{si} = Hb \times tlvl_i \tag{14.12}$$

where $tlvl_{\mathrm{i}}$ is the t-level value of task i, which is calculated as the largest sum of processing time of the task path from the entry task to task i. In addition, to discuss the performance of the algorithms under different situation, urgency ratio of tasks is set to control the deadlines and calculated as $T'_v = T_v/Loose$.

For each *CCR*, 500 independent DAGs and corresponding deadline values are produced and for each DAG instance, 3 *Loose* values are used. Thus, $500 \times 4 \times 3 = 6000$ instances are used to validate the efficiency of the proposed algorithm.

14.6.3 Compared with the Heuristic Method

The comparison results of modified heuristic method and our proposed algorithm are listed as Table 14.1. The relative percentage deviation (*RPD*) value and the *p*-values of two sample paired t-tests under 95% confidence level are listed. The *RPD* value is calculated as follows where U^{uEDA} and U^{blvl} represents the utility sum obtained by uEDA and the modified heuristic method. And the alternative hypothesis is the sum of task utility of heuristic method is less than our solution. If the hypothesis is accepted, the "Sig" appears to be "Y," otherwise appears "N."

$$RPD = \frac{U^{uEDA} - U^{blvl}}{U^{blvl}} \times 100\% \tag{14.13}$$

From Table 14.2, it could be seen that sum of utility of uEDA is larger than the heuristic solutions under 95% confidence level. Our algorithm performs much better when the deadline is tight and *CCR* is large. Considering the comparison results, it could be seen that our proposed algorithm is an efficient algorithm for this problem compared to the modified heuristic method.

In addition, based on the results, two chats are presented as follows. Figure 14.2 presents the trends of average *RPD* value overall *Loose* situation of either "*RPD*>0" or "*RPD*<0." In Figure 14.3, the blue, yellow, and green column each presents the percentage of "*RPD* < 0," "*RPD*=0," and "*RPD* > 0" overall *Loose* situation.

It could be seen from Figure 14.2 that for each *CCR* situation, the absolute value of average positive *RPD* is much larger than negative *RPD*, which denotes that

Table 14.2 Comparison results of utility between heuristic and uEDA.

	1.0			1.5			2.0		
	RPD	***p*-Value**	**Sig**	**RPD**	***p*-Value**	**Sig**	**RPD**	***p*-Value**	**Sig**
0.1	0.694	0.000	Y	3.769	0.000	Y	3.613	0.000	Y
0.5	0.458	0.000	Y	3.446	0.000	Y	3.491	0.000	Y
1.0	3.232	0.000	Y	14.599	0.000	Y	14.113	0.000	Y
5.0	78.139	0.000	Y	62.214	0.000	Y	56.982	0.000	Y

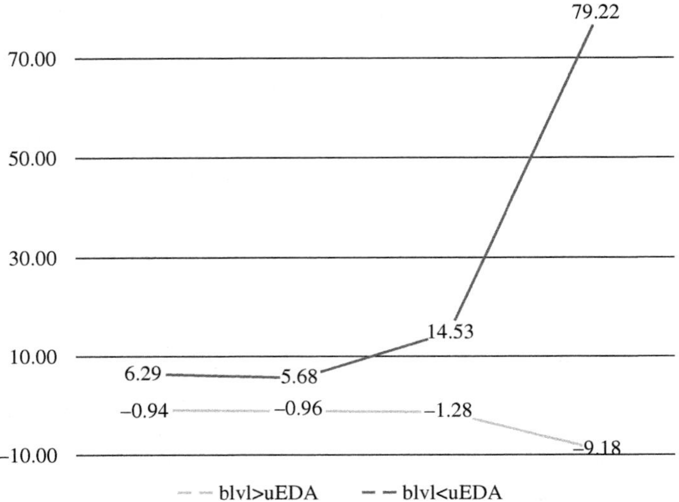

Figure 14.2 Line chat with average value markers of different *CCR* values.

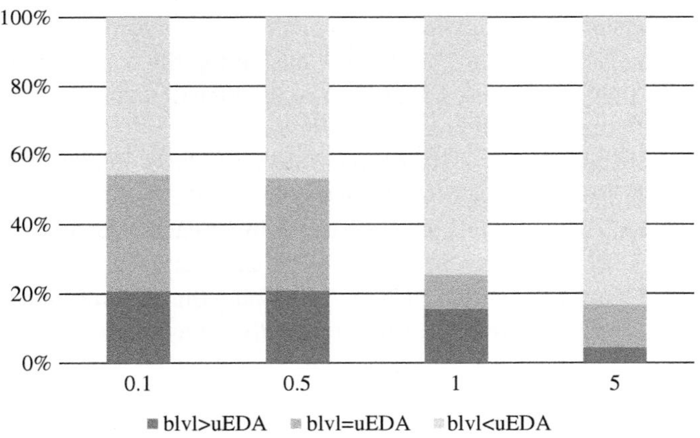

Figure 14.3 100% stack column chat of different *CCR* values.

the superiority of uEDA is much larger than the modified heuristic algorithm. In addition, as *CCR* grows, the superiority increases. And from Figure 14.3, it could be seen that the solutions obtained by uEDA dominate the modified heuristic method in most cases. And when *CCR* grows, the superiority of uEDA turns to be more obvious.

14.7 Conclusion

In this work, a problem specific uEDA is proposed to address the latency optimization of IoT applications under the three tier systems. The task utility is adopted to measure the performance of the task scheduling algorithm. Our proposed uEDA consists of the basic EDA, problem-specific decoding method and local search procedure. Based on the comparison results of the modified heuristic algorithm, it could be proved the efficiency of our proposed algorithm on this problem. In the future, we plan to consider the computing node capacity, as well as the robustness performance of the scheduling algorithm during the optimization. In addition, the prototype system is needed to achieve realistic benchmark instances and to test our algorithm.

References

1 Mattern, F., Floerkemeier, C., Sachs, K. et al. (2010). From the Internet of computers to the Internet of Things. In: *From Active Data Management to Event-Based Systems and More*, vol. 6462 (eds. K. Sachs, I. Petrov and P. Guerrero), 242–259. Berlin, Heidelberg: Springer.

2 D. Chu, Toward immersive mobile virtual reality, Proceedings of the 3rd Workshop on Hot Topics in Wireless (HotWireless '16), ACM, New York, October 3–7, 2016.

3 Tran, T.X., Pandey, P., and Hajisami, A. (2017). Collaborative multi-bitrate video caching and processing in mobile-edge computing networks. In: *Annual Conference on Wireless On-demand Network Systems and Services (WONS)*, 165–172. IEEE Congress on Jackson.

4 M. Gorlatova and M. Chiang, Characterizing task completion latencies in fog computing, ArXiv.org: 1811.02638, (2018).

5 Bonomi, F., Milito, R., Zhu, J., and Addepalli, S. (2012). Fog computing and its role in the internet of things. In: *Proceedings First Edition of the MCC Workshop on Mobile Cloud Computing*, 13–16. Helsinki, Finland, August 17,: ACM.

6 Kwok, Y.K. and Ahmad, I. (1999). Static scheduling algorithms for allocating directed task graphs to multiprocessors. *ACM Computing Surveys* 31: 406–471.

7 Yang, J., Xu, H., and Jia, P. (2013). Efficient search for genetic-based machine learning system via estimation of distribution algorithms and embedded feature reduction techniques. *Neurocomputing* 113: 105–121.

8 Armananzas, R., Inza, I., and Santana, R. (2008). A review of estimation of distribution algorithms in bioinformatics. *Biodata Mining* 1 (1): 6–12.

9 Wang, S.Y. and Wang, L. (2016). An estimation of distribution algorithm-based memetic algorithm for the distributed assembly permutation flow-shop scheduling problem. *IEEE Transactions on Systems, Man, and Cybernetics: Systems* 46: 139–149.

10 Wu, C.G. and Wang, L. (2018). A multi-model estimation of distribution algorithm for energy efficient scheduling under cloud computing system. *Journal of Parallel and Distributed Computing* 117: 63–72.

11 Wu, C.G., Li, L., Wang, L., and Zomaya, A. (2018). Hybrid evolutionary scheduling for energy-efficient fog-enhanced internet of things. *IEEE Transactions on Cloud Computing* 43 (8): 1575–81.

12 Wu, C.G. and Wang, L. (2019). A deadline-aware estimation of distribution algorithm for resource scheduling in fog computing systems. In: *IEEE Congress on Evolutionary Computation (CEC)*, 1–8. IEEE.

13 Yousefpour, A., Fung, C., Nguyen, T. et al. (2019). All one needs to know about fog computing and related edge computing paradigms: a complete survey. *Journal of Systems Architecture* 98: 289–330.

14 Ye, D., Wu, M., Tang, S., and Yu, R. (2016). Scalable fog computing with service offloading in bus networks. In: *Cyber Security and Cloud Computing (CSCloud), IEEE Conference on Beijing*, 247–251.

15 Pinedo, M. and Hadavi, K. (1992). Scheduling: theory, algorithms and systems development. In: *Operations Research Proceedings 1991* (eds. W. Gaul, A. Bachem, W. Habenicht, et al.). Berlin, Heidelberg: Springer.

16 Koulamas, C. (1994). The total tardiness problem: review and extensions. *Operations Research* 42: 1025–1041.

17 Sen, T., Sulek, J.M., and Dileepan, P. (2003). Static scheduling research to minimize weighted and unweighted tardiness: a state-of-the-art survey. *International Journal of Production Economics* 83: 1–12.

18 Liu, J. (2000). *Real-Time Systems*. Prentice Hall.

19 Li, W., Santos, I., Delicato, F.C. et al. (2017). System modelling and performance evaluation of a three-tier Cloud of Things. *Future Generation Computer Systems* 70: 104–125.

15

Reliable and Power-Efficient Machine Learning in Wearable Sensors

Parastoo Alinia and Hassan Ghasemzadeh

Washington State University, Pullman, WA, USA

15.1 Introduction

Metabolic equivalent of task (MET) is an approximation of energy expenditure and an indicator of the intensity of physical activities. This measurement is commonly used to assess performance of physical activity interventions associated with many chronic illnesses such as coronary heart disease, type-2 diabetes, and cancer [1]. Healthy lifestyle changes such as diet control and exercise, which maintain a balance between dietary intake and calories burned, are key approaches in reducing complications due to these diseases [2]. This requires real-time tracking of physical activities that individuals at high risk of chronic diseases perform daily [3]. There are several approaches to calculate food intake and level of physical activity, including traditional self-reported questionnaires, indirect calorie meters, doubly labeled water techniques, and electrocardiographs [3, 4]. In recent years, however, accelerometers, gyroscopes, pressure sensors, and heart rate monitors have been used for physical activity detection and energy expenditure calculation [5–7] due to their small size, portability, low-power consumption, and low cost [3, 4].

Accelerometers have been widely used to estimate energy expenditure and MET of physical activities [3–5, 8]. Although, the current approaches for estimation of MET values using wearable sensors have proven be to accurate [5, 8], they do not take two important issues into consideration for deployment in real-world settings. First, users would naturally tend to add new wearable sensors to the network as new sensors such as smart watches, ankle bracelets and necklaces, become available. Current research requires collection of new labeled training data for the purpose of algorithm development when a new sensor is added to the system. Second, users prefer to carry their mobile devices on various body locations, resulting in a displacement of the sensor [9]. Therefore, claimed accuracy

Fog Computing: Theory and Practice, First Edition.
Edited by Assad Abbas, Samee U. Khan, and Albert Y. Zomaya.

of current Metabolic equivalent of task estimation systems (MES) is dependent on adhering to the deployment protocols; for example, users must wear sensors on predefined body locations. These requirements are limiting practical use and potentially imposing discomfort for end-users. In order to make wearables of the futures more reliable and reconfigurable, the underlying MET estimation model needs to be updated upon changes in the wearable network [10]. This chapter presents a framework to address reliability and reconfiguration challenges of wearable sensor networks. The first part of this chapter addresses the unreliability due to the change of on-body sensor location while taking into account the computation complexity of the devised sensor localization algorithms. The second part of this chapter presents a novel transfer learning algorithm to adopt the knowledge of existing nodes in a new configuration of the network. The result is a reliable, and reconfigurable MES that allows users to change the location of the sensors or add new sensors to the network in real-time without any need for new data collection or retraining of the underlying signal processing and machine learning algorithms. The contributions of this chapter are: (i) a framework for estimating MET numbers to address unreliability and reconfigurability challenges in wearables; (2) a sensor localization algorithm based on machine learning techniques to automatically detect the location of the wearable sensors; (3) a transductive learning approach to transfer the knowledge of existing trained wearable sensors to a newly added untrained sensor; (4) regression-based algorithms for estimating the MET values of physical activities without need for collecting new training data; (v) Assessing the performance of the individual algorithms as well as the entire framework using real-data collected in two experiments involving both daily physical activities and fitness movements.

15.2 Preliminaries and Related Work

MET is the ratio of work metabolic rate to the resting metabolic rate. One unit of MET is defined as $1\,\text{kcal}\,\text{kg}^{-1}\,\text{h}^{-1}$, which is approximately equal to the energy cost of sitting quietly, equivalent to $3.5\,\text{mL}\,\text{kg}^{-1}\,\text{min}^{-1}$. In other words, MET is an energy expenditure measurement that demonstrates the intensity of physical activities [1]. The literature review in this section includes a discussion on how gold standard MET values are computed in clinical practice, followed by the state-of-the-art on estimating MET numbers using wearable sensors, and related research on sensor localization and transfer learning.

15.2.1 Gold Standard MET Computation

In recent studies, two approaches have been used to compute the gold standard MET values of physical activities. First, users can look up the MET values from the Compendium, which contains MET value of almost 300 daily activities in a table [11, 12]. In spite of the ease of use and wide range of physical activities in the Compendium, there are several limitations to this approach. The major problem is that the users need to perform physical activities in a fully controlled environment with knowledge on time and intensity associated with each activity. Therefore, the true energy cost for an individual may or may not be close to the stated mean MET level presented in the Compendium. In the second approach, researchers use a metabolic cart to compute MET values. They measure volume of oxygen breathed into the lungs while performing various activities. The following equation has been used in previous studies to compute the actual MET corresponding to each activity [3]:

$$MET = \frac{VO_2}{f \times m} \tag{15.1}$$

where $VO_2 \left(\frac{ml}{min}\right)$ and m denote the oxygen uptake volume and the mass of the user in kilograms, respectively. The symbol f represents a factor that depends on the general fitness features of the group participated in the experiment [13]. The main problem users may encounter while using this approach is that the metabolic cart is a heavy and expensive device and an oxygen uptake mask need to be worn while performing the activities. Thus, the utility of the metabolic cart is limited to in-door usage and constrained activities.

15.2.2 Sensor-based MET Estimation

To address the dificulties in calculating gold standard MET values, in recent studies, accelerometers have been used to estimate energy expenditure and MET of physical activities [3–5, 8]. This approach requires a training phase to develop an MET estimation model. Usually, several accelerometers are placed on different locations on the body and an estimation model (e.g. a regression model) is developed based on the features extracted from the acceleration signals and the gold standard MET numbers. In one application, researchers developed regression models to estimate MET values when playing a soccer game [3]. It demonstrates that the MET value of soccer exergaming movements can reach a value of 7, which is a standard value for actual casual intensity soccer. The authors in [5] compare a wearable multisensor with a single-sensor approach for energy expenditure estimation. The results show that a wearable multisensor approach outperforms the single-sensor solution using ActiGraph GT3X+ and linear regression. Another study proposes two MET estimation methods, one traditional single and multiple regression models, and one mono-exponential MET estimation method [4]. In all the aforementioned studies, the location of the wearable sensors is fixed, and a retraining of the MET estimation model is needed if a new sensor is added to the network.

15.2.3 Unreliability Mitigation

There have been several research efforts on sensor localization for wearable sensors. Those studies can be divided into three categories as follows: (1) sensor placement on different body locations (e.g. back pocket of trousers versus side pocket of jacket) [14–19]; (2) sensor displacement within a given coarse location (e.g. shifting from top upper arm to middle upper arm) [20, 21]; and (3) changes in the orientation of the sensors [22, 23]. In particular, there exist several recent studies that develop localization techniques based on machine learning algorithms [10, 14–17, 20, 21, 24, 25]. However, none of these studies investigated sensor localization in the context of MET estimation. Sensor misplacement can dramatically decrease the accuracy of MET estimation. There has been minimal effort in developing sensor localization algorithms for MET calculation. In one study, authors use a wearable sensor to estimate energy expenditure without relying on the prior knowledge about the location of the sensor [26]. The system, however, uses a single sensor on three predefined locations on the body and three different models are built for these body-locations. This chapter aims to present a global model of sensor localization where the developed machine learning algorithms can be used uniformly on all the sensors independent of their on-body location. This chapter presents only one model for all the activities users perform while wearing multiple sensors at the same time. The localization accuracy and MET estimation model in the prior research is based on one activity (i.e. walking).

15.2.4 Transfer Learning

To date, there has been no prior study on developing transfer learning approaches for MET estimation. However, researchers have conducted much research on transfer learning in the field of machine learning and artificial intelligence. Transfer learning approaches can be divided into three subcategories of inductive, transductive, and unsupervised based on different conditions of the source and target domains and the knowledge extraction tasks [9]. In inductive transfer learning [27, 28], the target and source tasks are different from each other and labeled data are available in the target domain. In transductive transfer learning [29–31], the source and target tasks are the same, while their domains are different. In this case, labeled data exist in source domain but not in the target view. In unsupervised transfer learning, similar to inductive transfer learning setting, the target task is different but related to the source task, and there exist no labeled data available in either source or target domain [32–34].

In [31], authors derived efficient transductive transfer learning algorithms based on a support vector machine (SVM) paradigm of a large margin hyperplane classifier on a feature space. In another study [29], authors introduce a maximum entropy-based technique, called Iterative Feature Transformation (IFT), and show that it achieves comparable performance with the state-of-the-art transductive SVMs in solving the problem of transfer learning for protein name extraction.

However, the claimed accuracy of the conventional MET estimation methods, using wearable sensors network, is strictly dependent on adhering to a specific deployment protocol. For example, the users are required to wear a fixed number of the sensors on predefined body locations for the conventional systems to maintain a high accuracy, all the time. Moreover, real-time monitoring of the users is not possible when these machine learning models do not consider power optimizations in the design of their algorithms. This chapter presents an MET estimation system that addresses these unreliability challenges that wearables might face when deployed in real-world dynamics. This chapter proposes a power-aware sensor localization algorithm that allows users to relocate the sensors on different body locations, and a transductive transfer learning technique to enable addition of a new wearable sensor to the network of the existing sensors while maintaining the high accuracy without the need to collect new data and retrain the under lying estimation models.

15.3 System Architecture and Methods

This section overviews the structure of the reliable and reconfigurable framework, which aims to address the challenges of sensor misplacement, and automatic sensor addition in wearable networks, with a special focus on MET estimation applications. First part of this section discusses a reliable MET calculation technique that estimates the MET values of physical activities without requiring the sensors to be placed on predefined locations while minimizing the energy consumption due to feature extractions. Second part introduces a novel transfer learning algorithm to allow for addition of new sensors without retraining of the MET estimation model. For this purpose, a system configuration $C(k, l_0, l_1, \ldots, l_k, t)$ is defined as the number of the sensors k, their location on the body $l_0, l_1, \ldots, l_k$ and whether the MET values of physical activities are known ($t = 1$) or not ($t = 0$). Each configuration of the system illustrates a different situation. For example, when a user wears two trained accelerometer sensors, one on the chest, and other on the waist, the configuration of the MES will be $C(2, \textit{chest}, \textit{waist}, 1)$. Since in MES, the MET labels are independent of the location of the sensors, any configuration with mixed type of untrained and trained sensors is not possible.

15.3.1 Reliable MET Calculation

Reliability in the context of MET calculation focuses on two properties of the system: (1) changing the location of the sensors should not have a significant effect on the accuracy of MES; and (2) the system needs to be power efficient in terms of sensing and feature computation complexity. The goal of the location-independent MET calculation is to demonstrate how detecting the location of the sensors before estimating the MET values would increase the accuracy of the underlying linear regression equation in estimating MET values of physical activities. The power-aware feature selection selects the set of optimal features taking into account their information relevance and computation complexity. The proposed reliable MET estimation model contains five main steps: (1) feature extraction, which computes a set of representative features from the signals captured using wearable accelerometers; (2) sensor localization, which is a constructed machine learning algorithm based on the optimal selected features; (3) label assignment, which assigns predicted location labels to the sensor readings based on the output of the sensor localization algorithm; (4) linear regression, which uses the extracted features from wearable sensors to estimate the MET values. As shown in Figure 15.1, in the training phase, the extracted features from the accelerometers labeled with their locations are fed into a classifier to build a location detection model. A linear regression model is then constructed on the magnitude of the accelerometer signals to estimate MET numbers. (5) Activity level classification, which classify the features extracted from the sensor signals into three groups of light, moderate, and vigorous activities based on mapping the exact MET numbers from the Pate model.

15.3.1.1 Sensor Localization

Several time-domain features can be extracted from human activity data collected from different type of sensors (e.g. accelerometers, and gyroscopes). Previous research studies such as the one in [35] have shown that statistical features, such as mean value, maximum va minimum of a signal, are positively effective in human activity recognition. However, the optimal features for sensor node localization have not been explored, to date. One goal to the sensor localization module is to identify the most effective features that can extracted from human activity signal, for node localization. For this purpose, an exhaustive set of potentially useful features is extracted from sensor data; then the number of features for each sensor is minimized to increase the performance of the model.

Table 15.1 lists the set of features extracted from individual sensor streams. This list includes 10 statistical features and 10 morphological features for each data segment.

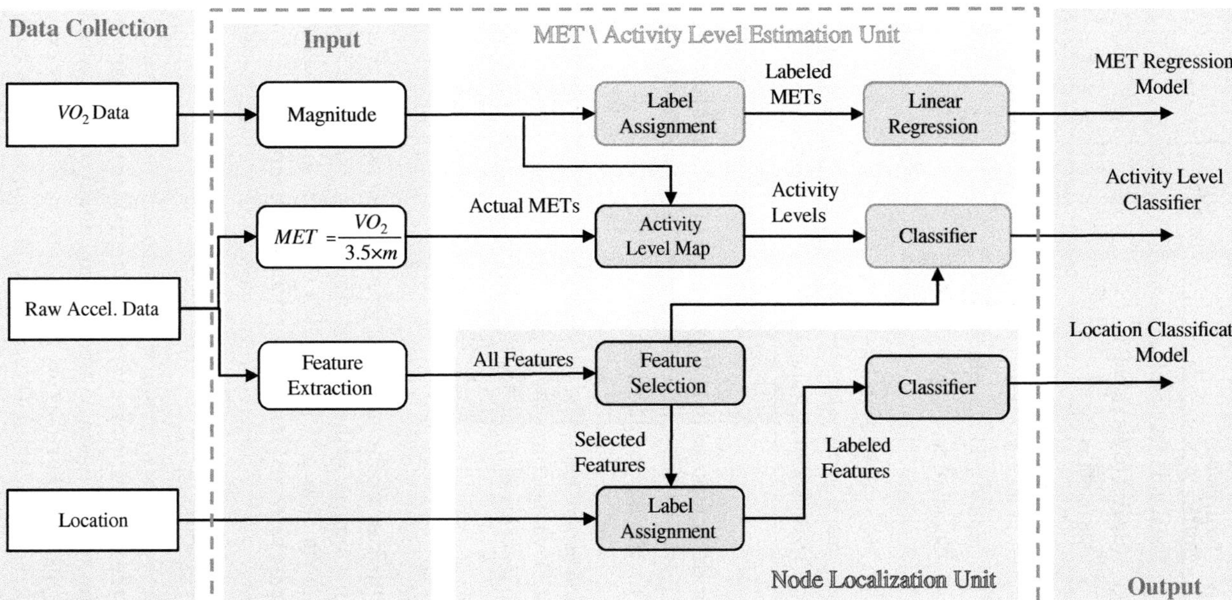

Figure 15.1 The process of developing location-independent MET estimation models.

Table 15.1 Extracted features from sensor signals.

Feature	Description
AMP	Amplitude of signal segment
ED	Median of the signal
MNVALUE	Mean of the signal
MAX	Maximum value of signal
MIN	Minimum value of signal
P2P	Peak to peak amplitude
STD	Standard deviation
VAR	Variance
RMS	Root mean square power
S2E	Stand to end value
MORPH	Morphological features

The greedy algorithm proposed in [10], is applied to extracted features to select the optimal set of features. The output of this module is a subset of features that are most effective to the node localization task.

15.3.1.2 MET Value Estimation

To estimate the MET values from the sensor data, a linear regression model is trained to a line on features as input and the MET values as output as given by:

$$L = \sum_{i=0}^{k-1} A^i f^i \tag{15.2}$$

where A denotes the corresponding coefficients for one feature f_1 (i.e. "magnitude"). The symbol k represents the number of sensors in the network. The result is a linear regression module, which accurately estimates the MET values given a set of features that are extracted from sensor data, as the input.

15.3.2 The Reconfigurable MET Estimation System

The objective of the reconfiguration unit is to transfer knowledge from the existing trained sensors in the network to a newly added (untrained) and adapt to the new configuration of the system. Based on the common analogy in machine learning domain, the new sensor is referred as the target sensor, and all other existing sensors in the network, are called the source sensors. The proposed approach, shown in Algorithm 15.1, transfers the MET labels through two main phases. (1) Sensor selection, which identifies the most similar sensor among the source sensors to the target sensor. (2) MET values transfer, which transfer the MET values from the selected sensor in the source view to the target. The proposed algorithm in Algorithm 15.1 is implemented to select the similar sensor and transfer the MET values. Parameter S_{source} denotes the set of input feature datasets with MET labels extracted from the source sensors. Symbol s_{target} is the target sensor feature set without MET labels. In the first phase, to avoid the possible negative transfer learning effects, the function *MaxCorrelation* selects most similar source sensor to the target sensor (s_{target}) using a correlation factor. The value of this metric ranges from 0, for entirely dissimilar sensors, to 1, for sensor with highest correlation. Function γ computes the correlation factor between two arbitrarily sensors in the algorithm. Equation (15.3) calculates the correlation factor between two arbitrarily sensors s_1 and s_2:

$$\gamma(s_1, s_2) = \sum_{k=0}^{n-1} U(s_1.f_k, s_2.f_k) \tag{15.3}$$

where U is symmetric uncertainty measure between two feature instance and n denotes the number of the features in the datasets. Symbols f_k represents k-the sample in the dataset from each sensor. The symmetric uncertainty between two random variables is given by:

$$U(X, Y) = \frac{2I(X, Y)}{H(X) + H(Y)} \tag{15.4}$$

where $H(X)$ and $H(Y)$ represents the entropy of random variables of X and Y, respectively, and $I(X, Y)$ denotes the information gain between them and is computed by:

$$I(X, Y) = H(X) - H(Y) \tag{15.5}$$

In the second phase, for each instance in the feature set of target sensor, s_{target}, Nearest-Neighbor classification is used to detect the closest instance from the feature set in the selected sensor s_{selected}. Finally, all instances in dataset of the target sensor are assigned with MET labels with the labels from the selected sensor feature set, based on the k-NN classification model, which links each instance in s_{target} to the closest instance in s_{selected} in a Euclidean feature space given by Eq. (15.6).

$$D(I_1, I_2) = \sqrt{\sum_{k=1}^{n} (I_1.f_k - I_2.f_k)^2} \tag{15.6}$$

where I_1 and I_2 are two arbitrarily feature vectors from the sensor datasets. Symbol f_k denotes the kth feature in feature vector instances when there are n feature vectors in the dataset. The output of the algorithm is the dataset of the target sensor s_{target} containing the transferred MET labels.

Algorithm 15.1 Transfer Label Algorithm

```
Input: S_target feature set of the new untrained sensor (without MET labels) which has
       been added to the network of a trained sensors on the body, S_source set of the
       trained sensors in the body sensor network with MET, MET_source
minDist ← +Inf; /* the minimum distance among the pairs of the instances
    in the feature set of the target and source sensors                    */
s_selected ← MaxCorrelation(s_target, S_source) foreach sample d_target in s_target do
    foreach sample d_selected in s_selected do
        if Distance(d_target, t_selected) < minDist then  minDist ← D(d_target, t_selected)
        d_target.met ← d_source.met ;

Output: s_target /* the dataset of the new sensor with transfered MET
    labels                                                                 */
Function MaxCorrelation(s_1, S)
    maxCor ← −Inf; /* the maximum correlation value between the s_1 and
        the sensors in set of sensors S                                    */
    foreach sensor s in S do
        if Correlation(s_1, s) > maxCor then  maxCore ← γ(s_1, s);
        s_selected ← s ;
    return s_selected
```

15.4 Data Collection and Experimental Procedures

This section investigates the effect of sensor localization and the effectiveness of the proposed transfer learning approach in improving the accuracy of the system in computing MET numbers. The proposed approaches are validated on two datasets: (1) exergaming movements [3], and (2) walking on the treadmill. The first dataset is collected from six subjects wearing two accelerometer-based sensors during a clinical experiment. The second dataset includes accelerometer data from three sensors located on the body of 15 participants while walking at three different speeds on a treadmill. Both experiments were approved by Institutional Review Board (IRB).

15.4.1 Exergaming Experiment

The exergaming dataset was collected from six male subjects, aged between 21 and 30 during a clinical trial. In this trial, two data collection modalities were used. Two GCDC $\pm 2g$ tri-axial accelerometers sampling at $50\,Hz$ were attached to the hip and ankle of participants. A metabolic cart that measured the volume of oxygen breathed into the lungs while doing activities was attached to the mouth of the users to compute ground truth MET values. Since the breathing pattern of each individuals is not unique, the average oxygen uptake of participants during a 30-second period was reported. Equation (15.1) is utilized to compute the actual value of MET, corresponding to each activity using the oxygen volume from the metabolic cart. Participants were asked to follow a protocol including six physical activities (e.g. running, sprint, pass, chip, medium shot, and full powered shot) and a simulated game. The duration of the physical activities was selected as 3 minutes to achieve a steady state in breathing. Subjects were allowed to have their own desired intensity while performing the physical activities. Sensor Localization and MET calculation models are developed on the accelerometer sensor data collected from the participants. Intensity of each action was computed based on the work in [3], which required repeating an action for three minutes. The magnitude feature of the tri-axial accelerometer sensor data is computed as a feature indicating the intensity of each action given by [36]:

$$M = \sqrt{x^2 + y^2 + z^2} \tag{15.7}$$

where M refers to the magnitude of each accelerometer data segment.

To synchronize the accelerometer data with VO_2 data, the output signal is averaged over a moving window of 3,000 samples with one sample overlap. As a result, each axis of the accelerometer signal is averaged over 30 seconds, which is the same as the VO_2 data. The magnitude feature of accelerometer data collected from each sensor (e.g. ankle and hip accelerometer sensors) is calculated. The peak of the magnitude signal, which were collected over a time span of 3 minutes, matched to MET value points computed from the metabolic data. Since, accelerometer sensor data is prone to environmental and collection device noises, an approach similar to Ousto's proposed in [7, 37], is required to detect the peaks accurately. This method of peak detection is implemented based on nonlinear neighbor filtering, in which a sliding neighboring is defined and centered on the signal where the output vector is the local maximum (e.g. peak). The peaks of the magnitude of both ankle and hip accelerometers are identified using the equation below:

$$\Phi(t-1) < \Phi(t) > \Phi(t+1) \tag{15.8}$$

where $\Phi(t)$ denotes the feature (magnitude) from accelerometer signal at time t. Figure 15.2 shows the output of the cross-correlation function on magnitude of ankle accelerometer signal from one participant. Seven peaks are detected from each participant corresponding to seven physical activities performed by subjects during the experiments.

15.4.2 Treadmill Experiment

The treadmill experiment was performed during a two-month experimental study on 15 young healthy adults, 5 females, and 10 males, aged from 21 to 33. Each participant was asked to walk at three different speeds (e.g. 2.5, 5.0, and 8.0 km h^{-1}) for

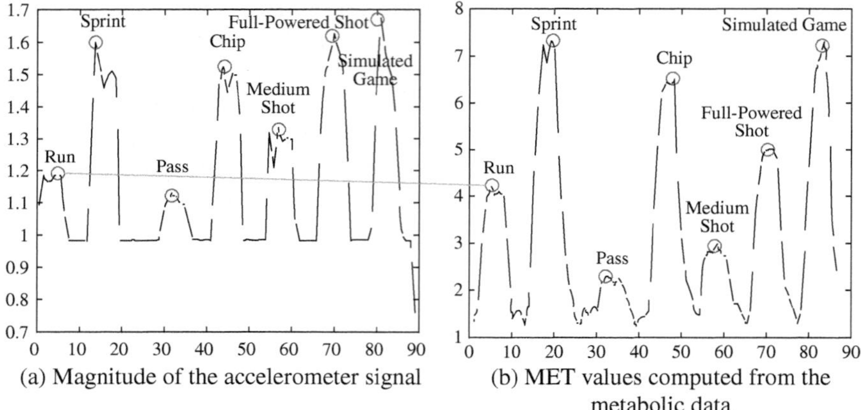

(a) Magnitude of the accelerometer signal

(b) MET values computed from the metabolic data

Figure 15.2 The result of the cross-correlation function on magnitude (a) and MET (c) signals of ankle accelerometer on one subject.

a duration of five minutes each. Three accelerometer-based motion sensors were used to collect three-axis accelerometer signals from the subjects during the experiment. Participants were asked to wear sensor nodes on specific on-body locations (right hand, chest, and left jacket pocket) as shown in Figure 15.3.

Based on what observed in previous research studies, age and gender merely impact the energy expenditure in common activities of daily living such as walking [38]. Therefore, the gold standard MET values of walking with different intensities were determined from the Compendium data. The MET values of 7.0, 3.5, and 2.0 were selected for walking at speed of 8.0, 5.0, and 2.5 km h^{-1}, respectively [11, 12] (Table 15.2).

15.5 Results

This section evaluates the reliability, reconfigurability and robustness of the MET estimation model. The accuracies for sensor localization and MET estimation are reported in the following sections.

15.5.1 Reliable MET Calculation

This section compares the performance of the MES with a baseline MET calculation when the sensor localization is not applied to the sensor data before estimating MET values. First part of the result sections provides results of the sensor localization algorithm. In the second part, the performance of linear regression

Figure 15.3 Three smart phones are placed on three different locations of each participant's body (chest, right hand, and left jacket pocket) to collect accelerometer signals while walking on the treadmill.

Table 15.2 Energy consumption of various configurations, where computation is in *nJ* and sensing and total are in *mJ*.

	Accuracy		Energy			
Feature Selection	**5-NN**	**ANN**	**Computation**	**Sensing**	**Total**	**Power saving**
FFS	90%	97.7%	56 777	5.28	8.11	34.7%
Ranker	93.5%	95.01%	48 757	5.28	7.71	31.4%
Greedy stepwise	91.1%	94.2%	8637	5.28	9.6	44.8%
Power-aware	92%	97.3%	270	5.28	5.29	—

model to estimate MET values, is investigated. Third part evaluates the impact of sensor localization in the accuracy of MET estimation. The last part reports the results on the performance of the proposed models when a new untrained sensor is added to the existing trained network of sensors.

15.5.1.1 Sensor Localization

The sensor localization module aims to identify the location of the sensor on the body of the user using classification algorithms. For this purpose, the localization algorithm is applied to a subset of 6 optimal features out of 40, which were extracted from a moving window of 100 samples (e.g. equivalent to 2s of data) on the data. The proposed localization algorithm is applied to the data collected from both treadmill and exergaming experiments. The localization algorithms are developed by WEKA machine learning tool [39] to classify the signals into two classes of on-body locations (e.g. ankle and hip).

Table 15.3, shows the accuracy of Artificial Neural Network (ANN) and K-Nearest Neighbor (e.g. k = 5) in identifying the location of the sensors in treadmill and exergaming experiments. The average classification accuracy on the sensor signal segments are 94.6% in exergaming experiment and 86.2% in treadmill experiment using 10-fold cross validation method. Between the classifiers, five-NN shows better superior performance with 97.3% and 99.6%, in the two experiments.

15.5.1.2 MET Value Estimation

To estimate the MET numbers from sensor data, a linear regression model is implemented on sensor data. The regression model is developed on all possible combinations of the data collected from sensors on ankle and hip for exergaming dataset, to identify the most accurate model in computing MET values. A recent study in [3] applied the sum of the magnitude feature of the accelerometer signals from different locations on the body as the input identifier to the linear regression model, which is compared against other variations of fusing data collected from sensors on ankle and hip.

Table 15.4 reports the R^2 statistics and error of different fusion on the sensors such as using a single sensor ankle or hip, using summation of both ankle and hip sensors and using ankle, and hip sensors separately together. As shown, using

Table 15.3 Accuracy of sensor localization.

	Exergaming				Treadmill			
	10-fold		LOSOCV		10-fold		LOSOCV	
Classifier	Accuracy	RMSE	Accuracy	RMSE	Accuracy	RMSE	Accuracy	RMSE
5-NN	97.3%	11.7%	87.2%	39.6%	99.6%	4.1%	88.1%	35.6%
ANN	92%	20.9%	86.2%	27.5%	99.6%	4.3%	84.8%	23%
Average	94.6%	16.3%	86.2%	33.5%	99.6%	4.2%	86.5%	29.3%

Table 15.4 The comparison between the accuracy of the regression on different combination of ankle and hip sensors.

	Ankle		Hip		Ankle+Hip		Ankle, Hip	
Sub.	R^2	**RMSE**	R^2	**RMSE**	R^2	**RMSE**	R^2	**RMSE**
1	0.11	3.25	0.71	1.85	0.71	1.58	0.83	1.61
2	0.22	2.5	0.84	1.25	0.53	1.94	1.84	1.25
3	0.63	1.58	0.84	1.02	0.86	0.96	0.92	0.79
4	0.23	2.01	0.60	1.43	0.71	0.23	0.71	1.37
5	0.60	1.02	0.53	1.11	0.80	0.71	0.82	0.75
6	0.76	1.23	0.44	1.17	0.72	1.88	0.79	1.23
All	0.60	1.38	0.32	1.81	0.63	1.34	0.76	1.09
Average	0.45	1.85	0.61	1:37	0:71	1:37	0:80	1:15

ankle and hip signals, improves the average R^2 value from 0.71 to 0.80 and reduces the amount of the error from 1.37 to 1.15 on average compared to the approach used in [3].

Due to differences in the training pattern of each participant, the results are reported based on a leave-one-subject-out validation. This validation approach provides a generalized and unbiased estimation of the performance of the proposed model and can be implemented very efficiently in the case of linear regression [40]. With six subjects in the experiment, in each iteration, a leave-one-subject-out validation is performed, then the root mean squared error (RMSE) and normalized mean squared error (NMSE) are used as validation measurements using following equation:

$$NMSE = \sqrt{\frac{\sum_{i=1}^{n} (M\hat{E}T_i - MET_i)^2}{n}} \times \frac{1}{\lambda} \tag{15.9}$$

where MET and $M\hat{E}T$ are the actual and estimated MET values, respectively. Symbols n and λ, respectively, are the size and range of the measured value. Table 15.5 lists the result of LOSOCV. The NMSE values in all cases are smaller than 0.12, which demonstrates the acceptable performance of the MET estimation model for a separate dataset from the training data.

15.5.1.3 The Impact of Sensor Localization

A linear regression model is developed on the magnitude feature of the data collected from accelerometer sensor and MET numbers that are computed based on

Table 15.5 Leave-one-subject-out cross validation test.

Subject	MSE	NMSE
1	0.41	0.08
2	0.29	0.06
3	0.33	0.05
4	0.59	0.12
5	0.39	0.11
6	0.58	0.09
Average	0.43	0.08

the VO_2 data. In real-life scenarios, there is no guarantee that users wear sensors on predefined locations. To address this issue, sensor localization is integrated with MET calculation to develop a robust MET calculation algorithm. In order to compare the result of MET value estimation, first the location of each signal segment is assumed to be known using the sensor localization algorithm presented in Section 15.5.1.1. Second, one of the most probable mistakes in which user might wear the hip accelerometer sensor on ankle and ankle accelerometer sensor on hip is considered. Such sensors replacement can be simulated by swapping the detected labels for each corresponding signal segment in ankle and hip accelerometers. As a result, two datasets including the baseline with estimated MET values for the ankle and hip, and the other one with the same features but swapped MET values for the ankle and hip accelerometers.

Table 15.6 contains the linear regression model for each participant. The regression model is applied accelerometer data collected from each participant to

Table 15.6 R^2 values from linear regression on MET vs Ankle and Hip Accelerometers.

Subject	R^2	Linear equation
1	0.83	$11.7\mu_a + 16.9\mu_h - 25.9$
2	0.84	$-2.3\mu_a + 30.3\mu_h - 24.8$
3	0.92	$2.7\mu_a + 7.3\mu_h - 7.4$
4	0.71	$2.2\mu_a + 17.9\mu_h - 18$
5	0.82	$1.8\mu_a + 1.7\mu_h - 2.5$
6	0.79	$3.0\mu_a + 2.5\mu_h - 3.2$
All	0.76	$4.9\mu_a + 4.0\mu_h - 8.0$

develop a linear formula that assumes the location of the sensors are known for the model based on the result of the proposed localization algorithm. Table 15.6 reports the R^2 values corresponding to each linear regression model. There exists variability among the model corresponding to each subject, which could be originated from that fact that different individual exhibit different intensities and patterns during the same physical activity. For instance, some people may move their hip during walking while others do not. Based on the results, the linear regression model represents all the subjects with R^2 value of 0.76.

The R^2 correlation coefficient of a regression model only reports how well the line fits to the training data. Therefore, an estimation error (EE) parameter is defined to more accurately estimate the performance of the proposed model and compare it against the baseline technique. EE is defined as the absolute difference between the actual MET values computed from the metabolic data (MET) and estimated MET values ($M\hat{E}T$) from the regression model, and is shown by ε in Equation (15.10).

$$\epsilon = | M\hat{E}T - MET | \tag{15.10}$$

Table 15.7 reports the R^2 metric of the linear model fitted on the magnitude feature of accelerometer sensor data and actual MET numbers. It compares the R^2 metric when sensor localization is applied to data with the case when not. The R^2 of proposed linear regression models (e.g. with sensor localization) ranges from 71% to 92%, while it dramatically decreases down to negative values when the location of the sensors is unknown (e.g. without sensor localization). A negative value for R^2 correlation for a linear regression model, which is fitted on the data samples demonstrates a poorly fitted model on the dataset. As a conclusion based on the average estimation error results, locating the position of the sensor nodes on different body parts improves the R^2 of the MES with a factor of 2.3 comparing to when there is no prior knowledge on the location of the sensors.

Table 15.7 Comparing R^2 values and error of linear regression on MET vs Ankle and Hip Accelerometers with and without sensor localization.

Subject	R^2 with NL	EE with NL	R^2 without NL	EE without NL
1	0.83	0.29	0.68	0.30
2	0.84	0.12	−19.30	0.34
3	0.92	0.09	−0.10	0.34
4	0.71	0.18	−14.41	0.96
5	0.82	0.13	0.34	0.15
6	0.79	0.20	0.50	0.37
Average	0.80	0.18	−4.98	0.41

15.5.2 Reconfigurable Design

The performance of the transfer learning approach is evaluated using both exergaming and treadmill datasets during two main system configurations. (1) Assume the user is wearing a set of sensors on different on-body locations while the MET values are available to the system through user annotation. The system could train an accurate linear model based on the data collected from each sensor and actual MET numbers. We call this configuration the source view, and the data collected data during this scenario the source dataset. (2) Then, the user decides to wear a new untrained sensor on a different location of the body, while the actual MET numbers to the physical activities are unknown to the system. This section aims to examine the ability of the proposed transfer learning approach in accurate estimation of MET values, despite the fact that the MET labels are unknown. We call this configuration and the collected data the target view and target dataset.

The MET estimation accuracy of the proposed transfer learning method is compared against an upper-bound and a baseline approach. The upper-bound method assumes the actual MET numbers to the physical activities are available in the target view. Therefore, it estimates the MET labels by fitting the linear regression model on the actual MET labels of the second configuration. The baseline approach merely uses the model trained on the source dataset collected from the source view for MET label estimation in the target view.

15.5.2.1 Treadmill Experiment

In the treadmill experiment, participants wore three accelerometers on the right hand, chest, and left jacket pocket. In the source view, users wear two sensors on two of the locations while walking on the treadmill. The MET numbers are available from the Compendium, in this configuration (e.g. source view). The sensors on the chest and in the left jacket pocket were selected arbitrarily for the first configuration ($C(2; \textit{chest}, \textit{pocket}, 1)$). In the second configuration (e.g. target view), the users hold a new untrained sensor in the their right hand while walking on the treadmill, while the actual MET values are not available ($C(3, \textit{chest}, \textit{pocket}, \textit{waist}, 0)$, 0 shows the unavailability of the MET values). Figure 15.4 compares the R^2 value of MET estimation during three different configuration of the baseline, proposed transfer learning and upper-bound.

Based on the MES architecture, the transfer learning algorithm identifies the most similar sensor from existing set of source sensors in the first configuration (e.g. chest and left jacket pocket sensors) to the target sensor (right hand sensor) based on a correlation factor. The target sensor (e.g. right-hand sensor) is more correlated to the chest rather than the left pocket sensor. Therefore, the MET values for the target view are estimated based on the similarity of the target data and source data collected from the chest sensor.

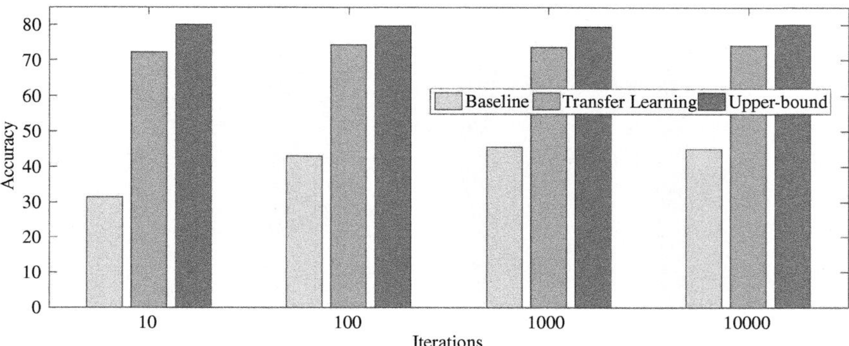

Figure 15.4 Performance comparison of the proposed transfer learning approach versus the base-line and upper bound regression model estimation in treadmill dataset.

The proposed transferred learning algorithm is run for 10, 100, 1,000, and 10,000 iterations, where in which iteration the dataset is shuffled before partitioning into source and target datasets. The average accuracy value of upper-bound, transfer learning and baseline methods during all the iterations have been reported in Figure 15.4. The upper bound on the average R-squared value of the linear regression on the treadmill dataset, in 1,000 iterations, reaches to 79.9%. In the baseline, existing models from the source view are applied to estimate the MET values in the target view. In this case, the R^2 value drops significantly to 44.9%. Applying the proposed transfer learning algorithm, the average R^2 value increases to 74.0%, which is only 7.3% drop comparing to the case when the actual MET values for the activities in the experiment are known (e.g. upper-bound).

15.5.2.2 Exergaming Experiment

In the exergaming experiment, the participants wore two accelerometer sensors, one on the ankle and other on the hip. The R^2 value of estimating MET values of the aforementioned three approaches are compared, as shown in Figure 15.5. In the first configuration, the user wears only one sensor on the ankle while performing the exergaming movements, with known MET values from the metabolic cart ($C(1, Ankle, 1)$). Later, in the second configuration, user adds another sensor on the hip during the experiment. However, the actual MET values of the activities user is performing are unknown ($C(2, Ankle, Hip, 0)$).

As one can observe in Figure 15.5, using the exergaming dataset of two sensors and actual MET values from the metabolic cart, the average R^2 value of MET estimation was 74% in 1,000 iterations in the proposed transfer learning algorithm, which is an upper-bound to the demanding R^2 value for the transfer learning approach. Utilizing the model from the first configuration compensate the drop in R^2 value by 20%. However, with the proposed transfer learning approach, the R^2 value reached to 62%.

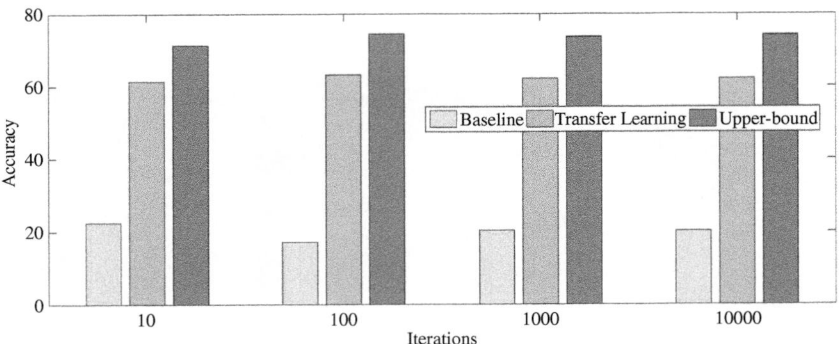

Figure 15.5 Performance comparison of the proposed transfer learning approach versus the base-line and upper bound regression model estimation in exergaming dataset.

15.6 Discussion and Future Work

This chapter illustrates how the presence of unreliability in the system (e.g. sensor misplacement) can affect the result of MET estimation using wearable sensors. First, different locations on the body might have different accuracy in MET estimation. Moreover, by what observed in Table 15.4, hip is a better wearing site for accelerometer sensor to estimate MET values comparing to the ankle. Looking at the individual results, when the sensor is worn on the hip, we observe higher R^2 value for the linear regression models compared to the ankle, except for the last two subjects. Comparing the results in Tables 15.4 and 15.6, the coefficient factor corresponding to the magnitude feature extracted from the sensor worn on the optimal location (e.g. hip) has a bigger value than the coefficient corresponding to the other sensor. For example, for subjects 1–4 in for whom the hip sensor is more accurate, the coefficient is bigger than the one for the ankle. Since there is not a consistent pattern for the optimal sensor location for all the subjects, we fused the magnitude features from ankle and hip, therefore achieved an estimation model with higher R^2 value than the either of the sensors separately ($R^2 value = 0.82$). Second, as shown in Table 15.7, changing the location of a sensor on the body results in average 2.3 times increase in the error of the MET estimation model, which developed on the last known location.

However, sensor misplacement is not the only concern regarding the use of wearables. There are several other unreliability factors such as displacement of the sensors. One future direction is to develop new algorithms to resolve the issue of sensor displacement of wearables in the context of MET estimation. Adding a new sensor to the network of wearable sensors can decrease the R^2 value of MET estimation significantly if the new sensor does not learn MET labels properly.

Therefore, R^2 value is increased by transferring knowledge about MET numbers of the most relevant sensor among the already trained sensors in the network. The proposed transfer learning algorithm is developed based on 1-Nearest Neighbor classification using Euclidean distance as the measure of similarity. One way to improve the performance of this approach is to investigate other classification techniques to transfer knowledge and other distance measures to compute the instance-based similarity among the features. Furthermore, other than adding a new sensor to the network of sensors, people tend to remove existing sensors in the wearable sensor network while doing physical activities which might decrease the R^2 value of the base model significantly. In order to address this issue, the future work involves studying dynamic reconfiguration of underlying signal processing models to compensate for sensor removal.

15.7 Summary

A reliable and reconfigurable design for a MET estimation system is presented in this chapter. The first aim was to maintain robustness and reliability cross different sensor locations. Moreover, it reconfigures the underlying estimation models when a new untrained sensor is added to the network of existing sensors in the system. MES identifies the location of the sensors using classification algorithms, which use the time-domain features extracted from wearable sensors data as their input. To evaluate the reliability of MES cross different sensor locations, a comparison is made between the MET estimation accuracies in two different scenarios. (1) The location of the data collecting sensor is estimated by the proposed sensor localization algorithm, and therefore is known to the system. (2) The location of the sensor nodes are unknown to the system. The second aim was reconfiguration of the underlying estimation model when the user adds a new untrained sensor to the network, using an algorithm based on the transductive transfer learning approach. Using the transfer learning algorithm presented in Algorithm 15.1, the R^2 value when a new sensor was added to the wearable sensor network is improved significantly, when MET values were unknown, in both exergaming and treadmill datasets. MES achieved average accuracy of 66% and 74% in MET estimation, respectively, in exergaming and treadmill experiments. These results demonstrate 60% improvement in R^2 value on average in both experiments, which proves the effectiveness of the proposed transfer learning approach in this chapter.

MES deployed a regression model to determine the MET values corresponding to soccer exergaming movements. Based on the results, the average error of estimating MET values corresponding to exergaming movements with sensor localization, is 2.3 times less than the case with no sensor localization. This chapter presented a localization algorithm based on the K-Nearest Neighbors (KNN) classifier to detect the location of on-body sensors with an accuracy of 92%. The proposed models could estimate the MET values of several exergaming movements with R^2 value 74% using two accelerometer sensors on hip and ankle, and the MET values computed from the metabolic cart.

References

1 Warburton, D.E.R., Nicol, C.W., and Bredin, S.S.D. (2006). Health benefits of physical activity: the evidence. *Canadian Medical Association Journal* 174 (6): 801–809.

2 N. Hezarjaribi, R. Fallahzadeh, and H. Ghasemzadeh, A machine learning approach for medication adherence monitoring using body-worn sensors. Proceedings of the 2016 Design, Automation & Test in Europe Conference & Exhibition (DATE), ICC, Dresden, Germany, March 14–18, 2016.

3 Mortazavi, B., Alsharufa, N., Lee, S.I. et al. (2013). MET calculations from on-body accelerometers for exergaming movements. In: *2013 IEEE International Conference on Body Sensor Networks (BSN)*, 1–6.

4 Altini, M., Penders, J., Vullers, R., and Amft, O. (2015). Estimating energy expenditure using body-worn accelerometers: a comparison of methods, sensors number and positioning. *IEEE Journal of Biomedical and Health Informatics* 19 (1): 219–226.

5 B., D., Biswas, S., Montoye, A., and Pfeiffer, K. (2013). Comparing metabolic energy expenditure esti-mation using wearable multi-sensor network and single accelerometer. In: *Engineering in Medicine and Biology Society (EMBC), 2013 35th Annual International Conference of the IEEE*, 2866–2869.

6 R. Fallahzadeh, S. Aminikhanghahi, A. N. Gibson, and D. J. Cook, Toward personalized and context-aware prompting for smartphone-based intervention,in the 38th Annual International Conference of Engineering in Medicine and Biology Society (EMBC). IEEE, 2016.

7 Ma, Y., Fallahzadeh, R., and Ghasemzadeh, H. (2015). Toward robust and platform-agnostic gait analysis. In: *2015 IEEE 12th Inter-National Conference on, Wearable and Implantable Body Sensor Networks (BSN)*, 1–6. *IEEE*.

8 Kawahara, Y., Ryu, N., and Asami, T. (2009). Monitoring daily energy expenditure using a 3-axis accelerometer with a low-power microprocessor. *e-Minds*: 1–1.

9 Pan, S.J. and Yang, Q. (2010). A survey on transfer learning. *IEEE Transactions on Knowledge and Data Engineering* 22 (10): 1345–1359.

10 Saeedi, R., Schimert, B., and Ghasemzadeh, H. (2014). Cost-sensitive feature selection for on-body sensor localization. In: *Proceedings of the 2014 ACM*

International Joint Conference on Pervasive and Ubiquitous Computing: Adjunct Publication, UbiComp '14 Adjunct, 833–842. New York: ACM.

11 Ainsworth, B.E., Haskell, W.L., Herrmann, S.D. et al. (2011). 2011 compendium of physical activities: a second update of codes and met values. *Medicine and Science in Sports and Exercise* 43 (8): 1575–1581.

12 Ainsworth, B.E., Haskell, W.L., Leon, A.S. et al. (1993). Compendium of physical activities: classification of energy costs of human physical activities. *Medicine and Science in Sports and Exercise* 25 (1): 71–80.

13 Brzycki, M. (1995). *A Practical Approach to Strength Training*. Masters Press.

14 Amini, N., Sarrafzadeh, M., Vahdatpour, A., and Xu, W. (2011). Accelerometer-based on-body sensor localization for health and medical monitoring applications. *Pervasive and Mobile Computing* 7 (6): 746–760.

15 Kunze, K. and Lukowicz, P. (2007). Using acceleration signatures from everyday activities for on-body device location. In: *2007 11th IEEE International Symposium on Wearable Computers*, 115–116. IEEE.

16 Kunze, K. and Lukowicz, P. (2014). Sensor placement variations in wearable activity recognition. *Pervasive Computing, IEEE* 13 (4): 32–41.

17 Kunze, K., Lukowicz, P., Junker, H., and Troster, G. (2005). Where am I: recognizing on-body positions of wearable sensors. In: *Location-and Context-Awareness*, 264–275. New York: Springer.

18 A. Vahdatpour, Unsupervised techniques for time series mining and information extraction in medical informatics, PhD dissertaion, University of California at Los Angeles, 2010.

19 Vahdatpour, A., Amini, N., and Sarrafzadeh, M. (2011). On-body device localization for health and medical monitoring applications. In: *2011 IEEE International Conference on, Pervasive Computing and Communications (PerCom)*, 37–44. *IEEE.*

20 Banos, O., Toth, M.A., Damas, M. et al. (2014). Dealing with the effects of sensor displacement in wearable activity recognition. *Sensors* 14 (6): 9995–10023.

21 Sagha, H., Bayati, H., Millan, J.d.R., and Chavarriaga, R. (2013). On-line anomaly detection and resilience in classifier ensembles. *Pattern Recognition Letters* 34 (15): 1916–1927.

22 Banos, O., Damas, M., Pomares, H., and Rojas, I. (2012). On the use of sensor fusion to reduce the impact of rotational and additive noise in human activity recognition. *Sensors* 12 (6): 8039–8054.

23 Jiang, M., Shang, H., Wang, Z. et al. (2011). A method to deal with installation errors of wearable accelerometers for human activity recognition. *Physiological Measurement* 32 (3): 347.

24 Saeedi, R., Amini, N., and Ghasemzadeh, H. (2014). Patient-centric on-body sensor localization in smart health systems. In: *48th Asilomar Conference* *on Signals, Systems and Computers, ACSSC 2014, Pacific Grove, CA, USA*, 2081–2085.

25 Saeedi, R., Purath, J., Venkatasubramanian, K., and Ghasemzadeh, H. (2014). Toward seamless wearable sensing: automatic on-body sensor localization for physical activity monitoring. In: *Engineering in Medicine and Biology Society (EMBC), 2014 36th Annual International Conference of the IEEE*, 5385–5388. *IEEE*.

26 B., D., Montoye, A., Biswas, S., and Pfeiffer, K. (2014). Metabolic energy expenditure estimation using a position-agnostic wearable sensor system. In: *2014 IEEE, Healthcare Innovation Conference (HIC)*, 34–37.

27 Croonenborghs, T., Driessens, K., and Bruynooghe, M. (2008). Learning relational options for in-ductive transfer in relational reinforcement learning. In: *Proceedings of the 17th International Conference on Inductive Logic Programming, ILP'07*, 88–97. Berlin, Heidelberg: Springer-Verlag.

28 Raykar, V.C., Krishnapuram, B., Bi, J. et al. (2008). Bayesian multiple instance learning: automatic feature selection and inductive transfer. In: *Proceedings of the 25th International Conference on Machine Learning*, 808–815.

29 Arnold, A., Nallapati, R., and Cohen, W.W. (2007). A comparative study of methods for transduc-tive transfer learning. In: *Seventh IEEE International Conference on Data Mining Workshops, 2007. ICDM Workshops 2007*, 77–82.

30 Bahadori, M.T., Liu, Y., and Zhang, D. (2014). *Knowledge and Information Systems* 38 (1): 61–83.

31 Quanz, B. and Huan, J. (2009). Large margin transductive transfer learning. In: *Proceedings of the 18th ACM Conference on Information and Knowledge Management, CIKM '09*, 1327–1336. New York: ACM.

32 Bengio, Y. (2012). Deep learning of representations for unsupervised and transfer learning. *Unsupervised and Transfer Learning Challenges in Machine Learning* 7 (19).

33 Mesnil, G., Dauphin, Y., Glorot, X. et al. (2012). Unsupervised and transfer learning challenge: a deep learning approach. In: *JMLR W& CP: Proceedings of the Unsupervised and Transfer Learning Challenge and Workshop*, vol. 27 (eds. I. Guyon, G. Dror, V. Lemaire, et al.), 97–110.

34 Raina, R., Battle, A., Lee, H. et al. (2007). Self-taught learning: transfer learning from unlabeled data. In: *Proceedings of the 24th International Conference on Machine Learning, ICML '07*, 759–766. New York: ACM.

35 Olgun, D. and Pentland, A.S. (2006). Human activity recognition: accuracy across common locations for wearable sensors. In: *Proceedings of 2006 10th IEEE International Symposium on Wearable Computers, Montreux, Switzerland*, 11–14. *Citeseer*.

36 Mortazavi, B., Rokni, A., Ghasemzadeh, H. et al. (2015). *IEEE*.

37 Glasbey, C.A. (1993). An analysis of histogram-based thresholding algorithms. *CVGIP: Graph, Models Image Process* 55 (6): 532–537.

38 Yue, A.S.Y., Woo, J., Ip, K.W.M. et al. (2007). Effect of age and gender on energy expenditure in common activities of daily living in a Chinese population. *Disability and Rehabilitation* 29 (2): 91–96.

39 Hall, M., Frank, E., Holmes, G. et al. (2009). The weka data mining software: an update. *ACM SIGKDD Explorations Newsletter* 11 (1): 10–18.

40 Cawley, G.C. and Talbot, N.L.C. (2008). Efficient approximate leave-one-out cross-validation for kernel logistic regression. *Machine Learning* 71 (2–3): 243–264.

16

Insights into Software-Defined Networking and Applications in Fog Computing

Osman Khalid, Imran Ali Khan, and Assad Abbas

Department of Computer Sciences, COMSATS University Islamabad, Abbottabad, Pakistan

16.1 Introduction

Software-defined networking (SDN) is an emerging implementation model for wired networking that decouples the data plane (system that forwards data to destination) from control plane (system that makes decision about where to send data). The aim of SDN is to simplify the network management. In traditional networks, the control and data planes are merged into the same networking device due to which the flexibility to accommodate new services is reduced. Moreover, devices manufactured by one vendor could not interoperate with the devices of other vendors. As a result the businesses were locked to a particular vendor specific hardware tool. SDN decouples data plane and control plane. With this separation the networking companies can use equipment from various vendors and can apply policies on the entire network. The switches are only used to forward data while the centralized SDN controller is used to implement control logic [1]. SDN provides multiple opportunities to traditional networks. The management of the network becomes efficient and easy with centralized control. These frameworks are well suited for end-to-end communication across different wireless technologies.

The Open Network Foundation (ONF) has defined architecture for SDN as illustrated in Figure 16.1. The ONF categorized SDN into three layers [2]:

1. The application layer consists of various end-users' business applications that utilize the communication services of SDN. Northbound application programming interface (API) provides the connection between application layers and the control layer.
2. The control layer controls the network forwarding decision through centralized logic.

Fog Computing: Theory and Practice, First Edition.
Edited by Assad Abbas, Samee U. Khan, and Albert Y. Zomaya.

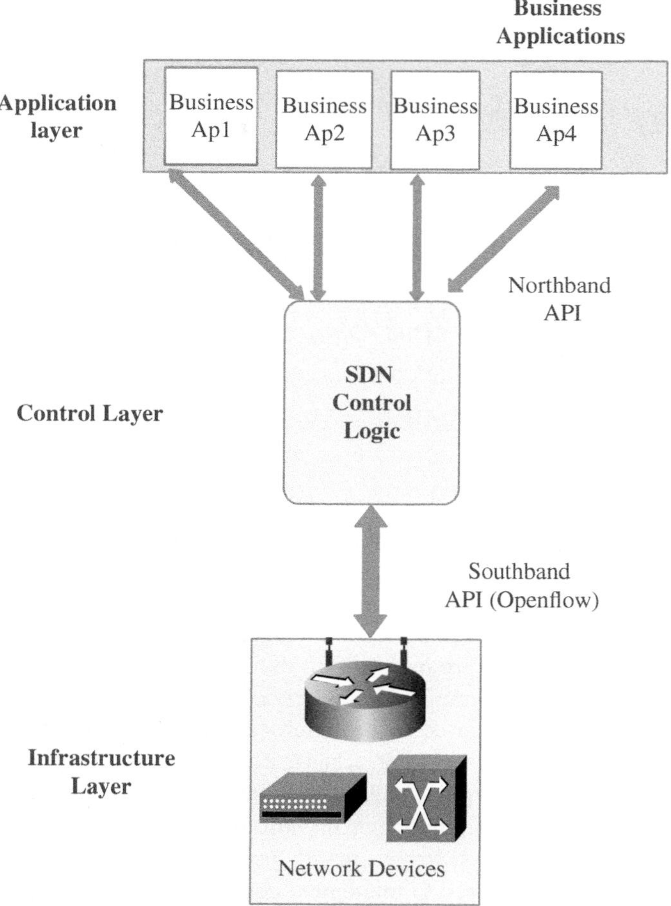

Figure 16.1 Architecture of software-defined network (SDN). (*See color plate section for the color representation of this figure*)

3. The infrastructure layer is composed of multiple network nodes. These nodes provide packet switching and forwarding from source to destination. OpenFlow communication protocol provides interaction between control layer and infrastructure layer. OpenFlow is the required interface for the SDN architecture.

To support sophisticated services in wireless frameworks, mobile operators should design devices that could manage growing traffic in these paradigms. Cellular interference in these frameworks is high due to small cell size. Mobile operators have to reuse frequency to increase capacity. Numerous wireless technologies

3G, 4G, Wi-Fi, and Bluetooth are being supported in today's mobile devices. To maintain a mechanism to migrate users from one technology to another is a challenging task. Seamless provision of services across different technologies is also a challenge for mobile operators. These challenges could be overcome by adding OpenFlow-based SDN communication protocol in wireless networks [2]. Centralized logic control in wireless SDN provides efficient collaboration of base stations and controller, hence avoiding inter-cellular interference. The logical centralized control in SDN has the potential to control any OpenFlow-based mobile nodes that are associated with any vendor. New network services and capabilities are expected to introduce quickly. This logical controller can also provide efficient use of resources, such as optimal use of the radio frequency spectrum. OpenFlow with SDN provides potentials to network operators to customize the behavior of the network. This makes efficient decisions of off-load flow to manage network traffic [3].

The use of SDN in mobile wireless networks not only simplifies networking but also provides simplicity in the deployment of protocol and application by providing collaboration between access points [1]. The use of multiple controllers provide the user with the flexibility to connect with any access point regardless of the operator it belongs to. Wireless SDN provides the flexibility of link sharing to network operators. For example, in case of sudden excessive traffic in a link the Internet service provider could off-load its traffic to other collaborative network technology. SDN monitoring mechanism also provides a clear vision of network activities. Any abnormal behavior could be recognized easily. Wireless SDN has the potential to provide user localization information even in the absence of GPS with the help of the controller that is collecting information from access points.

Besides these potentials, there are multiple challenges associated in wireless-based SDN networks (SDWN) [4]. Wireless networks have link separation challenges. SDN and OpenFlow were basically designed for wired networks and utilizing them in wireless infrastructures is difficult. Interference among the nodes is the main issue in these frameworks that limit the numbers of channels reused. The main types of interferences in these architectures are: (1) the receivers could not decode the signal due to physical interference because it is receiving some other strong signals, and (2) intrusion due to simultaneous transmission of multiaccess protocol [5]. Moreover, segmentation is required in wireless networks for isolation of communication channels to minimize interference between these segments. For wired and infrastructure networks, detachment of the link is simple by implementing different wavelengths in cables. In wireless networks, time synchronization (such as in time division multiplexing) and bandwidth wastage (in frequency division multiplexing) have to be taken care of. Load estimation on a wireless medium is handy due to the frequent change of state in the medium. Handling frequent handoff is another big challenge in implementing SDN. However, overall advantage of SDN dominates its challenges [1].

The rest of the chapter is organized as follows. Section 16.2 describes the OpenFlow protocol and architecture of OpenFlow switch. Section 16.3 outlines the different SDN-based research works. Section 16.4 surveys different state-of-art of SDN in fog computing. Section 16.5 discusses the applications of SDN in wireless mesh networks. Section 16.6 presents SDN in wireless sensor networks, and finally Section 16.7 concludes the chapter.

16.2 OpenFlow Protocol

OpenFlow is a standardized interface between the control and forwarding layers of SDWN architecture. ONF provides standards for OpenFlow [1]. It is the most common SDN interface manufactured by a number of vendors including IBM [6]. NOX, Floodlight, and Maestro are the examples of OpenFlow interface [6]. OpenFlow provides the integration of heterogeneous devices in a common way. It provides seamless subscriber mobility that works across different cellular technologies.

16.2.1 OpenFlow Switch

In SDN, OpenFlow switch is a hardware device used to forward packets. OpenFlow switch consists of OpenFlow secure channel, an external controller, and one or more flow tables. The basic function of OpenFlow switch is packet forwarding to the external controller. The switch management is performed by controller through OpenFlow protocol. The basic function of the controller is to add and remove flow entries from the flow table interface [6]. The controller performs the required operations, such as addition, deletion, and update, on the flow. The architecture of the OpenFlow switch is depicted in the Figure 16.2.

A set of flow entries is associated with every flow table in the switch. Every flow entry contains the following fields: (1) counters to notify against matching packets, (2) matching field to compare against packets. This consists of an entrance port and packet header with optional metadata provided by the previous table, and (3) instructions to apply to matching packets [7]. Matching of flow starts from the first flow table and proceeds to the next. The priority-based mechanism is used by flow entries. First, the matching entry in the table is found. After finding the matching entry, the associated commands with particular flow entry are performed. If no matching entry is found, then the packet may be dropped, or forwarded to the next flow table. Commands associated with each flow table indicate packet refinement, packet forwarding, pipeline computing, or group table computing. Pipeline computing allows instructions to the succeeding flow table for further executions. If the instruction set connected with a matching flow entry

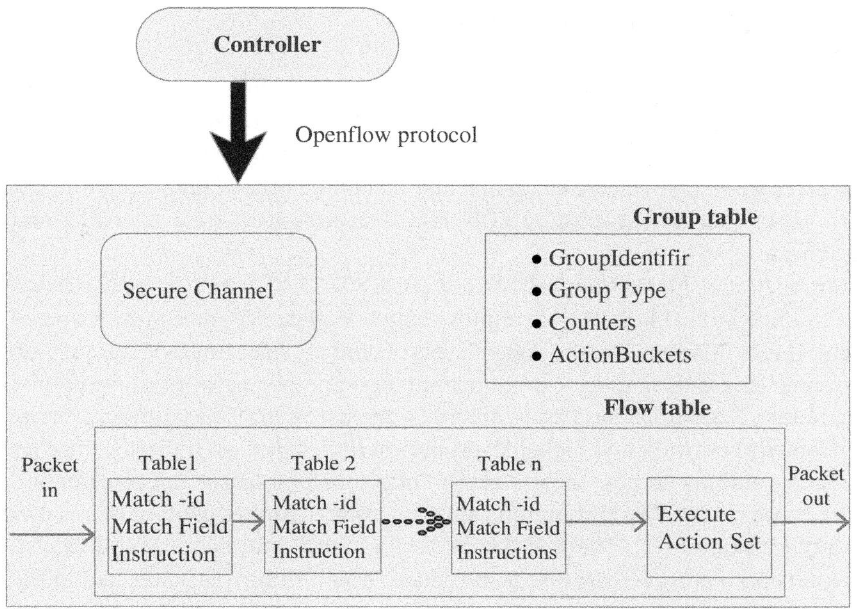

Figure 16.2 An OpenFlow switch communicate with controller over a secure connection using the OpenFlow protocol.

does not identify the next table, then the table pipeline computing is stopped. At this stage, the packets are either modified or forwarded. Packets may be forwarded to physical or virtual ports defined by the switch.

Group entries are part of group table. The capability for a flow to direct to a group empowers OpenFlow to use additional techniques of forwarding (e.g. multicast, broadcast). Each group entry consists of the following fields: (1) group identifier to uniquely identify a group, (2) group types, (3) counters that are updated after the processing of packets in group, (4) action buckets that contain the ordered list of actions. Each action bucket consists of some specific action to be executed.

The group types are named as: all group, select group, and indirect group. All group types process all buckets in the group and are used for broadcasting or multicast forwarding. Select group types process one bucket in the group. Indirect groups only process a single defined bucket in the group. The OpenFlow switch and controller are connected through an interface called OpenFlow channel. The controller manages the switch and receives messages from switch through this interface [6]. In the next section we present the research that applies SDN in the wireless environment.

16.3 SDN-Based Research Works

By using SDN in wireless networks, the control functionalities of network devices (base station [BS] and gateways) are migrated to centralized logical control. The centralized logical control enables the underlying physical infrastructure resources to be abstracted from global and dynamic traffic management. In this section, we discuss the existing SDN infrastructures with their strengths and weaknesses.

Costanzo et al. [3] have applied SDN on IEEE 802.15.4 low-rate wireless personnel area networks (LR-WPAN) to improve network efficiency and control management. Due to differences at the higher layers of protocol stack, the devices applying the same access technology cannot operate into another network in traditional technology. Such problems can be solved by the use of SDN, as the functionality performed at network and higher layers are controlled by a centralized controller (CC). The authors propose the protocol architecture for a sensor and actuator network based on LR-WPAN communication nodes with limited computational and energy capabilities [3]. The nodes consist of a micro-controller and transceiver. The network controller runs on a sink node, which in turn is connected to the embedded system. The proposed architecture is composed of generic node and sink node. All generic nodes form a peer-to-peer topology and run the basic physical and mandatory access control (MAC) layer functionalities of IEEE 802.15.4 standard. Packets from MAC layers are provided to the forwarding layer that identifies the type of the packets. The control packets are sent to network operating system (NOS), whereas the data packets are subjected to the required functionality of modification, broadcasting, and copying. A sink node's architecture is divided into two sublayers. The lower layer runs on the same type of hardware as that of generic node. The upper layer provides the additional high-level computational and communicational functionalities that are executed in the Linux base operating system. Devices and embedded systems are connected through communication interfaces like USB and RS232. The adaption layer in the embedded system is responsible for formatting messages. Besides the application layer, sink node consists of one or more controller(s) that implement network management. The topology information in 802.15.4 is collected by NOS. This information is needed by every node for communication with the sink node. Actions on the packets are performed according to packet format, which is described in the packet header. The proposed technique optimizes the system and shows effectiveness in the defined domain by utilizing lesser resources, and exhibits shorter response time for the end-users [3]. Moreover, the proposed system is generalized enough to allow customization. However, the security mechanism of the system is not that efficient and the system is not interoperable [8].

Sun et al. propose SDN wireless architecture for the next generation mobile communication (5G) to simplify the network management [9]. A multitier cloud controller architecture is designed for SDN based on 5G networks. The architecture of system prototype is based on OpenFlow interface. Proposed architecture is deployed with two basic entities: (1) edge controller (EC), which is used to process the event within a single network domain, and (2) global controller (GC), which processes the event within at least two heterogeneous networks. Event processing is implemented through online transaction processing (OTLP) and online analytical processing (OLAP). OTLP executes low-level events with real-time measurements. High-level events such as load balancing are performed through OLAP. Multiple virtual operators share the resources and network through network function virtualization (NFV) in SDWN-based 5G. OpenFlow data plane forwards the data packets. Two logical interfaces, security link (SSL) channel for event reporting and Simple Network Management Protocol (SNMP) for monitoring, are supported by the common Ethernet card. Cloud controller, OpenFlow switch, and network simulators are included in current prototype. For virtualization techniques, one GC and two EC are implemented in the cloud controller for this prototype. The extended monitoring of localization, energy management, topology discovery, and cloud radio reactive handover are the four new applications supported for the 5G network in this prototype. Software architecture for the controller and simulator consists of following layers: (a) the communication layer is responsible for message exchange function, (b) logical control layer performs the responsibility of wireless-related logic control functions, and (c) persistence layer provides user-friendly APIs and performs data operations. The messages and communication among the virtual switches, controller, and network simulator is modeled in Wi-simu-comm an OpenFlow wireless network simulator. The main contribution of the authors is the provision of global cloud radio access for 5G mobile infrastructures. In addition, the network status is monitored in real time. In case of network bottleneck or congestion at data, forwarding command is sent based on quality of service (QoS) parameters. The research proposed a layered cloud-based scheme having two components EC and GC for multiple controllers. The significance of this scheme is the balancing of load for cloud controller and reducing the response latency [10].

Sood et al. highlight the emergence of SDNs and describe the recent developments in the wireless environment to integrate SDN and Internet of Things (IoT) [11]. With the emergence of IoT, mobile networks will have to handle enormous data and increased network traffic with devices having limited bandwidth. IoT devices connect to serve heterogeneous applications where no single wireless standard could be enough. How to connect devices wirelessly to gain potentials of IoT is a challenging task. To cope with these types of challenges SDN is a good alternate to increase bandwidth of the networks. Information gathering, analysis of gathered information, decision making, and implementation have become simple by using SDN in IoT. The merging of SDN on IoT will provide the potential of exchanging bandwidth between users and devices and visibility of network resources. SDN smartly directs traffic with efficient use of network resources to prepare the network for massive data of IoT. Security and scalability are the main challenges in the integration of IoT with SDN. If hackers have the access to the controller they can attempt security breaches. A strong detection mechanism to detect malicious devices trying to access the network is necessary. In SDN, a logically CC controls the switch that evaluates each incoming packet. Each switch maintains a limited size of the flow table and could not scale beyond few hundred entries. The type and length of flow while planning scalability solutions is necessary. The larger the flow length, the more time is taken by the particular flow, hence there are more chances of collision and delay in WLAN networks. Therefore, IoT networks can be made more scalable with SDNs [11].

Managing, controlling, and evaluating a large number of access points (APs) in IEEE 802.11 WLAN frameworks is a challenging task. To overcome interference due to shared channels in wireless paradigms, the coordination between APs is necessary to increase performance. To address such issues, the authors proposed an SDN-based architecture where the management of APs is the responsibility of the access controller (AC) [12]. Communication between AC and AP is provided by Control and Provisioning of Wireless Access Point (CAPWAP) protocol that provides potentials for seamless roaming. In traditional networks, when a client leaves an AP range, the AP notifies the CC that directs the client to the new AP with the strongest signals. The new AP confirms the client's presence by four-handshake protocol. However, such handovers inculcate delays due to reconfiguration. Monin et al. describe a system that integrates SDN and WLAN to enable speedy roaming policy [12]. Roaming is started with a connection establishment between the client and new AP. The CAPWAP notifies SDN about the migration of client. The experimental results showed that SDN-based WLAN have 70% faster roaming than traditional frameworks with no additional delays.

Kumar et al. discuss the challenges of interferences among nodes that limit the numbers of channels being reused [5]. The authors propose SDN-based WLANs to cope with these kinds of interferences. The work also proposes a load-aware handoff algorithm for these infrastructures. The algorithm checks the signal intensity at wireless clients and reviews the state of congestion on APs. Load unbalancing among APs is resolved and seamless mobility is provided. The load on the network is periodically checked at the CC. To reduce delay between handoffs, Channel Switch Announcement (CSA) message in (Beacon) frames is used. The CC is the main component of the proposed SDN based WLANs. CC periodically records network traffic and reconfigures the network optimally. Assistive agent (AA) is another module of the defined architecture that is installed on AP. The

AP communicates all necessary parameters (congestion on AP, load of each AP, handoffs) with CC. OpenFlow support is provided by integrating APs with Open vSwitch. Handoff decisions by CC are notified by the APs. CC measures load on APs. The described algorithm runs on CC. Initial signal thresholds' value on APs is fixed, then it is dynamically changed on APs depending on their congestion that provides good QoS. Network state provision and parameters for load-aware handoff decisions are challenging in these frameworks. Experimental results shows that in proposed algorithms, both static and mobile users could be associated with less loaded APs and cleverly utilize bandwidth of the network. The proposed architecture could also be efficiently implemented with multichannel WLANs frameworks.

Chaudet et al. describe the challenges and opportunities posed by SDN in wireless infrastructure [4]. Implementation of SDN is relatively easy in wired networks as compared to wireless frameworks that pose link separation challenges. SDN and OpenFlow were basically designed for wired networks and incorporating those in wireless infrastructures is difficult. Segmentation is required in wireless networks for isolation of communication channels to minimize interference between these segments. For wired and infrastructure networks, detachment of links is simple by implementing different wavelength in cables. Moreover, in wireless networks, time synchronization and bandwidth wastage have to be taken care of. Handling of frequent handoff is another challenge in implementing SDN.

Bernardos et al. illustrates the advantages of using SDN model to mobile wireless networks [1]. By referencing the 3GPP Evolved Packet System, the paper proposes an architecture of SDWN that is implemented by using two models: (1) Evolutionary Model: new model is developed by incrementing new strategies in the existing network, and (2) Clean State Model: using an API between virtual operator and SDN controller, control plane functions are directly programmed on the SDN controller.

A lot of effort has been put forward in SDN architecture implementation in multihop networks [13]. These multihop networks included wireless mesh networks (WMNs) and wireless sensor networks. In the next section we analyze the use of SDN in these multihop networks.

16.4 SDN in Fog Computing

In recent years,fog computing has emerged as a technology to augment cloud computing. The term fog computing was first coined by Cisco, and is aimed to bring the storage and computing resources closer to the end-user. Fog computing is suitable for the applications that require real-time response, e.g. real-time gaming, augmented reality, real-time healthcare, and traffic management. The same applications when run on the cloud may experience delays due to far-off deployment of cloud nodes and congestion on the backbone network. Despite its benefits, fog computing may face numerous challenges in managing heterogeneous devices, e.g. large numbers of IoT devices connected with fog [14]. SDN can be integrated into fog computing in order to increase network performance [15]. In SDNs, the network intelligence is moved to a logically centralized SDN controller, which maintains a global view of the network, interacts with data-plane devices, and provides a programming interface for network management applications. Using this concept, the SDN controllers are usually deployed along fog networks to orchestrate the traffic engineering and resource management more efficiently in a centralized manner by having insight into applications' requirements and available resources. This helps in alleviating resource contention in IoT environments and improves overall IoT performance. Besides fog orchestration, SDN controller performs traffic control by using OpenFlow and connectivity management for IoT devices. As the size of fog-enabled IoT platforms increases, efficient routing mechanisms are required in fog computing to forward the ubiquitous IoT data to related servers with low latency, low bandwidth, and/or high security. Therefore, SDN can be employed to optimize the routing process in fog-enabled IoT platforms.

In the subsequent text, we mention some of the existing researches on SDN-based fog computing.

Nobre et al. proposed an architecture combining fog computing and SDN for vehicular ad hoc networks (VANETs) [16]. An SDN controller is located in the middle of fog and cloud layers and is responsible for fog orchestration and resource management. A vehicular fog node downloads the global traffic conditions from cloud and uploads the local traffic conditions using network infrastructure, such as vehicle-to-vehicle V2V or vehicle-to-infrastructure V2I. The cloud controls the fog nodes in a centralized fashion. The SDN-enabled VANET architecture enables content dissemination to efficiently accommodate a large number of vehicular users with any kinds of communication technologies and devices.

Khan et al. proposed a hierarchical 5G next generation VANET architecture using SDN and fog computing framework [17]. The authors propose a fog computing framework with the control functionality of controllers divided in a hierarchical manner to reduce the overhead on the CC. This helps in reducing the transmission delay and overhead in the proposed framework.

The authors illustrate a fog node–based distributed blockchain cloud architecture for IoT in [17]. The architecture utilizes SDN and fog computing to implement a distributed security scheme based on block chaining. At the fog layer, the authors propose the use of blockchain-based distributed secure SDN controller network for the fog node. All the SDN controllers are connected in a distributed fashion at the fog node using block chaining. Each SDN controller has an analysis function and packet migration function to secure the network during attacks. The network edge is deployed with multi-interface BSs equipped with an SDN switch to facilitate IoT-based wireless communication technologies. The BS act as a forwarding SDN switch for the fog controller collecting data from IoT devices and forwards to fog node controller. Each fog node is composed of distributed SDN controllers and uses the blockchain technique to provide scalable, reliable, and high-availability services. In [14], Yoshigoe et al. propose the use of SDN to alleviate resource contentions in the IoT environment to improve overall performance of IoT.

As the size of the fog-enabled IoT platforms increases, the efficient routing mechanisms are also required to forward the ubiquitous IoT data to related servers with low latency, low bandwidth, and/or high security. SDN has been employed by several proposals to optimize the routing process in fog-enabled IoT platforms [18]. Okay et al. propose a hierarchical SDN-based fog computing architecture for routing in fog-enabled IoT platforms [18]. The SDN-based fog controllers take the decisions locally for frequent evens, whereas the cloud controller takes decision globally for less frequent events. The authors evaluated the performance of the proposed scheme by varying the number of controllers and results indicated that hierarchical SDN framework has the potential to reduce routing delay and data transmission overhead.

16.5 SDN in Wireless Mesh Networks

16.5.1 Challenges in Wireless Mesh Networks

WMN consists of a mesh of routers. Few of these routers are connected to gateways. Mobile clients that are not connected to the Internet must route their traffic between mesh routers to reach gateway-connected routers. As there are only a limited number of routers that are providing Internet access, a bottleneck situation may arise. There is a need of routing the traffic to alternate gateways and routers to avoid congestion. In addition, mesh routers are configured in ad hoc mode and lack global knowledge of the network. Therefore, they provide poor performance of network resource allocation.

16.5.2 SDN Technique in WMNs

SDN can handle the aforementioned challenges in WMNs. Mesh routers in SDN architecture are OpenFlow enabled [19]. SDN provides simplified network management with the help of a global view of the WMN. It decouples the data plane and control plane as shown in Figure 16.3. Separation of data and control traffic is achieved via dividing the wireless spectrum. There is a logical CC that constitutes

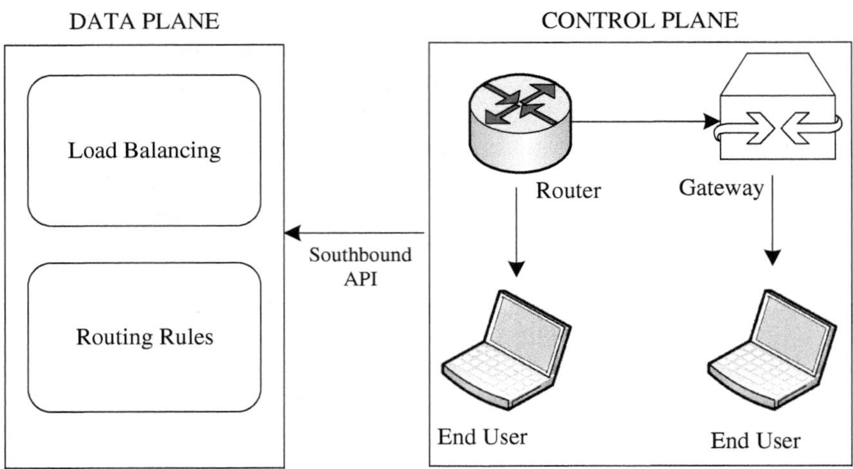

Figure 16.3 Architecture of SDN-based wireless mesh network.

Table 16.1 Comparison between centralized and distributed controller.

	Flexibility	Reliability	Global view of network	Complexity	Efficient response time	Cost	Load balancing	Control traffic congestion
Centralized controller	Yes	No	Yes	No	No	No	Yes	Yes
Distributed controller	No	Yes	No	Yes	Yes	Yes	No	No

the rules for data forwarding. Actual data forwarding is performed by numerous SDN mesh routers. Rules from the controller can be simple or complex. Simple rules match the incoming packet's port number while complex rules match MAC address for each packet. There are several benefits of a CC as in SDN over a distributed one but some parameters can also be compromised. Table 16.1 summarizes that SDN-based CC provides more flexibility to implement new services in WMNs and load balancing is achieved with this controller to avoid traffic congestion. However, at the same time SDN controller–based WMN requires a way for fault tolerance in case of failure. This suggests that distributed controller paradigm provides more reliability.

Three algorithms are proposed by Huang et al. that divide and assign the spectrum for data and control traffic to avoid congestion [13]. These algorithms are called Fixed-Band Non-Sharing (FB-NS) algorithm, Non-Fixed Band Non-Sharing algorithm (NFB-NS) and Non-Fixed-Band Sharing (NFB-S) algorithm. In FB-NS algorithm, the radio spectrum is divided in subbands. A fixed fraction of all subbands is allocated to control traffic. Whereas the remaining subbands are used to transmit data traffic. Fixed subband allocation in the FB-NS algorithm can lead to poor resource utilization. To overcome this problem another algorithm is proposed, called NFB-NS. In this algorithm, number of subbands can be selected freely, as per the need for data and control traffic; however, control traffic is given more priority. In addition, both the control and data traffic cannot share the same link, even if there is a remaining capacity in one of the links. This scenario decreases the spectrum utilization. To overcome this situation, NFB-S algorithm is introduced. It suggests that data traffic can be transmitted in same subband of control traffic if there is still some remaining capacity. Performance of all three algorithms is evaluated and compared. It is clear from the results that NFB-S algorithm has the highest throughput as compared to others, due to maximum utilization of spectrum and flexibility. However, there is also some overhead of its complex implementation.

16.5.3 Benefits of SDN in WMNs

Major contributions of SDN in WMNs are better mobility management and efficient allocation of resources. With flexible routing, load balancing is achieved and client mobility is made more efficient. In the Mesh Flow project, the authors introduced SDN-based mesh routers [20]. These mesh routers have multiple interfaces. With help of these interfaces, mesh routers can connect to the Internet or other routers. The physical interface of each router is further divided into two virtual interfaces. One virtual interface is used for data traffic and the other for control traffic. There is a unique service set ID (SSID) for each virtual interface.

Frequent topology changes in WMNs are handled with help of Optimized Link State Routing OLSR [21]. A large amount of control traffic is needed to update topology, which can cause a delay in control and data plane communication. Such a delay is minimized with the OLSR routing technique. However, OLSR routing without the OpenFlow controller only considers the shortest path for routing and does not optimize resource allocation. This scenario results in less throughput and congestion. In the Mesh Flow project, OLSR constitutes a daemon (program that runs in the background) named as OLSRD. This daemon program is used for IP routing and is connected to control interface. Whereas, data interface is connected to an SDN controller, which is based on OpenFlow architecture. The controller accesses the information related to topology changes from a local agent. This local agent does the network monitoring and is connected to each router. An end-user station also possesses a monitoring agent that gives information about handovers to the controller. These end stations are connected to mesh access points.

In Mesh Flow, the controller also constitutes a monitoring and control server (MCS) [20]. Handoffs are taken care of with the help of this MCS that provides an updated topology based on the information from the monitoring agents of routers. It helps the controller to create and maintain flow tables for routing. MCS also maintains information about the handovers for end-users. In this way, SDN-based WMNs can support effective load balancing and handovers.

Mesh Flow is extended further to support even more flexible routing schemes, such as *Gap* [22]. Gap also provides effective resource allocation in WMNs. This is achieved by setting effective routing rules by the gap controller. Multiple interfaces connect routers with data and control planes. A unique SSID is utilized by mesh routers for each virtual interface. There are four virtual interfaces. The first one utilizes Hybrid Wireless Mesh Protocol (HWMP), to control forwarding of data traffic. Second and third interfaces provide connection to the access network, while the fourth one serves as the backbone of WMN.

Yang et al. propose SDN architecture for effective load balancing [23]. To balance the traffic load on different routes and avoid congestion, a network management tool is utilized. This tool tracks the network traffic situation and dynamically updates the routing paths. An in-band control message is exchanged between mesh routers and CC to perform load balancing.

16.5.4 Fault Tolerance in SDN-based WMNs

One of the disadvantages of SDN-based WMNs is the single point of failure. As there is only a single controller, if it fails the whole network will suffer. There is no access to control the plane for the nodes in case of failure and hence routing rules cannot be retrieved by the nodes. To achieve fault tolerance, additional techniques are required that can detect failure of the CC. Mechanisms are also required to overcome the fault by activating an alternative controller, which is an extension of Mesh Flow and addresses the issue of fault tolerance in WMNs [24]. Detti et al. implement a distributed control mechanism in case of the failure of the controller. This supports continuity of the operations without interruption. The entries from the flow table of CC are removed and typical OLSR routing rules are implemented. This is achieved by an embedded flow table manager.

16.6 SDN in Wireless Sensor Networks

16.6.1 Challenges in Wireless Sensor Networks

Wireless sensor networks (WSN) constitutes of large number of spatially distributed sensors to measure a phenomenon of some interest like temperature and humidity [25]. As the sensors in WSNs are based on batteries, it is very important to reduce energy consumption. In addition, maintaining a topology that provides good coverage of survey area is another challenge for WSNs. Multihop topology is most suitable for WSNs as there are low-energy sensors with limited radio range. Large coverage area and tough physical situations can result in broken links. Due to these distorted links packets can fail to reach their destination. WSNs are data centric and application specific. In some scenarios, such networks can provide data for multiple applications having multiple types of sensors at the same physical area. To support such diversity of applications effective abstractions are required.

16.6.2 SDN in Wireless Sensor Networks

WSN architecture is related to SDN, as it consists of a centralized sink node that gathers the information from the network [26]. In this way, the sensor nodes are similar to data planes of SDN. These sensor nodes are only forwarding the data, so they have a simple architecture that consumes less energy. Routing and topology management is handled by the controller. As the controller utilizes the global view of the network, it can provide a better connectivity mechanism among different nodes. New protocols can be implemented very easily by updating the control plane and hence multiple applications can be supported. The architecture of SDN-enabled WSNs is shown in Figure 16.4. Control plane consists of routing and energy management components, whereas data plane consists of sensors.

Typical wireless networks implemented in SDN vary from WSNs, as in WSNs data fusion techniques that are applied to aggregate the traffic at the intermediate nodes [26]. Moreover, sensor nodes are embedded devices and are prone to faults. Therefore, the network topology is dynamic and unreliable. In addition, WSNs are data centric rather than address centric. Thus, SDN techniques must be applied in such a way that the aforementioned requirements of WSNs are

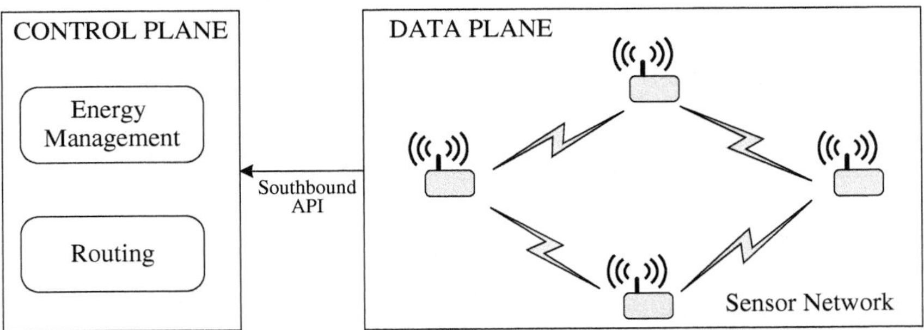

Figure 16.4 Architecture of SDN-enabled wireless sensor network.

carefully addressed. In short, WSNs should be able to support implementation of new architectures.

16.6.3 Sensor Open Flow

Sensor OpenFlow is the extension of OpenFlow standard, for WSNs [27]. Sensor OpenFlow introduces new forwarding techniques to meet the requirements of WSNs. It also enables multiple applications to operate simultaneously. In addition, duty-cycling is also controlled to provide less energy consumption [28]. As the traffic flow in WSNs is data centric rather than address centric, sensor OpenFlow is extended to utilize user-defined transport protocols. This is achieved by adding new functions that match the compact addresses and attributes. In SDN-enabled WSNs, the same network topology is used for both data and control traffic, rather than an individual control channel for communication with data plane. As the topology is changing frequently and there are only a few sinks, a lot of control packets flow from sensors toward the sinks. Time-out strategies are used by Sensor OpenFlow to address such scenarios.

16.6.4 Home Networks Using SDWN

SDN in wireless home networks is a novel research area. Many researchers have proposed different approaches to combine the home networks with SDN to provide virtualization [29, 30]. The techniques proposed in [29, 30] utilize NOX controllers and OpenFlow as a protocol to achieve simpler management, easy interface, and traffic isolation. There are differences in performing traffic isolation, and the management by various schemes. In [31], the authors discuss a DHCP- and DNS-based implementation that achieved enhanced control of home networks. New approaches have been introduced to add WSDN to home networks that focused on splitting the virtual layers into the slices [32]. Splitting the network layer will overcome the challenges of home networks as this will reduce the cost of deployment because multiple stakeholders with their own slice will share overall deployment cost. Moreover, in sliced networks, the management and data configuration is no more a challenging task. Slices distributed among different stakeholders have to be secured, which means that the data from one slice must be inaccessible to the other slice. However, the security, performance, customization, flexibility, and privacy cannot be achieved simultaneously, so still there is need for different strategies that can accomplish better balancing of aforementioned trade-offs.

16.6.5 Securing Software Defined Wireless Networks (SDWN)

Due to separation of data and control plane, SDWN are vulnerable to security attacks. Therefore, it requires new security mechanisms. Sood et al. describe various potential threats to SWDN [33]. Both forwarding devices and controller could be attacked with denial of service (DoS) attacks. The most severe attack is on CC. A compromised controller could be misused by the malicious user to corrupt the entire network. SDWN requires additional security mechanisms compared to traditional networks. Use of Transport Layer Security/Secure Socket Layer (TLS/SSL) in OpenFlow for mutual authentication between controller and forwarding devices could mitigate malicious threats. Confidentiality in SDWN could be achieved by encrypting communication medium between forwarding devices and controller. In case of DoS attacks, the data, devices, and services to authorized user could be made available by the controller by restoring the network routes dynamically. Message authentication code is used in SDWN to make sure that the message has not been changed by a malicious user. To achieve consistency in SDWN and to avoid conflicting rules, an intermediary between controller and application is used. The security mechanism of SWDN should be fast enough to tackle threats. There should be a proper monitoring mechanism in SDWN to deal with mobility and dynamic network states.

16.7 Conclusion

SDN introduces multiple opportunities in the network architecture space. The use of centralized intelligence has revolutionized the whole network infrastructure. SDN improved the resource virtualization, resource sharing among different vendors, and reduced the cost. However, the use of the technology in different networks is still an issue. Challenges, such as link separation in wireless domains, interference among the nodes due to limited numbers of channels, and load estimation on a wireless medium due to frequent change of state in the medium are the limitations of these networks.

References

1 Bernardos, C. and Oliva, A. (2014). Architecture for software defined wireless networking. *IEEE Wireless Communications* 21 (3): 52–61.

2 Open Networking Foundation, OpenFlow Enabled Mobile and wireless Networks, ONF Solution Brief [online]: https://www.opennetworking.org/wp-content/uploads/2013/03/sb-wireless-mobile.pdf. Accessed May, 2019.

3 Costanzo, S., Galluccio, L., Morabito, G., and Palazzo, S. (2012). Software defined wireless networks unbridling SDNs. In: *Proceedings of the 2012 European Workshop on Software Defined Networking*, 59–60. IEEE.

4 Chaudet, C. and Haddad, Y. (2013). Wireless defined network: challenges and opportunities. In: *Proceedings of IEEE International Conference on Microwaves Communication, Antennas and Electronic Systems*, 1–5. IEEE.

5 Rangisetti, A.K., Baldaniya, H.B., Kumar, P., and Tamma, B.R. (2014). Load-aware hands-offs in software defined wireless LANs. In: *IEEE 10th International Conference on Wireless and Mobile Computing, Networking and Communications (WiMob)*, 8–10. IEEE.

6 McKeown, N., Anderson, T., Balakrishnan, H. et al. (2008). OpenFlow: enabling innovation in campus networks. *ACM SIGCOMM Computer Communication Review Archive* 38 (2): 69–74, ACM: New York,.

7 Open Networking Foundation, OpenFlow Switch Specification Version 1.5.0, [online]: https://www.opennetworking.org/images/stories/downloads/sdn-resources/onf-specifications/openflow/openflow-switch-v1.5.0.noipr.pdf, accessed May 2019.

8 Sadiq, S. and Shaukat, F. (2015). A survey on wireless software defined networks. *International Journal of Computer and Communication System Engineering* 2 (1): 316–325.

9 Sun, G. and Liu, F. (2014). Software defined wireless network architecture for the next generation proposal and initial prototype. *Journal of Communication* 9 (12): 143–151.

10 S. K. Routray and K. P. Sharmila, Software defined networking for 5G, 4th International Conference on Advanced Computing and Communication Systems (ICACCS), Vol. 26, No. 1, pp. 82–92, 2015.

11 Sood, K. and Yu, S. (2016). Software defined wireless networking opportunities and challenges for internet-of-things: a review. *IEEE Internet of Things* 3 (4): 453–463.

12 S. Monin and A. Shalimov, Chandella: smooth and fast Wi-Fi roaming with SDN/Openflow, https://www.usenix.org/sites/default/files/ons2014-poster-monin.pdf, 2014.

13 H. Huang and S. Guo, Software-Defined Wireless Mesh Networks: Architecture and Traffic Orchestration, IEEE Network, Vol. 29, No. 4, 2015.

14 Yoshigoe, K., Maljevic, I., and Radusinovic, I. (2017). Software-defined fog network architecture for IoT. *Wireless Personal Communications* 92 (1): 181–196.

15 Ola, S., Imad, E., Ali, C., and Kayssi, A. (2018). IoT survey: an SDN and fog computing perspective. *Computer Networks* 143: 221–246.

16 Nobre, J.C., de Souza, A.M., Rosário, D. et al. (2019). Vehicular software-defined networking and fog computing: integration and design principles. *Ad Hoc Networks* 82: 172–181.

17 A. A. Khan, M. Abolhasan, and W. Ni, 5G next generation VANETs using SDN and fog computing framework, 15th IEEE Annual Consumer Communications & Networking Conference (CCNC), Las Vegas, NV, January 12–15, 2018.

18 Okay, F.Y. and Ozdemir, S. (2018). Routing in fog-enabled IoT platforms: a survey and an SDN-based solution. *IEEE Internet of Things Journal* 5 (6): 4871–4889.

19 P. Dely and A. Kassler, OpenFlow for wireless mesh networks, Proceedings of the 20th International Conference of Computer Communications and Networks (ICCCN), 2011.

20 B. Anderson, F. Farias, V. Nascimento, and A. Abelém, OpenMesh: OpenFlow in wireless mesh networks, 1st Workshop on Experimental Research on the Future Internet, December 2016.

21 S. Salsano and G. Siracusano, Controller selection in a Wireless Mesh SDN under network partitioning and merging scenarios, arXiv preprint arXiv:1406.2470, 2014.

22 V. Nascimento, M. Moraes, R. Gomes et al., Filling the gap between software defined networking and wireless mesh networks, in 10th International Conference on Network and Service Management (CNSM) IEEE, November 17–21, 2014.

23 F. Yang and V. Gondi, OpenFlow-based load balancing for wireless mesh infrastructure, in The 11th Annual IEEE Consumer Communications and Networking Conference (CCNC), January 10–13, 2014.

24 A. Detti, C. Pisa, S. Salsano, and N. Blefari-Melazzi, Wireless Mesh Software Defined Networks (wmSDN), 2013 IEEE 9th International Conference on Wireless and Mobile Computing, Networking and Communications (WiMob), Lyon, France, October 7–9, 2013.

25 Tilak, S., Abu-Ghazaleh, N.B., and Heinzelman, W. (2002). A taxonomy of wireless micro-sensor network models. *ACM SIGMOBILE Mobile Computing and Communications* 6 (2): 28–36.

26 C. Intanagonwiwat, D. Estrin, R. Govindan, and J. Heidemann, Impact of network density on data aggregation in wireless sensor networks, in Proceedings 22nd International Conference on Distributed Computing Systems, July 2–5, 2002.

27 Luo, T. and Tan, H.P. (2012). Sensor OpenFlow: enabling software-defined wireless sensor networks. *IEEE Communications Letters* 16 (11): 1896–1899.

28 W. Ye and J. Heidemann, An energy-efficient MAC protocol for wireless sensor networks, in Proceedings of IEEE International Conference on Computer Communications (INFOCOM), New York, 2002.

29 Philip Venmani, D., Gourhant, Y., Reynaud, L. et al. (2012). Substitution networks based on software defined networking. In: *Ad Hoc Networks*, 242–259. Berlin: Springer.

30 Chaudet, C. and Haddad, Y. (2013). Wireless software defined networks: challenges and opportunities. In: *IEEE International Conference on Microwaves, Communications, Antennas and Electronic Systems*, 1–5. ACM.

31 N. Soetens, J. Famaey, M. Verstappen, and S. Latre, SDN-based management of heterogeneous home network, 11th International Conference on Network and Service Management (CNSM), November 9–13, 2015.

32 Yiakoumis, Y., Yap, K., Katti, S. et al. (2011). Slicing home networks. In: *Proceedings of the 2nd ACM SIGCOMM Workshop on Home Networks*, 1–6. ACM.

33 He, D., Chan, S., and Guizani, M. (2016). Securing software defined wireless networks. *IEEE Communications Magazine* 54 (1): 20–25.

17

Time-Critical Fog Computing for Vehicular Networks

Ahmed Chebaane[1], Abdelmajid Khelil[1], and Neeraj Suri[2]

[1] Department of Software Engineering, Landshut University of Applied Sciences, Landshut, Germany
[2] Department of Computer Science, Lancaster University, UK

17.1 Introduction

Over the past few years, Internet of Things (IoT) has undoubtedly become an integral part of our quotidian lives connecting objects such as vehicles, machines, and products with various users through the Internet.

Cloud computing has been proposed as a promising approach for IoT applications/services to virtualize things and to deal with analytics, storage, and computation of data generated from IoT devices. This approach is viable in ample cases such as executing the application in the cloud for saving battery lifetime, offering on-demand data storage to the end-users, etc. In the automotive field, cloud computing and IoT support building smart vehicles by facilitating communication across vehicles, infrastructures, and other connected devices, which may improve road safety as well as traffic efficiency [1].

Dew computing, mobile cloud computing (MCC), and vehicular cloud computing (VCC) are variations of cloud computing. When cloud applications assume an active participation of the end devices along with the cloud in the execution of services and applications, cloud computing is also referred to as dew computing [2]. MCC [3] has been introduced to combine both paradigms mobile computing and cloud computing in order to overcome the mobile devices constraints. VCC [1] brings the MCC paradigm to vehicular networks by providing public services such as parking systems, traffic information, etc.

However, knowing that cloud data centers are geographically highly centralized, resulting in a large (network hop) distance between the end device/vehicle and the cloud/data center, usually leading to unpredictable communication delays [4], it is complicated if not impossible to enable time-critical applications with

Fog Computing: Theory and Practice, First Edition.
Edited by Assad Abbas, Samee U. Khan, and Albert Y. Zomaya.

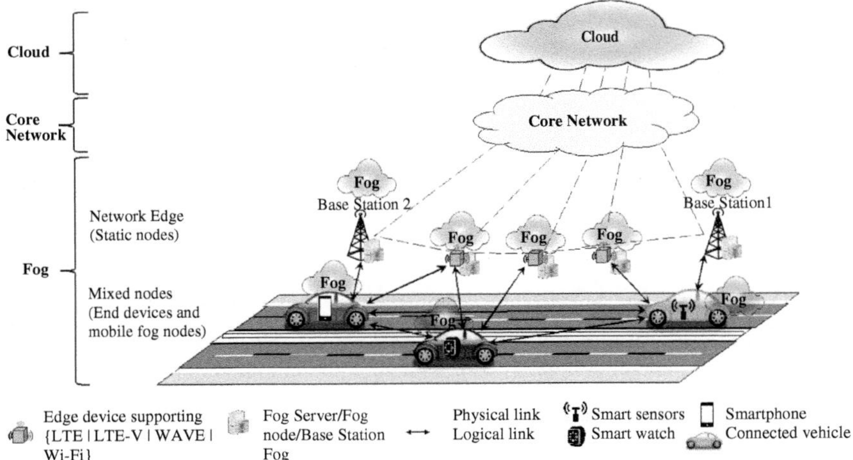

Figure 17.1 Fog computing for vehicular applications.

short-lived data across vehicles and clouds. Constraints, failures, and attacks in the heterogeneous computing environment are the typical perturbation of the timeliness in VCC. Therefore, there is a need to manage data very close to the point of use, i.e. end devices, in order to enable low-latency applications.

Fog computing was introduced in 2012 by both industry and academia [5, 6] to address the challenge of delay-sensitive IoT applications (Figure 17.1). Fog computing brings cloud computing to the edge of the network. Instead of centrally running analysis, processing, and storage functions in the cloud, they are now decentralized and running on gateways/fogs very close to the end-user devices, thus, decreasing latency, rendering it more predictable, and preserving data locality and privacy. Accordingly, this network architecture is more suitable for time-critical applications such as vehicle-to-vehicle (V2V), vehicle-to-device (V2D) [7], and vehicle-to-infrastructure (V2I) communication or autonomously driving cars, whose data processing must happen in a delay-sensitive manner. Provided a careful coping with perturbations, fog computing is indeed a highly promising architecture to make finally vehicular networked application into a reality after a long and intensive research over the last two decades.

Edge computing [8] pushes computation/intelligence even closer to the things, i.e. things may play the role of gateways/fogs [5].

We categorize fog computing either in delay-tolerant or delay-critical. Delay-tolerant fog computing addresses distributed applications that may tolerate either high or fluctuating timeliness. Compared to cloud computing, delay-tolerant fog computing still reduces network traffic and gives the data owner more localized control on own data. The main category of delay-tolerant vehicular applications are those that provide entertainment and infotainment for drivers and passengers. Usually, they are not safety-critical and accordingly not delay-critical. In this chapter, we focus on delay-critical vehicular applications such as detection of immediate obstacles on the road, cooperative, or platoon driving. We detail the target scenarios and their requirements on fog computing in Section 17.2.

Vehicular fog computing (VFC) [9, 10] has gained a significant focus in the last few years. In addition, VFC plays a significant role to support high mobility, rise computational capability and decrease communication latency, which suits well delay-sensitive applications. Static fogs for vehicular applications may be implemented on top of available architectures such as traffic lights, traffic signs, street lighting or cellular base stations, toll collect infrastructure, bridges. New architectures, known as mobile fog computing, aim to model vehicles as fog nodes for communication and computation, thus integrating fog computing, edge computing, and Vehicular Ad-hoc NETworks (VANETs) to a homogeneous architecture (Figure 17.1).

VANET computing [11] restricts communication among vehicles using a short-range communication (such as dedicated short-range communication [DSRC], IEEE 802.11p, D2D) to enable vehicles to analyze and to share information between each others. This information could be safety-relevant, e.g. accident prevention, traffic jams, or general information, e.g. position, weather, in order to enhance safety on the road. Compared to fog computing, VANET encompasses only vehicles for computation and sharing information to the neighboring vehicles that are referred to as V2V [12]; however, fog computing includes V2I [12] for the purpose of increasing computation capability and exchange of information for vehicles via infrastructure elements, i.e. fog nodes.

The interworking of cloud, fog, and VANET computing is illustrated in Figure 17.1. This interworking will be the common architecture to address all vehicular applications. However, delay-critical applications will rely either on VANET or fog computing or a combination of both. In the latter case, one may consider vehicles as either end-device or a mobile fog. Accordingly, mobile fog computing is an emerging architecture, where fogs may be mobile.

Though there are surveys on fog computing [4, 13, 14] and several recent papers presenting the VFC architectures, algorithms, etc. for enabling delay-sensitive applications, there is no survey of this emerging field. This chapter aims at filling this gap and presenting a comprehensive survey.

In this chapter, we comprehensively survey the literature on delay-critical fog-based vehicular applications. Nonetheless, we also briefly survey the fog computing support for other delay-critical application domains such as smart grid, industry 4.0 and IoT, while pointing to the potentials of the adoption of the available techniques to the field of vehicular networks.

In Section 17.2, we present the focused applications and their timeliness requirements. In addition, we survey the key perturbations that hinder fulfilling these timeliness requirements. In Section 17.3, we introduce the existing research works to cope with the perturbation in order to meet the timeliness requirements. In Section 17.4, we address the research gaps achieved in this survey and we provide a future research direction In Section 17.5, we conclude the paper.

17.2 Applications and Timeliness Guarantees and Perturbations

In this section, we present various scenarios of time-critical applications and their timeliness requirements. Next, we survey the perturbations that may complicate meeting the desired deadline.

17.2.1 Application Scenarios

We first detail a representative application scenario, i.e. obstacle detection that clearly shows the need for delay-critical fog computing in vehicular networks. Next, we survey a broad class of application scenarios that emphasize these needs.

Consider a (autonomous) vehicle driving with 200 km h^{-1} on the highway and suddenly an obstacle appears on the road. The obstacle detection application includes the type of obstacle and its velocity is a prerequisite for a suitable decision making. For example, depending on whether the obstacle is a human being, a large object, or something harmless, and depending on the surrounding vehicles, the vehicle should ignore, avoid, or overdrive it. This application could execute directly on the on-board computers of prime and modern vehicles. However, most of vehicles will not be able to do the processing on-board while meeting the deadline due to limited resources. For this purpose, the application (and its data) should be partitioned and selected parts of it should be transmitted with minimal delay and highest reliability to the surrounding fog infrastructure. A cloud solution is not suitable because the latency of the data transfer is not deterministic and may become intolerable (Figure 17.2).

Usually, fogs from the same provider are (directly) interconnected and can exchange data, balance loads among each other, and execute similar measures for reliable and delay-aware computing. When data is processed, the result should be transmitted back to the initiator to take appropriate decisions. For instance, if the obstacle is a human being, an immediate collision avoidance maneuver must be initiated. If it is just a harmless object, such as a tire part, it can be run over. Thus, nobody is endangered by an unnecessary lane change maneuver. This gained knowledge may be then communicated to other affected vehicles, such as those immediately following the considered vehicle.

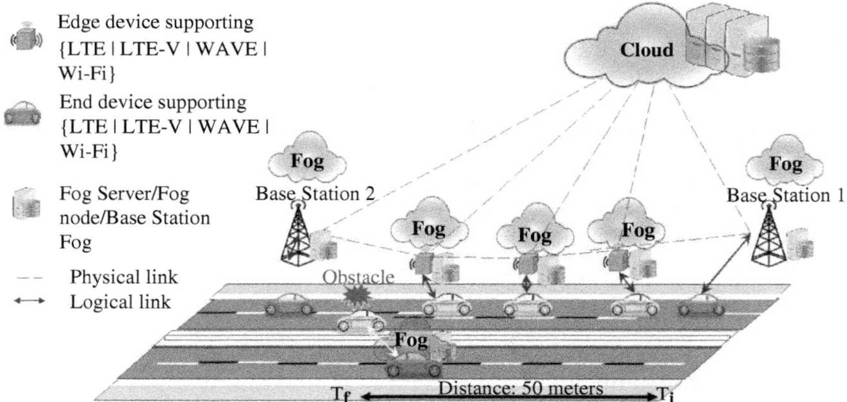

Figure 17.2 Obstacle detection as an example of delay-critical application scenarios.

In order to make this scenario possible, data processing and distribution should not violate the tolerable delay. At a speed of 200 km h^{-1}, a vehicle covers a distance of about 50 m in 900 ms. Therefore, the decision must be taken in maximum 90 ms, which is the time to execute the application. Therefore, we refer to such applications as short-lived ones.

The aforementioned representative application scenario clearly motivates the necessity of fog computing support to provide for such crucial applications. In the following, we illustrate further delay-critical application classes that require fog computing support.

A cooperative perception class is based on swapping/fusing data from different sensors sources/vehicles and/or infrastructures using wireless networks in order to cooperatively perceive an important context. This information should be treated as a map-merging problem. See-through, lifted-seat, or satellite view are some use cases from the cooperative perception class [15].

A cooperative driving class enables maneuvers to review, share, plan, coordinate, and apply information concerning driving trajectories among vehicles in a safe way including negotiation and optimization of trajectories. The possible cases in this class are lane change warning, lane merge, etc. [15].

The cooperative safety class addresses the presence of vulnerable road user (VRU), where affected vehicles and/or infrastructure entities should interchange the VRU information to improve safety on the road. Moreover, VRU information acquired is processed and analyzed by the on-board unit of the vehicles or external system. The alert message generated is transmitted to the drivers or to the autonomous driving system to take applicable and corrective decisions in order to provide safety. Obstacle detection, collision warning, network-assisted vulnerable pedestrians, and bike driver protection are a possible scenario in this class [15].

Autonomous navigation classes target the building of self-governing real-time intelligent high definition maps of the surrounding area. Precisely, the information comes from the cooperative perception and a well-defined map that provides accurate and optimum performance in achieving autonomous navigation, e.g. high-definition local map acquisition [15].

Autonomous driving classes enable self-driving vehicles through wireless communication that allows the control of the major vehicle component from outside the vehicles to facilitate remote driving, which requires information about the perception layer and infrastructure. An example of the use case is self-driving in the city [15].

17.2.2 Application Model

We now follow an application model that is commonly used in fog computing as well as in other distributed embedded systems communities. An application is represented as a directed acyclic flow graph of tasks [16]. The edges specify data dependencies between tasks. We differentiate two kinds of tasks: execution tasks and communication tasks. An execution task is a composite of code and data. A communication task is the transmission of an execution task from one node to another on a certain communication path.

An application has a certain priority that applies to all its tasks. Each task has specified execution times on selected computing nodes. An application has a timeliness requirement, which is usually based on executing the entire application while meeting a certain deadline. An application deadline is the maximum tolerable delay. The root task is executed on the application initiator (the vehicle that starts the application). The rest of the tasks can be either executed locally or on surrounding fogs. A task is usually represented by an application container along with its dependencies, the task execution time, and priority.

In order to efficiently deploy an application while meeting its timeliness requirements, we usually need a set of middleware building blocks. The goal of these building blocks is to find an assignment of tasks to nodes, and communication tasks to communication links. The key building blocks are resource monitoring and task scheduling (Section 17.2.5).

17.2.3 Timeliness Guarantees

A timeliness guarantee is a fundamental quality level for providing a service delivery that satisfies the application quality of service (QoS) requirements. We identify three main timeliness guarantee levels in the literature: hard real-time (RT), soft RT, and firm RT [17] requirement classes. We survey explicitly the existing efforts to address these requirements.

A hard RT application is defined as follows: any delay in completing application execution within deadline means system failure, which can lead to catastrophic damage on the road and a violation of security requirements. Hard RT requirement uses a preventive version to prioritize tasks for scheduling.

A soft RT application is tolerant with the deadline, which is based on three requirements types: number of deadline misses in an interval of time, tardiness, and probabilistic bounds [18]. Soft RT allows the system to fail respecting the deadline even many times while the tasks are performed correctly. In this case, the result still is useful for the end-user but its utility degrades after passing the desired deadline. Soft RT requirement uses a nonpreemptive version to prioritize tasks for scheduling.

A firm RT application [19] is tolerant to skipping some tasks but still meeting the deadline [20] (also known as weakly hard RT). Unlike soft RT, firm RT applications are not considered to have failed but the result of the request is useless once the system fails to reach the deadline.

In summary, soft RT is soft with the respective deadline, and the result is useful after missing the deadline. In contrast, for hard RT applications, missing the deadline may lead to catastrophic damages. Firm RT is between soft RT and hard RT, as it is strict with the deadline, so the result is useless but no harm happens when missing the deadline. It is noteworthy to mention that RT systems require clock synchronization across multiple networked entities. In vehicular networks, we assume vehicles and fogs are equipped with GPS receivers and therefore all clocks are synchronized with the GPS global clock.

17.2.4 Benchmarking Vehicular Applications Concerning Timeliness Guarantees

We now benchmark the application classes with respect to their timeliness guarantee requirements. As illustrated in Table 17.1, most of the applications require RT communication and computation but depending on the concrete application scenario and context, various RT classes may be required. For example, autonomous navigation and cooperative perception tolerate passing the deadline so that they belong to the soft RT application classes. Cooperative safety and autonomous driving require meeting the deadline. Because of the critical nature of the situation, a deadline in few tens of milliseconds needs to be met in order to avoid a fatal damage. Consequently, they are hard RT class. Cooperative driving mostly requires rm RT requirement due to the necessity to get the result within the deadline in order to enhance safety in the road but nothing critical happens when the execution time has exceeded the deadline.

To guarantee safety and satisfy the service delivery of all these application scenario classes, fog computing is a suitable candidate for computing architecture

Table 17.1 Benchmarking of application classes.

App class	Possible scenario	Need for fog computing	RT class	Timeliness guarantees (deadline)	Fog architecture
Cooperative driving	Lane change warning Lane merge	Timely communication Online and offline analysis Privacy preserving, authenticity and integrity Information sharing among V2V and V2I to enhance the QoS and enable the integration of legacy vehicles for calculation.	Firm RT requirements	Few 100–1000 ms	VANET Fog computing
Cooperative safety	Neighbor collision warning Obstacle detection Network assisted vulnerable pedestrian protection	High computation to process the presence of VRU in ultra low-latency Real time communication Security and Reliability Delay critical for deadline Information sharing among V2V and V2I.	Hard RT requirements	Few 10 ms	VANET Fog computing Locally (OnBoard)
Cooperative perception	see-through Life-seat Satellite view	Capacity (for analyze and localize of the detected object)	Soft RT requirements	Few seconds	VANET Fog computing

		Heterogeneity among vehicles (computing and communication)			Cloud computing
		Sharing information about localization and relative position need the communication through the infrastructure in addition to the V2V communication.			
Autonomous driving	Self driving in the city	Very low latency in communication and computation	Hard RT requirements	Few 10 ms	VANET
					Fog computing
		Almost require 100% reliability			Locally (OnBoard)
		Efficient security			
		Information sharing among V2V and V2I.			
Autonomous navigation	High-definition local map acquisition	Centralized and decentralized computation in real time	Soft RT requirements	Few seconds	Fog computing
					Cloud computing
		Real time distribution of map information			

that enables very low communication and computation delays among vehicles and infrastructures. If needed and possible, fog computing can be seamlessly integrated with cloud computing, which can support tolerant applications such as smart parking with higher computational capability.

Fog computing plays a fundamental role in the cooperative driving class, which can improve cooperation among drivers and enhance safety on the road by applying VANET to coordinate with the infrastructure in order to analyze and exchange information close to the vehicles.

17.2.5 Building Blocks to Reach Timeliness Guarantees

To ensure real-time computation in a distributed, mobile, and heterogeneous infrastructure, fog computing is considered as a viable solution for vehicular applications and services [21]. Resource monitoring, scheduling, RT computation, and RT communication are the key functional blocks to ensure service delivery deadline respecting the QoS of the system [22].

Resource monitoring creates awareness of current and future available networks and computation resources. Accordingly, it is fundamental for task scheduling. The network resource monitoring collects important network indicators such as available channels and their bandwidth in a current or future vehicle location. The computational resource monitoring maintains indicators concerning the available processing and storage resources in time. An integral part of the monitoring is to consider the impact of failures and the availability of the resources.

The resource monitoring function usually is distributed across multiple entities. It is hard to have one central entity that monitors all available resources in a multitenant vehicle environment.

Scheduling tasks target an effective planning of the application tasks depending on available resources and application requirements. The requirements specify the task dependencies, task priority, and the deadline. The resulting schedule is given as an assignment of starting times to every task and communication activity. Automatic rescheduling is also adapted to resolve the issues of the incoming and outgoing fog nodes around the vehicle. The scheduler should guarantee the deadline according to the existing resources.

Similar to resource monitoring, the scheduler functionality is usually shared with multiple nodes.

The RT task computation requires a careful resource management on the node executing a task (Section 17.3.1.2), in order to ensure the overall RT computation of the complete application across vehicles and fog nodes. On node-level this is a well-investigated topic in the literature. On the application level an RT scheduler is indispensable.

RT communication plays an effective role to assure application-level RT computation in fog computing by selecting a suitable fog link to distribute

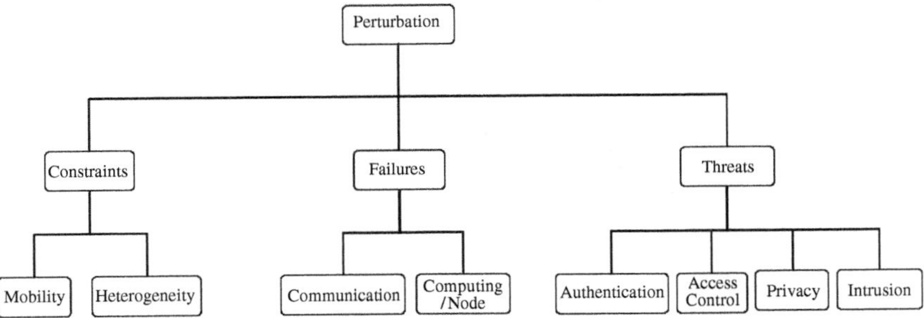

Figure 17.3 Timeliness perturbations.

communication tasks, i.e. to off-load/migrate execution tasks for processing in convenient networking delays, which enable RT communication between sensors/fog nodes to fog nodes in vehicular networks. Ge et al. [23] implement the 5G SDN vehicular network paradigm to enhance latency-aware communication time. Therefore, a good management of the network (Section 17.3.1.1) resources can enhance RT communication in VFC.

A communication task usually consists of migrating a task/container such as Docker [24] from the initiator vehicle to the selected fog nodes such as vehicles and roadside units (RSU), while keeping a high fidelity of the applications.

Docker enables application virtualization through containers. These containers combine individual application parts (tasks) together with all necessary auxiliaries. Therefore, Merkel [25] often refers to lightweight virtualization in terms of containers.

On the other hand, perturbation can break communication or computation between vehicles and/or infrastructures, which is a serious problem in such a scenario that needs ultralow latency in communication and computation. We elaborate more on timeliness perturbations in the subsequent section.

17.2.6 Timeliness Perturbations

After defining the applications, their requirements on latency and the building blocks that allow to fulfill these requirements, we now survey the perturbations, i.e. constraints, failures, and threats that complicate the design of delay-critical VFC (Figure 17.3).

17.2.6.1 Constraints

Mobility. Due to mobility, wireless network characteristics change frequently. For example, the effective available bandwidth is highly dynamic. This depends on the wireless technology (DSRC, 802.11p, WLAN, satellite, LTE-V,

D2D, etc.), access coverage, and a number of vehicles that have to share the wireless medium. Other key characteristics of the wireless links are latency and communication costs. These characteristics lead to considerably varied reliability/availability and connectivity of vehicles.

Heterogeneity. In the vehicular fog environment we usually observe a strong heterogeneity of nodes and links. Nodes strongly vary in computational resources. Prime class vehicles may have sufficient resources to run complex applications/tasks onboard. Other older or lower class vehicles may have highly limited resources. Also, fog nodes may significantly vary with their computational resources.

High mobility and strong heterogeneity in nodes and links obviously complicate monitoring and scheduling, thus fulfilling the timeliness requirements.

17.2.6.2 Failures

Communication failures. These constitute the majority of failures in the vehicular environment. We distinguish between two types of communication failures:

1. *Message loss*. Messages exchanged between the vehicle and the fog are highly vulnerable to loss due to the high bit error rate of wireless links, network congestion, and collisions. Message loss probably occurs in vehicular environments and needs to be explicitly taken into consideration.
2. *Network disconnection (or link disruption)*. Given its mobile nature, a vehicle can enter a geographical area out of coverage of any fog node so that it loses its connection to the network. The vehicle is said to be disconnected from the rest of the network. While disconnected from the network, the vehicle is not able to send or receive messages. As network disconnection is a common occurrence in mobile scenarios, it needs to be explicitly considered.

Communication failures usually lead to delays in the distributed application execution and subsequently to the violation of timeliness requirements.

Computing/node failures. The failure of computing resources may lead to delays and subsequently to the violation of timeliness requirements. Examples of computing failures: node failures, storage overflow, processor overload, etc.

17.2.6.3 Threats

Authentication threats. Vehicles continuously join and leave different fog nodes, which mostly interrupt the service continuity of the initiator. In addition, due to the limited resources in the connected vehicle, the initiator fails to authenticate to some fog layers using the traditional authentication (certificates and public-key infrastructure [PKI]).

Access control threats: In such a scenario, ensuring access permission among the fog and cloud becomes untrustworthy in the vehicular network.

Moreover, access the control system fails to control capacity utilization of the resource-constrained nodes as a result power or resources get empty.

Privacy threats. Usually, the vehicle interacts every time with multiple fog servers. Therefore, most of the fog servers know information about the vehicle, driver, position, etc., which is a risk to share this sensitive data to the other fog nodes. More than that, fog platforms are faced with many threats; as an example, Man-In-The-Middle (MITM) attacks can easily exploit unmanned aerial vehicles (UAVs)-based integrative IoT fog platform [26] to discover sensitive data (e.g. location, fog node identity) [27].

Intrusion threats. Intruders may harm computation and communication and therefore represent a high risk to violate the timeliness requirements of vehicular application.

17.3 Coping with Perturbation to Meet Timeliness Guarantees

We now survey the available research efforts to cope with the perturbations and still efficiently meet the timeliness requirements despite the perturbations.

17.3.1 Coping with Constraints

17.3.1.1 Network Resource Management

In VFC, network management includes vertical federation that depends on the physical partitioning (ex. 5G, LTE, WIFI, 802.11p), [15, 28–35] and horizontal federation that depends on the time portioning and bandwidth (e.g. slicing, software defined networking [SDN] to manage vehicular neighbor groups [VNGs], network function virtualization [NFV]) [30, 35, 36].

Mobility of vehicles and fogs as well as the heterogeneity of network nodes and links result in continuously changing network resources. Accordingly, an efficient network resource management is indispensable for VFC. In order to enable seamless handover among different fog nodes, Bao et al. [31] develop a follow-me fog (FMF) framework to reduce the latency of the handover scheme in fog computing. In addition, Palattella et al. [37] describe the gap of connectivity and security of the handover in the vehicle-to-everything (V2X) and propose a proactive cross-layer, cross-terrestrial-technology, and cross-slices handover approach based on fog computing, including 5G, to achieve zero-latency handover in the vehicular network. For security they aim to enable a quick authentication and re-authentication handover based on SDN and fog. A detailed overview on this research field as well a proposal for proactive handover can be found in [38].

17.3.1.2 Computational Resource and Data Management

From the point of view of an application initiator (root task), mobility and heterogeneity lead to a permanently changing pool of available and useful computational resources. Accordingly, an efficient resource management is crucial for VFC. The selected resource management technique has an effective impact on enhancing the organization and the optimization of resource allocation among fog nodes ensuring the QoS and minimizing the execution time and cost. Scalability in fog computing enables both horizontal and vertical extensibility of fog resources [39] to cover regularly the high demands of resources for vehicular networks. Resource management involves computation management and data management as shown in Figure 17.4. As defined in Section 17.2, we survey the existing literature that we judge useful/applicable for the VFC.

Computation management aims to provide proper resources to the application tasks and usually includes resource provisioning, resource consumption, and resource monitoring.

Resource provisioning allows preparing resources for computation to guarantee the application requirements. Resource provisioning includes resource estimation, fog node selection, and workload allocation.

Resource estimation allows estimating the necessary resources to allocate in the involved nodes. The estimation should consider the heterogeneity and availability of fog nodes. Aazam et al. [40] propose a flexible resource estimation mechanism based on user characteristics by estimating suitable resources to be allocated. They provide a mathematical model relying on resource relinquish probability to save resources and minimize costs taking into account the application quality. In [41], the authors implement a dynamic resource estimation by reducing the underutilization of resources and enhancing the QoS based on the characteristics of the services reaching devices. In [42], El Kafhali et al. develop a mathematical model by analyzing fog resources, estimating the right amount of fog nodes under any offered workload and the ability for a dynamic scaling following the incoming workload. Malandrino et al. [43] focus on estimating the server utilization and latency in relation with the extension of deployed mobile edge omputing (MEC). These efforts could support the realization of the considered time-critical scenarios for VFC as they allow for proactive decision making.

Fog node selection. After estimating the required resource, fog node selection assures the selection of the suitable fog nodes for the considered tasks. Li et al. [44] propose two schemes: (1) fog resource reservation that selects and reserves some fog resources to the respective vehicle based on traffic flow prediction methods, and (1) fog resource reallocation of resources depending on the application/task priority. Gedeon et al. [45] implement a brokering mechanism based on helping the client to select the relevant accessible surrogate to off-load

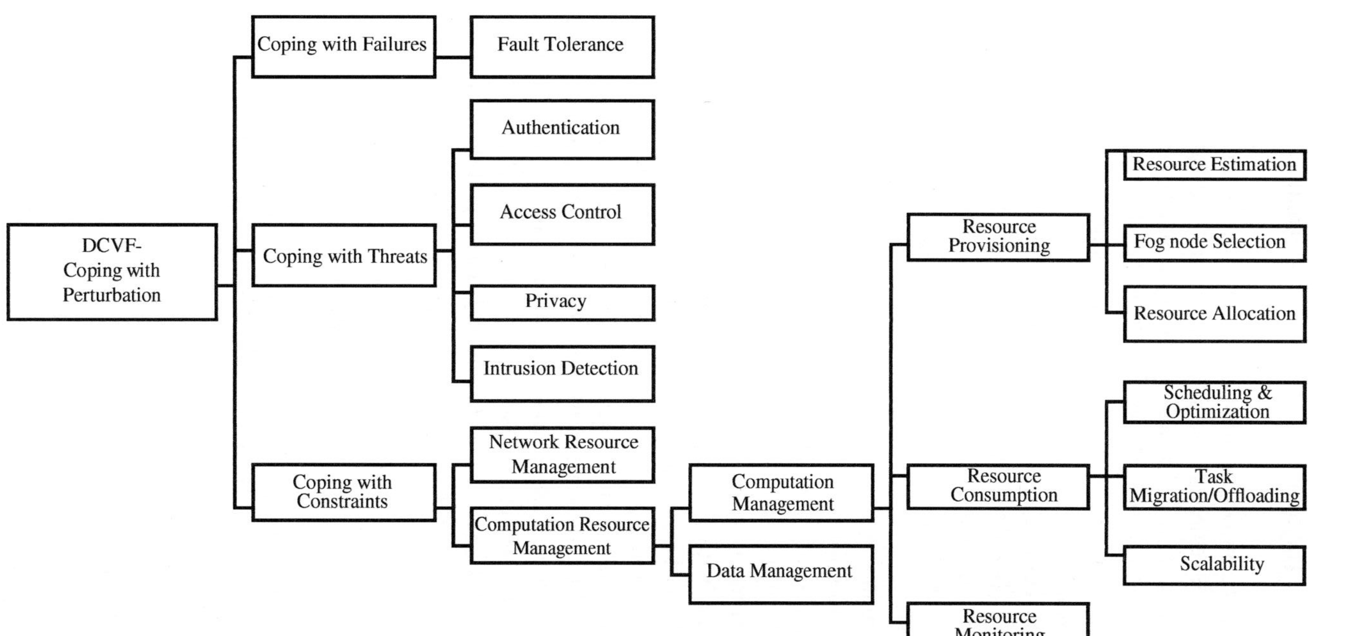

Figure 17.4 Coping with perturbation in DCVF.

computation tasks. The broker is responsible for updating information of all surrogates.

Resource allocation. Resource allocation is the assignment of the needed resources to the concrete application on the selected nodes. Load balancing across the different fog nodes is crucial to assure the availability of required resources, e.g. by delaying or deactivating lower priority tasks where needed. Sutagundar et al. [46] propose a game theory approach in fog-enhanced vehicular services to perform the resource allocation taking into account the resource estimation by predicting the required number of resources at the time of the workload allocation. Deng et al. [47] develop a workload allocation framework by balancing power consumption, delay, and workload allocation in fog–cloud computing environments. The authors [44, 48, 49] address workload allocation taking into account the QoS.

Resource consumption includes how to schedule a set of tasks and optimize resources, how to off-load a task from the fog node of the requester to the proper workload of the selected nodes, and to scale resources by extending them across vehicular nodes, fogs, and clouds.

Scheduling and optimization aim to control and optimize resources first and then arrange a set of jobs in the workload while respecting the QoS and the critical nature of the tasks. Park et al. [50] address the problem of critical delay services for the connected vehicles by providing an optimal scheduling algorithm based on Reinforcement Learning Data Scheduling. They focus on the problem proposed by the Markov decision process, which is a reinforcement learning method to increase the transmission number of services within the deadline. Zhu et al. [46] design a Folo model for VFC to optimize the task allocation. The task allocation process is considered a joint optimization problem. A dynamic task allocation approach based on Linear Programming-based Optimization and Binary Particle Swarm-based Optimization has proposed to solve the issue. Other related contributions in this field [22, 48, 51].

Task migration/offloading is a crucial technique to migrate Virtual Machine (VM) or container from Fog node to another Fog node. H. Yao et al. [52] introduce a rod side Cloudlet (RSC) that enable the VM migration in the VCC to improve the response time and reduce network and VM migration cost during vehicle movement. A. Machen et al. [53] show the performance of the container compared to the VM in the proposed layered migration framework among mobile edge Cloud. Additional related research in the service migration where the authors focus on the container in mobile Fog Computing [54–56]. I. Farris et al. [57] aim to enhance the proactive migration of latency-aware applications in MEC by providing two Integers Linear Problem optimization schemas to guarantee the desired Quality of Experience (QoE) and decrease the cost of proactive replication. Wang et al. [58] provide a relevant survey on the service migration in MEC.

Off-loading intends to send tasks from the application initiator to other fogs in order to reduce energy consumption and service delays. This technique is typically used in MCC such as on smart phones. Wu et al. [59] develop a task off-loading strategy in VFC based on the proposed model Direction-based Vehicular Network Model (DVNM) to perform off-loading of the tasks within vehicles and RSUs. Zhang et al. [60] develop an efficient code partition algorithm for MCC based on depth-first search and a linear time-searching scheme to find the convenient points on a sequence of calls for off-loading and integration. This contribution could adapt effectively to the VFC. Zhou et al. [61] provide a good survey that explains the data off-loading techniques for V2V, V2I, and V2X through VANET.

Scalability. Due to the unlimited resources among cloud and fog, this technique must be able to adopt all kinds of occurrences predictable or unpredictable and maintain the quality of service. Yan et al. [62] propose a user access mode selection mechanism to perform the scalability of the application in F-RAN (fog computing radio access network) taking into account the different nodes locations and QoS requirement of the service accessing entities. Tseng et al. [63] extend the oneM2M platform from cloud to fog computing based on containerization of the oneM2M middle node as a Docker container to make the system highly scalable and resolve the latency issues for some critical applications.

Resource monitoring control and monitor the events and the performance of each service and then record the results to the concerning services. Monitoring includes resource discovery. Resource discovery aims to discover the available fog resources and provides the necessary information to the application initiator such as location, availability, and the available capacity for use of the discovered fog nodes. In [45, 64, 65] the authors focus on discovering resources in fog computing. On the other hand, Lai et al. [66] provide an efficient schema called two-phase event-monitoring and data-gathering (TPEG) that collect data and monitor frequently the events of the fog nodes in VANET in order to select the suitable amount of data for decision making and avoid the useless messages transmissions based on two-level threshold adjustment (2LTA) algorithm.

Data management is very challenging because of the limited capacity of the vehicle. Data management needs an optimal separation between effective and ineffective data. The effective data is a sensitive data related to the safety, which must be stored onboard computer. The ineffective data is a soft data that should be moved into the cloud.

After processing the application correctly, the application initiator decides where to store the results/data. Most of the researchers are addressing optimization of the data for storage and then select the data that should be stored on the vehicle or on the cloud data centers [67, 68].

17.3.2 Coping with Failures

Fault tolerance is a fundamental solution to cope with failures. Some related research [69–72] addresses the fault tolerance in fog computing to meet the timeliness guarantees. Kopetz et al. [73] implement a fault-tolerance technique in VFC for the real-time application that well improved the allocation of the time-triggered virtual machine (TTVM) on different fog nodes systems.

In addition, fault tolerance is also highlighted in the connected vehicles to limit the constraint in message delivery infrastructure. Du [74] designs a distributed message delivery system by developing a prototype infrastructure using Kafka [75]. The results show the performance of the proposed prototype in the connected vehicle application, which is highly scalable, fault tolerant, and able to deliver in parallel a big amount of messages in a short time.

17.3.3 Coping with Threats

Coping with authentication threats. Fog node authentication, service migration authentication, etc. are indispensable for VFC. Dsouza et al. [76] implement a policy-driven security management framework in fog computing, which makes data, devices, instance, and data migration authentication more efficient and enhance the protection of the system for real-time services. Applying the physical contact for preauthentication in local ad-hoc wireless networks [77] and cloudlets authentication in near-field communication (NFC)-based mobile computing [78] on VFC can resolve the problem of authentication and enhance the security.

Coping with access control threats. In general, access control restricts the access to some services by defining specific rules to each node. Aazam et al. [79] implement a security layer in fog nodes as a smart gateway in order to control the access to such sensitive data that will be uploaded to the cloud. Salonikias [80] proposes a preliminary access control approach called attribute-based access control (ABAC) based on attributes and its security policy to resolve the problem of access control in intelligent transportation system architectures using fog computing for the required sensitive application.

Coping with privacy threats. Data generated from the vehicles may be very sensitive and require the protection of user privacy. It is necessary to ensure the confidentiality of vehicular fog in a distributed environment, but the subject still is a challenge for future research [9].

Data privacy. In order to ensure data privacy, the authors [81] provide a privacy preserving vehicular road surface condition monitoring using fog computing based on certificate less aggregate signcryption. Lu et al. [82] design an efficient privacy-preserving aggregation scheme based on homomorphic encryption to improve security and privacy preservation in smart grid communications.

Location privacy. This part needs proper attention to investigate in location privacy issues on the vehicular fog client, taking into account the high mobility of vehicles as a fog node.

Usage privacy. In this pattern, utilization of fog nodes services require a usage policy for the customer, taking an example, [83, 84] propose a privacy-preserving mechanism in the smart metering for the smart grid domain. But user usage still is not tackled efficiently in the VFC.

Network security. The authors [85–87] point on the network security SDN-based in terms of network monitoring and intrusion detection system, network resource access control and network sharing.

Coping with intrusion threats. The principal role of intrusion detection is to control and monitor all fog nodes from any kind of attacks, such as denial of service (DoS) attacks, insider attack, attacks on VM, and hypervisor [88]. For example, in VANET, Malla et al. [89] provide an efficient solution based on several lines of defense to counter DoS attacks. But mitigating attacks in vehicular networks using fog computing requires a high attention in further investigation.

17.4 Research Gaps and Future Research Directions

Our literature survey has shown that the following aspects are insufficiently addressed in the literature.

17.4.1 Mobile Fog Computing

Fog mobility is considered among the most crucial challenges in fog computing. Existing contributions usually consider static fog nodes and mobile or fixed user devices. Supporting mobility of fog nodes largely remains an open challenge due to the complication of resource and network management. In particular, in vehicular fog environments, the mobility is very high and even further hardens the application and system design. Accordingly, designing an efficient paradigm that can provide a wide range of scenarios and ensure mobility-aware management and coordination among mobile fog nodes is urgently needed.

Resource management due to the mobility of fog nodes, provisioning, virtualization, selection, and scheduling of resources need to be revised and optimized to cope with the continuously changing network topology and the fluctuating resource availability.

Resource provisioning. Discovering and selecting the relevant fog node in a high-mobility environment is challenging, and perhaps predicting resource discovery in mobile fog can be a suitable solution. In addition, migration of VM/container in other nodes and estimation of the execution time to meet the deadline taking into account heterogeneity of the computing systems. Continuously changing workloads on fogs and nonstatic fog servers will be more challenging.

Resource consumption. Regarding its heterogeneity and dynamic distributed nature, mobile fog computing requires deep attention to the scalability issues in the fog layer. In addition, efficient scheduling is needed to limit the difficulty in task off-loading. Task classification and optimization can be helpful to resolve these issues; as well as resource provisioning, resource consumption requires a careful investigation.

Virtualization. Inspired from the cloud data center approach, virtualization of fog computing is a big challenge that targets transforming a fog node into a small-scale data center at the edge network. This challenge can be a suitable solution for VM/container migration or task off-loading. The benefits are an improved QoS, reduced costs, and real-time communication and computation functions.

Decision. Taking a decision with respect to QoS and service delivery requirement is another research gap to be addressed.

Resource monitoring. The literature does not meet the needs for more proactiveness in monitoring computational and network resources in VFC. Accordingly, there is an urgent need for techniques that accurately estimate and forecast resource availability and needs in highly dynamic environments.

Network management. Mobile fog nodes frequently join and leave fog networks, which further complicates the network resource management. In particular, ensuring a seamless zero-delay handover is challenging. Discovering network resources in mobile fog networks is also an interesting and challenging research direction.

17.4.2 Fog Service Level Agreement (SLA)

Service level agreements (SLAs) are widely investigated for cloud computing, however, still need research efforts to adopt them for mobile fog computing with strict timeliness and bandwidth guarantees. This challenge should be addressed by defining and designing metrics and SLA enforcement techniques that are suitable for mobile fog computing. In addition to SLA, security LA (SecLA) represents a real challenge for VFC, due to the high mobility and geographically distributed vehicles that require a privacy-preserving, critical-data protection, and fast authentication.

Authentication. Due to the resource constraints in the connected vehicle, authentication with the traditional method using certificates and PKI to the fog layers is insufficient. In addition, ensuring service continuity despite mobile fog churns (node joins and leaves) is an open challenge.

Privacy preserving. Preserving privacy in fog computing is still challenging due to the shared resource of fog node with other nodes, which is a risk to expose some sensitive data.

Access control. Designing an efficient optimistic access control scheme along with a user permission policy that cover vehicle, fog, and cloud is another potential direction.

Intrusion detection. Intrusion detection should be carefully designed and implemented across vehicles, fogs, and clouds while ensuring efficient coordination between different levels. These challenges should be contributed taking into account the high mobility of vehicles and the distribution of the fog nodes.

17.5 Conclusion

In this chapter, we illustrate the different application scenario classes and their timeliness requirements. Next, we present the different perturbations that complicate the design of delay-critical VFC. Then, we survey literature on network management, resource management, security, and fault tolerance to cope with perturbation and to guarantee the timeliness requirements. For further investigation in delay-critical VFC, we point out key research gaps and challenges that require deep research attention.

References

1. Whaiduzzaman, M., Sookhaka, M., Gani, A., and Buyya, R. (2014). A survey on vehicular cloud computing. *Journal of Network and Computer Applications* 40: 325–344.
2. Skala, K., Davidovic, D., Afgan, E. et al. (2015). Scalable distributed computing hierarchy: cloud, fog and dew computing. *Open Journal of Cloud Computing (OJCC)* 2 (1): 16–24.
3. Dinh, H.T., Lee, C., Niyato, D., and Wang, P. (2013). A survey of mobile cloud computing: architecture, applications, and approaches. *Wireless Communications and Mobile Computing*.
4. Naha, R.K., Garg, S., Georgakopoulos, D. et al. (2018). Fog computing: survey of trends, architectures, requirements, and research directions. *IEEE Access* 6: 47980–48009.

5 OpenFogConsortium, Openfog reference architecture for fog computing 2017, [Online]. Available: https://www.openfogconsortium.org/ra, 2017.

6 Bonomi, F., Milito, R., Zhu, J., and Addepalli, S. (2012). Fog computing and its role in the internet of things. In: *Proceedings of the First Edition of the MCC Workshop on Mobile Cloud Computing*, 13–16. ACM.

7 A. Khelil and D. Soldani, On the suitability of device-to-device communications for road Traffic safety. 2014 IEEE World Forum on Internet of Things (WF-IoT), 2014.

8 Shi, W., Cao, J., Zhang, Q. et al. (2016). Edge computing: vision and challenges. *IEEE Internet of Things Journal* 3 (5): 637–646.

9 Hou, X., Li, Y., Chen, M. et al. (2016). Vehicular fog computing: a viewpoint of vehicles as the infrastructures. *IEEE Transactions on Vehicular Technology* 65 (6): 3860–3873.

10 Huang, C., Lu, R., and Choo, K. (2017). Vehicular fog computing: architecture use case and security and forensic challenges. *IEEE Communications Magazine* 55 (11): 105–111.

11 Al-Sultan, S., Al-Doori, M.M., Al-Bayatti, A.H., and Zedan, H. (2014). A comprehensive survey on vehicular ad hoc network. *Journal of Network and Computer Applications* 37: 380–392.

12 Hu, F. (2018). *Vehicle-to-Vehicle and Vehicle-to-Infrastructure Communications.* Boca Raton: CRC Press.

13 Mukherjee, M., Shu, L., and Wang, D. (2018). Survey of fog computing: fundamental network applications and research challenges. *IEEE Communications Surveys and Tutorials.*

14 Mouradian, C., Naboulsi, D., Yangui, S. et al. (2017). A comprehensive survey on fog computing: state-of-the-art and research challenges. *IEEE Communications Surveys and Tutorials.*

15 A. E. Fernandez, A. Servel, J. Tiphene et al., 5GCAR scenarios, use cases, requirements and KPIs, Fifth Generation Communication Automotive Research and Innovation, https://5gcar.eu/wp-content/uploads/2017/05/5GCAR_D2.1_v1.0.pdf, 2017.

16 Roig, C., Ripoll, A., and Guirado, F. (2007). A new task graph model for mapping message passing applications. *IEEE Transactions on Parallel and Distributed Systems.*

17 Gezer, V., Um, J., and Ruskowski, M. (2018). An introduction to edge computing and a real-time capable server architecture. *International Journal on Advances in Intelligent Systems* 11 (1 and 2): 105–114.

18 G. Lipari and L. Palopoli, Real-time scheduling: from hard to soft real-time systems. https://arxiv.org/abs/1512.01978, 2015.

19 Bernat, G., Burns, A., and Llamosi, A. (2001). Weakly hard real-time systems. *IEEE Transactions on Computers* 50 (4): 308–321.

20 T. Kaldewey, C. Lin, and S. Brandt, Firm real-time processing in an integrated real-time system, University of York, Department of Computer Science – Report, Vol. 398, p. 5, 2006.

21 Grover, J., Jain, A., Singhal, S., and Yadav, A. (2018). Real-time VANET applications using fog computing. In: *Proceedings of First International Conference on Smart System, Innovations and Computing*, 683–691. New York: Springer.

22 Mahmud, R., Ramamohanarao, K., and Buyya, R. (2017). Latency-aware application module management for fog computing environments. *ACM Transactions on Internet Technology (TOIT).*

23 Ge, X., Li, Z., and Li, S. (2017). 5G software defined vehicular networks. *IEEE Communications Magazine* 55 (7): 87–93.

24 Docker, www.docker.com.

25 Merkel, D. (2014). Docker: lightweight Linux containers for consistent development and deployment. *Linux Journal* 239.

26 Motlagh, N.H., Bagaa, M., and Taleb, T. (2017). UAV-based IoT platform: a crowd surveillance use case. *IEEE Communications Magazine* 55 (2): 128–134.

27 Mukherjee, M., Matam, R., Shu, L. et al. (2017). Security and privacy in fog computing: challenges. *IEEE Access.*

28 Virdis, A., Vallati, C., Nardini, G. et al. (2018). D2D communications for large-scale fog platforms: enabling direct M2M interactions. *IEEE Vehicular Technology Magazine.*

29 Froiz-Míguez, I., Fernández-Caramés, T.M., Fraga-Lamas, P., and Castedo, L. (2018). Design, implementation and practical evaluation of an IoT home automation system for fog computing applications based on MQTT and ZigBee-WiFi sensor nodes. *Sensors* 18: 2660.

30 Vinel, A., Breu, J., Luan, T.H., and Hu, H. (2017). Emerging technology for 5G-enabled vehicular networks. *IEEE Wireless Communications* 24 (6): 12.

31 Bao, W., Yuan, D., Yang, Z. et al. (2017). Follow me fog: toward seamless handover timing schemes in a fog computing environment. *IEEE Communications Magazine* 55 (11): 72–78.

32 Xiang, C., Rongqing, Z., and Liuqing, Y. (2019). *Introduction to 5G-Enabled VCN*. New York: Springer.

33 A. Soua and S. Tohme, Multi-level SDN with vehicles as fog computing infrastructures: a new integrated architecture for 5G-VANETs, 21st Conference on Innovation in Clouds, Internet and Networks and Workshops (ICIN), 2018.

34 A. A. Khan, M. Abolhasan, and W. Ni, 5G next generation VANETs using SDN and fog computing framework, 15th IEEE Annual Consumer Communications Networking Conference (CCNC), 2018.

35 Huang, X., Yu, R., Kang, J. et al. (2017). Exploring mobile edge computing for 5G-enabled software defined vehicular networks. *IEEE Wireless Communications* 24 (6): 55–63.

36 Truong, N.B., Lee, G.M., and Ghamri-Doudane, Y. (2015). Software defined networking-based vehicular Adhoc network with fog computing. In: *2015 IFIP/IEEE International Symposium on Integrated Network Management (IM)*, 1202–1207. IEEE.

37 M. R. Palattella, R. Soua, A. Khelil, and T. Engel, Fog computing as the key for seamless connectivity handover in future vehicular networks, in The Proceedings of the 34th ACM/SIGAPP Symposium on Applied Computing (SAC), 2019.

38 A. Khelil, M. R. Palattella, R. Soua, and T. Engel, Fog computing as the key for seamless connectivity handover in future vehicular networks, in The Proceedings of the 34th ACM/SIGAPP Symposium On Applied Computing (SAC), 2019.

39 Baccarelli, E., Naranjo, P.G.V., Scarpiniti, M. et al. (2017). Fog of everything: energy-efficient networked computing architectures, research challenges, and a case study. *IEEE Access* 5: 9882–9910.

40 M. Aazam and E. Huh, Dynamic resource provisioning through fog micro datacenter, IEEE International Conference on Pervasive Computing and Communication Workshops (PerCom Workshops), 2015.

41 M. Aazam, M. St-Hilaire, C. Lung, and I. Lambadaris, MeFoRE: QoE based resource estimation at fog to enhance QoS in IoT, 23rd International Conference on Telecommunications (ICT), 2016.

42 El Kafhali, S. and Salah, K. (2017). Efficient and dynamic scaling of fog nodes for IoT devices. *Journal of Supercomputing* 73 (12): 5261–5284.

43 Malandrino, F., Kirkpatrick, S., and Chiasserini, C.-F. (2016). How close to the edge? Delay/utilization trends in MEC. In: *Proceedings of the 2016 ACM Workshop on Cloud-Assisted Networking*, 37–42. ACM.

44 J. Li, C. Natalino, D. P. Van, L. Wosinska, and J. Chen, Resource management in fog-enhanced radio access network to support real-time vehicular services, IEEE 1st International Conference on Fog and Edge Computing (ICFEC), 2017.

45 J. Gedeon, C. Meurisch, D. Bhat et al., Router-based brokering for surrogate discovery in edge computing, IEEE 37th International Conference on Distributed Computing Systems Workshops (ICDCSW), 2017.

46 Zhu, C., Tao, J., Pastor, G. et al. (2018). Folo: latency and quality optimized task allocation in vehicular fog computing. *IEEE Internet of Things Journal*.

47 Deng, R., Lu, R., Lai, C. et al. (2016). Optimal workload allocation in fog-cloud computing towards balanced delay and power consumption. *IEEE Internet of Things Journal* 3: 1171–1181.

48 Zeng, D., Gu, L., Guo, S. et al. (2016). Joint optimization of task scheduling and image placement in fog computing supported software-defined embedded system. *IEEE Transactions on Computers* 65: 3702–3712.

49 Y. Chen, J. P. Walters, and S. P. Crago, Load balancing for minimizing deadline misses and total runtime for connected car systems in fog computing, IEEE International Symposium on Parallel and Distributed Processing with Applications and 2017 IEEE International Conference on Ubiquitous Computing and Communications (ISPA/IUCC), 2017.

50 Park, S. and Yoo, Y. (2018). Real-time scheduling using reinforcement learning technique for the connected vehicles. In: *IEEE 87th Vehicular Technology Conference (VTC Spring)*, 1–5. IEEE.

51 Lin, F., Zhou, Y., Pau, G., and Collotta, M. (2018). Optimization-oriented resource allocation management for vehicular fog computing. *IEEE Access* 6: 69294–69303.

52 Yao, H., Bai, C., Zeng, D. et al. (2015). Migrate or not? Exploring virtual machine migration in roadside cloudlet-based vehicular cloud. *Concurrency and Computation: Practice and Experience* 27 (18): 5780–5792.

53 Machen, A., Wang, S., Leung, K.K. et al. (2016). Migrating running applications across mobile edge clouds: poster. In: *Proceedings of the 22nd Annual International Conference on Mobile Computing and Networking (MobiCom)*, 435–436. ACM.

54 Montero, D. and Serral-Gracia, R. (2016). Offloading personal security applications to the network edge: a mobile user case scenario. In: *Proceedings of the International Wireless Communications and Mobile Computing Conference (IWCMC)*, 96–101. IEEE.

55 Saurez, E., Hong, K., Lillethun, D. et al. (2016). Incremental deployment and migration of geodistributed situation awareness applications in the fog. In: *Proceedings of the 10th ACM International Conference on Distributed and Event-Based Systems (DEBS)*, 258–269. ACM.

56 Wang, S., Urgaonkar, R., He, T. et al. (2017). Dynamic service placement for mobile micro-clouds with predicted future costs. *IEEE Transactions on Parallel and Distributed Systems* 28 (4): 1002–1016.

57 Farris, I., Taleb, T., Bagaa, M., and Flick, H. (2017). Optimizing service replication for mobile delay-sensitive applications in 5G edge network. *IEEE International Conference on Communications (ICC)*: 1–6.

58 Wang, S., Xu, J., Zhang, N., and Liu, Y. (2018). A survey on service migration in mobile edge computing. *IEEE Access* 6: 23511–23528.

59 Y. Wu, J. Wu, G. Zhou, and L. Chen, A direction-based vehicular network model in vehicular fog computing, IEEE SmartWorld, Ubiquitous Intelligence Computing, Advanced Trusted Computing, Scalable Computing Communications, Cloud Big Data Computing, Internet of People and Smart City Innovation (SmartWorld/SCALCOM/UIC/ATC/CBDCom/IOP/SCI), 2018.

60 Zhang, Y., Liu, H., Jiao, L., and Fu, X. (2012). To offload or not to offload: an efficient code partition algorithm for mobile cloud computing. In: *IEEE 1st International Conference on Cloud Networking (CLOUDNET)*, 80–86. IEEE.

61 Zhou, H., Wang, H., Chen, X. et al. (2018). Data offloading techniques through vehicular ad hoc networks: a survey. *IEEE Access* 6: 65250–65259.

62 Yan, S., Peng, M., and Wang, W. (2016). User access mode selection in fog computing based radio access networks. *IEEE International Conference on Communications (ICC)*: 1–6.

63 Tseng, C. and Lin, F.J. (2018). Extending scalability of IoT/M2M platforms with fog computing. In: *IEEE 4th World Forum on Internet of Things (WF-IoT)*, 825–830. IEEE.

64 Cho, J., Sundaresan, K., Mahindra, R. et al. (2016). Acacia: context-aware edge computing for continuous interactive applications over mobile networks. In: *Proceedings of the 12th International on Conference on Emerging Networking Experiments and Technologies*, 375–389. ACM.

65 Tanganelli, G., Vallati, C., and Mingozzi, E. (2018). Edge-centric distributed discovery and access in the Internet of Things. *IEEE Internet of Things Journal* 5 (1): 425–438.

66 Lai, Y., Yang, F., and Su et al., J. (2017). Fog-based two-phase event monitoring and data gathering in vehicular sensor networks. *Sensors*, http://www.mdpi.com/1424-8220/18/1/82.

67 Shi, H., Chen, N., and Deters, R. (2015). Combining mobile & fog computing using CoAP to link mobile device clouds with fog computing. In: *IEEE International Conference on Data Science and Data Intensive Systems*, 564–571. IEEE

68 Hassan, M.A., Xiao, M., Wei, Q., and Chen, S. (2015). Help your mobile applications with fog computing. In: *12th Annual IEEE International Conference on Sensing, Communication, and Networking – Workshops (SECON Workshops)*, 1–6. IEEE.

69 Wang, K., Shao, Y., Xie, L. et al. (2018). Adaptive and fault-tolerant data processing in healthcare IoT based on fog computing. *IEEE Transactions on Network Science and Engineering*.

70 Zhang, J., Zhou, A., Sun, Q. et al. (2018). Overview on fault tolerance strategies of composite service in service computing. *Wireless Communications and Mobile Computing* 2018, Article ID 9787503, 8 pp.

71 J. P. Araujo Neto, D. M. Pianto, and C. G. Ralha, An agent-based fog computing architecture for resilience on amazon EC2 spot instances, 7th Brazilian Conference on Intelligent Systems (BRACIS), 2018.

72 Xu, J.W., Ota, K., Dong, M.X. et al. (2018). SIoTFog: byzantine-resilient IoT fog networking. *Frontiers of Information Technology & Electronic Engineering*.

73 Kopetz, H. and Poledna, S. (2016). In-vehicle real-time fog computing. In: *Proceedings of the 2016 46th Annual IEEE/IFIP International Conference Dependable Systems and Networks Workshop (DSN-W)*, 162–167. IEEE.

74 Du, Y., Chowdhury, M., Rahman, M., and Dey, K. (2017). A distributed message delivery infrastructure for connected vehicle technology applications. *IEEE Transactions on Intelligent Transportation Systems.*

75 Kreps, J., Narkhede, N., and Rao, J. (2011). Kafka: a distributed messaging system for log processing. In: *Proc. NetDB*, 1–7. ACM.

76 Dsouza, C., Ahn, G.J., and Taguinod, M. (2014). Preliminary framework and a case study. Policy-driven security management for fog computing. In: *Proceedings of the 2014 IEEE 15th International Conference on Information Reuse and Integration*, 16–23. IEEE.

77 Balfanz, D., Smetters, D., Stewart, P., and Wong, H.C. (2002). Talking to strangers: authentication in ad-hoc wireless networks. In: *Proceedings of the Symposium on Network and Distributed System Security*, 23–35.

78 Bouzefrane, S., Mostefa, A.F.B., Houacine, F., and Cagnon, H. (2014). Cloudlets authentication in NFC-based mobile computing. In: *Proceedings of the 2nd IEEE International Conference on Mobile Cloud Computing, Services, and Engineering (MobileCloud)*, 267–272. Oxford, UK. IEEE.

79 Aazam, M. and Huh, E.N. (2014). Fog computing and smart gateway based communication for cloud of things. In: *Proceedings of the International Conference on Future Internet of Things and Cloud (FiCloud)*, 464–470. IEEE.

80 Salonikias, S., Mavridis, I., and Gritzalis, D. (2016). Access control issues in utilizing fog computing for transport infrastructure. In: *Proceedings of CRITIS*, 15–26. Springer.

81 Basudan, S., Lin, X., and Sankaranarayanan, K. (2017). A privacy-preserving vehicular crowdsensing-based road surface condition monitoring system using fog computing. *IEEE Internet of Things Journal* 4 (3): 772–782.

82 Lu, R., Liang, X., Li, X. et al. (2012). EPPA: an efficient and privacy-preserving aggregation scheme for secure smart grid communications. *IEEE Transactions on Parallel and Distributed Systems* 23 (9): 1621–1631.

83 A. Rial and G. Danezis, Privacy-preserving smart metering, in Proceedings of the ACM WPES, 2011.

84 McLaughlin, S., McDaniel, P., and Aiello, W. (2011). Protecting consumer privacy from electric load monitoring. In: *Proceedings of the 18th ACM Conference on Computer and Communications Security*, 87–98. ACM.

85 S. Shin and G. Gu, CloudWatcher: Network security monitoring using OpenFlow in dynamic cloud networks (or: How to provide security monitoring as a service in clouds), in Proceedings of the 20th IEEE International Conference on Network Protocols, 2012.

86 Klaedtke, F., Karame, G.O., Bifulco, R., and Cui, H. (2014). Access control for SDN controllers. In: *Proceedings of the Third Workshop on Hot Topics Software Defined Network*, 219–220. ACM.

87 K.-K. Yap, Y. Yiakoumis, M. Kobayashi et al., Separating authentication, access and accounting: a case study with OpenWiFi, OpenFlow Technical Report 2011-1, 2011.

88 Modi, C., Patel, D., Borisaniya, B. et al. (2013). A survey of intrusion detection techniques in cloud. *Journal of Network and Computer Applications* 36 (1): 42–57.

89 Malla, A.M. and Sahu, R.K. (2013). Security attacks with an effective solution for DoS attacks in VANET. *International Journal of Computers and Applications* 66 (22): 45.

18

A Reliable and Efficient Fog-Based Architecture for Autonomous Vehicular Networks

Shuja Mughal[1], Kamran Sattar Awaisi[1], Assad Abbas[1], Inayat ur Rehman[1], Muhammad Usman Shahid Khan[2], and Mazhar Ali[2]

[1] *Department of Computer Science, COMSATS University Islamabad, Islamabad Campus, Pakistan*
[2] *Department of Computer Science, COMSATS University Islamabad, Abbottabad Campus, Pakistan*

18.1 Introduction

In recent years, technological developments have revolutionized the automobile industry. Cars being manufactured today include features like auto-drive, cruise control, and parking assistance systems. These features allow the car to steer itself on the road or park itself. Major companies today have made a significant leap by creating autonomous vehicles (AVs) [1, 2]. These AVs are also called automated or self-driving vehicles. The AVs have the capability to navigate roadways and environmental contexts by themselves without human input. The AVs have the potential to change the transportation system by preventing deadly crashes, providing help to the elderly and disabled, and saving fuel as described in [2–5].

Cloud computing is a new technology in information technology that separates computing and data away from user devices into large data centers. Cloud computing deals with software, storages services, data access, and computation at a fixed physical location, from which is delivering the services [6]. The cloud comes from telecommunications systems where users access virtual private networks (VPNs) for data communications. With the usage of the Internet around the world in almost every field, cloud computing services are also being provided through various methods. The aim of cloud computing is to make an efficient and effective use of the distributed resources to achieve higher output and to solve complex computation problems. However, limitations of cloud computing have become evident for latency-sensitive applications that require nodes to meet their delay requirements [6–10].

Cloud computing is a centralized computing model where most of the computations are performed in the cloud [6]. Although the data-processing speed has

Fog Computing: Theory and Practice, First Edition.
Edited by Assad Abbas, Samee U. Khan, and Albert Y. Zomaya.

increased with time, the network bandwidth has not increased, which may result in higher latency. Some of the applications require a low latency and mobility support. The examples include AVs, traffic light system, emergency response systems, smart health care, and several other latency-sensitive applications. The delay caused by transferring data is not acceptable as it could be life-threatening. As safety of human life is on the top priority, therefore, no risks can be tolerated in this matter [11–16].

It is also worth mentioning that the aforementioned challenges are caused by the sudden growth of the Internet of Things (IoT) and are hard to address on the cloud model. Therefore, a new platform called fog computing has exhibited tremendous potential to overcome the challenges of low network bandwidth and latency [17]. A key benefit of fog computing is that it allows to perform the computations near the network edges and consequently the issues pertinent to the latency are resolved, particularly for delay-sensitive applications, such as the AVs. In fog environment, some of the decisions of applications can be made locally without having to be transmitted to the cloud, which greatly decreases the latency within the applications [14, 18–29].

This chapter proposes a fog-based architecture to increase the reliability of Autonomous Vehicular Networks (AVNs). To evaluate the effectiveness of the proposed architecture, we created a scenario for both the fog-based architecture and cloud-based architecture on iFogSim. We compared the latency, bandwidth, and scalability of both the fog-based and cloud-based architectures. Experimental results demonstrate that the proposed fog-based architecture not only minimizes the latency but also consumes less bandwidth as compared to the cloud environment.

To address the shortcomings in the existing literature, we take into consideration the following research questions:

RQ 1: *Does the proposed fog-based architecture minimize the latency as compared to cloud-based architecture?*

RQ 2: *Does the proposed architecture minimize the bandwidth as compared to cloud-based architecture?*

RQ 2: *Does the proposed architecture maximize the scalability as compared to cloud-based architecture?*

To answer the previously stated research questions, we implemented the proposed fog-based architecture, analyzed, and compared the execution with the cloud-based architecture. To test our proposed architecture, we used iFogSim to implement both fog-based and cloud-based architectures. We ran multiple simulations of the scenario created on both of the architectures. The results of the simulation presented that the proposed fog-based architecture is not only supportive in minimizing the latency, it also consumes less bandwidth and is more scalable than the cloud-based network.

Several research works have been carried out in the context of fog computing and the AVNs. Hajibaba et al. [6] reviewed modern distributed computing paradigms, such as cloud computing, jungle computing, and fog computing. Madsen et al. [23] worked on the reliability aspects of fog computing in the utility computing era. Truong et al. [25] researched on software-defined networking-based vehicular ad hoc networks (VANETs) with fog computing. The authors made software-defined networks with fog computing for VANETs. Hou et al. [28] offered a viewpoint of vehicles as the infrastructure by using vehicular fog computing, in which the authors described different infrastructures for vehicular fog computing. Moreover, various techniques, applications, and issues pertaining to fog computing in AVN have been described in [14, 18, 21–28, 30].

The rest of the chapter is organized as follows: Section 18.2 presents the proposed methodology, whereas Section 18.3 presents the hypothesis formulation. Section 18.4 presents the simulation design and experimental results, and Section 18.5 concludes the chapter.

18.2 Proposed Methodology

This chapter proposes a fog-based architecture to increase the reliability of AVNs.

Figure 18.1 shows the cloud-based architecture for the communication in the AVN. In the figure, we can see that all the nodes that are AVs in our case are communicating directly with the cloud. Some of the applications' decisions can be made locally without having to be transmitted to the cloud, but in this case every decision is being made by the cloud, which greatly increases the latency within the

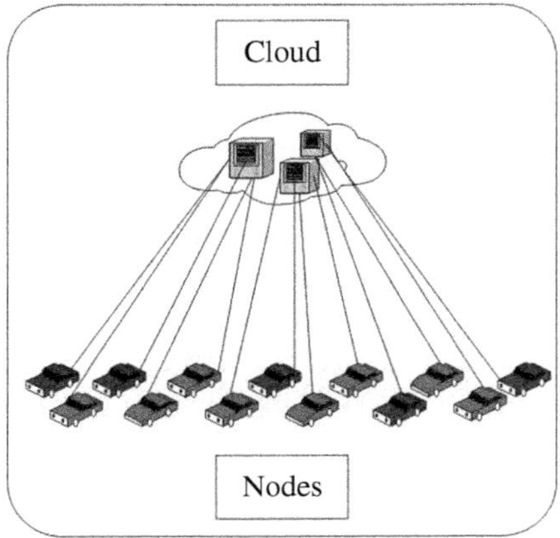

Figure 18.1 Cloud based architecture for the communication between autonomous vehicles.

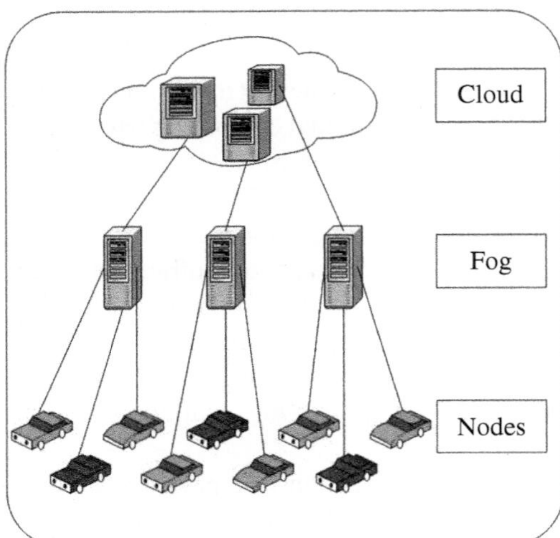

Figure 18.2 Proposed fog-based architecture for the communication among autonomous vehicles.

applications. This results in significant increase in the latency within the AVN, as all the decisions are being made at the cloud level, instead of being taken locally.

Figure 18.2 shows the proposed fog-based architecture for communication in the AVN. In the figure, we can see that all the nodes that are AVs in this case communicate with the fog instead of through direct communication with the cloud, because during the communication of AVs with each other, certain decisions need to be made instantly and, therefore, such processing is performed at the fog nodes in the proposed architecture. If the data cannot be handled by the fog node, then it is transmitted to the cloud. This will result in greatly decreasing the latency within the AVN. It is important to mention that the fog environment is not a replacement of the cloud. Instead the fog computing environment extends the capability of the cloud by pushing the processing close to the network's edges.

Figure 18.3 shows the realistic and practical view of the proposed fog-based architecture for communication in the AVN. In the figure, we can see that all the nodes that are AVs in our case are communicating with the towers that are the fog nodes. The fog nodes subsequently communicate with the data center, i.e. the cloud.

We implemented the proposed fog-based architecture, analyzed, and compared the execution with the cloud-based architecture. To test our proposed architecture, we used iFogSim to implement both the fog-based and the cloud-based architectures. We ran multiple simulations of the scenarios created on both architectures. It is revealed by the simulations that the proposed architecture not only minimized the latency of the network, but also consumed less bandwidth than the cloud-based network. The results also showed that by using the fog-based architecture, the reliability of the AVN is enhanced, as the AVs

Figure 18.3 Realistic and practical view of the proposed fog-based architecture for the communication among autonomous vehicles.

can communicate with each other in significantly less time than the cloud environment. Communication delays among AVs cannot be tolerated, as such delays may result in loss of human life.

18.3 Hypothesis Formulation

We formulated the following hypotheses:

H1: *μ(Latency of fog-based architecture) < μ(Latency of cloud-based architecture)*
H10: *μ(Latency of fog-based architecture) = μ(Latency of cloud-based architecture)*

Where H10 is the null hypothesis on which testing will be done, while H1 is the alternate hypothesis.

H2: *μ(Network usage of fog-based architecture) < μ(Network usage of cloud-based architecture)*
H20: *μ(Network usage of fog-based architecture) = μ(Network usage of cloud-based architecture)*

Where H20 is the null hypothesis on which testing will be done, while H2 is the alternate hypothesis.

H3: *μ(Scalability of fog-based architecture) > μ(Scalability of cloud-based architecture)*

H30: *μ(Scalability of fog-based architecture) = μ(Scalability of cloud-based architecture)*

Where H30 is the null hypothesis on which testing will be done, while H3 is the alternate hypothesis.

18.4 Simulation Design

We used iFogSim to implement both fog-based architecture and cloud-based architecture and designed simulation to run on both the architectures. iFogSim is a simulator that models IoT, cloud, and fog environments and also measures the impact of different resource management techniques like latency, network congestion, energy consumption, and cost [29]. During the simulations, we continuously varied the number of vehicles for both the architectures, to see variations in the latency and the bandwidth. For both architectures, the experiments were conducted for 10, 20, 30, 40, 50, 60, 70, 80, 90, and 100 vehicles in simulations.

Different results were seen with the different numbers of vehicles. The details of the results are explained in the next section.

18.4.1 Results and Discussions

This section presents the results for the proposed fog-based architecture and compares the results with the cloud-based implementation. We created a scenario in iFogSim and evaluated the performance of the fog-based and cloud-based implementation for latency and network usage.

Table 18.1 depicts the results of the latency in milliseconds (ms) for both the proposed fog-based and cloud-based architectures. It can be observed that even with 10 vehicles the latency of fog-based architecture is far less than the latency of the cloud-based architecture. Also, we can see a severe increase in the latency of cloud-based architecture with 40 vehicles. Such delays in latency in vehicular area networks, where autonomous cars need to communicate with each other frequently, are not tolerable and can result in fatal accidents. Therefore, the aforementioned results demonstrate the efficacy of the proposed fog-based architecture for the AVs and also enhances the reliability of the smart vehicular environments.

Table 18.2 depicts the log-transformed results of the latency for both proposed fog-based and cloud-based architectures. The reason to depict the log-transformed data here is that the latency values for the cloud are significantly higher and it is impractical to plot with high variations.

Figure 18.4 shows that there is a severe increase in the latency of cloud-based architecture as compared to tfog-based architecture. The y-axis shows the

Table 18.1 Results of the latency in the both architectures.

No. of vehicles	Latency in fog network	Latency in cloud network
10	0.457	7.03
20	0.81	8.76
30	1.171	11.48
40	1.528	1967
50	1.88	3049.5
60	2.243	3641.559
70	2.600	4000.1
80	2.95	4233.75
90	3.31	4394.35
100	3.671	4509.8

Table 18.2 Log-transformed data for latency.

No. of vehicles	Latency in fog network	Latency in cloud network
10	−0.34008	0.846955
20	−0.09151	0.942504
30	0.068557	1.059942
40	0.184123	3.293804
50	0.274158	3.484229
60	0.350829	3.561287
70	0.414973	3.602071
80	0.469822	3.626725
90	0.519828	3.642895
100	0.564784	3.654157

log-transformed latency, while the x-axis shows the two box plots. There are 100 vehicles, which is shown by the boxes, with whiskers for both architectures. Even the whiskers of both architectures do not overlap, which shows that fog-based architecture is far more reliable in terms of latency than cloud-based architecture, in the context of autonomous car environments.

Table 18.3 depicts the results of network usage for both the proposed fog-based architecture and the cloud-based architecture. As we can see from the results, even with 100 vehicles, network usage is lower in the fog-based architecture than in the

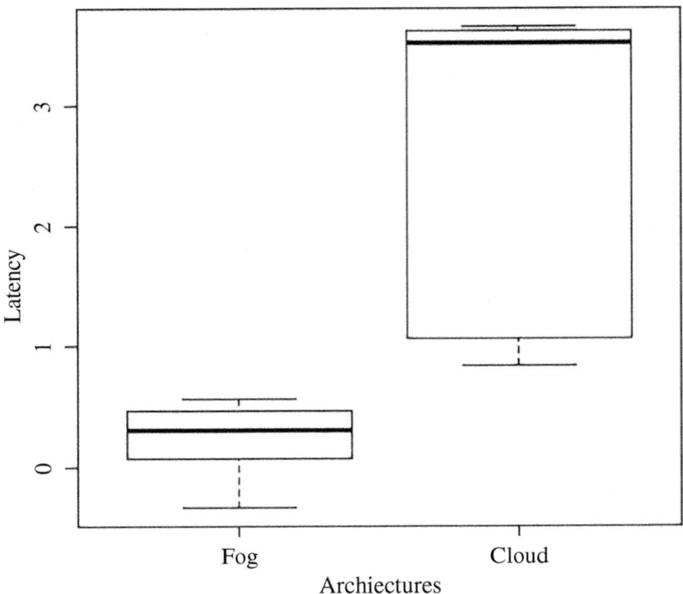

Figure 18.4 Box plot for the latency for both fog-based architecture and cloud-based architecture.

Table 18.3 Results of the network usage in the both architectures.

No. of vehicles	Network usage in fog network	Network usage in cloud network
10	509.49	22096.8
20	1018.98	48090.6
30	1528.47	78057.76
40	2037.96	83795.16
50	2547.45	89105.22
60	3056.94	95948.98
70	3566.43	103678.64
80	4075.92	111966.92
90	4585.41	120628.58
100	5094.9	129553.84

Table 18.4 Log-transformed data for network usage.

No. of vehicles	Network usage in fog network	Network usage in cloud network
10	2.707136	4.344329
20	3.008166	4.68206
30	3.184257	4.892416
40	3.309196	4.923219
50	3.406106	4.949903
60	3.485287	4.98204
70	3.552234	5.015689
80	3.610226	5.04909
90	3.661378	5.08145
100	3.707136	5.11245

cloud-based architecture. Moreover, it was also observed that network usage with a slightly higher number of vehicles, such as over 30 vehicles, increased severely and consequently reduced the efficacy of the cloud-based architecture.

Table 18.4 depicts the log-transformed results of network usage for both the proposed fog-based and the cloud-based architectures.

It can be seen in Figure 18.5 that there is a severe increase in network usage of the cloud-based architecture as compared to the fog-based architecture. The y-axis shows the log-transformed network usage while the x-axis shows the two box plots. The numbers of vehicles are 100 which are shown by the boxes, whiskers, and outliers for both architectures. Again, as in Figure 18.4 the whiskers in the box plots of both architectures do not overlap, which shows that fog-based architecture is significantly more reliable in terms of network usage when compared to cloud-based architecture in the given scenario.

18.4.2 Hypothesis Testing

For hypothesis testing, we will start from the first hypothesis, i.e. null hypothesis H10 and alternate hypothesis H1.

18.4.2.1 First Hypothesis

The null hypothesis H10 states:

H10: *μ(Latency of fog-based architecture) = μ(Latency of cloud-based architecture)*

We can see in Table 18.1 that, even with 100 vehicles, the latency of fog-based architecture is far less than the latency of the cloud-based architecture. Also, a

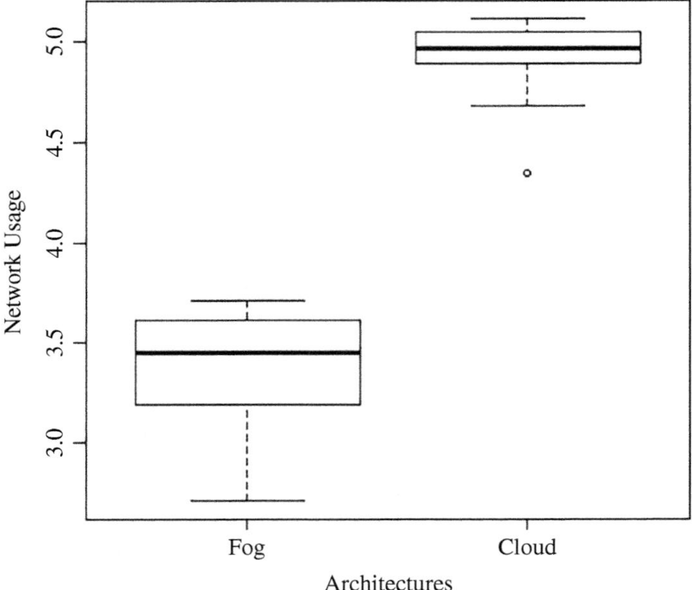

Figure 18.5 Box plot for network usage for both fog-based architecture and cloud-based architecture.

significant decrease in the latency of cloud-based architecture with 40 vehicles is observed. With the analyzed results, we can say that the fog-based architecture for the AVs has less latency when compared to the cloud-based architecture.

We can reject the null hypothesis H10, which stated that μ(Latency of fog-based architecture) = μ(Latency of cloud-based architecture).

As null hypothesis H10 is rejected, then the alternate hypothesis, which states that μ(Latency of fog-based architecture) < μ(Latency of cloud-based architecture), is accepted.

18.4.2.2 Second Hypothesis

The null hypothesis H20 states:

H20: *μ(Network usage of fog-based architecture) = μ(Network usage of cloud-based architecture)*

We can see in Table 18.3 that, even with 10 vehicles, network usage of fog-based architecture is far lower than network usage of the cloud-based architecture. Also, we can see a radical increase in network usage of cloud-based architecture with 30 vehicles. It can be seen that the fog-based architecture for the AVs has lower network usage when compared to the cloud-based architecture.

Figure 18.5 also shows us that network usage of cloud-based architecture when compared to the fog-based architecture is significantly higher. Even the whiskers in the box plots of both architectures don't overlap, which shows that fog-based architecture is far more reliable in terms of network usage when compared to cloud-based architecture.

We can reject the null hypothesis H20, which states that μ(Network usage of fog-based architecture) = μ(Network usage of cloud-based architecture).

Therefore, the alternate hypothesis H2, which states μ(Network usage of fog-based architecture) < μ(Network usage of cloud-based architecture), is accepted.

18.4.2.3 Third Hypothesis

The null hypothesis H30 states:

H30: *μ(Scalability of fog-based architecture) = μ(Scalability of cloud-based architecture)*

We can see in Tables 18.1 and 18.3 that, from 10 vehicles to 100 vehicles, the latency and network usage of fog-based architecture is far less than the network usage of the cloud-based architecture. Also, we can see a radical increase in both the latency and network usage of cloud-based architecture with an increase from 30 to 40 vehicles. With the analyzed results, we can say that the fog-based architecture for the AVs is more scalable when compared to the cloud-based architecture.

From Figures 18.4 and 18.5, it can be observed that the latency and network usage of cloud-based architecture when compared to the fog-based architecture is significantly higher. Also, the whiskers in the box plots of the both architectures do not intersect, which shows that fog-based architecture is far more scalable when compared to the cloud-based architecture.

We can reject the null hypothesis H30, which states that μ(Scalability of fog-based architecture) = μ(Scalability of cloud-based architecture).

Therefore, the alternate hypothesis H3, which states that μ(Scalability of fog-based architecture) > μ(Scalability of cloud-based architecture), is accepted.

18.5 Conclusions

In the past few years, due to technological advancements in the fields of artificial intelligence, robotics, sensor technologies, and self-driving vehicles, we are able to sense the surroundings of vehicles in real time. The coming of AVs has caused a radical increase in data traffic over networks. However, the AVs demand low latency and high bandwidth to efficiently communicate with the other vehicles. Connecting the AVs to the cloud only may result in a high rate of delays, which might result in accidents in smart vehicular environments. In this chapter, we proposed a fog-based architecture that not only minimizes the latency and bandwidth, but also increases the scalability as well. To validate our proposed architecture, we used a simulator called iFogSim. We used iFogSim to implement both the fog-based and cloud-based architectures. We performed multiple simulations of the scenario created on both of the architectures. The results of the simulation demonstrated that the proposed fog-based architecture not only minimizes the latency and network usage of the network, but also increases the scalability when compared to the cloud-based network. The results also are evidence of the reliability and effectiveness of fog-based applications for AVNs.

References

1 Fagnant, D., J. and Kockelman, K. (2015). Preparing a nation for autonomous vehicles: opportunities, barriers and policy recommendations. *Transportation Research Part A: Policy and Practice* 77: 167–181.

2 Campbell, M., Egerstedt, M., How, J.P., and Murray, R.M. (2010). Autonomous driving in urban environments: approaches, lessons and challenges. *Philosophical Transactions of the Royal Society A: Mathematical, Physical and Engineering Sciences* 368 (1928): 4649–4672.

3 Kevin Bullis, How vehicle automation will cut fuel consumption, MIT's Technology Review, October 24, 2011.

4 Chengalva, M.K., Bletsis, R., and Moss, B.P. (2008). Low-cost autonomous vehicles for urban environments. *SAE International Journal of Commercial Vehicles* 1, 2008-01-2717: 516–526.

5 Clements, L.M. and Kockelman, K.M. (2017). Economic effects of automated vehicles. *Transportation Research Record* 2606 (1): 106–114.

6 Hajibaba, M. and Gorgin, S. (2014). A review on modern distributed computing paradigms: cloud computing, jungle computing and fog computing. *Journal of Computing and Information Technology* 22 (2): 69–84.

7 Harauz, J., Kaufman, L.M., and Potter, B. (2009). Data security in the world of cloud computing. In: *IEEE Security & Privacy*. Copublished by the IEEE Computer and Reliability Societies.

8 Dikaiakos, M.D., Katsaros, D., Mehra, P. et al. (2009). Cloud computing: distributed Internet computing for IT and scientific research. *IEEE Internet Computing* 13 (5): 10–13.

9 Stantchev, V., Barnawi, A., Ghulam, S. et al. (2015). Smart items, fog and cloud computing as enablers of servitization in healthcare. *Sensors & Transducers* 185 (2): –121.

10 Yannuzzi, M., Milito, R., Serral-Gracià, R. et al. (2014). Key ingredients in an IoT recipe: fog computing, cloud computing, and more fog computing. In: *2014 IEEE 19th International Workshop on Computer Aided Modeling and Design of Communication Links and Networks (CAMAD)*, 325–329. IEEE.

11 Ken Laberteaux, How might automated driving impact US land use. 2014 Automated Vehicle Symposium, 2014.

12 Patrick Lin, The ethics of saving lives with autonomous cars is far murkier than you think. *WIRED*, July 30, 2013.

13 Vaquero, L.M. and Rodero-Merino, L. (2014). Finding your way in the fog: towards a comprehensive definition of fog computing. *ACM SIGCOMM Computer Communication Review* 44 (5): 27–32.

14 Bonomi, F. (2011). Connected vehicles, the Internet of things, and fog computing. In: *The Eighth ACM International Workshop on Vehicular Inter-networking (VANET)*. Las Vegas, USA.

15 Cao, Y., Chen, S., Hou, P., and Brown, D. (2015). FAST: A fog computing assisted distributed analytics system to monitor fall for stroke mitigation. In: *2015 IEEE International Conference on Networking, Architecture and Storage (NAS)*, 2–11. IEEE.

16 Arkian, H.R., Diyanat, A., and Pourkhalili, A. (2017). MIST: fog-based data analytics scheme with cost-efficient resource provisioning for IoT crowdsensing applications. *Journal of Network and Computer Applications* 82: 152–165.

17 Dastjerdi, A.V. and Buyya, R. (2016). Fog computing: helping the Internet of Things realize its potential. *Computer* 49 (8): 112–116.

18 Yi, S., Hao, Z., Qin, Z., and Li, Q. (2015). Fog computing: Platform and applications. In: *2015 Third IEEE Workshop on Hot Topics in Web Systems and Technologies (HotWeb)*, 73–78. IEEE.

19 Datta, S.K., Bonnet, C., and Haerri, J. (2015). Fog computing architecture to enable consumer centric internet of things services. In: *2015 International Symposium on Consumer Electronics (ISCE)*, IEEE.

20 Aazam, M. and Huh, E.-N. (2016). Fog computing: the cloud-iot\/ioe middleware paradigm. *IEEE Potentials* 35 (3): 40–44.

21 Sarkar, S. and Misra, S. (2016). Theoretical modelling of fog computing: a green computing paradigm to support IoT applications. *Iet Networks* 5 (2): 23–29.

22 Natraj, A. (2016). Fog computing focusing on users at the edge of Internet of things. *International Journal of Engineering Research* 5 (5): 1004–1008.

23 Madsen, H., Burtschy, B., Albeanu, G., and Popentiu-Vladicescu, F.L. (2013). Reliability in the utility computing era: towards reliable fog computing. In: *2013 20th International Conference on Systems, Signals and Image Processing (IWSSIP)*, 43–46. IEEE.

24 Intharawijitr, K., Iida, K., and Koga, H. (2016, 2016). Analysis of fog model considering computing and communication latency in 5G cellular networks. In: *IEEE International Conference on Pervasive Computing and Communication Workshops (PerCom Workshops)*. IEEE.

25 Truong, N.B., Lee, G.M., and Ghamri-Doudane, Y. (2015). Software defined networking-based vehicular adhoc network with fog computing. In: *2015 IFIP/IEEE International Symposium on Integrated Network Management (IM)*. IEEE.

26 Amendola, D., Cordeschi, N., and Baccarelli, E. (2016). Bandwidth management VMs live migration in wireless fog computing for 5G networks. In: *2016 5th IEEE International Conference on Cloud Networking (Cloudnet)*. IEEE.

27 Li, J., Jin, J., Yuan, D. et al. (2015). EHOPES: data-centered Fog platform for smart living. In: *2015 International Telecommunication Networks and Applications Conference (ITNAC)*, 308–313. IEEE.

28 Hou, X., Li, Y., Chen, M. et al. (2016). Vehicular fog computing: a viewpoint of vehicles as the infrastructures. *IEEE Transactions on Vehicular Technology* 65 (6): 3860–3873.

29 Gupta, H., Vahid Dastjerdi, A., Ghosh, S.K., and Buyya, R. (2017). iFogSim: A toolkit for modeling and simulation of resource management techniques in the Internet of Things, Edge and Fog computing environments. *Software: Practice and Experience* 47 (9): 1275–1296.

30 Yi, S., Li, C., and Li, Q. (2015). A survey of fog computing: concepts, applications and issues. In: *Proceedings of the 2015 Workshop on Mobile Big Data*. ACM.

19

Fog Computing to Enable Geospatial Video Analytics for Disaster-incident Situational Awareness

Dmitrii Chemodanov, Prasad Calyam, and Kannappan Palaniappan

Department of Electrical Engineering and Computer Science, University of Missouri Columbia, MO, USA

19.1 Introduction

> Computer Science is no more about computers than astronomy is about telescopes.
>
> –E.W. Dijkstra

Computer science advances go far beyond creating a new piece of hardware or software. The latest advances are increasingly fostering the development of new algorithms and protocols for a more effective/efficient use of cutting-edge technologies within new computing paradigms. What follows is an introduction to a novel function-centric computing (FCC) paradigm that allows traditional computer vision applications to scale faster over geo-distributed hierarchical cloud-fog infrastructures. We discuss salient challenges in realizing this paradigm and show its benefits through a prism of disaster-incident response scenarios. Based on innovative data collection and processing solutions, we illustrate how geospatial video analytics enabled by FCC can help coordinate disaster relief resources to save lives.

19.1.1 How Can Geospatial Video Analytics Help with Disaster-Incident Situational Awareness?

In the event of natural or man-made disasters, timely and accurate situational awareness is crucial for assisting first responders in disaster relief coordination. To this end, imagery data, such as videos and photographs, can be collected from numerous disaster-incident scenes using surveillance cameras, civilian mobile devices, and aerial platforms. This imagery data processing can be essential for

Fog Computing: Theory and Practice, First Edition.
Edited by Assad Abbas, Samee U. Khan, and Albert Y. Zomaya.

first responders to (1) provide situational awareness for law enforcement officials (e.g. by using online face recognition to reunite lost citizens [1]), and (2) inform critical decisions for allocating scarce relief resources (e.g. medical staff/ambulances or search-and-rescue teams [2]). Building dynamic three-dimensional reconstructions of incident scenes can increase situational awareness in a theater-scale setting (as large as two city blocks) of the incident scene. This can be achieved by fusing crowd-sources and data-intensive surveillance imagery [3]. Furthermore, tracking objects of interest in wide-area motion imagery (WAMI) can provide analytics for planning wide-area relief and law enforcement activities at the regional-scale of incident scenes (tens of city blocks) [4]. We use the term *geospatial video analytics* to refer to such an integration of computer vision algorithms implemented for different scales and types of imagery/video data collected from Internet of Things (IoT) devices, and processed through geo-distributed hierarchical cloud-fog platforms.

19.1.2 Fog Computing for Geospatial Video Analytics

To be effective for users (i.e. incident commanders, first responders), geospatial video analytics needs to involve high-throughput data collection as well as seamless data processing of imagery/video [5]. For instance, the user's quality of experience (QoE) expectations demand that the geospatial video analytics is resilient to any edge network connectivity issues for data collection, while also delivering low-latency access (e.g. in real time) to process large amounts of visual data using cloud-fog resources that run complex computer vision algorithms. The fog resources are augmented by cloud resources, as well as services closer to end-user's IoT devices, to allow computing anywhere within the IoT-to-cloud continuum. Thus, providing applications with options for fog computing reduces cloud service latencies enables computing closer to the IoT data sources at the network edges.

Figure 19.1 shows how cloud/fog resources are used for imagery data processing to realize a collection, computation, and consumption (3C) pipeline that is common for any geospatial video analytics use case. In this stage of data collection, it is possible that the network edge could have lost infrastructure, e.g. loss of cellular base stations in a disaster scene and/or intermittently accessible IoT devices (e.g. sensors, wearable heads-up display devices, Bluetooth beacons). Similarly, in the stage of data processing, it is quite likely that infrastructure edges are rarely equipped with high-performance computation capabilities to run computer vision algorithms. Moreover, cloud-processed data needs to be moved closer to users through content caching at fog resources for thin-client consumption and interactive visual data exploration. Thus, adoption of fog computing in conjunction with cloud computing platforms for relevant compute, storage, and network resource provisioning and management requires a new computing paradigm.

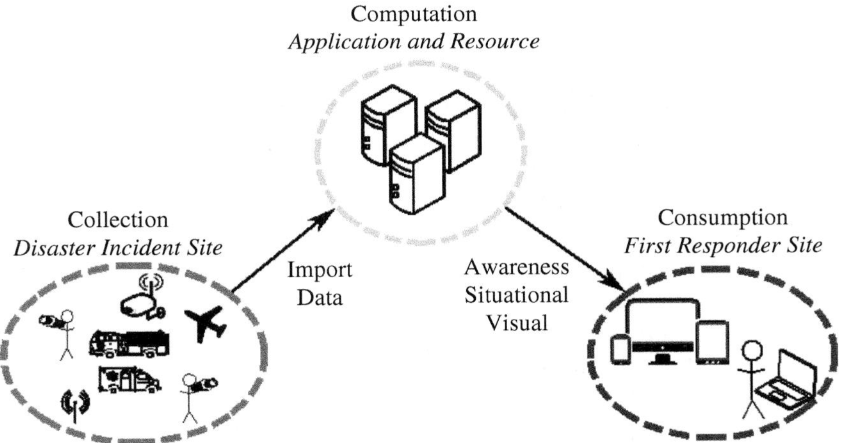

Figure 19.1 Illustrative example of a visual data computing at network edges (i.e. at fog) that needs to span geographically dispersed sites of data collection, computation, and consumption: fog computing resources here are linked with cloud computing platforms.

Moreover, the cloud-fog resource provisioning becomes an NP-hard problem to solve in large-scale visual data processing across distributed disaster incident scenes [6].

19.1.3 Function-Centric Cloud/Fog Computing Paradigm

To address both the data collection and data processing problems in geospatial video analytics, we prescribe a novel FCC paradigm that integrates computer vision, edge routing, and computer/network virtualization areas. Figure 19.2 illustrates the FCC paradigm that extends the basic 3C pipeline for data collection at the edge with preprocessing and human-computer interaction (HCI) analysis

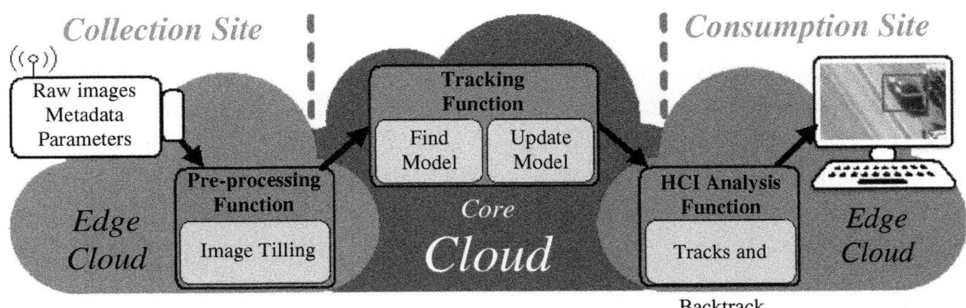

Figure 19.2 Illustrative example of the function-centric fog/cloud computing paradigm used for the real-time object tracking pipeline [5].

functions (i.e. "small instance functions") that can be placed for low-latency access on edge servers by using fog computing. For example, in this context, to cope with the potential loss of infrastructure at the collection site near a disaster scene, we can use mobile ad hoc wireless networks (MANETs) which need to be operational for collecting media-rich visual information from this scene as quickly as possible at the edge cloud gateway. In another example, for processing large-scale visual data sets using object tracking computer vision algorithms, the FCC involves the placement of computer-intensive tracking functions (i.e. "large instance functions") on a cloud server. Thus, applying the FCC paradigm can accomplish geospatial video analytics by suitable computer location selection that is based on decoupling of visual data processing functions for theater-scale or regional-scale applications. The decoupling can allow for cloud-fog resource orchestration that results in speed-up of traditional computer vision algorithms by several orders of magnitude [5]. In essence, FCC can be extended to existing computation offloading techniques found in mobile cloud computing literature, where virtualization principles are used for decoupling computers, storage, and network functions in resource allocation and management of cloud-fog resources [7, 8].

19.1.4 Function-Centric Fog/Cloud Computing Challenges

First, FCC assumes that data collection involves the ability to collect data even in lost infrastructure regions, e.g. loss of cellular base stations in a disaster scene, as well as in the presence of intermittently available and mobile IoT devices. For FCC to be operational in such edge network scenarios, we need to steer user traffic dynamically within MANETs to satisfy its throughput and latency requirements. To this aim, we cannot adopt full-fledged routing solutions from core networks due to mobility as well as power constraints of IoT devices. Instead, lightweight routing approaches, such as those based on geographic routing, are more suited. However, there is a lack of traffic steering techniques in MANETs that can provide sustainable high-speed delivery of data to a geo-distributed cloud infrastructure gateway component [9]. Specifically, this can be achieved by designing a more high-performing greedy-forwarding approach that is not subject to the local minimum problem due to (severe) IoT failures and mobility. This problem is caused by the lack of global network knowledge of greedy forwarding algorithms (see e.g. [10–13]), which can deliver packets to nodes with no neighbors closer to the destination than themselves. Second, FCC implementation requires refactoring computer vision applications that are typically developed with codes that are tightly coupled for ad hoc theater/regional task-specific solutions, and are not designed for function portability. For this, principles from the recent advances in serverless computing used in Amazon Web Services Lambda [14], Google Cloud Functions [15], Microsoft Azure Functions [16], and IBM OpenWhisk [17] can be used.

More specifically, FCC can mitigate application scalability limitations via use of microservices, where application code decoupling is performed via RESTful Application Programming Interfaces (APIs) [5] to manage load balancing, elasticity, and server instance types independently for each application function.

Third, FCC implementation that can satisfy geo-location and latency demands of geospatial video analytics needs to cope with node failures and congested network paths that frequent impact (QoS) requirements [18]. Especially in specific cases of natural or man-made disaster incidents, FCC will be subject to severe infrastructure outages and austere edge-network environments [19]. Moreover, computation and network QoS demands of computer vision functions can fluctuate, depending on the progress of the disaster incident response activities [20]. Thus, the FCC needs to use reliability-ensuring mechanisms to cope proactively with both potential computer vision function demand fluctuations [20] as well as possible infrastructure outages [18, 19].

Lastly, optimal placement of computer vision functions as shown in Figure 19.2 is a known "Service Function Chaining (SFC)" problem in the network function virtualization (NFV) area. The SFC has known approximation guarantees only in some special cases where chaining of service functions [21] and/or their ordering [22, 23] are omitted. In the general case, however, it requires solving of the NP-hard integer multi-commodity-chain flow (MCCF) problem to align flow splits with supported hardware granularity [24]. It is also necessary to support cases when service functions or their associated flows are nonsplittable. This subproblem has no known approximation guarantees and has been previously reported as the integer NFV service distribution problem [25]. Furthermore, its complexity can be exacerbated by incorporated reliability and geo-location/latency-aware mechanisms. The former aims to cope proactively with both possible infrastructure outages as well as function demand fluctuations, whereas the latter is needed to satisfy QoS demands of geo-distributed latency-sensitive function chains.

19.1.5 Chapter Organization

This chapter seeks to introduce concepts of fog computing related to enabling geospatial video analytics at theater and regional scales for disaster-incident situational awareness. The chapter will first discuss the natural decomposability of common computer vision applications (i.e. for face recognition, object tracking, and 3-D scene reconstruction) to a set of functions that motivates the need of FCC paradigm. Following this, we outline innovative state-of-the-art solutions to the "data collection" and "data processing" problems in geospatial video analytics that are based on theoretical and experimental research conducted by the authors in the Virtualization, Multimedia and Networking (VIMAN) Lab, and Computational Imaging and VisAnalysis (CIVA) Lab at University of Missouri-Columbia (supported in part by the Coulter Foundation Translational Partnership Program, and the National Science Foundation CNS-1647084 award). More specifically, we present a novel "artificial intelligence (AI)–augmented geographic routing approach" that can address the data collection challenges of geospatial video analytics at the wireless network edge within lost infrastructure regions. In addition, we present a novel "metapath composite variable approach" that can be used for a near-optimal and practical geo/latency-constrained SFC over fog/cloud platforms to enable data processing in geospatial video analytics. Last, we discuss the main findings of this chapter with the list of open challenges for adopting fog computing architectures in geospatial video analytics to effectively and efficiently deliver disaster-incident situational awareness.

19.2 Computer Vision Application Case Studies and FCC Motivation

In this section, we consider three common computer vision applications that operate on different data scales (e.g. theater-scale vs. regional-scale) and have different latency and geo-location requirements. In particular, we describe benefits of FCC for the following application case studies: (1) real-time patient triage status tracking with Panacea's Cloud incident command dashboard [26] featuring face recognition, (2) reconstruction of dynamic visualizations from 3-D light detection and ranging (LIDAR) scans [27], and (3) tracking objects of interest in WAMI [2]. Recall that we distinguish between theater-scale and regional-scale applications based on the geographical coverage of the incident and the nature of the distributed visual data – with theater-scale being small area (two city blocks) around a disaster incident site, and regional-scale being large areas (dozens of city blocks) distributed across multiple disaster incident sites. The salient contribution of our work is to transform exemplar computer vision applications (e.g. face recognition, 3-D scene reconstruction, and object tracking), with state-of-the-art solutions for increasing speed of "data collection" and scale of "data processing" using edge routing and SFC, as detailed in the following Sections 19.3 and 19.4, respectively.

19.2.1 Patient Tracking with Face Recognition Case Study

19.2.1.1 Application's 3C Pipeline Needs

Following medical triage protocols of hospitals during response coordination at natural or man-made disaster incident scenes is challenging. Especially when dealing with several patients with trauma, it is stressful for paramedics to verbally communicate and track patients.

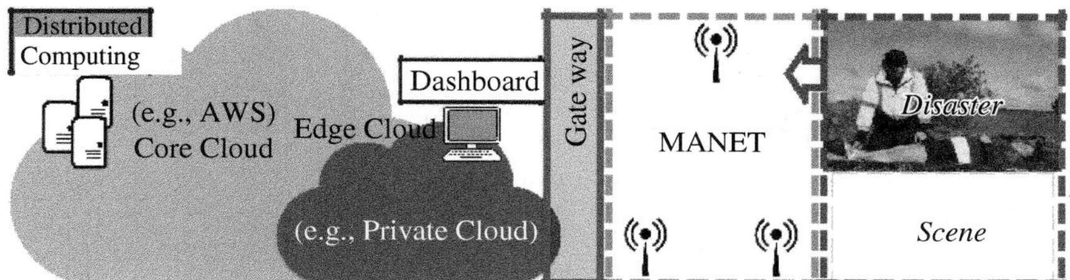

Figure 19.3 Illustrative example of the Panacea's Cloud setup: IoT device data sets generated on-site (e.g. near disaster scenes) need to be collected through a MANET at the edge cloud for further processing in conjunction with the core cloud.

In our first case study application, "Panacea's Cloud" [26], the FCC needs to support a incident command dashboard, as shown in Figure 19.3, that can handle patient tracking using real-time IoT device data streams (e.g. video streams from wearable heads-up display or smartphones, geolocation information from virtual beacons) from multiple incident scenes. Proper aggregation of the IoT data sets and intuitive user interfaces in the dashboard can be critical for efficient coordination between first responder agencies, e.g. fire, police, hospitals, and so on [28]. The dashboard supports real-time videoconferencing of paramedics with the incident commander for telemedical consultation, medical supplies replenishment communication, or coordination of ambulance routing at a theater-scale incident site. It can also facilitate patient triage status tracking through secure mapping of the geolocation information with a face recognition application, where the latter is used to recognize the person and, if possible, pull his/her relevant medical information [29]. Such a mapping is essential to ensure that the relevant patients at the incident sites are provided the necessary care, and follow-ups on the care can be scheduled with new responders or at a new location of the patient. The face recognition capability can also be used in the dashboard for "lost person" use cases, where first responders can detect children when attempting to reunite them with their guardians [30], or identify bad actors in crowds, who might raise public safety resource allocation decisions.

19.2.1.2 Face Recognition Pipeline Details

The visual data processing pipeline steps for face recognition are shown in Figure 19.4 and can be divided into two main classes according to the nature of the inherent functions, such as:

- *Small function processing*. Compression, preprocessing, and results verification
- *Large function processing*. Segmentation and face detection, features extraction and matching

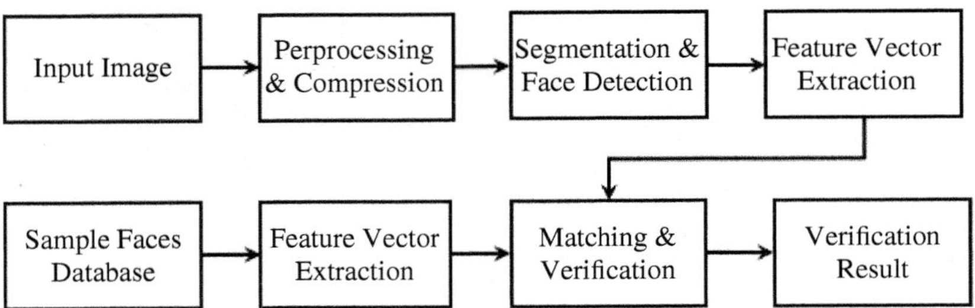

Figure 19.4 Overview of visual data processing stages in a facial recognition application used in patient triage status tracking.

Small function processing is mainly focused on pure pixel-level information. In contrast, large function processing is focused on both pixel and object level information. To work effectively, large functions typically require data preprocessing stages. Figure 19.4 shows facial recognition steps that are used for "Panacea's Cloud" patient tracking and involve the digital image at the client side and a larger image sample dataset at the server side. A preprocessing step is performed to first detect a human face within a small amount of time (i.e. with low latency). During this step, all input images are compressed to one quarter of their original size. Subsequently, the image is fed into a pretrained face classifier. This classifier is provided by Dlib [31] (an open-source library) and is based on [32] for facial recognition tasks. Training is done using a "very deep" convolutional neural network, which comprises a long sequence of convolutional layers that have recently shown state-of-the-art performance in related tasks.

19.2.2 3-D Scene Reconstruction from LIDAR Scans

19.2.2.1 Application's 3C Pipeline Needs

3-D scene reconstructions have been proven to be useful for a quick damage assessment by public safety organizations. Such an assessment is possible through highly accurate LIDAR scans at incident scenes that provide evidence at relevant locations from multiple viewpoints [33]. As Figure 19.5 shows, 3-D models of a scene can be created, e.g. by fusing a set of 2-D videos and LIDAR scans. This visual data can be obtained from civilian mobile devices as well as from surveillance cameras near (or at) incident scenes. In our second case study application, FCC registers 2-D videos with 3-D LIDAR scans collected at the theater scale. As a result, first responders can view sets of videos in an intuitive (3-D) virtual environment [27]. This simplifies the cumbersome task of analyzing several disparate 2-D videos on a grid display. However, commonly used data from LIDAR scans can be large in size – a typical resolution of about 1 cm for data collected at a range of up to 300 m with 6 mm accuracy. Thus, when collected from large-area incident scenes, this

Figure 19.5 Illustrative example of a 3-D scene reconstruction with use of LIDAR scans: a 2-D video frame (top left) is first projected onto LIDAR scan (bottom left) to then reconstruct a 3-D scene (right).

data can be computationally expensive to process. To help with this processing, FCC can help take advantage of this rich source of information and quickly provide the situational awareness.

19.2.2.2 3-D Scene Reconstruction Pipeline Details

Fusing a video with a LIDAR scan requires calculation of camera poses for 2-D video with respect to the 3-D point space. Figure 19.6 illustrates steps of matching a 2-D video frame to LIDAR scans with known 3-D correspondences, calculation of projection matrix for the camera, and segmentation of moving objects in the 3-D space. The steps of the visual data processing pipeline in this case study application can be classified as follows:

- *Small function processing*. Metadata data preprocessing, 2-D video frame background registration, rendering of 3-D
- *Large function processing*. Projection matrix calculation, segmentation of motion, positioning of dynamic objects

To better understand what each function computes, we describe one possible pipeline of the 3-D scene reconstruction using LIDAR scan and 2-D videos.

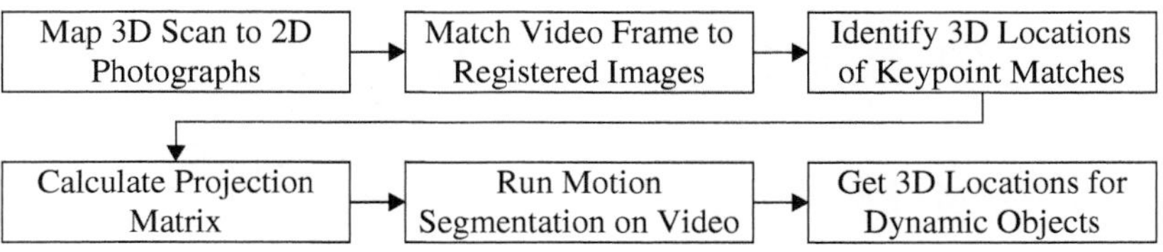

Figure 19.6 Overview of 3-D scene reconstruction stages with 2-D videos and LIDAR scans.

We start by estimation of the 2-D-3-D relationship between the LIDAR scan and its photographs. To this end, we map 2-D pixels to 3-D points during a preprocessing step using standard computer vision techniques described in [34]. The entire point space is projected onto each image exactly once. Subsequently, the 2-D-3-D mappings can be stored at a remote cloud server location. Identification of moving objects in the 2-D video is performed using the mixture of Gaussians (MOG) approach [35] that yields a binary image featuring the motion segmented from the background.

19.2.3 Tracking Objects of Interest in WAMI

19.2.3.1 Application's 3C Pipeline Needs

Tracking objects of interest is crucial for (intelligent) search-and-rescue activities after a large-area disaster incident. Object tracking has long shown potential for city-wide crowd surveillance by providing a hawk-eye's view for public safety organizations. Our third case study application that relates to search-and-rescue type of activities operates on the regional scale by identifying and tracking objects of interest in WAMI. This in turn can assist incident managers in studying the behaviors of particular vehicles [2]. The tracker's input is in the form of bounding boxes that indicate targets, as shown in Figure 19.7. Novel sensor technologies allow responders to capture high-resolution imagery data (ranging between 10 cm and 1 m) for wide-area surveillance. The processing is performed by our likelihood of features tracking (LoFT) framework, which utilizes WAMI frames with a high resolution of 25 cm ground sampling distance (GSD) coming at about one to four frames per second [36]. Utilizing FCC for tracking such imagery data can be challenging, for several reasons. First of all, the objects of interest are usually small with a (relatively) large motion displacement due to the inherent

Figure 19.7 Illustrative example of WAMI imagery ecosystem: tiled (TIFF) aerial images of 80 MB size and with a resolution of 7800 × 10 600 pixels.

low frame rate. Secondly, WAMI imagery itself is challenging for automated analytics due to multiple reasons including (but not limited to) variations in illumination, oblique camera viewing angles, tracking through shadows, occlusions from tall structures, blurring, and stabilization artifacts (e.g. due to atmospheric conditions). LOFT uses a set of imagery features, including gradient magnitude, histogram of oriented gradients, median binary patterns [37], eigenvalues of the Hessian matrix for shape indices, and intensity maps.

19.2.3.2 Object Tracking Pipeline Details

Typical object tracking in wide-area motion imagery comprises several visual data processing stages tested on (large-area) aerial data [38, 39]. Figure 19.8 outlines these stages, which can be classified according to the tracker functionality as follows:

- *Small function processing.* Raw data compression, data storage, metadata preprocessing, geo-projection, tiling, and stabilization
- *Large function processing.* Initialization of objects of interest, their detection, tracking, and analysis. A small function processing mainly operates on pure pixel level information, whereas a large function processing deals with information on both pixel as well as object levels. We remark that for most of large functions to work effectively, we need the preprocessing stages while applying our FCC approach. For example, common object trackers benefit from imagery stabilization, and hence registration becomes crucial during data preprocessing. Our LOFT case study utilizes a track-before-detect approach to significantly reduce the search space, which is especially helpful in large wide-area motion imagery, as most of the objects in WAMI look similar [2, 40]. However, a proper synchronization between large tracking and small preprocessing functions is needed for this approach. To this aim, latency constraints have to be imposed to avoid bottlenecks in the tracking pipeline. When the tracker is fully consolidated on a single server, it has a processing rate of three to four frames per second. Thus, it is also important to take into account bandwidth demands because even a single WAMI frame can be 80 MB in size; and at four frames per second, the maximum throughput required is up to 2.5 Gbps.

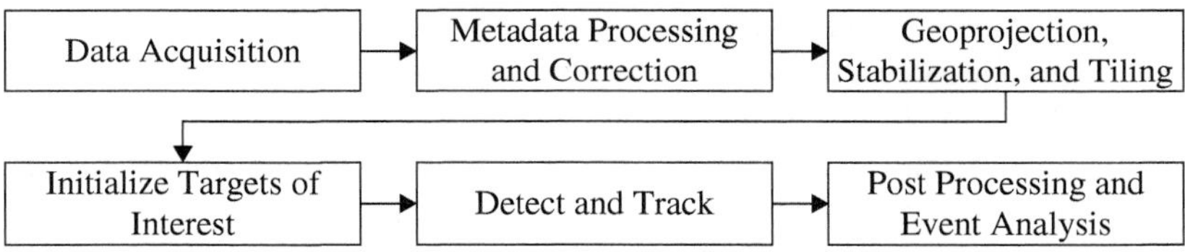

Figure 19.8 Overview of object tracking stages in a typical WAMI analysis pipeline.

19.3 Geospatial Video Analytics Data Collection Using Edge Routing

Fog computing relies on the use of wireless edge networks to facilitate data marshaling with IoT devices that are intermittently available and mobile. It is possible for IoT devices to experience scarce energy, high mobility, and frequent failures [41, 42], which consequently makes fog computing difficult. In extreme conditions (i.e. after a disaster or in remote areas), fog computing needs to be set up even within lost infrastructure regions without, e.g., cellular connectivity and operational in specialized wireless sensor network environments. In this section, we consider a special case of data marshaling at the network edge to collect data for geospatial video analytics at a gateway dashboard from disaster incident scenes over MANETs. However, MANETs are commonly subject to the high node-mobility as well as severe node failures (e.g. caused by intermittent energy supply). Consequently, there is a need for packet delivery solutions that utilize either network topology knowledge [42–45] or rely on a logically centralized network control [5, 8]. In this section, we detail our novel edge routing protocol based on our work in [28], in order to overcome throughput sustainability challenges of data collection when using existing solutions within lost infrastructure regions.

19.3.1 Network Edge Geographic Routing Challenges

Software-defined networking (SDN) and NFV have been recently adopted to overcome some of the challenges in data marshaling between IoT devices and the edge network gateway (e.g. to access cloud-fog resources) [5, 8]. For instance, a control plane can be used to dynamically find paths that satisfy IoT-based application low-latency and bandwidth demands [5, 46]. Existing edge routing solutions commonly rely on the network topology knowledge, such as spanning trees [44, 47], or on the network clusters knowledge [45]. These solutions may not be suitable in highly dynamic conditions, such as the one caused by severe node failures and/or high node mobility. To this aim, a geographic routing approach can be utilized to cope with existing MANET solution limitations in disaster incident scenarios. However, routing protocols that use this approach are not typically capable of providing sustained and high-throughput data rates for a satisfactory visual data delivery to the edge gateway [9]. This is mainly due to the local minimum problem that frequently occurs in the presence of nonarbitrary node mobility and failures. This problem manifests mainly owing to the lack of global network knowledge of the greedy forwarding algorithm [10–13]. As a result, packets may be delivered to nodes with no neighbors closer to the destination than themselves. This, in turn, makes it impossible to find the next hop during packet-forwarding actions.

Existing solutions only partially address the local minimum problem and can be divided into stateless and stateful solutions. Existing stateless greedy forwarding algorithms may not deliver packets even if a path exists [11, 12]. In addition, they can also stretch such paths significantly to desperately find a way to the destination [10]. On the other hand, existing stateful greedy forwarding solutions that rely on some network topology knowledge (e.g. partial paths [42] or spanning trees [44, 47]) are sensitive to high IoT device mobility and their frequent failures [42, 43, 48]. Given that such conditions are common in disaster incident scenarios, implementations that use these algorithms can have poor/unacceptable performance to deliver high-speed visual data to the network edge gateway in order to provide situational awareness.

Examples of routing protocols that guarantee packet delivery are the greedy perimeter stateless routing (GPSR) [11] and GFG [49]. Although both protocols can recover packets from a local minimum by using face routing, they are based on strong assumptions, such as planar and unit disk graphs. Their strong assumptions rarely hold in practice especially when: nodes are mobile, graphs have arbitrary shapes and physical obstacles manifest [50]. To overcome such limitations, the authors in [50] propose a solution that avoids use of planar graphs by proposing a cross-link detection protocol (CLDP) complication. However, it has been shown that CLDP requires an expensive signaling mechanism to first detect and then remove crossed edges [43]. As an alternative, authors in [44, 51] have proposed a greedy distributed spanning tree routing (GDSTR) protocol. This protocol uses (less expensive) distributed spanning trees to guarantee packet delivery and recover them from local minima. Another solution to the local minimum problem is to use a greedy embedding method that assigns "local minima-free" virtual coordinates based on spanning trees [47]. More recent works [42, 43, 48] have shown that spanning trees are also highly sensitive to network dynamics, such as node mobility and failures.

Owing to the above limitations, newer routing protocols have been proposed to cope with topology dynamics to some extent [42, 43]. For example, MTD routing protocol requires construction of Delaunay triangulation (DT) graphs for local minimum recovery [43]. When topology changes, nodes may lose their Delaunay neighbors, which are needed for recovery from a local minimum degrading MTD performance under node mobility and/or failures. The authors in [10] build their approach upon the work in [47]. In particular, they show how packet delivery can subject to a local minimum due to greedy embedding inaccuracies caused by network dynamics. As a solution, they propose a novel routing protocol, viz., gravity pressure greedy forwarding (GPGF) [10]. GPGF is shown to have guaranteed packet delivery on graphs of an arbitrary shape at the expense of packet header space [48] and can stretch paths significantly violating packets' time-to-leave (TTL) constraints. In contrast, our work titled

"Artificial Intelligence–Augmented Geographic Routing Approach (AGRA)" in [28] overcomes aforementioned challenges and delivers packets with sustained and high-speed throughput to the network edge gateway. In the remainder of this section, we detail our AGRA solution, which proactively avoids local minima by utilizing obstacles knowledge mined from satellite imagery.

19.3.2 Artificial Intelligence Relevance in Geographic Routing

Under severe node failures and mobility caused by disaster-incident scenarios, routing protocols cannot rely on the network topology knowledge, such as spanning trees, routing tables, etc. As a result, most of the geographic routing-based protocols (especially those designed for static sensor networks) today are poorly applicable for MANETs during the disaster-incident response application use cases described earlier in Section 19.2. At the same time, we can observe how local minimum of the greedy forwarding often happens near large physical obstacles (especially those of concave shapes). Examples of such obstacles include (but are not limited to) natural obstacles (e.g. lakes or ponds) or man-made obstacles (e.g. buildings). Figure 19.9a,b illustrate these obstacles over an example satellite imagery dataset.

As part of the geospatial video analytics pipeline, the information about physical obstacles can be mined directly from the satellite maps of the disaster-incident area. Figure 19.10 shows maps of Joplin, Missouri, before and right after tornado damages that occurred on May 22, 2011. We can see how information (such as location, size, etc.) of the Joplin Hospital (see Figure 19.10a,b) and the Joplin High School (see Figure 19.10c,d) buildings is mainly preserved after tornado damages. As a result, we can mine information about potential physical obstacles even from maps of the disaster-incident area that do not contain information

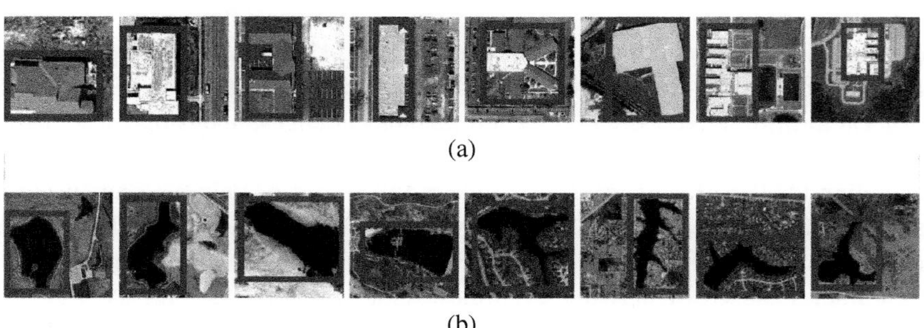

Figure 19.9 Various physical obstacles including both man-made e.g. buildings (a) and natural e.g. lakes or ponds (b) over an example satellite imagery dataset that can be used for training purposes.

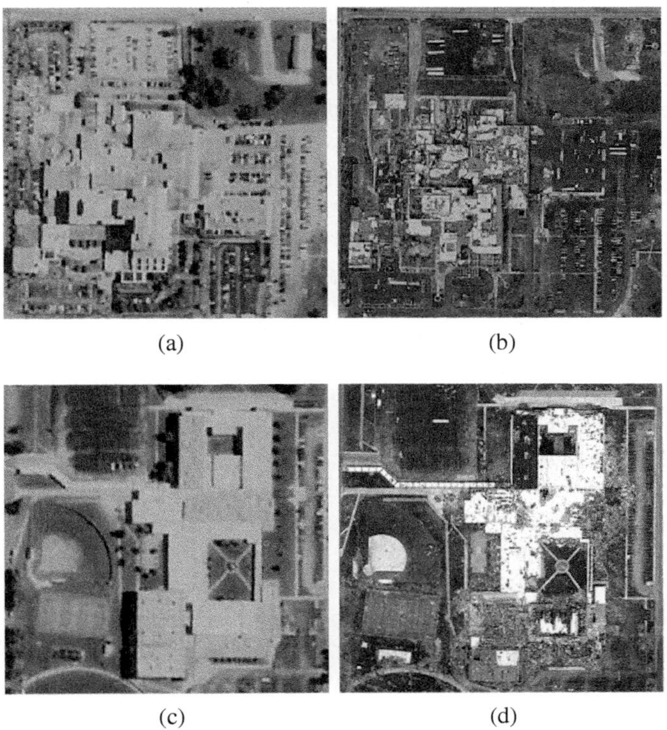

Figure 19.10 Joplin, MO satellite maps of Joplin Hospital (a, b) and Joplin High School (c, d) buildings before (a, c) and right after (b, d) tornado damages that occurred on May 22, 2011. This satellite imagery is taken from openly available source at [52].

about any marked damages. To this aim, we need to address the following problems: (1) how can information about potential obstacles be extracted from the satellite imagery of the disaster scene, and (2) how can we utilize this knowledge within a geographic routing-based protocol to improve the latter's application layer data-throughput. Thus, a well-designed cloud-fog computing integration with this extended geographical routing approach that uses satellite imagery knowledge can ensure that user QoE demands are met in terms of obtaining real-time situational awareness in disaster response coordination.

19.3.3 AI-Augmented Geographic Routing Implementation

To mine information about physical obstacles that can be utilized by the network edge geographic routing algorithm, one can manually label all such obstacles on the map that can cause potential local minima. However, due to the fact that time is critical for first responders, manually labeling physical obstacles on maps in a timely manner is not feasible for large-area incident scenes, and thus we aim to automate this labeling process. The AI techniques, particularly in the pattern recognition literature, can be ideal candidates to automate this process, as they include many approaches to detect objects (i.e. determine their size and location) in any given (satellite) imagery. These approaches can include nearest neighbor, support vector machines, deep learning and other techniques. However, present-day state-of-the-art detectors are based on deep learning approaches [53, 54]. For example, the "You Only Look Once" deep-learning-based technique can detect objects in images using only a single (26 layers) neural network – an easier task for the fog resources [53]. On the other hand, its performance can be worse than the one shown by more sophisticated deep learning-based detectors [54].[1]

The deep-learning-based detectors may misclassify obstacles or not find them at all, and thus, they still need some human assistance in sample labeling. To further address deep learning complexity limitations, we move its functions to the cloud resources, as shown in Figure 19.11 for our Panacea's Cloud application, and overcome resource limitations of fog storage and computation. Assuming that training samples can be collected and partly labeled at the fog (e.g. during prior incident responses), we can use them for (semi-) supervised deep learning [56] to improve object detection in the future. The available detector can be always (pre-)uploaded to the fog resources and used in an off-line manner to support edge routing within lost infrastructure regions of a disaster-incident. After detection, the information about physical obstacles is propagated by AGRA through services supported at the network edge gateway to the MANET.

To benefit from an obstacles information, AGRA features a conceptually different greedy forwarding mode that repels packets away from these obstacles [28]. This mode utilizes the electrostatic potential field calculated via Green's function, and to the best of our knowledge it is the first greedy forwarding approach that provides theoretical guarantees on a shortest path approximation as well as on a local minima avoidance. In [28], we also show how this mode can be augmented with the GPGF protocol gravity pressure mode to provide practical guarantees on local minima avoidance and recovery. To this end, we have proposed two algorithms, viz., attractive repulsive greedy forwarding (ARGF) and attractive repulsive pressure greedy forwarding (ARPGF). Both algorithms alternate attractive and repulsive packet greedy forwarding modes in 2-D or 3-D Euclidean spaces [28]. Our numerical simulations also indicate their superior performance in terms of a path stretch reduction as well as a delivery ratio increase with respect to GPGF. Thus, ARPGF can be suitable for routing in MANETs under

1 Note that an alternative to object detection can consider geographical object-based image analysis [55]. Such analysis uses the spectral information extracted from image pixels that require LIDAR hardware which may be not necessarily available at disaster incident scenes [5].

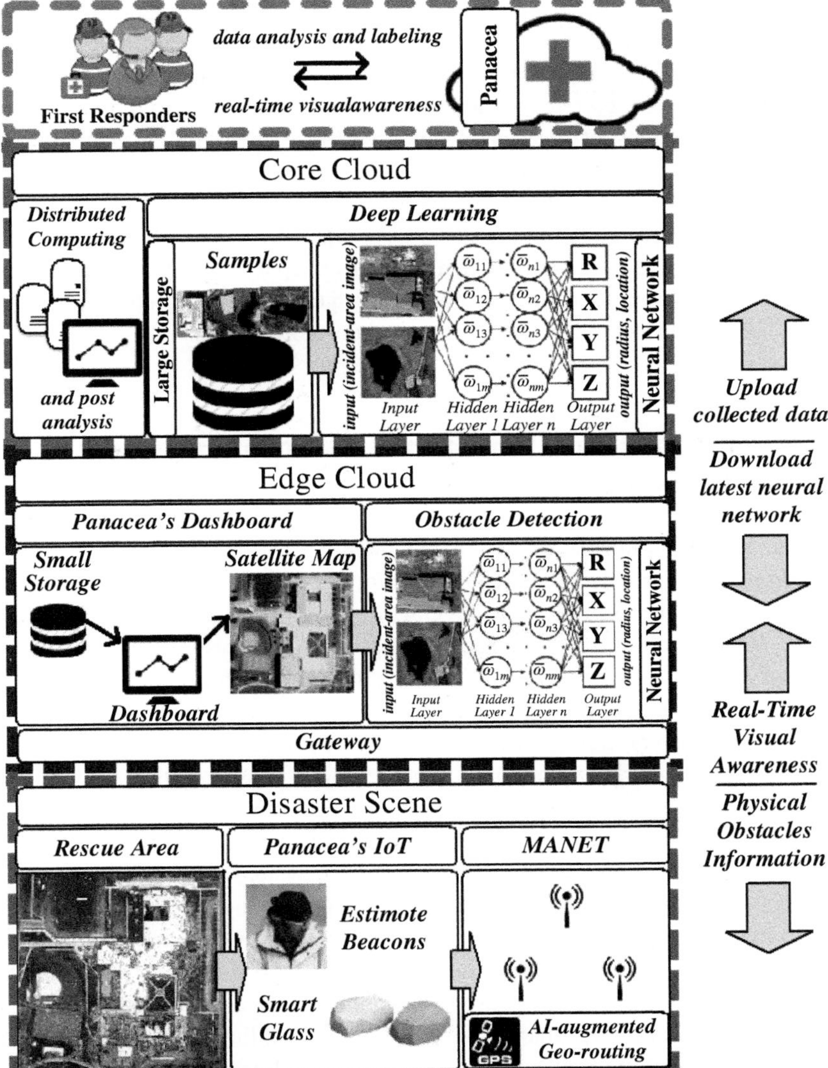

Figure 19.11 To cope with deep learning functions complexity of the obstacle detector, we move them to the core cloud. The up-to-date detector can be then pre-uploaded to the fog and used in an off-line manner to enhance edge routing.

challenging disaster-incident conditions. We remark that such improvements are due to the knowledge of obstacles that allow our improved routing protocols better cope with severe node failures and high mobility. As a result, both these routing protocols demonstrate overall greater application level throughput under challenging conditions of disaster-incident response scenarios, making them crucial for real-time (visual) situational awareness. Note that the performance of these routing protocols degrades with respect to their predecessor performances, i.e. ARGF to GF [12] and ARPGF to GPGF [10], with the object information quality degradation, e.g. due to miscalculations by deep learning-based detectors. Due to space constraints, we omit details of our edge routing protocols that utilize the AI-AGRA, and refer curious readers to [28] for more details.

19.4 Fog/Cloud Data Processing for Geospatial Video Analytics Consumption

Once data collection hurdles are overcome using AGRA, data processing within cloud-fog platforms needs to be orchestrated for the consequent geospatial video analytics. To ensure satisfactory user QoE in the consumption stages of the pipeline, fog computing needs to cope with geographical resource accessibility and latency demands of geospatial video analytics [57]. To this end, the fog computing APIs augment the core/public cloud APIs closer to the end user locations at the expense of computation/storage capabilities available locally (on-site). Correspondingly, emerging paradigms, such as the microservices introduced in Section 19.1, can help seamless data-intensive fog computing in conjunction with cloud computing to compensate the insufficient local processing capabilities within a geographical area of interest. An essential system aspect in the design to use microservices is to perform application code decoupling (as atomic services) to manage fog/cloud leased resources independently for each application function. Resulting geo-distributed latency-sensitive "service chains" have to be orchestrated as follows: they first have to be composed from the leased fog/cloud platform resources and then maintained throughout their lifetime. Composing an instance of a virtual function chain (or a virtual network in general) requires the QoS-constrained service chain to be mapped on top of a physical network hosted by a single infrastructure provider, or by a federation of providers. This problem is similar to NFV SFC, where traffic is redirected on-demand trough a "chain" of middleboxes that host specific network functions,, as firewalls, load balancers, and others to satisfy resource providers' policies [58]. In the rest of this section, we first describe SFC challenges related to data processing of geospatial video analytics. We then propose a novel constrained shortest path–based SFC composition and maintenance approach, the "Matapath-based composite variable approach" [6].

19.4.1 Geo-Distributed Latency-Sensitive SFC Challenges

SFC is traditionally used in NFV to place a set of middleboxes and chain relevant functions to steer traffic through them [59]. Existing SFC solutions either separate the service placement from the service chaining phase [21–23], or jointly optimize both of the two phases [20, 25].

19.4.1.1 SFC Optimality

In some special cases the optimal SFC is shown to have approximation guarantees [21–23]. For instance, authors in [22, 23] provide near-optimal approximation algorithms for the SFC problem without chaining and ordering constraints. Also, the authors in [21] propose the first SFC solution with the approximation guarantees, which admits ordering constraints, but still omits chaining constraints. The work in [60] shows approximation guarantees for SFCs with both ordering and chaining constraints, but only under assumptions that available service chaining options are of polynomial size. In the general case, however, when service functions need to be jointly placed and chained in a geo-distributed cloud-fog infrastructure with a corresponding compute/network resource allocation, possible SFC compositions are of exponential size. Thus, it becomes a linear topology virtual network embedding (VNE) [61, 62] and can be formulated as the (NP-hard) MCCF problem with integrality constraints with no known approximation guarantees [25]. To this aim, the authors in [25] propose a heuristic algorithm that relies on a number of active resources that can be used to consolidate flows. The preliminary evaluation results of this algorithm in a small-scale network settings (of 10 nodes) show promise for providing efficient solutions to the integer MCCF problem in practical settings.

19.4.1.2 SFC Reliability

With the advent of edge networking and a growing number of latency sensitive services, recent works also consider problems of geodistributed [63] and edge SFC [64]. Although these works mainly focus on the new load balancing and latency optimization techniques, they omit an important reliability aspect of geo-distributed latency-sensitive SFCs. The closest works related to ours are [20, 65]. Authors in [20] propose a prediction-based approach that proactively handles SFC demand fluctuations. However, their approach does not account for network/infrastructure outages that mainly cause service function failures [18]. At the same time, work in [65] proposes a SFC solution that ensures a sufficient infrastructure reliability, but neither proactively nor reactively handles SFC demand fluctuations.

19.4.1.3 Our Approach

We present the first (to our knowledge) practical and near-optimal SFC composition approach in the general case of joint service function placement and chaining in a geo-distributed cloud-fog infrastructure. Our approach also is the first to admit end-to-end network QoS constraints, such as latency, packet loss, etc. In particular, we describe a novel metapath composite variable approach that reduces a combinatorial complexity of the (master) integer MCCF problem. As a result, our approach achieves 99% optimality on average and takes seconds to compose SFCs for practically sized problems of US Tier-1 (300 nodes) and regional (600 nodes) infrastructure providers' topologies. In contrast, master MCCF problem solution takes hours using a high-performance computing cloud server [6]. Moreover, in contrast to [20, 65], the proposed metapath composite variable approach allows us to uniquely ensure reliability of geo-distributed latency-sensitive SFCs via use of chance-constraints and backup policies. Details of our metapath composite variable approach design and implementation to cope with both SFC demand fluctuations and infrastructure outages as described in [6] are summarized in the remainder of this section.

19.4.2 Metapath-Based Composite Variable Approach

To simplify the combinatorial complexity of the integer MCCF-based SFC composition problem, we present an abridged description of our novel metapath-based composite variable approach detailed in [6]. Similar to existing composite variable schemes [66], our goal is to create a binary variable that composes multiple (preferably close to optimal) decisions. To this end, we build upon a known result in optimization theory: all network flow problems can be decomposed into paths and cycles [67]. We first introduce our notion of metapath and its relevance to the constrained shortest-path problem [68–70]. We then use the constrained shortest metapaths to create variables with composite decisions for the SFC composition problem and discuss scalability improvements of this approach.

19.4.2.1 Metalinks and Metapaths

Before defining the metapath, it is useful to introduce the idea of "metalinks." Metalinks have been widely adopted in prior NFV/VNE literature to solve optimally graph matching problems [25, 71]. A metalink is an augmentation link in a network graph. In our case, it represents the (potential) feasible placement of some service a on some physical node A, as shown in Figure 19.12. Formally, we have:

Definition 19.1 ***Metalink*** A link si for SFC a 2 A belongs to the set of metalinks E_Ma if and only if the service s 2 $N_V{}^a$ of SFC a can be placed onto the node i 2 N_S.

Building on the definition of a metalink, we can define a metapath as the path that connects any two services through the physical network augmented with

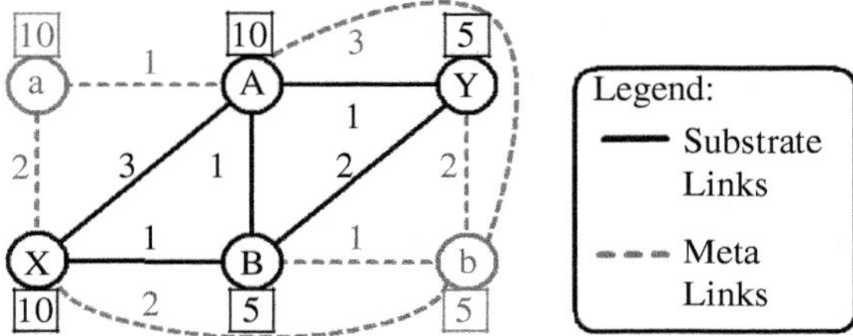

Figure 19.12 Illustrative example of the augmented with metalinks physical network which represent feasible service a, b, and c placements; numbers indicate fitness function values – red and black values annotate service placement and service chaining via some physical link, respectively.

metalinks. For example, consider following metapaths a–A–Y–b and a–A–B–b, as shown in Figure 19.12. Formally, we have:

Definition 19.2 ***Metapath*** The path P_{ijst} is a metapath between services s and t for SFC a 2 A if and only if 8kl 2 P_{ij}^{st}: kl 2 E_S _ kl 2 {si, tj}.

Intuitively, metapath P_{ijst} is formed by exactly two metalinks that connect *s* and *t* to the physical network and an arbitrary number of physical links kl 2 E_S.

19.4.2.2 Constrained Shortest Metapaths

Having defined metapaths, let us consider a simple case of the SFC composition problem – composition of a single-link chain (i.e. two services connected via a single virtual link): the optimal composition of a single-link chain can be seen as the constrained shortest (meta)path problem. This connects two services via the augmented physical network, where (1) all physical links have arbitrary fitness values of a service chaining (virtual link mapping), and (2) all metalinks have arbitrary fitness values of a service placement divided by the number of neighboring services (i.e. by 1 for a single-link chain). In our example, shown in Figure 19.3, the optimal single-link SFC a–b composition can be represented by the constrained shortest metapath a–A–Y–b, which satisfies all SFC composition constraints with the overall fitness function of 3.

19.4.2.3 Multiple-link Chain Composition via Metapath

While observing Figure 19.12, we can notice how using only a single constrained shortest metapath per a single-link segment of a multiple-link SFC a–b–c can lead to an unfeasible composition: as the optimal a–b composition is a–A–Y–b metapath. Also, the optimal b–c composition is b–B–X–c metapath, and the service b has to be simultaneously placed on Y and B physical nodes. Thus, we cannot stitch these metapaths, and we need to find more than one constrained

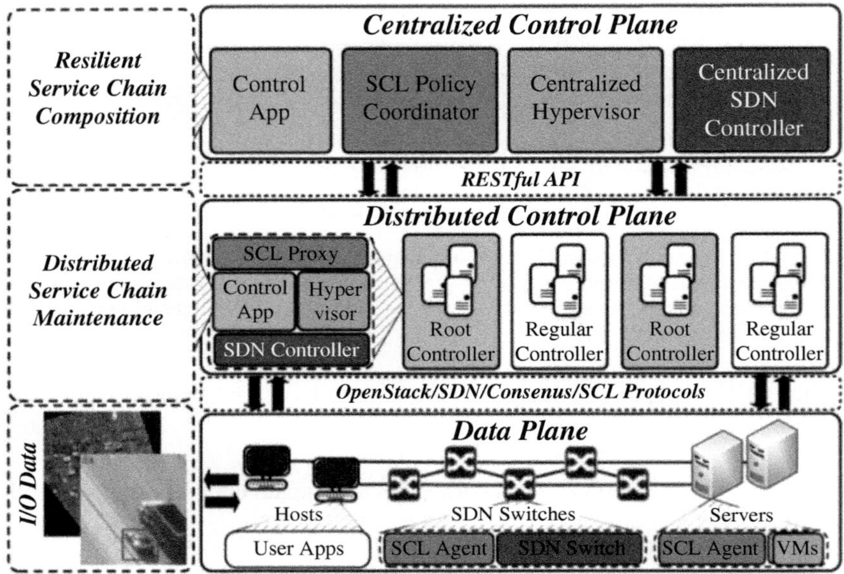

Figure 19.13 System architecture of our Incident-Supporting Service Chain Orchestration prototype includes four main logical components: (1) control application is responsible for service chain composition in a centralized control plane and its maintenance in a distributed control plane; (2) simple coordination layer (SCL) and root controllers are responsible for guaranteeing consistency in the distributed control plane; (3) SDN is responsible for traffic steering in the data plane; and (4) hypervisor is responsible for virtual machines (VMs) provisioning.

shortest metapath per a single-link chain. In our composite variable approach, we find k-constrained shortest metapaths (to create k binary variables) per each single-link segment of a multilink service chain. To find metapaths, any constrained shortest path algorithm can be used [68–70]. However, we build our metapath composite variable approach upon the path finder proposed in our prior work that is an order of magnitude faster than recent solutions [72].

To further benefit from constrained shortest metapaths and simplify the chain composition problem, we offload its constraints (either fully or partially) to either metalinks or the path finder. Specifically, geo-location and an arbitrary number of end-to-end network (e.g. latency) QoS constraints can be fully offloaded to metalinks and to the path finder, respectively. At the same time, capacity constraints of the SFC composition problem are global and can be only partially offloaded. Once k-constrained shortest paths have been found for each single-link service chain segment, we can solve the GAP problem [73] to assign each single-link chain segment to exactly one constrained shortest metapath and stitch these metapaths as described below. Aside from solving the NP-hard GAP problem directly, we can also solve the GAP using its polynomial Lagrangian relaxation by compromising both its optimality and feasibility guarantees [73]. Finally, found metapaths can be used subsequently to migrate failed SFC segments during their maintenance. The details of the metapath-based composite variable approach as well as the main MCCF problem integer programming formulation are omitted due to space constraints, and can be found in our detailed related work in [6].

19.4.2.4 Allowable Fitness Functions for Metapath-Based Variables

In general, fitness functions qualify for our metapath composite variable approach if they comprise either additive or multiplicative terms. The above requirement fits for most SFC objectives [59], and other objectives can also qualify if well-behaved (e.g. if their single-link chain fitness values can be minimized by a path finder).

19.4.2.5 Metapath Composite Variable Approach Results

We found that our approach achieves 99% optimality on average and takes seconds to compose SFCs for practically sized problems of US Tier–1 (300 nodes) and regional (600 nodes) infrastructure providers' topologies. For the same setup, we found that the master problem solution takes hours using a cloud server tailored for high-performance computing. As a result, our metapath approach can secure up to 2 times more SFCs in comparison to the state-of-the-art NFV/VNE approaches under challenging disaster incident conditions. Moreover, supported policies allow efficient trade-off between a SFC reliability and its composition optimality [6].

19.4.3 Metapath-Based SFC Orchestration Implementation

In this subsection, we describe the architecture of our incident-supporting metapath-based SFC orchestration implementation prototype shown in Figure 19.13. Our prototype architecture includes four main logical components: (1) a controller application, used to compose and maintain service chains; (2) our implementation of the Simple Coordination Layer (SCL), used to guarantee consistency of the distributed control plane and root controllers; (3) an SDN-based system with (4) a hypervisor to allocate mapped physical resources. The prototype source code for this incident-supporting metapath-based SFC orchestration implementation is publicly available under a GNU license at [74]. In the remainder of this section, we describe additional details for each of the four prototype components.

19.4.3.1 Control Applications

We have two main types of control applications that use our metapath-based composite variable approach. The first type is responsible for the resilient service-chain composition in the centralized control plane. The second type is responsible for maintenance of composed service chains. Note that we move the service-chain maintenance to the distributed control plane to avoid both a single point of failure as well as congestion in the centralized control plane.

19.4.3.2 SCL

To guarantee consistency in the distributed control plane and avoid various related violations (e.g. looping paths, SLO violations, etc.), one can use a SCL [75], or alternatively any other existing comparable consensus protocols. SCL includes three main components: SCL agent running on physical resources, SCL proxy controller running on controllers in the distributed control plane, and SCL policy coordinator running in the centralized control plane. The agent periodically exchanges messages with corresponding proxy controllers and triggers any changes in the physical resources. Proxy controllers send information to the service chain maintenance control application and periodically interact with other SCL proxy controllers. Finally, all policy changes (e.g. in control applications, in SCL, etc.) are committed via a two-phase commit [75] that is directed by the policy coordinator.

19.4.3.3 SDN and Hypervisor

The last two logical components of our prototype implementation are commonly used SDN and hypervisor systems being used in cloud-fog deployments. Guided by control applications, both SDN and hypervisor are responsible for traffic steering and virtual machines (VMs) provisioning in the data plane, respectively. In our prototype implementation, we use OpenFlow as our main SDN system [76] and Docker containers [77] as our hypervisor system to place services on the physical server.

19.5 Concluding Remarks

19.5.1 What Have We Learned?

In this chapter, we have learned how geospatial video analytics using fog/cloud resources can benefit from the FCC paradigm. We described how FCC advances current knowledge of the parameterization of application resource requirements in the form of small and large instance visual-data processing functions. The decoupling into small and large instance functions enables new optimizations of cloud/fog computation location selection utilizing network virtualization to connect the network edges (wired and wireless) in the fog with the core/public cloud platforms. To demonstrate the potential of FCC in cloud-fog infrastructures, we have considered three computer vision application case studies for disaster incident response and showed how these applications can be decoupled into a set of processing functions. We have also learnt how FCC is subject to challenges of: (1) "data collection" at the wireless edge within lost/austere infrastructure regions, and (2) the consequent "data processing" within geo-distributed fog/cloud platforms that satisfies geo/latency QoS constraints of a geospatial video analytics consumption with satisfactory user QoE.

To cope with challenges of data collection, we have seen how we can utilize geographic edge routing protocols and mobile ad hoc networks (MANETs). To provide sustainable high-speed data delivery to the edge cloud gateway, we have to overcome limitations of geographic routing approaches, such as the local minimum problem. We discussed one potential solution to this problem that considered an AI-AGRA that uses satellite imagery datasets corresponding to the disaster incident area(s). In this approach, we saw how recent deep-learning solutions can be used to augment geographic routing algorithms with the potential local minimum knowledge retrieved from satellite imagery.

To cope with challenges of data processing, we discussed how one can apply the microservices paradigm using natural decomposability of computer vision applications. Using microservices, we can utilize the NFV SFC mechanisms to satisfy geo/latency requirements of geospatial video analytics as well as overcome insufficient local processing capabilities within data collection regions. As the optimal SFC in a general case is basically the NP-hard MCCF problem, we presented a near-optimal and practical metapath composite variable approach. This approach achieves 99% optimality on average and allows us to significantly reduce combinatorial complexity of the master MCCF problem, i.e. it takes just a few seconds for SFC over US Tier-1 and regional infrastructure topologies, whereas the MCCF solution in the same case takes several hours to compute.

19.5.2 The Road Ahead and Open Problems

We conclude this chapter with a list of open challenges for adopting fog computing architectures in geospatial video analytics. Addressing these challenges is essential for a variety of computer vision applications, such as face recognition in crowds, object tracking in aerial wide-area motion imagery, reconnaissance, and video surveillance that are relevant for delivering disaster-incident situational awareness.

- *Consideration of energy-awareness.* To improve overall data collection performance from IoT devices at the wireless edge within lost infrastructure regions, edge routing protocols should not only account for a routing accuracy, but also take into account residual energy levels on routers/IoT devices [29], as well as optimize traffic based on user behavior (e.g. mobility pattern of first responders) to improve the overall edge network utilization as well as satisfy user QoE demands.

- *Consideration of virtual topologies of functions.* In some cases when handling diverse spatio-temporal data sets, computer vision applications may not be easily decomposable to a chain of functions; hence other (more complex) virtual topologies of functions should be addressed in potential cases where benefits of the proposed metapath composite variable approach may no longer hold.
- *Consideration of processing states.* When computer vision applications for video analytics are not stateless, e.g. tracking objects of interest may require maintenance of an object model, and state information needs to be preserved across functions. This in turn adds an additional layer of complexity that needs to be handled when considering how we can better handle noisy data for processing and compose chains of such functions.
- *Consideration of hierarchical control architectures.* To fully benefit from the FCC paradigm and deliver disaster-incident situational awareness, existing fog/cloud platforms may need to be extended with a support of hierarchical software-defined control planes to avoid single point of failure/congestion and enable efficient utilization of fog/cloud resources.
- *Consideration of distributed intelligence for IoT.* There will undoubtedly be impressive advances in wearables and other smart IoT devices (with higher data resolutions/sizes), 5G wireless networks, and flexible integration, as well as management of fog/cloud platforms occur; better modeling, suitable architectures, pertinent optimizations, and new machine-learning algorithms (particularly deep learning and/or reinforcement learning, real-time recommenders) for geospatial video analytics will need to be considered to adapt these advances for disaster incident–supporting computer vision applications.

Solving the above open issues in future explorations is significantly important to advance the knowledge to further the area of geospatial video analytics. Investigations to find solutions to these open issues can lead to a more helpful and accurate disaster-incident situational awareness that can foster effective and efficient disaster relief coordination to save lives.

References

1. Thoma, G., Antani, S., Gill, M. et al. (2012). People locator: a system for family reunification. *IT Professional* 14 (3): 13–21.
2. Pelapur, R., Candemir, S., Bunyak, F. et al. (2012). Persistent target tracking using likelihood fusion in wide-area and full motion video sequences. In: *IEEE International Conference on Information Fusion*, 2420, 2427.
3. Schurr, N., Marecki, J., Tambe, M. et al. (2005). The future of disaster response: humans working with multiagent teams using DEFACTO. In: *AAAI Spring Symposium: AI Technologies for Homeland Security*, 9–16.

4 Klontz, J.C. and Jain, A.K. (2013). A case study on unconstrained facial recognition using the Boston marathon bombings suspects, Michigan State University. *Technical Report* 119 (120): 1.

5 Gargees, R., Morago, B., Pelapur, R. et al. (2017). Incident-supporting visual cloud computing utilizing software-defined networking. *IEEE Transactions on Circuits and Systems for Video Technology* 27 (1): 182–197.

6 Chemodanov, D., Calyam, P., and Esposito, F. (2019). A near-optimal reliable composition approach for geo-distributed latency-sensitive service chains. In: *IEEE International Conference on Computer Communications (INFOCOM).*

7 Dinh, H.T., Lee, C., Niyato, D., and Wang, P. (2013). A survey of mobile cloud computing: architecture, applications, and approaches. *Wireless Communications and Mobile Computing* 13 (18): 1587–1611.

8 Fernando, N., Loke, S.W., and Rahayu, W. (2013). Mobile cloud computing: a survey. *Future Generation Computer Systems* 29 (1): 84–106.

9 Burchard, J., Chemodanov, D., Gillis, J., and Calyam, P. (2017). Wireless mesh networking protocol for sustained throughput in edge computing. In: *IEEE International Conference on Computing, Networking and Communications (ICNC)*, 958–962. IEEE.

10 Cvetkovski, A. and Crovella, M. (2009). Hyperbolic embedding and routing for dynamic graphs. In: *IEEE International Conference on Computer Communications (INFOCOM)*, 1647–1655.

11 Karp, B. and Kung, H.-T. (2000). GPSR: Greedy perimeter stateless routing for wireless networks. In: *ACM International Conference on Mobile Computing and Networking*, 243–254. ACM.

12 Kranakis, E., Singh, H., and Urrutia, J. (1999). Compass routing on geometric networks. In: *IEEE Conference on Computational Geometry.*

13 Sukhov, A.M. and Chemodanov, D.Y. (2013). The neighborhoods method and routing in sensor networks. In: *IEEE Conference on Wireless Sensor (ICWISE)*, 7–12. IEEE.

14 AWS lambda. https://aws.amazon.com/lambda. Accessed February 2018.

15 Google Cloud Functions. https://cloud.google.com/functions. Accessed February 2018.

16 Microsoft azure functions. https://azure.microsoft.com/en-us/services/functions. Accessed February 2018.

17 IBM OpenWhisk. https://www.ibm.com/cloud/functions. Accessed February 2018.

18 Potharaju, R. and Jain, N. (2013). Demystifying the dark side of the middle: a field study of middlebox failures in datacenters. In: *Proceedings of Internet Measurement Conference*, 9–22. ACM.

19 Eriksson, B., Durairajan, R., and Barford, P. (2013). Riskroute: a framework for mitigating network outage threats. In: *ACM Conference on Emerging Networking Experiments and Technologies*, 405–416.

20 Fei, X., Liu, F., Xu, H., and Jin, H. (2018). Adaptive VNF scaling and flow routing with proactive demand prediction. In: *IEEE International Conference on Computer Communications (INFOCOM)*, 486–494.

21 Tomassilli, A., Giroire, F., Huin, N., and Prennes, S. (2018). Provably efficient algorithms for placement of SFCs with ordering constraints. In: *IEEE International Conference on Computer Communications (INFOCOM).*

22 Cohen, R., Lewin-Eytan, L., Naor, J.S., and Raz, D. (2015). Near optimal placement of virtual network functions. In: *IEEE International Conference on Computer Communications (INFOCOM)*, 1346–1354.

23 Sang, Y., Bo, J., Gupta, G.R. et al. (2017). Provably efficient algorithms for joint placement and allocation of virtual network functions. In: *IEEE International Conference on Computer Communications (INFOCOM)*, 1–9.

24 Jain, S., Kumar, A., Mandal, S. et al. (2013). B4: experience with a globally-deployed software defined WAN. *ACM SIGCOMM Computer Communication Review* 43: 3–14.

25 Feng, H., Llorca, J., Tulino, A.M. et al. (2017). Approximation algorithms for the NFV service distribution problem. In: *IEEE International Conference on Computer Communications (INFOCOM)*, 1–9.

26 Gillis, J., Calyam, P., Bartels, A. et al. (2015). Panacea's glass: mobile cloud framework for communication in mass casualty disaster triage. In: *IEEE International Conference on Mobile Cloud Computing, Services, and Engineering (MobileCloud)*, 128–134. IEEE.

27 Morago, B., Bui, G., and Duan, Y. (2014). Integrating LiDAR range scans and photographs with temporal changes. In: *IEEE Conference on Computer Vision and Pattern Recognition Workshops (CVPRW)*, 732–737.

28 Chemodanov, D., Esposito, F., Sukhov, A. et al. (2019). Agra: AI-augmented geographic routing approach for IoT-based incident-supporting applications. *Future Generation Computer Systems* 92: 1051–1065.

29 Trinh, H., Chemodanov, D., Yao, S. et al. (2017). Energy-aware mobile edge computing for low-latency visual data processing. In: *IEEE International Conference on Future Internet of Things and Cloud (FiCloud).*

30 Chung, S., Christoudias, C.M., Darrell, T. et al. (2012). A novel image-based tool to reunite children with their families after disasters. *Academic Emergency Medicine* 19 (11): 1227–1234.

31 King, D.E. (2009). Dlib-ml: a machine learning toolkit. *Journal of Machine Learning Research* 10: 1755–1758.

32 Parkhi, O.M., Vedaldi, A., Zisserman, A. et al. (2015). Deep face recognition. In: *British Machine Vision Conference*, vol. 1, 6.

33 Kwan, M.-P. and Ransberger, D. (2010). LiDAR assisted emergency response: detection of transport network obstructions caused by major disasters. *Computers, Environment and Urban Systems* 34 (3): 179–188.

34 Hartley, R. and Zisserman, A. (2010). *Multiple View Geometry*. Cambridge University Press.

35 Stauffer, C. and Stauffer, E.G. (1999). Adaptive background mixture models for real-time tracking. In: *IEEE Conference on Computer Vision and Pattern Recognition (CVPR)*, 246–253.

36 Palaniappan, K., Rao, R., and Seetharaman, G. (2011). Wide-area persistent airborne video: architecture and challenges. In: *Distributed Video Sensor Networks*, 349–371.

37 Hafiane, A., Seetharaman, G., Palaniappan, K., and Zavidovique, B. (2008). Rotationally invariant hashing of median patterns for texture classification. In: *Lecture Notes in Computer Science (LNCS)*, 619–629.

38 Aliakbarpour, H., Palaniappan, K., and Seetharaman, G. (2015). Robust camera pose refinement and rapid SfM for multiview aerial imagery without RANSAC. In: *IEEE Geoscience and Remote Sensing Letters (GRSL)*, 2203–2207.

39 Hafiane, A., Palaniappan, K., and Seetharaman, G. (2008). UAV-video registration using block-based features. In: *IEEE International Symposium on Geoscience and Remote Sensing (IGARSS)*, 1104–1107.

40 Palaniappan, K., Bunyak, F., Kumar, P. et al. (2010). Efficient feature extraction and likelihood fusion for vehicle tracking in low frame rate airborne video. In: *IEEE Conference on Information Fusion (FUSION)*, 1–8.

41 Akhtar, F., Rehmani, M.H., and Reisslein, M. (2016). White space: definitional perspectives and their role in exploiting spectrum opportunities. *Telecommunications Policy* 40 (4): 319–331.

42 Król, M., Schiller, E., Rousseau, F., and Duda, A. (2016). Weave: efficient geographical routing in large-scale networks. In: *EWSN*, 89–100.

43 Lam, S.S. and Qian, C. (2013). Geographic routing in d-dimensional spaces with guaranteed delivery and low stretch. *IEEE/ACM Transactions on Networking (TON)* 21 (2): 663–677.

44 Leong, B., Liskov, B., and Morris, R. (2006). Geographic routing without planarization. In: *NSDI*, vol. 6, 25.

45 Umer, T., Amjad, M., Afzal, M.K., and Aslam, M. (2016). Hybrid rapid response routing approach for delay-sensitive data in hospital body area sensor network. In: *ACM International Conference on Computing Communication and Networking Technologies*.

46 Esposito, F. (2017). Catena: a distributed architecture for robust service function chain instantiation with guarantees. In: *IEEE Conference of Network Softwarization*, 1–9.

47 Kleinberg, R. (2007). Geographic routing using hyperbolic space. In: *IEEE International Conference on Computer Communications (INFOCOM)*, 1902–1909.

48 Sahhaf, S., Tavernier, W., Colle, D. et al. (2015). Experimental validation of resilient tree-based greedy geometric routing. *Computer Networks* 82: 156–171.

49 Bose, P., Morin, P., Stojmenović, I., and Urrutia, J. (2001). Routing with guaranteed delivery in ad hoc wireless networks. *Wireless Networks* 7 (6): 609–616.

50 Kim, Y.-J., Govindan, R., Karp, B., and Shenker, S. (2005). Geographic routing made practical. In: *Proceedings of the 2nd Conference on Symposium on Networked Systems Design & Implementation*, vol. 2, 217–230. USENIX Association.

51 Zhou, J., Chen, Y., Leong, B., and Sundaramoorthy, P.S. (2010). Practical 3-D geographic routing for wireless sensor networks. In: *ACM Conference on Embedded Networked Sensor Systems*, 337–350.

52 National Oceanic and Atmospheric Organization. https://geodesy.noaa.gov/stormarchive/storms/joplin/index.html. Accessed March 2018.

53 Redmon, J., Divvala, S., Girshick, R., and Farhadi, A. (2016). You only look once: unified, real-time object detection. In: *IEEE Conference on Computer Vision and Pattern Recognition*, 779–788.

54 Gidaris, S. and Komodakis, N. (2015). Object detection via a multi-region and semantic segmentation-aware CNN model. In: *IEEE International Conference on Computer Vision*, 1134–1142.

55 Blaschke, T. (2010). Object based image analysis for remote sensing. *ISPRS Journal of Photogrammetry and Remote Sensing* 65 (1): 2–16.

56 Weston, J., Ratle, F., Mobahi, H., and Collobert, R. (2012). Deep learning via semi-supervised embedding. In: *Neural Networks: Tricks of the Trade*, 639–655. Springer.

57 Bonomi, F., Milito, R., Zhu, J., and Addepalli, S. (2012). Fog computing and its role in the internet of things. In: *ACM Workshop on Mobile Cloud Computing*, 13–16.

58 Gil Herrera, J. and Botero, J.F. (2016). Resource allocation in NFV: a comprehensive survey. *IEEE Transactions on Network and Service Management* 13 (3): 518–532.

59 Bhamare, D., Jain, R., Samaka, M., and Erbad, A. (2016). A survey on service function chaining. *Journal of Network and Computer Applications* 75: 138–155.

60 Guo, L., Pang, J., and Walid, A. (2018). Joint placement and routing of network function chains in data centers. In: *IEEE International Conference on Computer Communications (INFOCOM)*, 612–620.

61 Yu, R., Xue, G., and Zhang, X. (2018). Application provisioning in fog computing-enabled Internet-of-Things: a network perspective. In: *IEEE International Conference on Computer Communications (INFOCOM)*, 783–791.

62 Chowdhury, M., Rahman, M.R., and Boutaba, R. (2012). Vineyard: virtual network embedding algorithms with coordinated node and link mapping. *IEEE/ACM Transactions on Networking* 20 (1): 206–219.

63 Fei, X., Liu, F., Hong, X., and Jin, H. (2017). Towards load-balanced VNF assignment in geo-distributed NFV infrastructure. *IEEE International Symposium on Quality of Service (IWQoS)*: 1–10.

64 Cziva, R., Anagnostopoulos, C., and Pezaros, D.P. (2018). Dynamic, latency-optimal VNF placement at the network edge. In: *IEEE International Conference on Computer Communications (INFOCOM)*, 693–701.

65 Spinnewyn, B., Mennes, R., Botero, J.F., and Latré, S. (2017). Resilient application placement for geo-distributed cloud networks. *Journal of Network and Computer Applications* 85: 14–31.

66 Barnhart, C., Farahat, A., and Lohatepanont, M. (2009). Airline fleet assignment with enhanced revenue modeling. *Operations Research* 57 (1): 231–244.

67 Ahuja, R.K., Magnanti, T.L., and Orlin, J.B. (2014). *Network Flows*. Elsevier.

68 Lozano, L. and Medaglia, A.L. (2013). On an exact method for the constrained shortest path problem. *Computers & Operations Research* 40 (1): 378–384.

69 Chen, X., Cai, H., and Wolf, T. (2015). Multi-criteria routing in networks with path choices. In: *IEEE International Conference on Network Protocols*, 334–344.

70 Van Mieghem, P. and Kuipers, F.A. (2004). Concepts of exact QoS routing algorithms. *IEEE/ACM Transactions on Networking* 12 (5): 851–864.

71 Mijumbi, R., Serrat, J., Gorricho, J.-L., and Boutaba, R. (2015). A path generation approach to embedding of virtual networks. *IEEE Transactions on Network and Service Management* 12 (3): 334–348.

72 Chemodanov, D., Esposito, F., Calyam, P., and Sukhov, A. (2018). A constrained shortest path scheme for virtual network service management. *IEEE Transactions on Network and Service Management* 16 (1): 127–142.

73 Fisher, M.L. (1981). The lagrangian relaxation method for solving integer programming problems. *Management Science* 27 (1): 1–18.

74 Metapath-based service chain orchestration repository. https://goo.gl/TMKZxj. Accessed February 2018.

75 Panda, A., Zheng, W., Hu, X. et al. (2017). SCL: simplifying distributed SDN control planes. In: *NSDI*, 329–345.

76 McKeown, N., Anderson, T., Balakrishnan, H. et al. (2008). Openflow: enabling innovation in campus networks. *ACM SIGCOMM Computer Communication Review* 38 (2): 69–74.

77 Docker containers. https://www.docker.com. Accessed February 2018.

20

An Insight into 5G Networks with Fog Computing

Osman Khalid[1], *Imran Ali Khan*[1], *Rao Naveed Bin Rais*[2], *and Asad Waqar Malik*[3]

[1] *Department of Computer Science, COMSATS University Islamabad, Campus, Abbottabad, Pakistan*
[2] *College of Engineering and Information Technology, Ajman University, Ajman, UAE*
[3] *Department of Computing, School of Electrical Engineering and Computer Science, National University of Sciences and Technology, Islamabad 44000, Pakistan*

20.1 Introduction

4G has seen significant progress over the past few years, and with passage of time it has spread worldwide. A question remains about the future of wireless technologies. Current wireless technology generations have adopted a huge model shift, as shown in Figure 20.1. In the past few years, cellular communication experienced a tremendous increase in data traffic. Due to the popularity of smart devices, multimedia contents, video streaming, and voice services, higher data transmission rates are required [1]. Current 3G and 4G can handle continuously increasing mobile data traffic for next few years but will not be able to handle large data and rapidly growing devices [2]. Big data is getting bigger with the advancement of new and quick methods for gathering, storing, and processing data [3]. Higher capacity, low latency, improved transfer rate of data, and enhanced quality of service (QoS) are primary challenges that need to be handled. Continuously growing demand for 3D ("Device," "Data creation," and "Data transfer") services has motivated manufacturers to develop a new wireless generation called 5G [4, 5]. Currently, 4G is not fulfilling all the above-mentioned requirements of cellular network, and, consequently, a new generation is being developed.

5G is the emerging standard beyond 4G and it will fulfill the current demands of wireless cellular communication [6–8]. Table 20.1 presents a comparison of 5G with existing technologies [9, 10]. What are the appropriate technologies that will

Fog Computing: Theory and Practice, First Edition.
Edited by Assad Abbas, Samee U. Khan, and Albert Y. Zomaya.

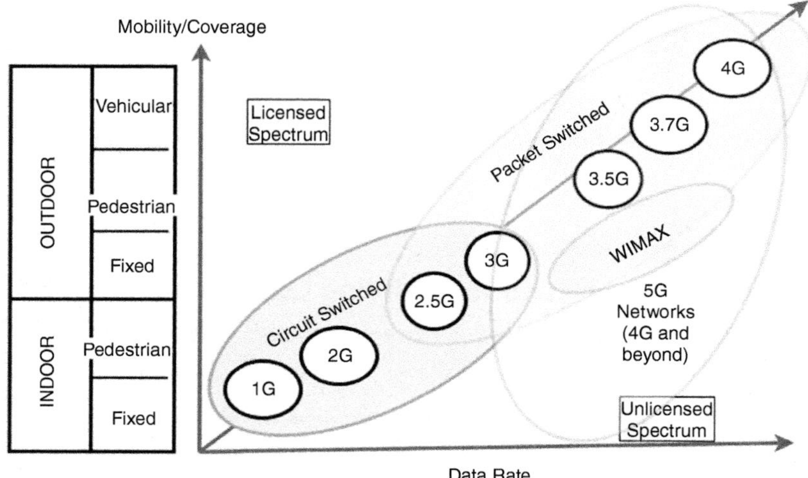

Figure 20.1 Evolution of wireless technologies.

Table 20.1 Comparison of wireless technologies.

Generations	1G	2G	3G	4G	5G
Deployment	1980s	1990s	1998–1990s	2008–2009s	2020
Bandwidth	2 kbps	14.4 kbps	2 Mbps	200 Mbps–1 Gbps	1 Gbps and higher
Web Standard	—	www	www (IPv4)	www (IPv4)	wwww (IPv6)
Switching	Circuit	Circuit	Packet	Packet	Packet
Limitations	Less security	Less security	Less support for Internet	Unable to support max. devices	—
Services	Mobile telephony	Digital voice/short SMS etc.	High quality audio/video	Dynamic access information, wearable devices	Dynamic access, wearable device with AI abilities

define 5G? Will it just be the extension of 4G or it will be entirely a new technology? These are some of the questions that are the topic of hot debate among researchers. The following aspects are discussed encompassing 5G [8]:

1. Device-centric architecture review, e.g. uplink/downlink and controlling all the channels to make routing information among nodes

2. Massive multiple-input multiple-output (MIMO) requires some modifications such as design of macro-station
3. IEEE 802.11 has set a standard for mmWave (millimeter wave) because of bandwidth limitations in backhaul in small cell
4. Machine-to-machine (M2M) communication must have very low latency
5. Heterogeneous Network (HetNet) design and support
6. 5G must provide a seamless user experience [1]

In general, 5G copes with issues/challenges that include network deployments, capacity, data rate, minimization of end-to-end latency, devices' connectivity, capital reduction, operational cost, QoS, smart antenna system, wider area coverage, free movement from one technology to another, choice among various wireless systems, and data encryption [11, 12]. It is believed that 5G will fulfill all the growing demands of data-centric applications due to its physical layer security design and mmWave spectrum, and because of the use of multiple antennas. It will fulfill the demand of exceeding data rate, wide-area radio coverage, excessive amount of nodes, and less latency [2]. Another main challenge that 5G needs to address is data caching [13, 14].

The rest of the chapter is organized as follows. Section 20.2 discusses the vision of 5G. Section 20.3 discusses fog computing with 5G. Section 20.4 presents 5G network architecture. Section 20.5 illustrates the growing technologies based on 5G. Section 20.6 presents the 5G applications, and Section 20.7 discusses challenges of 5G. Finally, Section 20.8 concludes the chapter.

20.2 Vision of 5G

4G technology is practically implemented in few countries, but full-scale deployment is yet to be seen. 5G is conceptualized as a real network, and it must support all the existing applications supported by previous technologies. Now 5G technology is in its evolution and it will provide an affordable wireless connectivity on a very high speed [15]. According to Siddiqui et al. 5G has two aspects: revolutionary and evolutionary [9]. The revolutionary view states that 5G technology is able to connect the whole world by reducing the limits to wireless communication. Application can be robots with artificial intelligence. In addition, according to the evolutionary view, the 5G systems will be able to support WWWW, and will allow a flexible network, for instance, dynamic ad hoc wireless network (DAWN). Requirements change from app to app, which will have to fulfill certain configurations. For example, driverless cars or public security apps require solutions to the main issues of delay and reliability. Alternatively, high data rate apps, such as streaming high-quality videos, may exhibit delay and reliability.

20.3 Fog Computing with 5G Networks

20.3.1 Fog Computing

Fog computing is an emerging technology that extends the cloud capabilities, such as computation, storage, and networking services from the remote data center to closer to the end-user [16–18]. By locally processing the requests, data traffic is reduced on the core network, thereby reducing the processing time, latency, and network cost. The fog computing architecture is implemented on multiple layers of a network's technology [19, 20]. By adding layers of fog nodes, applications can be partitioned to run at the optimal network level [21, 22]. Fog computing specifically aims the time-critical applications, e.g. real-time gaming, virtual reality, emergency response system, vehicular networks, real-time health care, Internet of Things (IoT), and so on, to name a few [23, 24].

20.3.2 The Need of Fog Computing in 5G Networks

The future 5G networks aim for ultra-low latency and high bandwidth. 5G communication will be driven by high-frequency millimeter waves, and a 5G cellular network will connect mobile users to the base station (BS) connected to the core network. The low latency radio interfaces in the 5G network aim to reduce the delays between a user and a base station. However, the forwarding of user application requests from core network to the cloud would significantly increase latency. Consequently, 5G networks require moving processing power closer to the end-user, which can be accomplished with fog computing. Fog computing can be enabled in 5G by leveraging both the small and macro-based stations to achieve the ultra-low latency demanded in 5G networks. A fog computing–based radio access network (F-RAN) could provide 5G with reduced latency and high energy efficiency.

20.4 Architecture of 5G

20.4.1 Cellular Architecture

In 5G, each device (cell phone) has a constant "home IP address," in additional to a "care-of" address. Care-of address represents actual current location whereas IP identifies network interface and location addressing. Before communicating with a mobile, a computer passes a message packet to its IP address. Then, it is moved to the care-of address by directory server through a tunnel. The directory tunnel replies back to the computer and informs it about the correct address, and then further communication starts. As compared to IPV4 addressing, a connectionless protocol is being used in the switched layer. A total of 128 bits are used, which are subdivided into four categories, dedicated for different functions. The first 32 bits are used for the "home" IP address, and the other 32 are specifically used for the "care-of" address. The second part is for communication among cell phones and other devices (e.g. computer). The third part is used for tunneling to establish interconnection among wired and wireless networks. In the third part, the directory server uses the IP address of the mobile phone to create a communication channel. The last 32 bits of IPv6 address may be used for local addressing, e.g. for VPN (virtual private networks) sharing [25]. Figure 20.2 shows a general 5G cellular network architecture [10].

5G phone is designed to be an open source and includes every layer from physical layer to the end-user layer (application layer). Open wireless architecture, which was proposed for 4G mobiles phone, is adapted in this conceptual phone, and some changes in network layer and other layers up to application layer are made. The proposed concept divides the network layer into further two sublayers: upper and lower network layers. The lower layer is for each interface and the upper layer is for mobile terminal. The middleware between the upper and lower layer translates the IP addresses [26]. The protocol stack for the proposed 5G mobile phone concept is illustrated in Table 20.2. To allow the maximum optimized usage of specific wireless implementations of transport protocols, the open transport layer is proposed in this concept. The proposed

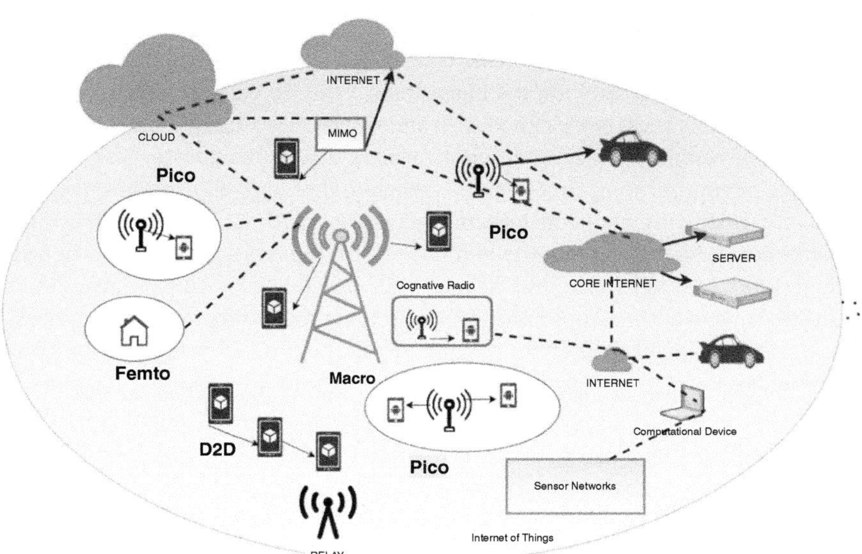

Figure 20.2 5G cellular architecture.

Table 20.2 5G protocol stack.

Application layer	Application (services)	
Session layer	Open transport protocol (OTP)	
Network layer	Upper network layer	Lower network layer
Data link layer	Open wireless architecture (OWB)	

concept will open a way to select many wireless technologies on different QoS and price constraints for different services, e.g. voice over Internet protocol (VoIP), streaming Internet protocol television (IPTV), web, messaging, and gaming. To fulfill this, 5G mobiles will create a database that will gather and provide information for all the services and technologies in the cellular device. Development modes will provide the best QoS and cost-to-use available wireless technologies within a 5G phone. Figure 20.3 explains the model, proposing a 5G architecture that is all-IP based and mobile network interoperability composed of nodes and independent radio access technologies (RATs). It is anticipated that 5G will change the concept of station-based-centric architecture of mobile networks, and methods of uplink/downlink control and data channels, to make routing and the flow of information in a better way. Currently, mobile systems are not letting devices communicate with each other directly. Every communication works through a base station. Tehrani et al. proposed a two-tier cellular network that allows a base-to-device communication ,and device-to-device communication, as well [27]. In a two-tier mobile system, as users' data is passed by other users' cell phones, safety and security are the biggest concerns. To ensure breaches do not occur, the systems must be designed with smart interface strategies and managed allocation of resources. The traditional macro-cell tier includes a base station for device communication. In 5G-based device-to-device (D2D) communication, the device is directly connected to another device without a base station and transmission is directly among the two nodes with the assistance of other nodes. Base stations are still there, but devices will be allowed to create an ad-hoc mesh network. Another important thing is pricing and charge of D2D services. Pricing models are needed to be designed to let devices to participate in D2D communication [11].

20.4.2 Energy Efficiency

The main goal of 5G is to reduce energy. Therefore, energy efficient infrastructures play a vital role in 5G networks. Rowell et al. consider a usage of combined spectral efficiency and optimization of energy [28]. In a user-centric approach, a user's equipment is able to select uplink and downlink channels from various BS,

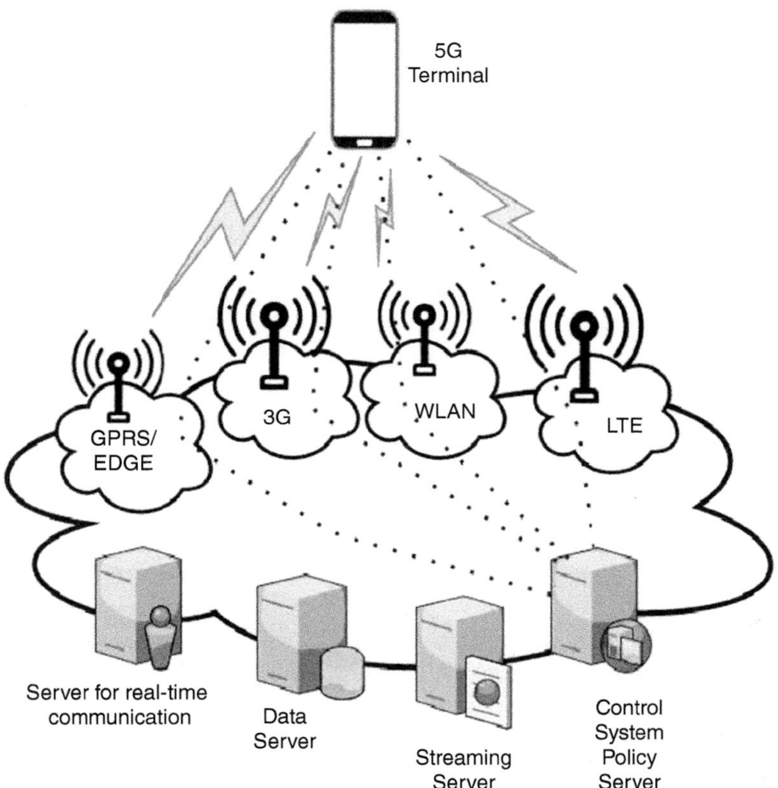

Figure 20.3 5G IP-based architecture.

depending upon amount of load, condition of channel, applications, and services requirements. Direct association cannot be performed entirely with user's equipment (UEs); it must consider both the channels at the time when they make an association with base stations (BSs). In this way, demodulation of signals and data is beneficial to save energy. An example of this scenario is a mobile base station (MBS), which may initiate signaling of a BS, and a server base station (SBS) to serve all data requests. If there is no data request for that time duration, we can hibernate the SBS. Zhang et al. proposed the same technique for decoupling of data in those UEs that are connected to SBS, according to the functionality provided by MBS, and output the result, which ensures less energy consumption and a lower chances of interference [29]. Hu et al. proposed C-RAN architecture for energy efficiency in a manner that UEs and Remote Radio Head (RHHs) serve almost the same number of times [30]. They also proposed an approach to interface management to decrease power consumption of SBSs and MBSs.

20.4.3 Two-Tier Architecture

Another proposed architecture for 5G is two-tier architecture. In two-tier architecture, MBS works at a higher tier and SBSs work under MBS, known as the lower tier. Higher and lower tiers share the same frequency band. There exists a macro cell, which covers all the small cells. These small cells include micro, pico, and femto cells. Small cells enhance the services and coverage area of macro cells. Two-tier architecture is enhanced to multitier architecture by using D2D and cognitive radio network (CRN)–based communication.

To separate the indoor and outdoor users is a big challenge. Wang et al. proposed a solution to this challenge by enhancing the MBS's responsibilities [14]. They deployed a huge array of antennas, having some of its elements distributed across macro cells, and they are connected by fiber optics to MBS. For communicating with MBS, a large array of antennas and SBS are deployed in buildings. SBS or Wi-Fi connects UEs with all other UEs in a building.

Wang et al. also proposed that a mobile (small-cell) placed inside a bus wants to communicate with internal UEs and an antenna that is located outside of the bus with MBS [14]. Thus, inside the bus, all UEs appear as a signal group or a single unit with corresponding MBS. SBS appear there as an MBS for all UEs inside the bus.

There are two basic issues while deploying two-tier architecture in small cells. Interference management and Backhaul data transfer. The interference may include inter-tier and intra-tier interference. The inter-tier interference will be as a result from MBS to SBS and from MBS to SBS's UEs. In intra-tier interference, resultant interference will be from SBS's to other SBS's UEs. In backhaul data transfer, an opaque placement of small cells in a network requires large data transfer rate and cost-efficient architecture. There are different models proposed by Nisha et al. for backhaul data transfer. They are wireless point-to-mMultipoint (PTMP), wired optical fiber (WPF), and wireless point-to-point (PTP). WPF backhaul data transfer is modeled through establishing the web objects framework (WOF) link from every BSS to MBS. It is an expensive and time-consuming process. PTMP deploys a wireless PTMP-base station on MBS. MBS then communicates with SBS and transfers backhauled data to the core network. In PTP, antennas are deployed in line-of-sight and have been proved to supply data at a high rate and capacity at low cost.

20.4.4 Cognitive Radio

A CRN is based on a two-tier architecture. It considers either MBS or SBS to implement cognitive radio properties at different mediums. Xuemin et al. [31] mention types of CRN-based architecture as (1) cooperative CRNs and (2) noncooperative CRNs. Cooperative CRNs use only cognitive and licensed bands/channels due to

QoS. A licensed channel is used to serve UEs by a SBS, and the opportunistic access to the licensed channel is used to transfer backhauled data to the MBS. A noncooperative approach establishes a multi-RAT system, including radio interface and operates at temporary and licensed channels by programmable unit (PU), known as cognitive channel (CC). The source unit (SU) runs only on CC to form a CRN. It overlaps on an existing cellular wireless network. As there is connection between two networks, they can be integrated in the upper layer and must separate in the physical layer. The end-users use a cognitive and licensed channel to connect with a nearer MBS, and the communication between SBS and MBS occurs through a licensed channel.

In 5G, SBS works as an SU to monitor all activities and work on temporarily used frequency bands of small cells [31]. This band is called spectrum hole [32]. The PU provides services by micro cell to their UEs while having a minimum disturbance to macro cell functions. For the small cells, a dynamic pricing model based on theoretical framework is suggested in [28]. Due to this model, high data rates are achieved by UEs as compared to macro-cell and UEs pricing model is different as compared to macro-cell. Multitier architecture consists of different type of small cells, broadcasting, and serving D2D communication to serve users with QoS that are required for spectrum efficiency and energy efficient manners [33].

20.4.5 Cloud-Based Architecture

McKeown et al. defines cloud computing as an infrastructure providing simple, on demand, scalable, and less costly access to configurable resources that are at shared pool without worrying about resource management [34]. Mobile cellular communication and cloud's integration can provide tremendous beneficial growth to communication system. Here we briefly discuss cloud-based radio odes (C-RANs) for 5G as mentioned by [35]. Beyond C-RAN, the aim is to execute MBS functions in cloud hence; MBS's functionality is divided into two layers: control layer and data layer. Data layer contains heterogeneous physical resource and perform signal-processing task. Control layer performs resource management and baseband processing. C-RAN provides dynamic service without costly charges to network devices. Two basic components of MBS are baseband unit (BBU) and RRH. BBU implements baseband processing by utilizing baseband processor. Whereas, RRH is used for performing radio functionalities. C-RAN provides scalable and flexible architecture due to most of the BBUs placed in clouds, and RRHs are in MBSs.

Checko et al. proposed two-layered C-RAN architecture, in which two C-RAN are used based on MBSs [35]. The MBSs functionalities are divided into two groups: full centralized C-RAN and partially centralized C-RAN. In full-centralized C-RAN, BBU, and all high-level core functionalities of MBS are

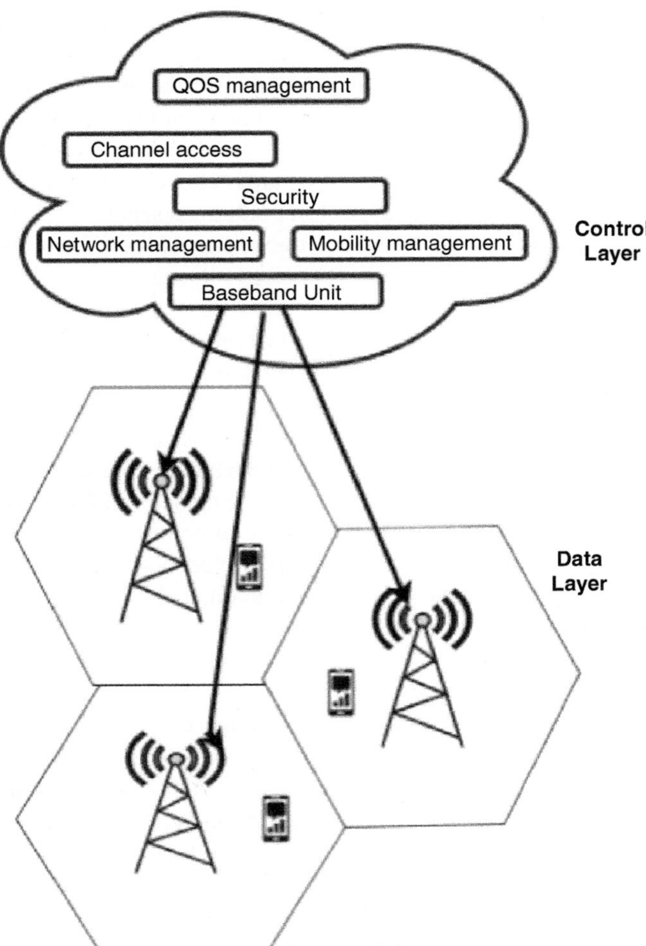

Figure 20.4 Cloud-based architecture.

placed at the cloud level. While RHH and some other core functionalities are located in MBS, the high-level functionalities are still in the cloud, as mentioned in Figure 20.4.

20.5 Technology and Methodology for 5G

Integration of current and previous Wi-Fi and other cellular standards allows 5G to provide an extraordinarily high data rate with decreased delays for end-users. 5G enhances technologies like MIMO, mmWave reference, and ultra-densification. Despite all these benefits, an official standard of 5G is still needed, and the absence of such a standard means that 5G has endless possibilities. 5G is expected to fulfill and fix all the flaws of 4G, and will be divided into several sublayers for an all-in-one IP connectivity. 5G technology will enable us to create such systems, to address safety and security problems, and to improve the capacity of the systems. 5G literature framework is providing researchers with directions to address: security, technology, network implementation, and application problems for innovative development [36].

20.5.1 HetNet

Humans use services such as social networking, VoIP, video conferencing, video streaming, and multiplayer gaming. Such forms of data communication are called human-type communication (HTC). When machines communicate with each other without human interference, then this form of data communication is called machine-type communication (MTC). When humans and machines communicate with each other, then this form of communication is termed as combined-type communication (CTC). CTC and MTC are important for IoT and cyber physical systems (CPS). 5G networks are presented in different use cases to establish the aforementioned communication mechanisms. The use cases are categorized in three ways: HTC, MTC, and CTC.

HTC Use Case

1. *Inelastic.* This use case is to satisfy user's demand for high-definition multimedia content at anytime, anywhere with a range of mobility and communication standards.
2. *Elastic.* This handles the large number of users and availability of Internet anytime, anywhere with the mobility communication standard.

MTC Use Case

1. Inelastic:
   a) *Intelligent roads*: A set of sensor and video cameras are arranged along a network of roads.
   b) *Inter-vehicular communication.* Autonomous driving, intelligent intersections, and traffic management need inter-vehicular communications.
   c) *Industrial process automation.* Sensors monitor automated industrial processes, and report their readings to a controller.
   d) *Theft control.* Sensor-based systems can be used to detect product theft in supply chain and/or in stores.

2. Elastic:
   a) *Home automation*. Different devices inside a home collaborate to accomplish a task.
   b) *Intelligent parking*. Sensors attached to parking spaces can communicate with online servers to provide information about available parking spaces in a particular area.

CTC Use Case

1. *Inelastic*. Critical control of remote devices and remote network management.
2. *Elastic*. Controlling and monitoring different home appliances from within a house or over the Internet.

5G research community still face issues, such as a scalable network architecture, QoS framework, intelligent network selection, and inter-network handover algorithms to satisfy diverse HTC, MTC, and CTC requirements. 5G supports a large amount of data initiated by HTC, MTC, and CTC [37].

20.5.2 Beam Division Multiple Access (BDMA)

5G technology uses CDMA and beam division multiple access (BDMA). BDMA allows data rate up to 100 Mbps with high mobility and at low mobility it is up to 1 Gbps. It uses an orthogonal beam, which is allocated to each mobile station. To increase the capacity of the system, an antenna beam is used that divides the band allocated to the base station. Mobile stations and the base stations are in the line of sight; both transmit the beam for the proper communication and reducing interference [38]. BDMA is shown in Figure 20.5.

20.5.3 Mixed Bandwidth Data Path

The architecture of 5G combines pico cell, micro cell, and macro cell, and can utilize bandwidth based on WLAN and CDMA2000. The WLAN network covers a small area, whereas CDMA2000 covers a large area. Moreover, in 5G, data requests can be sent from one network (MN, BS, PDSN, CN) and replies can arrive from the other network (CN, PDSN, AP, MN). The bandwidth selection is made with the exchange of messages between the networks, as depicted in Figure 20.6 [39].

20.5.4 Wireless Virtualization

5G supports network virtualization. Virtualizing mobile networks and sharing of their resources will provide more efficient utilization of wireless networks. Network virtualization can reduce the amount of base station equipment, as well as reduce energy consumption [40].

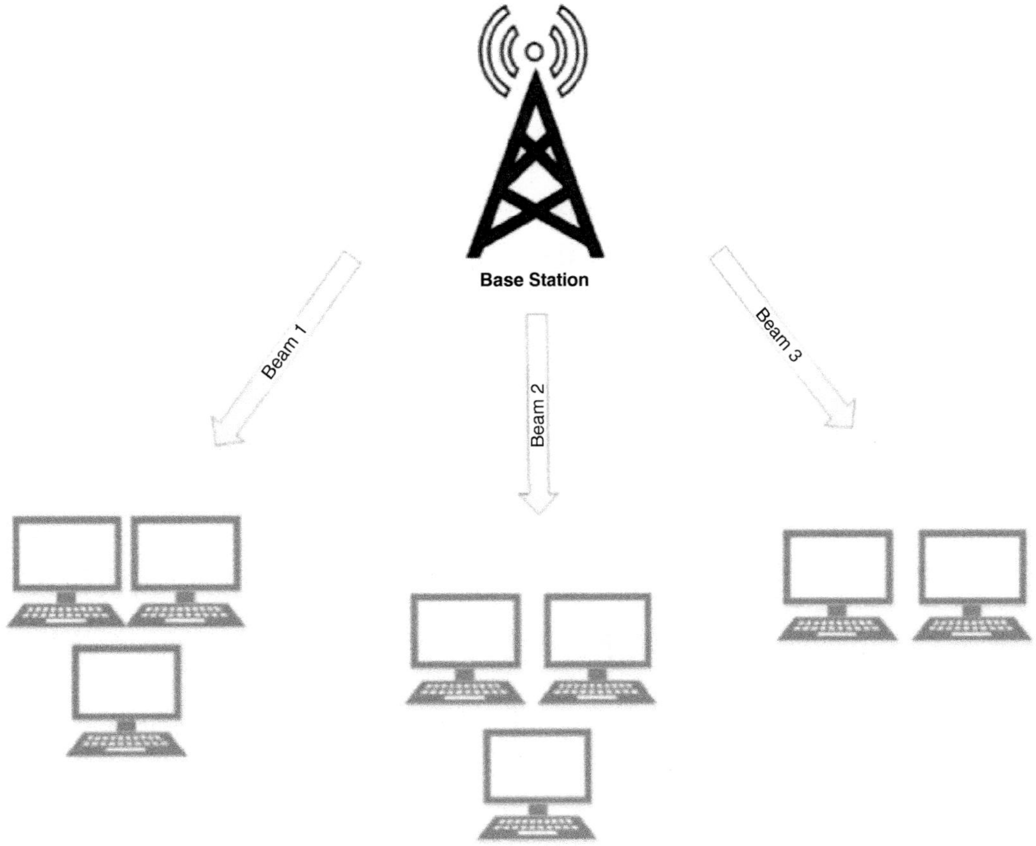

Figure 20.5 Beam division multiple access (BDMA).

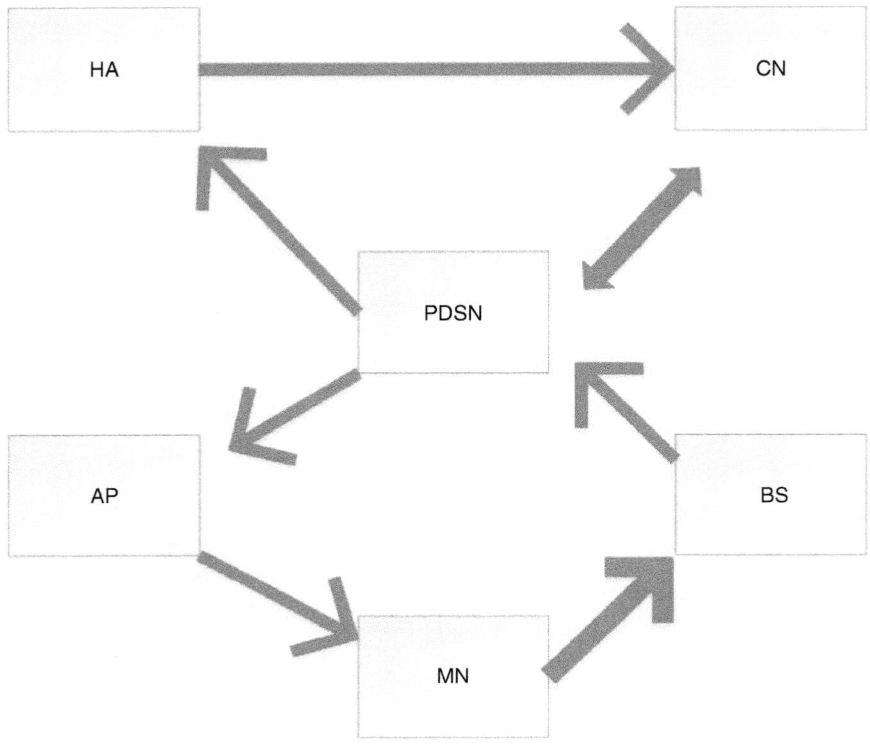

Figure 20.6 Mixed bandwidth data path.

20.5.5 Flexible Duplex

5G allows very flexible and dynamic test-driven design (TDD) transmission resource. Existing TDD-based mobile technologies include TDD-LTE for restriction on uplink/downlink configuration. The TDD requires interference like direct base-station-to-base-station and device-to-device that is similar to base-station-to-device and device-to-base-station [41].

20.5.6 Multiple-Input Multiple-Output (MIMO)

Multiple-input multiple-output (MIMO): Multiple antennas are used to multiplex messages for multiple nodes, keeping the energy radiations toward the specific directions, which decreases the mobile interference. 5G systems do not encourage the idea of keeping complete control at the infrastructure side and take advantage of intelligence at the device side among the different layers of the stack protocol. The 5G idea requires an enhancement at the device level (component change);

it also has implications at the architectural level. As wireless technology is emerging and data traffic is rapidly growing, the demand for swift and reliable base stations has increased. A new term, *massive MIMO*, describes a system that is emerging to fulfill this need. There are many benefits of this technique that are being realized by using large antenna arrays at the transmitters of receivers [2]. Massive MIMO and advanced code technology is a new concept in wireless paradigm for multi-users with multi-antenna systems [42]. It is able to provide better improvement in radiated energy efficiency and bandwidth. Advanced code refers to polar code, which are the first-ever codes that achieved the capacity of symmetric binary-input discrete memory-less channels (B-DMCs). They are regarded as the most promising code candidates for 5G, due to having advantages in performance, complexity, and flexibility. A MIMO has a number of receivers and transmitters and channels by which they communicate. Massive MIMO with deployment of multiple antennas is represented in Figure 20.7.

20.5.7 M2M

In 5G, M2M includes three basic requirements: Support for a large number of devices with low rate of transfer, sustenance of low data rate in non-real-time situations, and lesser delay in data transfer rate. Different research directions are discussed, which can lead toward basic changes in mobile networks design. Technologies like mmWave, MIMO, device-centric architectures can lead to form the basic pillars of 5G.

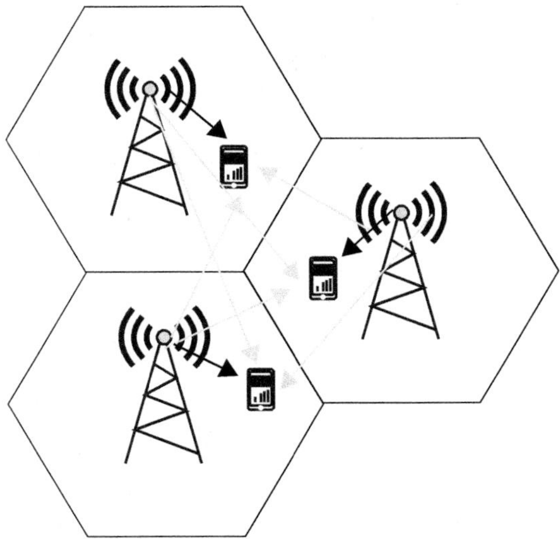

Figure 20.7 Cellular network with the deployment of massive MIMO.

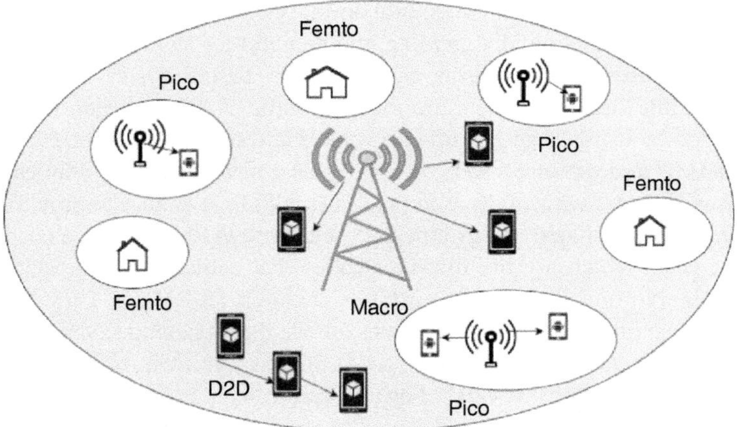

Figure 20.8 Heterogeneous network.

Yang et al. proposed the technique of heterogeneous network (HetNet), which creates a multinode topology in which many nodes of same attributes are deployed. In this perspective, the core components in HetNet are spatial modeling of nodes, associativity of cell users, and direct connection among devices. Figure 20.8 illustrates HetNet [2].

20.5.8 Multibeam-Based Communication System

MmWave has propagation limitations that are overcome by technologies like MIMO and beam-mmWave system. The basic structure for 5G is a multibeam-based communication system. Performance can be measured by the cell's capacity. Antenna array of massive MIMO and advanced beam technology is used to handle propagation limits of MMave band. 5G's basic structure for mobile network is known as a multibeam communication system. Advanced beam-forming technology has been introduced to evaluate cell capacity and disruption probability in mmWave systems with massive MIMO.

20.5.9 Software-Defined Networking (SDN)

Software defined networking (SDN) is one of the well-known techniques used by 5G. This technique partitions two functions of network: control and data-forwarding. Functions used by SDN are programmable and handle applications infrastructure, including services provided by a network. SDN is divided into three subgroups. The first is a software controller that is used to control network functions, such as a network's application programming interface (APIs), operating system, and global network view. The second group is southbound, and is dedicated to interface and protocols in the system's network. Third is the northbound group, which is responsible to provide an interface to applications of SDN and its controller.

20.6 Applications

With minimal latency and high data transfer rate, and all other salient features, 5G is supposed to serve a wide range of applications, including health care, personal usage, the tactile internet, automation, industrial usage, smart cities, virtualized homes, smart grids, etc. Some of the possible applications for 5G are discussed below.

20.6.1 Health Care

One of the most important and beneficial uses of 5G is in health care, by providing reliable and secure mobile communication for frequent data transmission from patient to cloud or to and from a remote health facility where the database for a specific patient has been maintained. A patient can get urgent and reliable medical service in the shortest possible span of time rather than by following the traditional procedures.

20.6.2 Smart Grid

5G can be pivotal in implementing smart grid technology, which focuses on better utilization of energy consumption by decentralizing energy distribution. Technologies used in 5G networks will help in smart distribution of power, providing efficient energy utilization.

20.6.3 Logistic and Tracking

5G is useful for location-based tracking. In future mobile communication systems, the tracking is easier. One of the practical examples of this application is radio frequency identification (RFID) tags.

20.6.4 Personal Usage

For heterogeneous networks, one of the key application of 5G is personal usage. It provides maximum use of UEs and meets high demand of not only text data but high-resolution video calls, voice calls, web surfing, and multimedia content, as well as providing QoS.

20.6.5 Virtualized Home

One of the most interesting applications of 5G is called virtualized home. They can be used by advances in wireless sensor networks. It will use C-RAN architecture with low-cost UEs with services of physical and data link layers.

20.7 Challenges

5G networks must address a few mentioned challenges that are not discussed for traditional LTE networks [11]. Table 20.3 briefly illustrates these challenges, and Table 20.4 shows the current research projects to address these challenges. Another big challenge that the 5G network faces is caching, which will need to meet high multimedia content demands, due to the rapid growth of smart devices' users. Moreover, currently used wireless networks cannot cope with this usage and fail to tackle the explosively growing traffic, due to its centralized architecture. Intermediate servers or routers use content caching so that the same content can be loaded easily and swiftly without the need of retransmission from remote servers, and redundant or duplicated traffic could be eliminated. The traditional caching used by both 3G and 4G LTE networks has been shown to reduce mobile data traffic by one to two thirds. However, there are still some issues including

Table 20.3 Challenges of 5G.

Challenges	Enable to address challenges	5G design principles
Capacity ×1000 >70% indoor	• Spectrum • Massive 3D MIMO • New air interface • Optical network	UHF and other spectrum options, e.g. pooling.
Data rate ×10–100	• Spectrum • Massive 3D MIMO • New air interface • All optical • Caching prefetching technique • Small cell • Local off-load	Optical transmission and switching wireless possible.
Massive number of connections ×10–100	• Spectrum • Local off load • New air interface • New SDN cloud • Energy efficient hardware • Energy management techniques	Minimize number of network beam and pool resource as much as possible.

Table 20.4 Research projects on 5G.

Research project/ institutions Research groups	Research area	URL
5G NOW (5th Generation Non Orthogonal Waveforms for asynchronous signaling)	Nonorthogonal waveform	http://www.5gnow.eu
5G PPP (5G Infrastructure Public Private Partnership)	Next generation of communication networks, ubiquitous super-fast connectivity	http://5g-ppp.eu/
COMBO (Convergence of fixed and Mobile Broadband access/aggregation networks)	Fixed/mobile converged (FMC) broadband access/aggregation networks	http://www.ict-combo.eu
(Interworking and Joint Design of an Open Access and Backhaul Network)	RAN-as-a-Service, radio access based upon small cells, and a heterogeneous backhaul	http://www.ict-ijoin.eu
MAMMOET (Massive MIMO for Efficient Transmission)	Massive MIMO	http://www.mammoet-project.eu
METIS (Mobile and wireless Communications Enablers for Twenty-twenty)	Provide a holistic framework 5G system concept	https://www.metis2020.com
MCN (Mobile Cloud Networking)	Mobile network, decentralized computing, smart storage	http://www.mobile-cloud-networking.eu/site
MOTO (Mobile Opportunistic Traffic Offloading)	Traffic off-loading architecture	http://www.ict-ras.eu/index.php/ras-projects/moto
PHYLAWS (Physical Layer Wireless Security)	Security approaches for handsets and communications nodes	http://www.phylaws-ict.org
TROPIC (Traffic Optimization by the Integration of Information and Control) 5GrEEn	Femto cell networking and cloud computing environmentally friendly 5G mobile network	http://www.ict-tropic.eu
University of Edinburgh	Indoor wireless communications capacity	www.ed.ac.uk
University of Surrey 5G Innovation Centre (5GIC)	Lowering network costs, anticipating user data needs to preallocate resources, dense small cells, device-to-device communication	www.surrey.ac.uk/5gic

where to cache, e.g. at evolved packet core (EPC) or radio access network (RAN); and what to cache and how to cache, e.g. least recently used (LRU), LFU, or FIFO. EPC and RAN caching can minimize the redundant traffic to servers. Also, caching on the network edge balances the traffic for longer periods. By studying the traffic reduction and cost estimation, it is concluded that EPC and RAN caching techniques can enhance current 3G and 4G wireless networks and could be useful in 5G [14]. 5G systems will have different technologies and standards in one device and platform and will properly combine the CDMA-OFDMA together that will enhance the quality and performance. This new reuse division system of HMA will increase the capacity of the nodes [43, 44].

20.8 Conclusion

This article has discussed and observed a complete and detailed survey of 5G's future. The discussion involved basics of existing technologies, their comparison and evolution. 5G wireless requirements are defined as capacity, data flooding, and energy efficiency, low latency, high frequency, delay tolerant, and QoS in future cellular networks. Further, we discussed different works on wireless network architecture with amendments required for 5G that are energy efficient, cloud-based, and data-centric, and cognitive RAN-based architecture. A complete survey of 5G architectures is defined, along with all growing technologies and their architectures such as BDMA, mmWAVE, HetNet, flexible duplex, and more. The article concludes with the main issues faced by 5G and research areas.

References

1 Zheng, M., Quan, Z.Z., Guo, D.Z. et al. (2015). *Key Techniques for 5G Wireless Communications: Network Architecture, Physical Layer, and MAC Layer Perspectives*. Berlin, Heidelberg: Science China Press/Springer-Verlag.

2 Yang, N., Wang, L., Geraci, G. et al. (2015). Safeguarding 5G wireless communication networks using physical layer security. *IEEE Communications Magazine* 53 (4): 20–27.

3 Kiran, P., Jibukumar, G.M., and Permkumar, V.C. (2016). Resource allocation optimization in LTE-A/5G networks using big data analytics. In: *Proceedings of the 2016 International Conference on Information Networking (ICOIN)*, 254–259. Washington, DC: IEEE.

4 Lin, Z., Du, X., Chen, H. et al. (2019). Millimeter-wave propagation modeling and measurements for 5G mobile networks. *IEEE Wireless Communications* 26 (1): 72–77.

5 Yaacoub, E., Husseini, M., and Ghaziri, H. (2016). An overview of research topics and challenges for 5G massive MIMO antennas. In: *2016 IEEE Middle East Conference on Antennas and Propagation (MECAP), Beirut, Lebanon.*

6 Addad, R., Bagaa, M., Taleb, T. et al. (2019). Optimization model for cross-domain network slices in 5G networks. *IEEE Transactions on Mobile Computing* https://doi.org/10.1109/TMC.2019.2905599.

7 Elayoub, S.E., Brown, P., Deghel, M., and Galindo-Serrano, A. (2019). Radio resource allocation and retransmission schemes for URLLC over 5G networks. *IEEE Journal on Selected Areas in Communications* 37 (4): 896–904.

8 Yao, M., Sohul, M., Marojevic, V., and Reed, J.H. (2019). Artificial intelligence defined 5G radio access networks. *IEEE Communications Magazine* 57 (3): 14–20.

9 Siddiqui, M.M. (2011). Vision of 5G communication. In: *Communications in Computer and Information Science, Springer Nature*, 252–256. Springer.

10 Gupta, A. and Jha, R.K. (2015). A survey of 5G network: architecture and emerging technologies. *IEEE Access* 3: 1206–1232.

11 Agyapong, P.K., Iwamura, M., Staehle, D. et al. (2014). Design considerations for a 5G network architecture. *IEEE Communications Magazine* 52 (11): 65–75.

12 Modi, H. and Patel, S.K. (2013). 5G technology of mobile communication. *International Journal of Electronics and Computer Science Engineering* 18 (3): 1265–1275.

13 Qazi, F., Khalid, O., Rais, R.N.B. et al. (2019). Optimal content caching in content-centric networks. *Wireless Communications and Mobile Computing* 2019, Art. No: 6373960: 15. doi: 10.1155/2019/6373960.

14 Wang, X., Chen, M., Taleb, T. et al. (2014). Cache in the air: exploiting content caching and delivery techniques for 5G systems. *IEEE Communications Magazine* 52 (2): 131–139.

15 Sapakal, R.S. and Kadam, S.S. (2013). 5G Mobile technology, International Journal of Advanced Research in Computer. *Engineering & Technology* 2 (2): 568–571.

16 Wang, Y., Wang, K., Huang, H. et al. (2019). Traffic and computation co-offloading with reinforcement learning in fog computing for industrial applications. *IEEE Transactions on Industrial Informatics* 15 (2): 976–986.

17 Ning, Z., Huang, J., and Wang, X. (2019). Vehicular fog computing: enabling real-time traffic management for smart cities. *IEEE Wireless Communications* 26 (1): 87–93.

18 Nobre, J.C., Souza, A.M., Rosárioc, D. et al. (2019). Vehicular software-defined networking and fog computing: integration and design principles. *Ad Hoc Networks* 82: 172–181.

19 Mutlag, A.A., Ghania, M.K.A., Arunkumar, N. et al. (2019). Enabling technologies for fog computing in healthcare IoT systems. *Future Generation Computer Systems* 90: 62–78.

20 Yao, X., Kong, H., Liu, H. et al. (2019). An attribute credential based public key scheme for fog computing in digital manufacturing. *IEEE Transactions on Industrial Informatics* 15 (4): 2297–2307.

21 Bilal, K., Khalid, O., Erbad, E., and Khan, S.U. (2018). Potentials, trends, and prospects in edge technologies: fog, cloudlets, mobile edge, and micro data centers. *Computer Networks* 130: 94–120.

22 Qayyum, T., Malik, A.W., Khattak, M.A.K. et al. (2018). FogNetSim++: a toolkit for modeling and simulation of distributed fog environment. *IEEE Access* 6: 63570–63583. https://doi.org/10.1109/ACCESS.2018.2877696.

23 Ghosh, A., Khalid, O., Rais, R.N.B. et al. (2019). Data offloading in IoT environments: modeling, analysis, and verification. *EURASIP Journal on Wireless Communications and Networking* 2019 Art. No: 53 https://doi.org/10.1186/s13638-019-1358-8.

24 Barzegaran, M., Cervin, A., and Pop, P. (2019). Toward quality-of-control-aware scheduling of industrial applications on fog computing platforms. *Proceedings of the Workshop on Fog Computing and the IoT, Montreal, Quebec, Canada, April* 15.

25 Andrews, J.G., Buzzi, S., Choi, W. et al. (2014). What will 5G be? *IEEE Journal on Selected Areas in Communications* 32 (6): 1065–1082.

26 J. Sheetal, Architecture of 5G technology in mobile communication, Proceedings of the 18th IRF International Conference, Pune, India, 2015.

27 Tehrani, M.N., Uysal, M., and Yanikomeroglu, H. (2014). Device-to-device communication in 5G cellular networks: challenges, solutions, and future directions. *IEEE Communications Magazine* 52 (5): 86–92.

28 Rowell, C.L.I.C., Han, S., Xu, Z. et al. (2014). Toward green and soft: a 5G perspective. *IEEE Communications Magazine* 52 (2): 66–73.

29 Zhang, X., Zhang, J., Wang, W. et al. (2015). Macro-assisted data-only carrier for 5G green cellular systems. *IEEE Communications Magazine* 53 (5): 223–231.

30 Hu, R.Q. and Qian, Y. (2014). An energy-efficient and spectrum efficient wireless heterogeneous network framework for 5G systems. *IEEE Communications Magazine* 52 (5): 94–101.

31 Hong, X., Wang, J., Wang, C., and Shi, J. (2014). Cognitive radio in 5G: a perspective on energy-spectral efficiency trade-off. *IEEE Communications Magazine* 52: 46–53.

32 Akyildiz, F.I., Lee, W.Y., and Chowdhury, K.R. (2009). CRAHNs: cognitive radio ad hoc networks. *Ad Hoc Networks* 7 (5): 810–836.

33 Hossain, E., Rasti, M., Tabassum, H., and Abdelnasser, A. (2014). Evolution toward 5G multitier cellular wireless networks: an interference management perspective. *IEEE Wireless Communications* 21 (3): 118–127.

34 McKeown, N., Anderson, T., Balakrishnan, H. et al. (2008). OpenFlow: 1197 enabling innovation in campus networks. *IEEE Computer Communication Review* 38 (2): 69–74.

35 Checko, A., Christiansen, H.L., Yan, Y. et al. (2015). Cloud RAN for mobile networks – A1099 technology overview. *IEEE Communication Surveys and Tutorials* 17 (1): 405–426.

36 Felita, C. and Suryanegara, M. (2013). 5G key technologies: identifying innovation. In: *2013 International Conference on Quality in Research, QiR 2013*, 235–238. IEEE.

37 M. O. Farooq, C. J. Sreenan, K.N. Brown, Research challenges in 5G networks: a HetNets perspective, Proceedings of the 19th International Conference on Innovations in Clouds, Internet, and Networks (ICIN 2016), 2016.

38 Goyal, P. and Buttar, A.S. (2015). A study on 5G evolution and revolution. *International Journal of Computer Networks and Applications* 2: 2.

39 Aryaputra, A. and Bhuvaneshwari, N. (2011). *5G – The future of Mobile Network, Proceedings of the World Congress on Engineering and Computer Science 2011*, 843–847. San Francisco, CA.

40 Le, L.B., Lau, V., Jorswieck, E., Dao, N., Haghighat, A., Kim, D.I., Le-Ngoc, T., "Enabling 5G mobile wireless technologies," EURASIP Journal on Wireless Communications and Networking, no. 218, (2015). https://doi.org/10.1186/s13638-015-0452-9.

41 E. White, 5G radio access [online]: http://www.5gsummit.org/seattle/docs/slides/Kumar-5GSummit-Radio-Access-revA3.pdf, 2019 (accessed December 2019).

42 Liu, L. and Zhang, C. (2015). Circuits and systems for 5G network: massive MIMO and advanced coding. In: *IEEE 11th International Conference on ASIC (ASICOON)*, 34–55.

43 Wang, L. and Rangapillai, S. (2012). A survey on green 5G cellular networks. In: *International Conference on Signal Processing and Communications (SPCOM)*, 232–245. IEEE.

44 Ge, X., Cheng, H., Guizani, M., and Han, T. (2014). 5G wireless backhaul networks: challenges and research advances. *IEEE Network* 28 (6): 6–11.

21

Fog Computing for Bioinformatics Applications

Hafeez Ur Rehman, Asad Khan, and Usman Habib

Department of Computer Science, National University of Computer and Emerging Sciences, Peshawar, Pakistan

21.1 Introduction

Bioinformatics is an interdisciplinary field having a myriad of applications in the area of health and life sciences. Bioinformatics algorithms generally have high computational and storage requirements [1, 2]. Currently, the most apt preference for bioinformatics researchers is to use cloud computing–based services to meet their high computational requirements within a limited budget. However, the scientists need to port their data and algorithms to the cloud environment in order to perform analysis. In recent days, due to the emerging trend of "on-the-fly solutions" for bioinformatics problems, the computational resources need to be brought closer to the users [3–6]. Fog computing, as an extension of cloud computing, brings the services to the edge of the network. This essentially brings the advantages and power of the cloud closer to the place where the data is actually generated, benefiting the resource demanding bioinformatics algorithms.

In the field of bioinformatics, the high-throughput technologies such as next-generation sequencing result into a diverse deluge of sequencing data [7, 8]. These raw sequences are called omics sequences, which include DNA sequences, RNA sequences, and protein sequences, among others. The study of different types of omics sequences produced different areas of research; for example, the study of gene sequences established the field of genomics, likewise the study of protein sequences resulted in the field proteomics, and so on [9–11]. In each subfield, the scientists try to explore the effect of omics sequences in understanding biological organism as a completely engineered system. In addition to sequencing data, one of the promises of bioinformatics research, in general, is to develop personalized medicine. The area that focuses on personalized medicine

Fog Computing: Theory and Practice, First Edition.
Edited by Assad Abbas, Samee U. Khan, and Albert Y. Zomaya.

is called pharmacogenomics, the study of genetic variations in an organism due to the application of a certain drug. An emergent area in pharmacogenomics is single nucleotide polymorphism (SNP), the study of variations in nucleotides at a specific position in a genome. SNPs in pharmacogenomics are used to select and optimize drugs based on genetic profiles of different patients.

Advancement in high-throughput technologies has made it easier to obtain the genetic profiles of patients [2, 12]. Next-generation sequencing technologies have reduced the cost of sequencing genomes at record lows [8]. Nowadays such sequencing tools are extensively used in pharmacogenomics and genomics studies alike. These technologies, when applied to huge populations in carrying out clinical experiments, result in a deluge of data. This abundance of data, along with opportunities, present some challenges for researchers. First, the storage of such data due to a real-time increase requires a lot of space, and provision of space for increasing data is becoming cumbersome [1, 2]. Second, the analysis of omics data also requires specialized bioinformatics software and tools that need extensive computational resources, and provision of those resources is a challenging task.

A primitive solution to these problems emerged in the form of bioinformatics tools, often delivered in the form of web services, to help manage and analyze data stored in biological databases in distant geographical locations [4, 13]. This solution had its disadvantages; for example, the storage of data in distributed databases made the process slower. To overcome this problem, cloud computing came to the rescue [14–17].

Cloud computing is a service-oriented computing model specifically designed for problems requiring large-scale data and high computational resources. Due to its appropriateness for across-the-board resource availability for resource-hungry algorithms, the cloud-computing model has gained fame in the research community and the concept has spread swiftly in recent years. The main aim of cloud computing is to supply hardware and software resources to users through network links.

The resources that are provided by cloud computing may vary from cloud to cloud. These resources may include memory, CPU, storage, specific applications, operating systems, etc. The cloud resources have dynamic scalability, virtualization, and accessibility rendered over the Internet [18]. The model offers new facilities for users that require scalable and massive storage, computing resources, applications, virtual technologies, etc. on demand [19]. Therefore, cloud computing can perform a significant role in various stages of a bioinformatics analysis pipeline, like data storage, preprocessing, sharing, integration, and exploration as well as visualization.

Regardless of numerous merits of cloud computing, it is facing several issues related to security, technology, management, and ethics. The security of data (especially privacy), legal responsibilities, and geographical localization of data are open challenging problems for cloud-computing platforms. For example, in the context of bioinformatics, what will be legal responsibilities in data leakages of sensitive information related to patients in the time of data uploading and processing in the cloud? In addition to these problems, researchers need to port their data and algorithms to the cloud environment in order to perform any analysis. In recent days, due to the emerging trend of on-the-fly solutions for bioinformatics problems, computational resources need to be brought closer to users [5, 6]. Fog computing, as an extension of cloud computing, brings the services to the edge of the network. This essentially brings the advantages and power of the cloud closer to the place where the data is produced, thus, helping and speeding up "on-the-fly solutions" for bioinformatics applications.

In this chapter, we first discuss the suitability of cloud computing for bioinformatics problems, alongside highlighting its limitations. Furthermore, we discuss the appropriateness of fog computing as an extension of clouding computing to solve bioinformatics problems. The overall chapter is organized as follows: In Section 21.2, we give an overview of cloud computing along with the overall cloud service model. Section 21.3 presents a complete review of existing use of cloud computing for bioinformatics application development. It also highlights the key bioinformatics projects that utilize cloud computing to make personalized, preventive as well as precision medicines. In Section 21.4, we present the concept of fog computing with an emphasis on its key properties (such as low jitter, low latency, improved security, etc.) that distinguish it from the cloud paradigm. In Section 21.5, we discuss the suitability of the fog computing paradigm and its potential for data and resource extensive bioinformatics applications, while elaborating it with a real time microorganism detection example. In the last section, the chapter is concluded.

21.2 Cloud Computing

Cloud computing is a computing paradigm in which different computers are configured to provide on-demand services at a higher level, thus freeing the user from underlying hardware, storage, and other issues. The authors have defined the term "cloud computing" with around 22 excerpts, by evaluating the common properties of cloud computing [20]. The authors have emphasized on the significance of service level agreements (SLAs) in order to make the cloud environment more reliable. In addition, the virtualization property of the cloud is the key enabler of service provisioning. Generally, the cloud computing paradigm has two fundamental models, i.e. the service model and the delivery model. Each of the models, along with types, is discussed in the proceeding sections.

21.2.1 Service Models

The cloud service models describe the level of services in which customers interact with the cloud. There are three type of service models.

1. *Infrastructure as a Service, abbreviated as IaaS*. As its name suggests, this model provides an infrastructure to customers as a service that has high computational power and a large storage space. This is lower model; i.e. the customer/organization will deploy its platform on its infrastructure and also its application software's according to the needs of the customer/organization. These models support virtual machine server facilities that can be managed and configured according to need. These types of service models can be acquired by large-scale organizations for achieving required goals. The common example is Amazon's Elastic Compute Cloud (E2C), which has virtual machine facility that users can manage according to his/her requirements. Another example is Amazon's Simple Storage Service (S3) for storing and retrieving data through a web interface.
2. *Platform as a Service, abbreviated as PaaS*. In this service model, users have no control on infrastructure, which means that the cloud service provider installs the platform in advance and users can configure it according to their need. The platform provides users the facility of creating, testing, and installing an application according to requirements. The user program can be written or created using already available libraries, programming languages, and tolls in the platform. The Google apps engine is a common example of a Pass model, in which users can create programs of Java and Python using the available software development kit (SDK) of both languages.
3. *Software as a Service, abbreviated as SaaS*. As its name suggests, users have no control on infrastructure and platform but use already available software in the cloud. The customers can use these services through a web interface. In some scenarios,customers can manage and configure the application program according to their need. Dropbox is a common example of the SaaS model.

21.2.2 Delivery Models

Cloud services can be offered to end-users in using one of the three delivery models, i.e. public, private, and hybrid clouds. The following is a short introduction of different types of delivery models:

Public Cloud. This type of cloud is publicly available. The customer can use the hardware and software resources of a data center. These resources can be acquired from a vender organization through credit cards. Amazon, Azure, and Google apps are common examples of public clouds.

Private Cloud. These types of cloud can be used only by employees of a specific organization. The software and hardware resources can be configured by users and the cloud administrator according to the requirements of specific users. Several open sources and commercial software available on the Internet to establish these types of clouds. OpenStake, Open Nebula, and VMware Cloud are common examples of software used to create private cloud.

Hybrid Cloud. This type of cloud is a combination of two clouds of different delivery models that are connected through specific technology in order to handle data and application portability.

21.3 Cloud Computing Applications in Bioinformatics

Bioinformatics research normally involves large datasets that are usually downloaded from publicly available repositories and then performing experiments using an available in-house infrastructure. The main problem faced by bioinformatics researchers is the lack of sufficient computing resources to perform experiments in a limited time. A dedicated high computing system for all researchers is not feasible. As an alternative, researchers tend to utilize the services offered by cloud computing platforms. Cloud computing has scalable computational power that can be ordered on demand to run experiments within a certain time and in a cost-effective way. Researchers can choose different cloud models according to their requirements for storing and processing data. There is no need for cloud users to install an operating system or application software; they can log in to the cloud with their account and use the available platforms and applications [21].

The three service models of cloud computing are: (1) Software as a Service (SaaS), (2) Platform as a Service (PaaS), and (3) Infrastructure as a Service (IaaS). In the following subsection, we discuss each model, along with the tools that are built on the basis of these models.

21.3.1 Bioinformatics Tools Deployed as SaaS

More recently, researchers have developed different cloud-based tools to properly execute bioinformatics tasks [1]. These tools can solve problems related to sequence alignment, mapping applications, and gene expressions [2]. Some examples related to SaaS bioinformatics tools are presented in the following paragraphs, highlighting the pros and cons of each.

In ref. [3], the authors developed the STORMSeq (Scalable Tools for Open-Source Read Mapping) tool based on cloud computing. The tool has a graphical user interface (GUI) for read mapping, read cleaning, and variant calling and annotation to process personal genomic data. STORMSeq charges $2 for processing a full exome sequence in 510 hours and $30 to process a whole genome in 38 days. The services have open access and open source resources and GUI and can be accessed using Amazon's EC2 cloud.

CloudBurst [4] and CloudAligner [13] are cloud-based tools designed for mapping next-generation sequences of the human genome and other species. Both of them use the MapReduce platform of Hadoop for parallelizing execution on different nodes. CloudAligner is preferred over cloudBurt because it can process long sequences. Crossbow [22] is another tool for alignment and single nucleotide polymorphism (SNP) detection; it can detect whole human genome SNPs and align them in a day using short read aligner.

The Variant Annotation Tool (VAT) [14] was developed for transcription-level annotation of variants from multiple personal genomes and use it to identify summarized sets of traits of individual populations. VAT also provides visualization of details from many genetic aspects, including comparative analysis, gene and allele gene occurrence and frequencies, which are based on the data collected from a wide range of individuals. These visualizations can further be used to draw additional conclusions, for example by constructing phylogenetic trees. The tool provides both web-based and command-line interfaces for researchers to use.

For RNA-Seq analysis, the FX [15] tool has been developed; it uses the cloud computing infrastructure for gene-level expression estimation and variant calling. The FX tool uses a web-based interface to take advantage of the cloud computing infrastructure. Myrna [16] is another elegant tool that calculates gene expression from large datasets. The tool uses short read alignment and combines it with interval calculation, normalization and aggregation, and statistical modeling to speed up execution time. Once aligned, the calculation step for genes exon coverage with deferential expression is done using both parametric and nonparametric permutation tests. Both Amazon's Elastic MapReduce and Hadoop can be used for performing these operations.

PeakRanger [17] is software that has been developed for the chromatin immunoprecipitation sequencing (ChIP-seq) technique. This technique is related to the next-generation sequencing (NGS) platform and has the capability of studying the interactions between proteins and DNA. PeakRanger is a peak caller package that can be executed in a parallel cloud computing environment to get tremendously high performance on huge datasets.

Cloud4SNP [23] is a novel tool based on cloud computing infrastructure. The tool can preprocess pharmacogenomics SNP microarray data and can also perform the statistical analysis. Different statistical tests are done using the partition of data sets on cloud virtual servers. Additionally, different types of statistical corrections such as false discovery rate, bonferroni correction, etc. can be done on the cloud in parallel, permitting the user to select different statistical models according to their requirements.

21.3.2 Bioinformatics Platforms Deployed as PaaS

Nowadays the Galaxy cloud is one of the most used PaaS-based platforms for bioinformatics applications; it uses the Galaxy cloud-based platform to analyze large-scale data. It has the facility for every user to run a private Galaxy installation on the cloud, without sharing resources to other users. Using the Galaxy cloud, users have full control of customizing software resources, deployment as well as launching its instances. Galaxy Cloud has another advantage over other platforms, which is portability – data as well as results can be moved from one cloud to another. This portability makes the Galaxy Cloud an ideal choice. Another publicly available Galaxy Cloud platform is CloudMan [24]. It is part of the Amazon Web Services cloud, yet it is compatible with Eucalyptus and other clouds [25]. CloudMan provides the facility of deployment and customization, as well as sharing the entire environment of analysis framework like data, tool, and configuration.

In [26] the researchers developed Eoulsan, a scalable modular framework based on the Hadoop cloud infrastructure and the MapReduce algorithm for the analysis of high-throughput sequences. The Eoulsan framework is rapidly scalable and provides facilities to generate clusters that further help in the analysis of several samples at once by using various software solutions available. The Eoulsan framework is implemented in Java and is supported on Linux systems. The framework is distributed under the LGPL license.

21.3.3 Bioinformatics Tools Deployed as IaaS

Cloud computing also provides the infrastructure to be used as a service. Bionimbus was introduced as an IaaS platform to handle the processing of genomic and phenotypic data [27]. It uses OpenStack for virtual machines. OpenStack manages a virtual machine on an on-demand basis to acquire computational resources. Bionimbus uses the famous ClusterFS file system in its clusters, which makes it the ideal candidate for applications with a lots of read/write operations. The other key features are that it contains Tukey, which works like a portal and is associated with middleware. It also contains Yates for automatic installation as well as for configuration and maintenance required for software infrastructure. Table 21.1 shows the summary of the bioinformatics applications utilizing different service models, including IaaS.

Cloud Virtual Resource, named CloVR [28], is a novel desktop-based software application that utilizes cloud computing resources to handle the analysis of

Table 21.1 A summary of existing cloud computing based major bioinformatics applications.

Cloud computing–based bioinformatics applications			
Project name	**Service model**	**Task**	**Reference**
STORMSeq	SaaS	Genome sequencing	[3]
CloudBurst	SaaS	Short read aligner for whole genome	[4]
CloudAligner	SaaS	Short read aligner for whole genome	[13]
Crossbow	SaaS	Genome sequencing	[22]
VAT	SaaS	Genome variant annotation tool	[14]
FX	SaaS	RNA sequencing tool	[15]
Myrna	SaaS	RNA sequencing tool	[16]
PeakRanger	SaaS	Chip sequencing tool	[17]
Cloud4SNP	SaaS	SNP analysis tool	[23]
CloudMan	PaaS	Galaxy-based framework for bioinformatics	[25]
Eoulsan	PaaS	High-throughput sequencing analysis framework	[26]
Bionimbus	IaaS	A infrastructure base on cloud for genome database sharing, analyzing, and management	[27]
CloVR	IaaS	Automatic and portable virtual machine for microbial sequences	[28]
CloudBioLinux	IaaS	Provide resources for genomes analysis	[29]

genomic data automatically. CloVR has portable VM machines for metagenomics, microbial genomics, and the whole genome. The CloVR virtual machine can be run on PCs. It utilizes local resources and has minimal installation requirements, although it also CloVR has support for using remote cloud computing resources for large-scale sequence processing, which improves the performance.

Another IaaS example is Cloud BioLinux [29], which is a virtual machine that is publicly available for the worldwide research community, aimed at expediting bioinformatics research. The BioLinux platform has both command line and graphical user interfaces for a variety of preconfigured software applications, along with documentation. There are more than 100 state-of-the-art bioinformatics packages to perform sequence operations, including sequence alignment, clustering of sequences, sequence assembly, sequence feature visualization, customization, and phylogenetic analysis. In addition, BioLinux is integrated with Amazons EC2 cloud, through the Eucalyptus cloud platform. Users have the convenience of accessing the desired bioinformatics tools as well as scalable cloud resources by using the EC2 cloud on a local machine.

The difficulties encountered by bioinformatics researchers in order to carry out their research in a cost-effective and fast manner is resolved, to some extent, with the help of the cloud computing services described earlier. However, there are growing privacy concerns, longer delays, and jitter, and on top of that researchers need to port their data and algorithms to the cloud environment in order to perform analysis and retrieve results. Recently, the emerging trend of on-the-fly solutions for bioinformatics problems, as well as the demand for more and more real-time applications, requires reducing the remoteness between computing platforms and users' data by bringing the computational resources closer to the users. Fog computing, as an extension of cloud computing, brings the services to the edge of the network. In the next section we give a thorough introduction of fog computing and highlight its appropriateness for the bioinformatics community.

21.4 Fog Computing

Presently, the Internet of Things architecture has the ability to connect physical devices, i.e. things, to analytics and machine learning (ML) applications. These applications take decisions without human intervention from the data generated by these devices [30]. The fundamental parameters on which the Internet of Things (IoT) performance is measured is fast data processing, scalability in analytics, and quick response.

Currently, centralized cloud-based architectures are facing problems with meeting these requirements. Thus, constructing efficient and logical spots in the middle of the data source and the cloud has been proposed. This paradigm is known as fog computing [5, 6]. The main aim of this decentralized model is to bring devices and software to the edge of the network where the data is being produced. The key purpose of fog computing is to reduce the volume of data that is transferred for processing and analysis to cloud data centers. It is also improving security, a main concern in the IoT industry [5].

The fog layer is like a junction point where there are sufficient amounts of networking, computing, and storage resources to handle the local ingestion of data, which can be acquired quickly and produce quick results. In most circumstances, low-power system-on-chip (SoC) devices are used, for the reason that they are designed to maintain the trade-off between computing performance and power consumption. On the other side, the cloud servers have the horsepower to perform sophisticated analytics and machine learning jobs to integrate time series

Table 21.2 Comparison between the cloud computing and fog computing paradigms based on desirable properties.

Property	Cloud computing	Fog computing
Latency	High latency	Low latency
Security	Usually undefined or difficult to define	Can be defined as having control of edge device
Delay jitter	Multiple systems involved, so high delay jitter	It has very low delay jitter
Location of service	Within the Internet	At the edge of the local network
Distance between client and server	Usually multiple hops	One hop
Attack on data in route	High probability	Low probability
Location awareness	No	Yes
Geo-distribution	Centralized	Distributed
Number of server nodes	Few	Large number
Mobility support	Limited	Supported
Type of last connectivity	Leased line	Wireless

formed by a number of heterogeneous or mixed types of things. Table 21.2 shows the benefits of fog computing over cloud computing.

A reference architecture for fog computing can be seen in Figure 21.1. The concept for the proposed architecture has been discussed in [6]. Generally, fog systems use the two kinds of programming models, i.e. sense-process-actuate and stream-processing. After sensing some real-life phenomena such as temperature, heart rate etc., sensors send the stream data to IoT networks. Moreover, the applications running on fog devices subscribe to and process the received data. The results obtained gives an insight that can be further translated into actions and sent to actuators. The proposed architecture has a layer where fog systems can dynamically discover and use application programming interfaces (APIs) in order to build complex functionalities.

The resource monitoring service provides information at the resource-management layer in order to track the state of available cloud, fog, and network resources, and thus can be helpful in identifying the best candidates for processing the incoming tasks. The resource-management components can prioritize the tasks by using the multitenant applications. In order to communicate between the edge and cloud resources, the machine-to-machine (M2M) standards such as Message Queuing Telemetry Transport (MQTT) and the Constrained Application

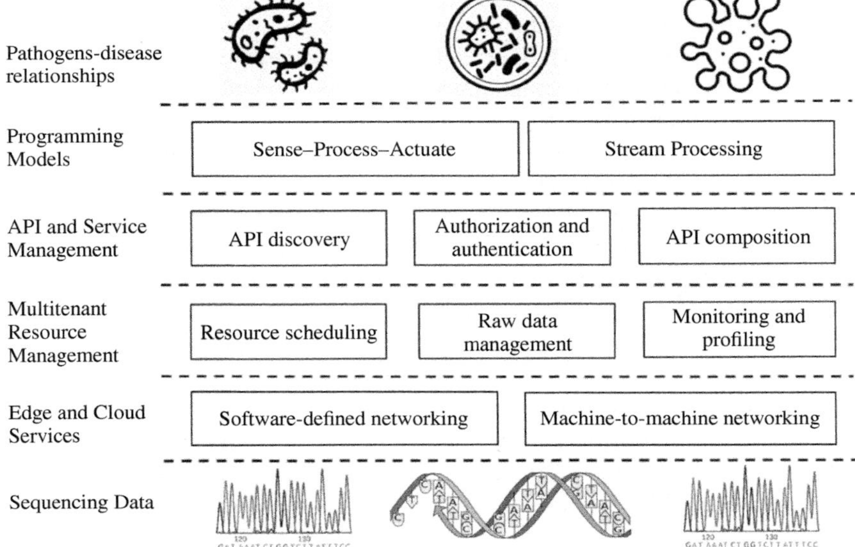

Figure 21.1 Fog computing architecture tailored for bioinformatics sequencing data.

Protocol (CoAP) are used. The efficient management of heterogeneous fog networks can be accomplished while using software-defined networking (SDN).

21.5 Fog Computing for Bioinformatics Applications

In comparison to cloud computing, fog computing provides better guarantees of quality of service (QOS) for latency-sensitive applications along with provisioning of improved security and privacy. Many computing-extensive areas, in particular bioinformatics (having high storage and computing requirements), can take advantage of this new paradigm. For example, one of the objectives of bioinformatics research (specifically the Human Genome Project) is to make personalized as well as preventive medicines [9–11]. An important problem for many algorithms aimed at making personalized and preventive medicines is to search for similar sequences (also called homolog sequences) when given a set of uncharacterized query sequences. The homology-based searches require computationally extensive algorithms (e.g. FASTA or BLAST [31]) as well as large sequence databases (take an example of the NCBI's database for proteins). To solve these problems using the fog computing paradigm, their relevant data can be stored locally; as additionally, the algorithms can be run in the vicinity of users, thus providing a significant boost to achieving the aim of personalized medicines. Once achieved, prescribing the most fitting medicines to patients, in accordance with their genetic blueprints, will be possible. Thus, personalized medicines will ultimately improve the presently unsatisfactory situations of drug safety and efficacy, which are the primary contributors of soaring medical conditions (including life losses) to society from adverse drug reactions (ADRs).

Another example of using the fog computing platform would be to enhance our understanding of complex biological organisms, with the use of computer simulations, as completely engineered systems. Biological cells have many components that work together to form a living cell, including genes, mRNAs, tRNAs, miRNAs, proteins, etc. Computer simulations are often performed to understand the behavior of such systems. However, due to the large number of components and inherent complexity that emerge from the interaction of such components, we require larger platforms with high computational capabilities to simulate and understand the working of such organisms. Fog computing, due to provisioning of a scalable and reliable computing platform close to the users, is an appropriate choice to simulate such systems. Researchers worldwide working on simulating portions of cells, e.g. cancer pathways, metabolic pathways, signaling pathways, etc., can use the service-oriented architecture of fog computing to solve their problems as well as to share their findings.

In general, the use of fog computing to solve bioinformatics problems will pave ways for improved medical care, safer pharmacotherapy, and better health, as can be seen in Figure 21.2. In the following section we describe a practical use case, i.e. a real-time microorganism detection system, for which fog computing infrastructure can potentially be utilized to speed the pace of bioinformatics research for microorganism detection.

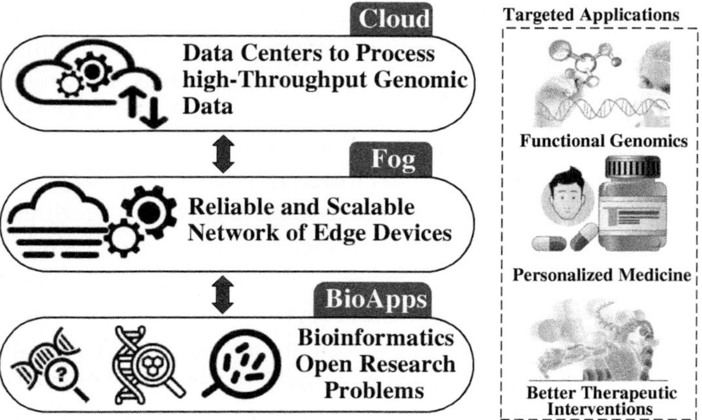

Figure 21.2 Fog computing architecture for bioinformatics applications.

21.5.1 Real-Time Microorganism Detection System

Microorganisms or microbes are microscopic organisms that surround us everywhere, i.e. in water, in soil, in air, and even in extreme conditions where normal life cannot prosper. These organisms live together and efficiently interact with each other for different purposes; these intricate interactions form complex entities called microbial communities. These tiny communities play a crucial role in keeping their eukaryotic hosts healthy as well as in the cycling of important life elements, such as carbon, phosphorus, nitrogen, etc. However, despite their biological importance, the factors that contribute to the functioning of microbial communities and their relationship with environmental changes are not very well understood. It is not yet clear how these microorganisms, along with environmental factors, perform very particular functions that have staggering complexities.

More recently, microorganism research has been sped up by advances in sequencing technologies. In particular, the genomes of microbes can be directly sequenced in real time by using the samples taken from different environments. Undeniably, with the dawn of metagenomics, microbe communities can be understood in previously unimagined ways, in reference to their genetic, taxonomical, structural, and functional relationships, both within and across microbe communities.

Nowadays, different types of bacterias in different environments can be monitored using different types of portal genome sequencing technologies; one such example is the Oxford Nanopore MinIoN device [32]. These devices can be utilized in the pharmaceutical industry to generate alarms upon finding the candidate microbial pathogens in the environment. They can also be used as bacterial monitoring devices, to filter air in different environments, such as food industries and hospitals, etc. [33, 34]. In addition, these applications can monitor microbe populations inside canals, lakes, rivers, and seas [35]. There are also applications that monitor bacteria populations inside cultivated soil and greenhouses [36, 37]. An interesting example is related to animals, where such devices can be used to detect the susceptibility to different diseases in animals using samples from different animal environments [38–40].

In almost all the applications discussed so far, the processes related to sequencing analysis of microbes are manually conducted, as these are spot analyses. With the passage of time, due to the emergence of myriad applications for microorganism detection, the detection systems needed to be utilized in different domains. As such devices will run in a local domain, where they will generate a lot of sequencing data, real-time analysis is a daunting task. The computationally extensive operations common to almost all the detection systems are mainly the following: (1) base calling, which is an operation to identify genomic sequence based on interpreting sequencer's signals, and (2) bacteria identification, which is the second step in classifying the type of bacteria based on sequence information.

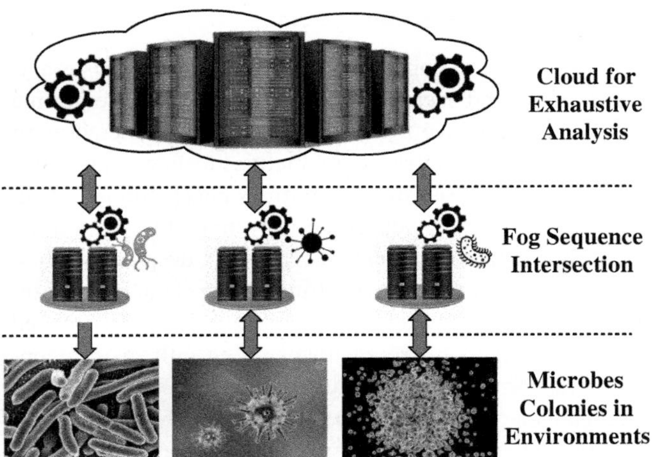

Figure 21.3 Fog computing for real time microorganism detection.

To process the microbe sequencing data in real time without delay, a computing paradigm with provisioning of storage and computing power close to the sequencing machines is the ideal choice. Fog computing, having the said properties, comes to the rescue and manages such data locally. The overall theme for the use of such a system is presented in Figure 21.3. The use of the fog paradigm will reduce data portability to the cloud's data centers, thus intrinsically reducing network delay overheads as well as privacy issues. Only significant results will be sent to the cloud for further processing.

There is another aspect of fog computing that makes it suitable for real-time microorganism detection. Oxford Nanopore MinIoN devices have an interesting feature according to a computational point of view: they send data streaming of sequences immediately without sending the data in bulk form to the other end for processing. This type of real-time streaming makes instantaneous analysis of data possible as soon as the sequencing results become available. This real-time streaming aspect of Oxford Nanopore MinIoN technology, when used alongside the benefits of fog computing, can result in quick identification of bacteria samples. As reported by some authors, the identification of bacteria and in response antimicrobial resistance is activated within 5–10 minutes [41], compared to much longer times using other platforms.

Some authors did similar work [42], in which they used the fog computing model to design a real-time system utilizing a network of MinION devices to generate sequencing data. This system showed the integration of MinION and SoC devices to produce and intelligently process the raw sequences in the fog environment. The system, although in a constrained environment, was able to process data stream (i.e. sequences) and perform the base calling as well as the identification of bacteria in real time using the fog computing paradigm. The system was able to successfully raise alarms when placed in environments with abnormal microbe populations.

Although a small amount of work has been done in the area of microorganism detection using fog computing, it remains an open research area. The metagenomics approach provides a lot of opportunities, but there are still many questions and challenges to be answered and solved by the fusion of bioinformatics tools and technologies alongside the benefits of the fog computing platform.

21.6 Conclusion

Bioinformatics is a data-rich field; many high-throughput technologies in the area of bioinformatics, such as next-generation sequencing, result in a deluge of sequencing data. Today, we have full genome datasets of different species readily available and many more are being sequenced. These genomic sequences are of the utmost importance in understanding the working of biological organisms and have myriad applications in our daily life. Processing this huge amount of data with conventional methods is a time-consuming task. In addition, if data is large, complex, and is coming from heterogeneous sources, then processing this type of data becomes more tedious and daunting. Analysis of such data might take hours or days to produce results and has caused current cloud computing paradigms to face numerous challenges (e.g. network overheads, data security including data provenance and data privacy, etc.).

To overcome the limitations of cloud computing, many researchers have proposed multiple paradigms with the unified aim of deploying the resources close to the edge of the network. Among the proposed paradigms, for scalable resource rendering, fog computing is widely used by researchers worldwide. Fog computing uses the cloud at the back end while extending the cloud-to-things continuum by bringing resources close to the edge of devices, thus overcoming many limitations of the cloud computing paradigm. Based on the desirable properties of fog computing, such as low jitter, low latency, improved security, etc., we argue that the fog computing paradigm has great potential for data- and resource-extensive bioinformatics applications.

References

1 Dai, L., Gao, X., Guo, Y. et al. (2012). Bioinformatics clouds for big data manipulation. *Biology Direct* 7 (1): 43.

2 Zhang, L., Gu, S., Liu, Y. et al. (2011). Gene set analysis in the cloud. *Bioinformatics* 28 (2): 294–295.

3 Karczewski, K.J., Fernald, G.H., Martin, A.R. et al. (2014). STORMSeq: an open-source, user-friendly pipeline for processing personal genomics data in the cloud. *PLoS One* 9 (1): e84860.

4 Schatz, M.C. (2009). CloudBurst: highly sensitive read mapping with MapReduce. *Bioinformatics* 25 (11): 1363–1369.

5 Bonomi, F., Milito, R., Zhu, J., and Addepalli, S. (2012). Fog computing and its role in the Internet of Things. In: *Proceedings of the First Edition of the MCC Workshop on Mobile Cloud Computing*, 13–16. ACM.

6 Dastjerdi, A.V. and Buyya, R. (2016). Fog computing: helping the Internet of Things realize its potential. *Computer* 49 (8): 112–116.

7 Anaparthy, N., Ho, Y.J., Martelotto, L. et al. (2019). Single-cell applications of next-generation sequencing. *Cold Spring Harbor Perspectives in Medicine*: a026898. https://doi.org/10.1101/cshperspect.a026898.

8 Romanel, A. (2019). Allele-specific expression analysis in cancer using next-generation sequencing data. In: *Cancer Bioinformatics. Methods in Molecular Biology*, vol. 1878 (ed. A. Krasnitz), 125–137. New York, NY: Humana Press.

9 Rehman, H.U., Benso, A., Di Carlo, S. et al. (2012). Combining homolog and motif similarity data with gene ontology relationships for protein function prediction. In: *2012 IEEE International Conference on Bioinformatics and Biomedicine*, 1–4. IEEE.

10 Benso, A., Di Carlo, S., Rehman, H.U. et al. (2013). Accounting for post-transcriptional regulation in Boolean Networks based regulatory models. In: *International Work-Conference on Bioinformatics and Biomedical Engineering*, 397–404. IWBBIO.

11 Benso, A., Di Carlo, S., Politano, G., and Savino, A. (2012). Using genome wide data for protein function prediction by exploiting gene ontology relationships. In: *Proceedings of 2012 IEEE International Conference on Automation, Quality and Testing, Robotics*, 497–502. IEEE.

12 Carter, M.D., Gaston, D., Huang, W.Y. et al. (2018). Genetic profiles of different subsets of Merkel cell carcinoma show links between combined and pure MCPyV-negative tumors. *Human Pathology* 71: 117–125.

13 Nguyen, T., Shi, W., and Ruden, D. (2011). CloudAligner: a fast and full-featured MapReduce based tool for sequence mapping. *BMC Research Notes* 4 (1): 171.

14 Habegger, L., Balasubramanian, S., Chen, D.Z. et al. (2012). VAT: a computational framework to functionally annotate variants in personal genomes within a cloud-computing environment. *Bioinformatics* 28 (17): 2267–2269.

15 Hong, D., Rhie, A., Park, S.S. et al. (2012). FX: an RNA-Seq analysis tool on the cloud. *Bioinformatics* 28 (5): 721–723.

16 Langmead, B., Hansen, K.D., and Leek, J.T. (2010). Cloud-scale RNA-sequencing differential expression analysis with Myrna. *Genome Biology* 11 (8): R83.

17 Feng, X., Grossman, R., and Stein, L. (2011). PeakRanger: a cloud-enabled peak caller for ChIP-seq data. *BMC Bioinformatics* 12 (1): 139.

18 P. Mell and T. Grance, The NIST definition of cloud computing. Special Publication SP 800-145, doi: 10.6028/NIST.SP.800-145, National Institute of Standards and Technology Computer Security Division, Information Technology Laboratory, 2011.

19 A. Fox, R. Griffith, A. Joseph et al., Above the clouds: A Berkeley view of cloud computing. Technical Report No. UCB/EECS-2009-28, Electrical Engineering and Computing Sciences, University of California at Berkeley, 2009.

20 Vaquero, L.M., Rodero-Merino, L., Caceres, J., and Lindner, M. (2008). A break in the clouds: towards a cloud definition. *ACM SIGCOMM Computer Communication Review* 39 (1): 50–55.

21 Shakil, K.A. and Alam, M. (2018). Cloud computing in bioinformatics and big data analytics: current status and future research. In: *Big Data Analytics*, 629–640. Singapore: Springer.

22 Langmead, B., Schatz, M.C., Lin, J. et al. (2009). Searching for SNPs with cloud computing. *Genome Biology* 10 (11): R134.

23 Agapito, G., Cannataro, M., Guzzi, P.H. et al. (2013). Cloud4SNP: distributed analysis of SNP microarray data on the cloud. In: *Proceedings of the International Conference on Bioinformatics, Computational Biology and Biomedical Informatics*, 468. ACM.

24 Afgan, E., Chapman, B., and Taylor, J. (2012). CloudMan as a platform for tool, data, and analysis distribution. *BMC Bioinformatics* 13 (1): 315.

25 Afgan, E., Baker, D., Coraor, N. et al. (2011). Harnessing cloud computing with Galaxy Cloud. *Nature Biotechnology* 29 (11): 972.

26 Jourdren, L., Bernard, M., Dillies, M.A., and Le Crom, S. (2012). Eoulsan: a cloud computing-based framework facilitating high throughput sequencing analyses. *Bioinformatics* 28 (11): 1542–1543.

27 Heath, A.P., Greenway, M., Powell, R. et al. (2014). Bionimbus: a cloud for managing, analyzing and sharing large genomics datasets. *Journal of the American Medical Informatics Association* 21 (6): 969–975.

28 Angiuoli, S.V., Matalka, M., Gussman, A. et al. (2011). CloVR: a virtual machine for automated and portable sequence analysis from the desktop using cloud computing. *BMC Bioinformatics* 12 (1): 356.

29 Krampis, K., Booth, T., Chapman, B. et al. (2012). Cloud BioLinux: pre-configured and on-demand bioinformatics computing for the genomics community. *BMC Bioinformatics* 13 (1): 42.

30 Maksimović, M. and Vujović, V. (2017). Internet of Things based e-health systems: ideas, expectations and concerns. In: *Handbook of Large-Scale Distributed Computing in Smart Healthcare*, 241–280. Cham: Springer.

31 Altschul, S.F., Gish, W., Miller, W. et al. (1990). Basic local alignment search tool. *Journal of Molecular Biology* 215 (3): 403–410.

32 Jain, M., Olsen, H.E., Paten, B., and Akeson, M. (2016). The Oxford nanopore MinION: delivery of nanopore sequencing to the genomics community. *Genome Biology* 17 (1): 239.

33 Alsan, M. and Klompas, M. (2010). *Acinetobacter baumannii*: an emerging and important pathogen. *Journal of Clinical Outcomes Management: JCOM* 17 (8): 363.

34 Peleg, A.Y. and Hooper, D.C. (2010). Hospital-acquired infections due to gram-negative bacteria. *New England Journal of Medicine* 362 (19): 1804–1813.

35 Tan, B., Ng, C.M., Nshimyimana, J.P. et al. (2015). Next-generation sequencing (NGS) for assessment of microbial water quality: current progress, challenges, and future opportunities. *Frontiers in Microbiology* 6: 1027.

36 Daniel, R. (2005). The metagenomics of soil. *Nature Reviews Microbiology* 3 (6): 470.

37 Castañeda, L.E. and Barbosa, O. (2017). Metagenomic analysis exploring taxonomic and functional diversity of soil microbial communities in Chilean vineyards and surrounding native forests. *PeerJ* 5: e3098.

38 A. Edwards, A. R. Debbonaire, B. Sattler et al., Extreme metagenomics using nanopore DNA sequencing: a field report from Svalbard, 78 N. Preprint bioRxiv, doi: https://doi.org/10.1101/073965, 2016.

39 Castro-Wallace, S.L., Chiu, C.Y., John, K.K. et al. (2017). Nanopore DNA sequencing and genome assembly on the International Space Station. *Scientific Reports* 7 (1): 18022.

40 McIntyre, A.B., Rizzardi, L., Angela, M.Y. et al. (2016). Nanopore sequencing in microgravity. *npj Microgravity* 2: 16035.

41 Greninger, A.L., Naccache, S.N., Federman, S. et al. (2015). Rapid metagenomic identification of viral pathogens in clinical samples by real-time nanopore sequencing analysis. *Genome Medicine* 7 (1): 99.

42 Merelli, I., Morganti, L., Corni, E. et al. (2018). Low-power portable devices for metagenomics analysis: Fog computing makes bioinformatics ready for the Internet of Things. *Future Generation Computer Systems* 88: 467–478.

Index

Fog Computing: Theory and Practice, First Edition.
Edited by Assad Abbas, Samee U. Khan, and Albert Y. Zomaya.

d

e

k

l

n

q

r

s

t

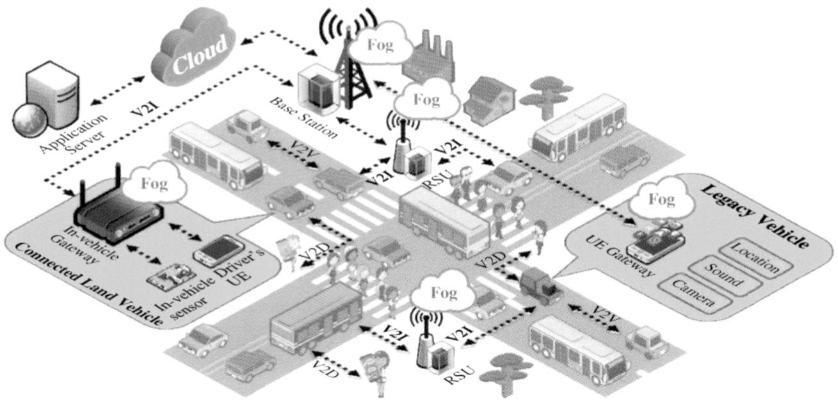

Figure 1.1 Land-vehicular fog computing examples.

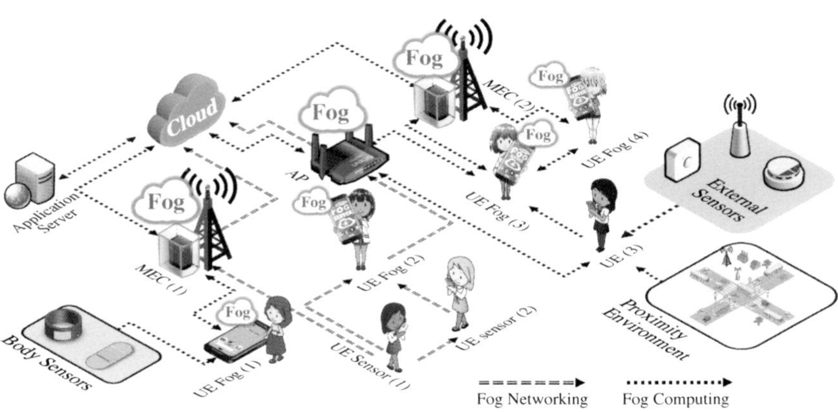

Figure 1.4 UE fog computing examples.

Fog Computing: Theory and Practice, First Edition.
Edited by Assad Abbas, Samee U. Khan, and Albert Y. Zomaya.

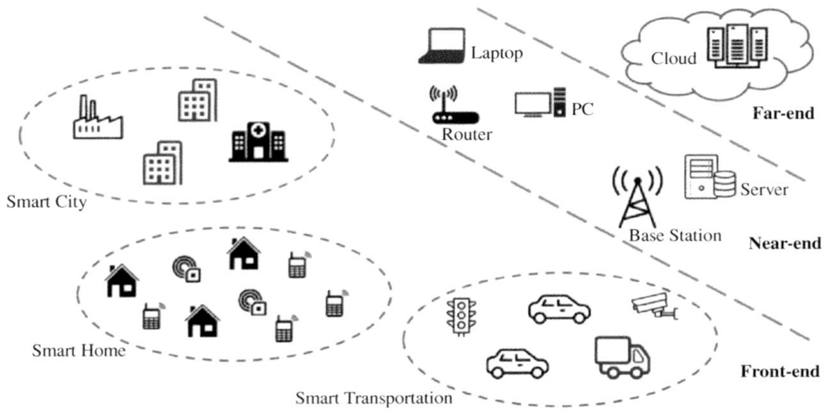

Figure 2.2 An overview of edge computing architecture [16].

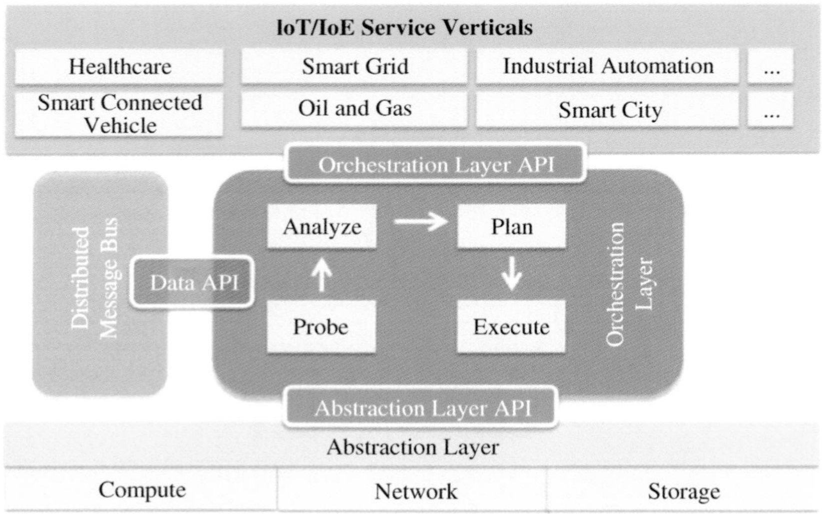

Figure 2.4 Fog computing architecture [10].

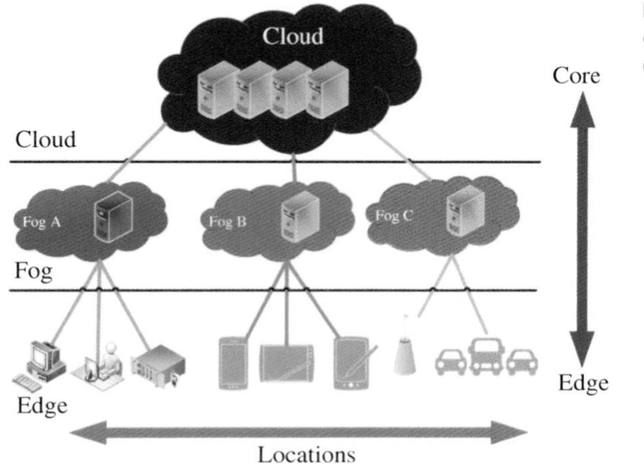

Figure 5.1 Fog computing enabled smart cities.

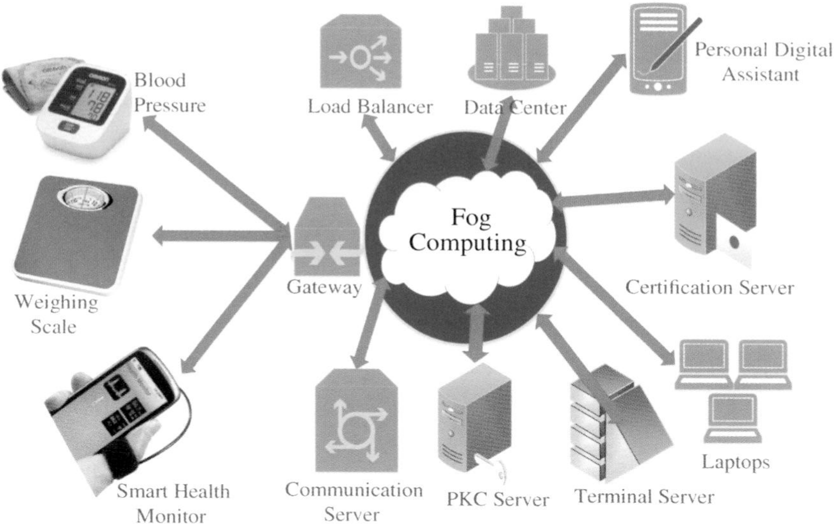

Figure 5.2 A generic fog enabled IoT environment.

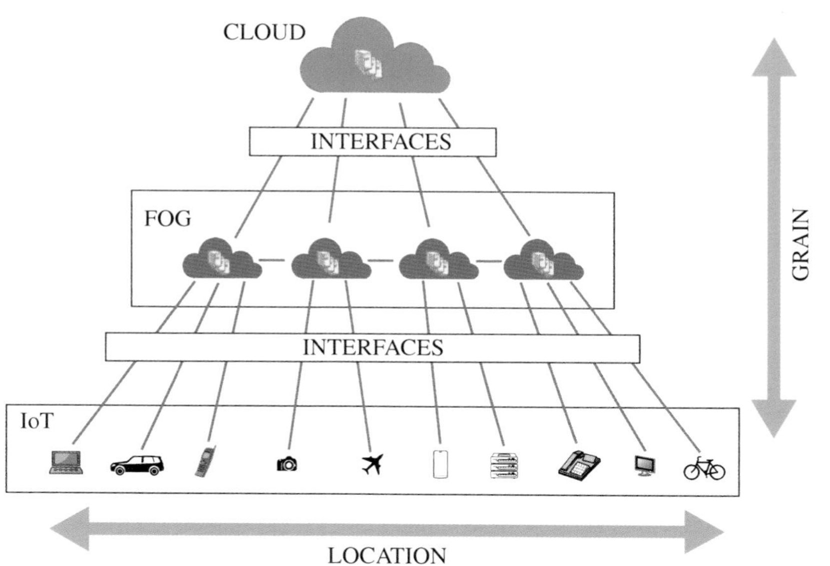

Figure 6.1 Cloud-fog-IoT architecture.

Harnessing The Computing Continuum

Science-driven Problems

e.g.: "Predict urban response to rainfall, trigger intelligent reaction..."

Goal-oriented Annotations

Notional trigger {flood_actuation, resident_warning}
Example: when {wx_prediction, sewer_model} implies (traffic_capacity < 70%) or (home_flooding > 5%)

Continuum Abstract Model and Runtime

Existing Resources and Services

Figure 7.2 Continuum Computing Research Areas: A pictorial depiction of the computer science areas that require research to successfully program the computing continuum. To address problems, such as the sciences examples given in Table 1.4, using existing resources and services, we need an abstract programming model with goal-oriented annotations, along with a run-time system and an execution model.

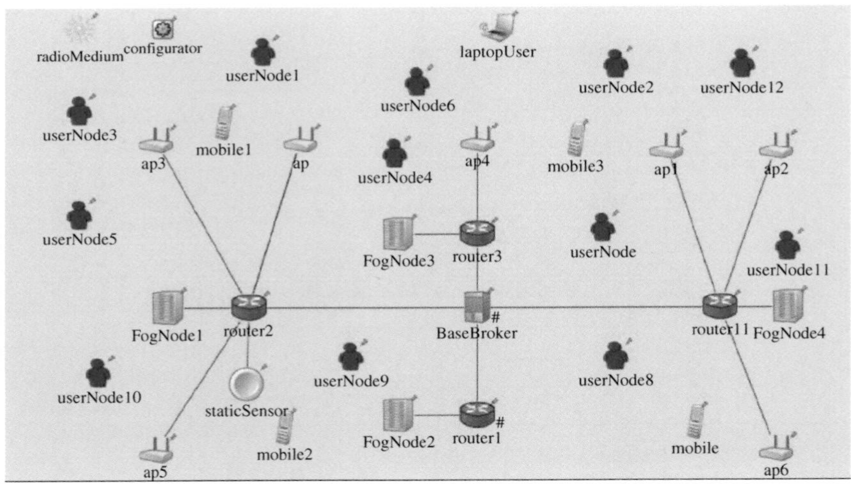

Figure 11.3 FogNetSim++: Graphical user interface, showing static, mobile, and fog computing nodes.

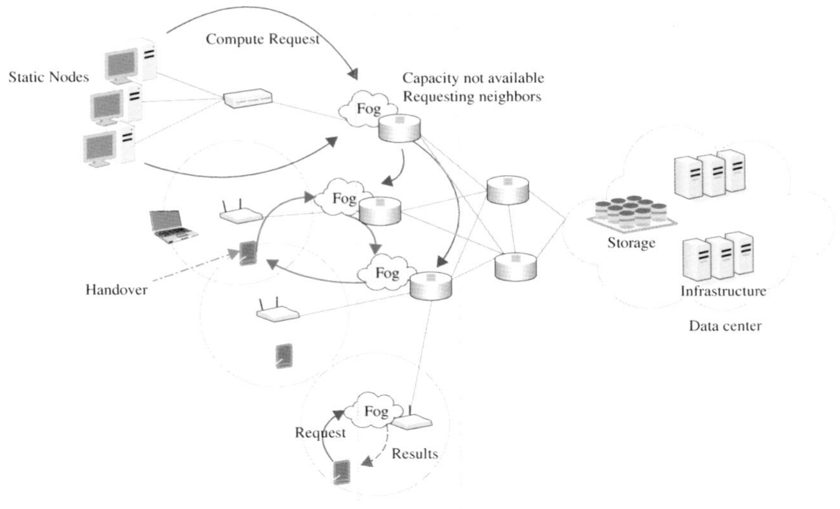

Figure 11.4 FogNetSim++: showing the handover features managed through single broker node.

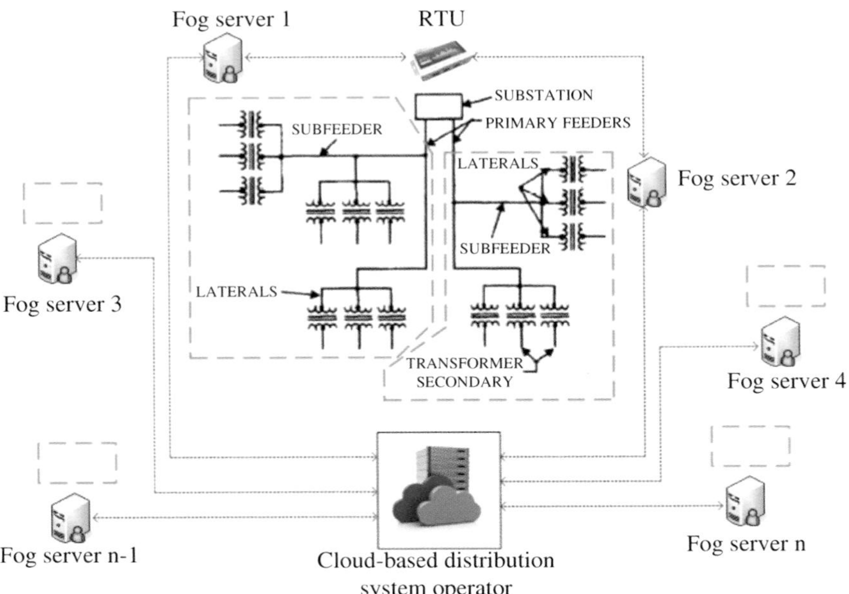

Figure 13.3 Feeder-based communication scheme for DMS using fog/cloud computing.

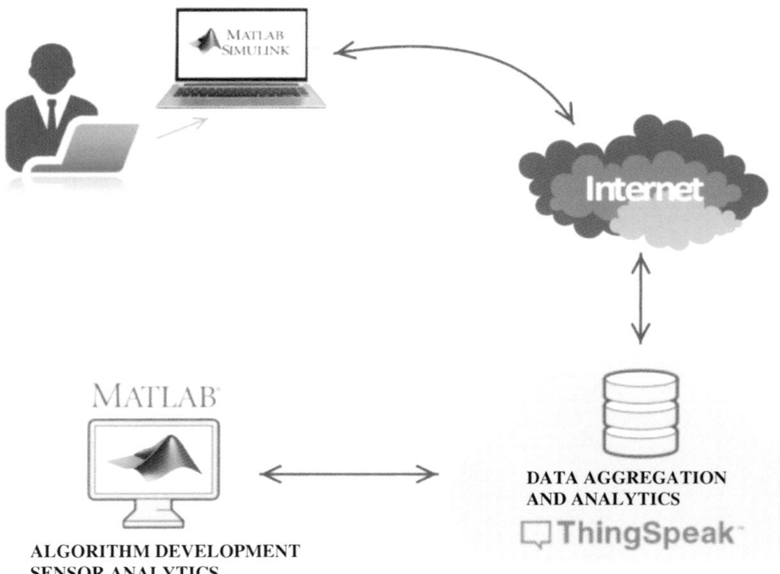

Figure 13.4 Simplified communication scheme connecting MATLAB/Simulink and ThingSpeak IoT platform used for DMS simulation.

Figure 15.3 Three smart phones are placed on three different locations of each participant's body (chest, right hand, and left jacket pocket) to collect accelerometer signals while walking on the treadmill.

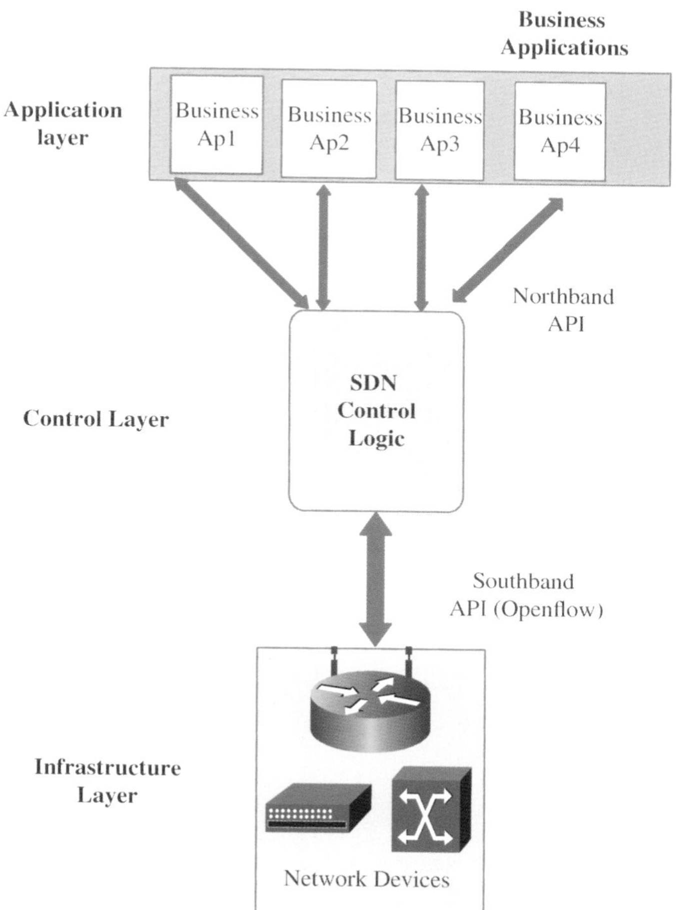

Figure 16.1 Architecture of software-defined network (SDN).

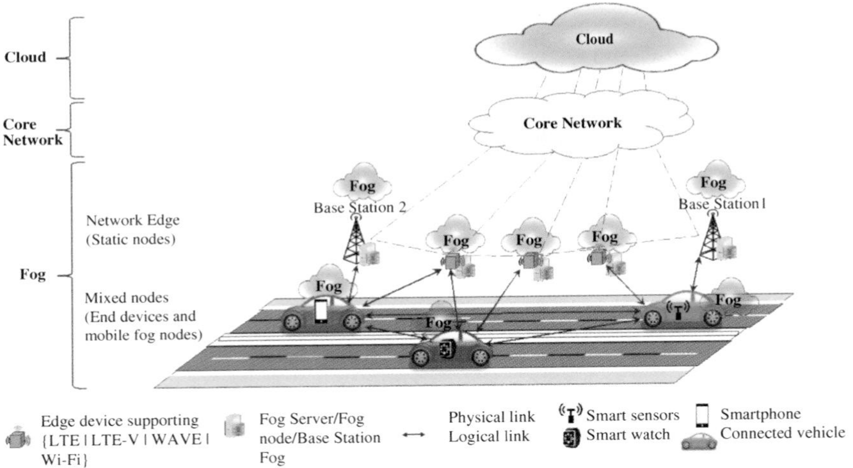

Figure 17.1 Fog computing for vehicular applications.

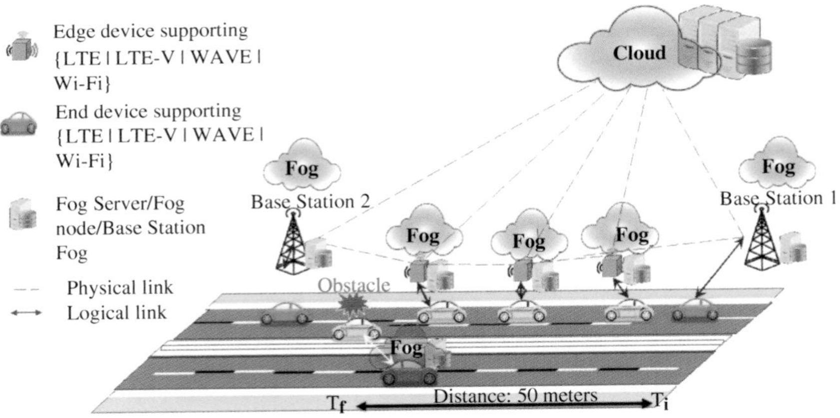

Figure 17.2 Obstacle detection as an example of delay-critical application scenarios.